ELECTRIC, OPTIC, AND ACOUSTIC INTERACTIONS IN DIELECTRICS

ELECTRIC, OPTIC, AND ACOUSTIC INTERACTIONS IN DIELECTRICS

DONALD F. NELSON
Bell Laboratories
Murray Hill, New Jersey

A WILEY-INTERSCIENCE PUBLICATION

JOHN WILEY & SONS, New York • Chichester • Brisbane • Toronto

Published by John Wiley & Sons, Inc.

Library of Congress Cataloging in Publication Data

Nelson, Donald Frederick, 1930-
Electric, optic, and acoustic interactions in dielectrics.

"A Wiley-Interscience publication."
Bibliography: p.
Includes index.
1. Dielectrics. I. Title.

QC585.N44 537.2'4 78-25964
ISBN 0-471-05199-3

Printed in the United States of America

10 9 8 7 6 5 4 3 2 1

PREFACE

This book is both a textbook and a research monograph. This is made possible by the special nature of the material covered—its newness and its pedagogical unity.

The book is a research monograph because the mode of treatment is based on papers published in the 1970s. In addition, a large number of the topics covered have either been discovered or greatly expanded by work of the past 15 years.

The book is a textbook because it treats a broad range of phenomena on a *first principles*, *unified*, *deductive* basis. This results because all of the material of the book is derived from a single Lagrangian that represents a dielectric crystal in interaction with the electromagnetic field. The phenomena covered consist of electromagnetism including linear and nonlinear optics, elasticity including linear and nonlinear acoustics at frequencies up through the hypersonic region, acoustooptics, the continuum mechanics of solids, and tensor properties of crystals for long wavelength phenomena (wavelengths large compared to unit cell dimensions). The Lagrangian treatment derives in a completely consistent manner all the constitutive properties of the crystalline medium along with the dynamical equations that govern the electromagnetic field and the long wavelength acoustic and optic modes of mechanical motion of the crystal. Thus the patchwork of phenomenologically assumed constitutive relations of so many treatments is avoided.

The book is organized into four parts: introductory material, the construction of the Lagrangian, the derivation of the equations of motion and general consequences of them, and applications to linear and nonlinear phenomena.

The introductory material is contained in the first three chapters. Chapter 1 briefly considers Lagrangian mechanics of lattices of discrete particles. Chapter 2 introduces continuum Lagrange mechanics, continuum limits of discrete particle systems, and the Lagrangian of the electromagnetic field. Chapter 3 considers the kinematics of deformation, the various measures of deformation, and the dualism of descriptions involving the spatial and material coordinate systems.

The construction of the Lagrangian occupies Chapters 4, 5, and 6. In Chapter 4 a discrete particle Lagrangian of the dielectric crystal in interaction with the electromagnetic field is first formed and then a continuum limit of it is taken. Chapter 5 considers the relation of invariances of physical laws to symmetry transformations, requirements on the Lagrangian needed to satisfy them, and the conservation laws that result from them. This chapter is the most abstract in the book and can be omitted on first reading. The invariance requirements on the Lagrangian, derived in Chapter 5, are restated in Chapter 6 and are quite plausible. The most general form of the stored energy consistent with the conservation laws is then found in Chapter 6. With the construction of the Lagrangian completed in Chapter 6 the remainder of the book is then deductively obtained from it.

The third section consisting of Chapters 7 and 8 derives the general equations of motion of the matter and the electromagnetic field and the associated boundary conditions. The general forms of the conservation laws are also obtained and certain consequences concerning the symmetry of the total stress tensor, exchange of angular momentum, the meaning of the group velocity, and so on are derived.

Applications to various physical phenomena are made in Chapters 9 to 18 with the exception of Chapter 16. These chapters are written in as self-contained a manner as possible in order to make them readily accessible to the casual user. Chapter 9 covers linear crystal optics; Chapters 10 and 11 cover linear elasticity and piezoelectricity including acoustics; Chapter 12 introduces three-field nonlinear optical interactions and Chapters 13, 14, and 15 then explore the elastooptic effect, the electrooptic effect, and optical mixing. Chapter 16 returns to a reformulation of the Maxwell equations in the material coordinate system. This general result is then used in Chapter 17 in the consideration of nonlinear electroacoustic phenomena that include acoustic wave mixing and electroacoustic parametric interactions. Chapter 18 concludes the book with a consideration of a few examples of higher order nonlinear optical and acoustooptic interactions.

The chapters on applications can be read in several sequences. A person interested in optical interactions could read Chapters 9, 12, 13, 14, 15, and 18, one interested in acoustic interactions could read Chapters 10, 11, 16, and 17, and one interested in acoustooptic interactions could read Chapters 9, 12, 13, and 18.

The great majority of the material in the main body of the text, Chapter 4 through Chapter 18, cannot be found elsewhere in book form. This is because the mode of treatment—deducing the results of the entire book from a single Lagrangian—has not been employed on such a broad segment of continuum physics before. It is also because of the choice of the applications made for detailed examination and the form in which they are presented. Even the chapter on linear crystal optics differs substantially from previous treatments.

The level of treatment is aimed at first year graduate students though the methods used are usually known by senior year science students. Most of the material of the book has been taught twice in the In-Hours Continuing Education Program at Bell Laboratories and once in a graduate level course at Princeton University.

Though the treatment is mathematical in nature, an effort is made to avoid intricate mathematical analysis in the applications by the consideration of simple geometries where single frequency excitation, plane wave propagation, homogeneous perturbations, and so forth are adequate representations of the fields. One exception to this is made in nearly every application presented: an arbitrary orientation of the crystalline anisotropy is allowed rather than adopting simple orientations. This is done for several reasons. Anisotropy is what makes so many of the interesting crystalline properties possible and so is basic to our treatment. Developing techniques to handle anisotropy in a general manner thus has considerable importance to this work. Handling anisotropy in a general manner is usually no longer, and sometimes shorter, than treating a special case and, of course, the final result has a far wider range of applicability.

It might be supposed that an extensive knowledge of group theory as applied to crystal symmetry would be needed to understand this treatment. This is not true, however, because the formulation applies to any crystal symmetry and so does not make use of particular symmetries. For applications to individual crystals the symmetry restrictions on the material interaction tensors must then be known. This is handled by an Appendix which lists the forms of the various tensors for the 32 crystal classes. For the interested reader several books on crystal symmetry are listed in the Bibliography.

Some of the mathematical objects needed to characterize interactions in crystals are complicated, being at the same time both multirank matrices and multirank tensors. The notation is thus necessarily somewhat complicated. The nature of the subject, however, requires this, and any simplification of the notation would either restrict the results or lead to ambiguity in the meaning. With a little effort applied early in the treatment it is hoped that one will see the mathematical objects unencumbered by their notation and so "see the forest through the trees."

Though the equations that govern nonlinear physical phenomena are in general mathematically nonlinear, the illustrative applications that we solve involve only solutions of linear differential equations. To do this, examples of optical, acoustic, and acoustooptic mixing are studied in the regime where depletion of input waves is negligible. In this way only a linear inhomogeneous wave equation needs to be solved in each case. Such solutions are obtained, either by the free-plus-forced-wave method applied to the wave equations or by the slowly-varying-amplitude method that leads to first order coupled mode equations.

In each of the chapters treating a specific interaction a few illustrative examples of that interaction are solved explicitly in a form useful to an experimenter. Space, however, does not allow treatment of all important examples. Many phenomena that are contained in the general equations, such as surface acoustic waves, excitonic polaritons, the first and second order Raman interactions, and solitons, could not be included. Each of these particular subjects, however, is introduced in the problems that follow the chapters and the treatment lays a sound basis for the study of still other applications.

Some of the subjects covered have a long history, and it is inappropriate here to attempt complete referencing of prominent advances in those subjects. Many of the subjects, however, are of recent origin, and references to the original literature concerning them serve a useful purpose. The references cited are thus to papers of recent times that originated subjects, papers from which recent experimental results are quoted, and papers on which the form of the treatment of certain subjects is based.

I wish to acknowledge with gratitude my colleagues who have collaborated with me on a number of the subjects discussed here. First and foremost is Professor Melvin Lax who collaborated on several papers that are basic to the treatment presented in this book. Second is Dr. David A. Kleinman who for a long period of time has acted as a sounding board and critic for my ideas. Both have contributed to shaping my thinking on many of the subjects of this book and thus to the form that the treatment takes. I also thank Dr. John R. Klauder for encouraging me to undertake writing this book while teaching the material at Bell Laboratories and Professor Walter C. Johnson for the invitation to teach a course on this material at Princeton University which offered an opportunity to revise and expand the manuscript. Last but not least, I acknowledge the patience, understanding, and support of my wife, Margaret, during the long time spent on preparing the manuscript.

DONALD F. NELSON

Murray Hill, New Jersey
February 1979

CONTENTS

ELECTRIC, OPTIC, AND ACOUSTIC INTERACTIONS IN DIELECTRICS

Introduction

The physics of dielectrics embodies a broad range of phenomena, which are usually taught in fragments in a variety of physics courses. These include elasticity, acoustics, electromagnetic theory, and optics, which as subjects of classical physics involve macroscopic properties of dielectrics. They also include subjects such as x-ray diffraction, band structure, and defect characteristics, which involve microscopic properties and need the concepts of quantum physics for their analysis. Though quantum physics has received more attention by physicists in recent decades, advances in the subjects of macroscopic physics have continually occurred and applications of them to technology have continually grown.

It is the purpose of this book to treat a broad segment of the macroscopic physics of dielectrics from a first principles, unified, deductive point of view. This has been made possible by a recent theory [Lax and Nelson, 1971], based on the methods of continuum mechanics, of the modes of motion (acoustic and optic) of a dielectric crystal and their interaction with the electromagnetic field. The complexity of phenomena of interest to us such as nonlinear acoustic or optical effects in crystals of low symmetry leads us to begin with the simplest and most basic point of view in order to avoid error. Thus we begin from a microscopic viewpoint even though we aim at macroscopic results. We use a Lagrangian formulation, since this requires only the construction of a single scalar quantity, the Lagrangian, formed from well known energies. Since macroscopic physics is our goal, we pass to a continuum limit of the solid at an early point of the development. Such a limit is adequate for phenomena, whether optic or acoustic, where wavelengths are long compared to the dimensions of a unit cell of the crystal structure. Because the continuum limit blurs the details of the microscopic structure, classical physics can be used to describe the microscopic starting point. The simplicity of this approach, in contrast to a quantum mechanical approach, allows us to present an entirely deductive theory beginning from first principles.

The result of this procedure is a completely unified treatment of all long wavelength motions (both optic and acoustic) of an arbitrary crystal in interaction with the electromagnetic field that may be at any frequency from 0 to $\sim 10^{15}$ Hz. The crystal can have any symmetry, any degree of anisotropy, any level of structural complexity, and any order of nonlinearity. Thus, besides *simple dielectrics*, which possess only even rank material tensor properties, dielectric crystals that possess odd rank material tensor properties are also included. Two important classes of crystals of this type are *piezoelectrics*, which possess a third rank tensor coupling between the strain and the electric field, and *pyroelectrics*, which possess a spontaneous polarization vector. The latter type of crystal includes *ferroelectrics*, which are reversible pyroelectrics, that is, crystals whose spontaneous polarization can be permanently reversed by the application of a sufficiently large electric field. *Ferroelastics* are another class of crystals included in the treatment. They are crystals that can exist in either of two mirror-image-related forms (*twins*) which can be reversibly transformed into one another by the application of a sufficiently large mechanical stress. If a reference configuration having zero strain and lying midway between the twins in structure is defined, then each twin has a *spontaneous strain* which is equal in magnitude and opposite in sign to that of the other twin. Of course, if one twin is chosen as the reference configuration, that twin has zero spontaneous strain.

Our method yields more than just the dynamical equations and associated boundary conditions governing the acoustic and internal motions of the crystal and those governing the electromagnetic fields (Maxwell's equations). It also yields deductively the constitutive relations which in so many treatments are added assumptions. Furthermore, the crystal group and interchange symmetries and the frequency dispersion of the various interaction tensors entering the constitutive relations and the dynamical equations follow from the development. When two or more lower order phenomena can combine to contribute to a higher order interaction (an indirect contribution), that too appears automatically from the development.

There are, of course, limitations on this treatment apart from its restriction to long wavelength phenomena. As we are interested in dielectrics, we exclude free charge distributions and free charge currents. We also ignore loss (damping) in any of the modes of motion of the crystal for simplicity. Since our object is not the treatment of ferromagnetic phenomena, we ignore the existence of electron spin. All magnetic effects resulting from the motion of bound charge, however, arise naturally in the treatment. Our interests here are directed toward organized mechanical phenomena such as static effects and wave motion. Thus we do not consider thermal phenomena. Strictly speaking our results apply only at the absolute zero of temperature. However, they may reasonably be applied at any temperature where the crystal structure is intact. Lastly, we point out that we present a nonrelativistic treatment that restricts the material velocities considered to values small compared to the velocity of light.

We conclude these introductory remarks by observing that, while the treatment to be presented has pedagogical qualities in its unifying of such diverse fields as elasticity, linear and nonlinear acoustics and ultrasonics, electromagnetic theory, linear and nonlinear optics, acoustooptics and continuum mechanics, it has equal qualities as a research tool. Application of these methods to the elastooptic effect [Nelson and Lax, 1970, 1971b] led to the discovery that the interaction tensor governing this effect had a lower interchange symmetry than believed since the work of Pockels in the 1890s. This resulted from finding that rotations could play as important a role as strains in strongly anisotropic crystals in causing light scattering or diffraction. The predicted contribution from rotations was numerically verified and its size shown to be comparable to that from strains in both rutile [Nelson and Lazay, 1970] and calcite [Nelson, Lazay, and Lax, 1972].

These methods were also applied to acoustically induced optical harmonic generation [Nelson and Lax, 1971c], a four-wave acoustooptic interaction in which an input acoustic wave and two input optical waves produce a phase matched optical output wave at a frequency displaced from the optical harmonic frequency by the acoustic frequency. The theory showed the susceptibility governing the interaction to be exceptionally complex with one direct interaction term, three two-step indirect interaction terms, and two three-step interaction terms. Coupling to both strains and rotations was found. Experiments were performed in GaAs [Boyd, Nash, and Nelson, 1970]. Multiple phase matching which the theory revealed was studied in $LiNbO_3$ in a triply phase matched five-wave interaction [Nelson and Mikulyak, 1972].

The application of this theory to nonlinear electroacoustics produced the first fully correct expression for the material interaction coefficient responsible for acoustic wave mixing in piezoelectric crystals and the analogous quantity responsible for the parametric interaction of two counter propagating acoustic waves and a homogeneous electric field [Nelson, 1978a, 1978b, 1978c]. A key technique in these derivations was the use of the material form of the Maxwell equations [Lax and Nelson, 1976b]. This theory also showed that the electrostriction tensor was not the low frequency limit of the elastooptic tensor, as long believed, but differs from that part of it coupling to strain by the Maxwell stress tensor of a polarizable medium.

Another useful aspect of this theory is the *three-eigenvector* formulation of crystal optics [Lax and Nelson, 1971]. Though it is convenient for proofs in linear crystal optics, it is particularly useful when studying forced electric waves that occur in nonlinear optics and in piezoelectricity. Performing perturbation theory on the eigenvector equation of crystal optics was found to be a convenient way of obtaining the general solution to the electrooptic effect [Nelson, 1975].

Further useful applications of the methods of this book can be expected in the future.

CHAPTER 1

Lattice Dynamics of Discrete Systems

Two simple examples of vibrations in discrete lattices are discussed in this chapter. A Lagrangian approach is used and the long wavelength limit is examined. The results illustrate phenomena that we find for realistic lattices in later chapters.

1.1 Vibrations of a Primitive One-Dimensional Discrete Lattice

Consider a one-dimensional array of identical pointlike atoms of mass M which are separated by a distance a. The *equilibrium position* of the m atom is then

$$X^m = ma. \tag{1.1.1}$$

Figure 1.1 illustrates the equilibrium configuration of the lattice. The *deformed position* of the m atom is

$$x^m = X^m + u^m \tag{1.1.2}$$

where u^m is the *displacement* of the m atom from its equilibrium position.

The dynamical equation for u^m can be found from the Lagrangian L of the system given by

$$L = T - V \tag{1.1.3}$$

where T is the kinetic energy and V is the potential energy. The kinetic energy is given simply by

$$T = \frac{M}{2} \sum_{m=-\infty}^{\infty} (\dot{u}^m)^2 \tag{1.1.4}$$

where the dot denotes the time derivative.

In order to construct the potential energy V we assume that the restoring force on any atom is due only to neighboring atom interactions

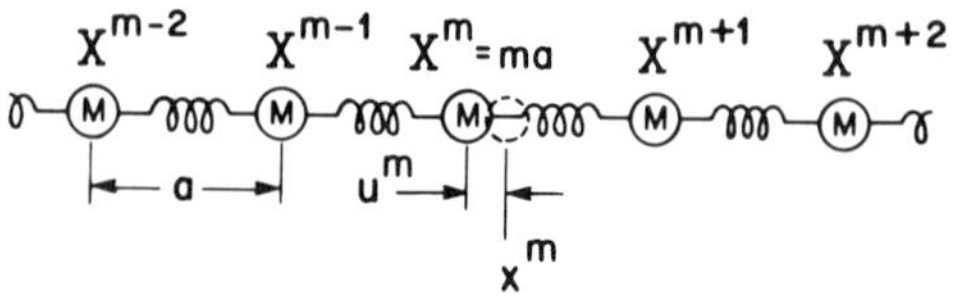

Fig. 1.1. Schematic representation of a one-dimensional lattice of identical atoms of mass M held together by Hooke's law "springs" of force constant K. The m atom is shown dotted in a deformed position denoted by x^m. The displacement u^m and the equilibrium position X^m are also shown.

that obey Hooke's law. This means that the restoring force is proportional to the change in the separation, that is, the elongation between neighboring atoms, and that the potential energy is quadratic in the elongation. The total potential energy of all the atoms associated with deformation is then

$$V=\frac{1}{4}\sum_{m=-\infty}^{\infty}K\left[(u^m-u^{m-1})^2+(u^{m+1}-u^m)^2\right]$$
$$=\frac{1}{2}\sum_{m}K(u^m-u^{m-1})^2 \tag{1.1.5}$$

where K is a constant.

The Lagrange equation for u^p is

$$\frac{d}{dt}\frac{\partial L}{\partial \dot{u}^p}=\frac{\partial L}{\partial u^p}. \tag{1.1.6}$$

Use of (1.1.3) to (1.1.5) then yields

$$M\ddot{u}^p=-K(2u^p-u^{p-1}-u^{p+1}), \tag{1.1.7}$$

the dynamical force equation for u^p. There is an infinite set of such equations corresponding to the infinite range of the discrete index p.

Let us consider the response of this one-dimensional lattice to a simple periodic wave (an *acoustic wave*) of the form

$$u^p=be^{i(kX^p-\omega t)} \tag{1.1.8}$$

where b is an amplitude constant, k the wavenumber, and ω the angular frequency. Substitution into (1.1.7) with the use of (1.1.1) yields

$$M\omega^2=2K(1-\cos ka)$$
$$=4K\sin^2\frac{ka}{2}. \tag{1.1.9}$$

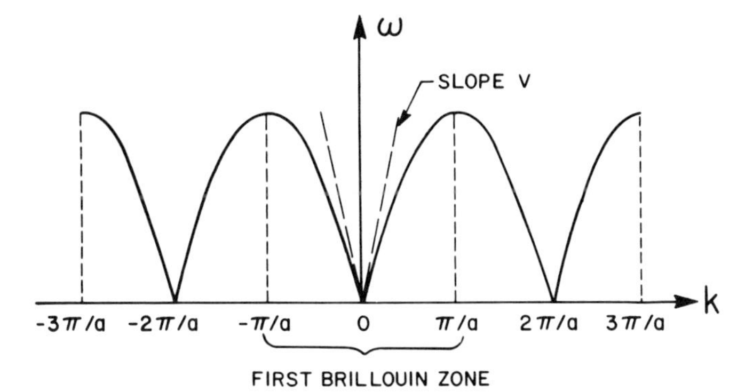

Fig. 1.2. The dispersion relation $\omega(k)$ for a primitive one-dimensional lattice. The long wavelength ($ka \ll 1$) velocity is the slope v of the dashed line.

A plot of this equation is shown in Fig. 1.2. Such a plot, or the equation it represents, is called a *dispersion relation* because it determines the dependence of the velocity of a wave on the frequency (or wavenumber) and thus how different frequency components of a disturbance disperse from one another. Note that the plot of $\omega(k)$ is periodic,

$$\omega\left(k+\frac{2\pi}{a}\right)=\omega(k), \tag{1.1.10}$$

and the function $\omega(k)$ is a repetition of its variation in the interval $-\pi/a<k\leqslant\pi/a$, called the *first Brillouin zone*. The phase velocity of the wave in general is

$$v\equiv\frac{\omega}{k} \tag{1.1.11}$$

which for this one-dimensional lattice results in

$$v=\frac{2}{k}\left(\frac{K}{M}\right)^{1/2}\sin\frac{ka}{2}. \tag{1.1.12}$$

The velocity of energy propagation of the wave (group velocity), as we show in Chapter 8, is

$$v_g\equiv\frac{\partial\omega}{\partial k} \tag{1.1.13}$$

which leads to

$$v_g=\left(\frac{Ka^2}{M}\right)^{1/2}\cos\frac{ka}{2}. \tag{1.1.14}$$

The long wavelength limit, $\lambda \gg a$, corresponds to $ka \ll 1$. In this limit the discrete lattice with cell spacing a acts as a continuum and the dispersion relation (1.1.9) becomes

$$\omega^2 = \left(\frac{Ka^2}{M}\right)k^2. \tag{1.1.15}$$

This is the dispersion relation of a long wavelength acoustic mode whose stiffness constant c is

$$c \equiv Ka \tag{1.1.16}$$

and whose mass density ρ is

$$\rho \equiv \frac{M}{a}. \tag{1.1.17}$$

In the long wavelength limit the phase and group velocities become equal,

$$v = \left(\frac{Ka^2}{M}\right) = \left(\frac{c}{\rho}\right) = v_g. \tag{1.1.18}$$

Note that in this limit neither velocity has any dispersion.

1.2 Vibrations of a Diatomic One-Dimensional Discrete Lattice

In order to illustrate an *internal* (or *optic*) *mode of motion*, as contrasted to the acoustic mode of motion, we consider a diatomic lattice, taking it as one-dimensional again for simplicity. The masses of the two types of atoms are denoted by M_1 and M_2. Figure 1.3 illustrates the diatomic lattice considered. The spacing of successive atoms of type 1 and 2, proceeding in a particular direction, is different from that of successive

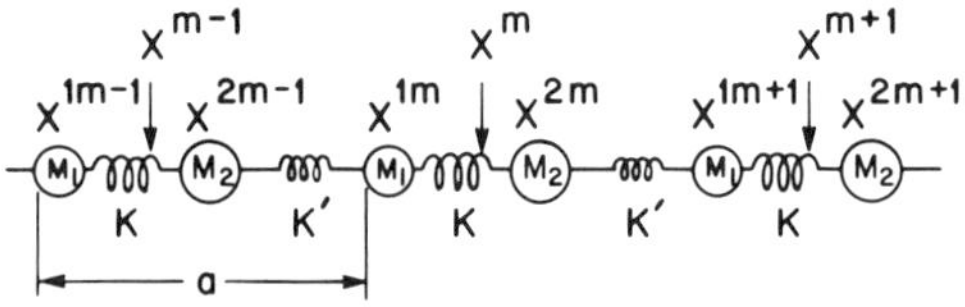

Fig. 1.3. Schematic representation of a one-dimensional diatomic lattice with only nearest neighbor Hooke's law forces shown. The equilibrium positions $X^{\alpha m}(\alpha = 1,2)$ of the atoms in the m cell and the equilibrium position X^m of the center of mass of the m cell are shown.

atoms of type 2 and 1. The Hooke's law spring constants can be different also. The cell size once again is denoted by a while the equilibrium positions of atoms are given by

$$X^{\alpha m} = ma + a^{\alpha}, \tag{1.2.1}$$

α denoting the atom type ($\alpha = 1,2$), m denoting the cell, and a^{α} being a constant different for each α. We express the deformed position $x^{\alpha m}$ of an atom in terms of a displacement $u^{\alpha m}$ from the equilibrium position of the atom according to

$$x^{\alpha m} = X^{\alpha m} + u^{\alpha m}. \tag{1.2.2}$$

We again construct a Lagrangian of the kinetic energy minus the potential energy. The kinetic energy is

$$T = \tfrac{1}{2}\sum_{m\alpha} M_{\alpha}(\dot{u}^{\alpha m})^2. \tag{1.2.3}$$

In constructing the potential energy this time, let us include the possibility of a Hooke's law interaction of an atom with every other atom in the lattice. Thus it takes the form

$$V = \tfrac{1}{4}\sum_{\alpha m, \beta n} K^{\alpha m \beta n}(u^{\alpha m} - u^{\beta n})^2 \tag{1.2.4}$$

where the symmetry of the quadratic form requires

$$K^{\alpha m \beta n} = K^{\beta n \alpha m}. \tag{1.2.5}$$

Homogeneity of the lattice requires

$$K^{\alpha m \beta n} = K^{\alpha (m+j) \beta (n+j)} = K^{\alpha\beta}(m-n), \tag{1.2.6}$$

that is, $K^{\alpha m \beta n}$ is a function only of the difference $m-n$. Proceeding as in the previous section, we find the Lagrange equation for $u^{\gamma p}$ to be

$$\begin{aligned} M_{\gamma}\ddot{u}^{\gamma p} &= \tfrac{1}{2}\sum_{\alpha m} K^{\alpha m \gamma p}(u^{\alpha m} - u^{\gamma p}) - \tfrac{1}{2}\sum_{\beta n} K^{\gamma p \beta n}(u^{\gamma p} - u^{\beta n}) \\ &= \sum_{\alpha m} K^{\alpha m \gamma p}(u^{\alpha m} - u^{\gamma p}) \end{aligned} \tag{1.2.7}$$

in which (1.2.5) is used. Thus we have now a doubly infinite set of dynamical equations.

The dispersion relation is found by substituting into (1.2.7) a wave of the form

$$u^{\gamma p} = b^{\gamma} e^{i(kX^p - \omega t)} \tag{1.2.8}$$

where X^p is the equilibrium position of the center of mass of the p cell given by

$$X^p \equiv \frac{\sum_\alpha M_\alpha X^{\alpha p}}{\sum_\alpha M_\alpha}. \tag{1.2.9}$$

The use of X^p rather than $X^{\gamma p}$ in (1.2.8) is not essentially different, since the factor $\exp ik(X^{\gamma p} - X^p)$ is independent of p and so can be regarded as having been absorbed in the definition of b^γ. The result is

$$M_\gamma \omega^2 b^\gamma = \sum_{\alpha m} K^{\alpha m \gamma p}\left(b^\gamma - b^\alpha e^{ik(X^m - X^p)}\right). \tag{1.2.10}$$

A solution for $b^{(1)}$ and $b^{(2)}$ exists provided

$$\det\left[\sum_{\beta m-p} K^{\beta\gamma}(m-p)\left(\delta^{\gamma\alpha} - \delta^{\beta\alpha} e^{ika(m-p)}\right) - M_\gamma \omega^2 \delta^{\gamma\alpha}\right] = 0 \tag{1.2.11}$$

where $\delta^{\gamma\alpha}$ is the Kronecker symbol and we use (1.2.6) and the fact that

$$X^m - X^p = (m-p)a \tag{1.2.12}$$

found from (1.2.9) and (1.2.1). Equation (1.2.11) is the dispersion relation $\omega(k)$ of the diatomic one-dimensional lattice.

In order to make the physical content of the dispersion relation more explicit, we specialize it to nearest neighbor forces only. The force constants then are

$$K^{1m2m} = K^{2m1m} = K, \tag{1.2.13a}$$

$$K^{1m2m-1} = K^{2m1m+1} = K', \tag{1.2.13b}$$

$$\text{all other } K^{\alpha i \beta j} = 0. \tag{1.2.13c}$$

The set of force equations (1.2.10) then consists of two ($\gamma = 1, 2$) equations

$$b^{(1)}\left(K + K' - M_1\omega^2\right) + b^{(2)}\left(-K - K'e^{-ika}\right) = 0, \tag{1.2.14a}$$

$$b^{(1)}\left(-K - K'e^{ika}\right) + b^{(2)}\left(K + K' - M_2\omega^2\right) = 0. \tag{1.2.14b}$$

The dispersion relation, the determinant of the coefficients of (1.2.14), can be put into the form

$$\omega^2 = \frac{(K+K')(M_1+M_2)}{2M_1M_2}\left[1 \pm \left(1 - \frac{16KK'M_1M_2}{(K+K')^2(M_1+M_2)^2}\sin^2\frac{ka}{2}\right)^{1/2}\right]. \tag{1.2.15}$$

The periodicity of $\omega(k)$ (1.1.10) is again apparent. The new effect that the diatomic structure of the lattice causes is the presence of two $\omega(k)$ curves. This is illustrated in Fig. 1.4.

The two modes of motion represented by the two solutions of (1.2.14) can be elucidated by considering the long wavelength limit, $ka \ll 1$. The limiting forms of (1.2.15) are

$$\omega^2 = \frac{KK'a^2k^2}{(K+K')(M_1+M_2)}, \tag{1.2.16}$$

$$\omega^2 = \frac{(K+K')(M_1+M_2)}{M_1M_2}. \tag{1.2.17}$$

The first of these is the acoustic mode having equal phase and group velocities

$$v = v_g = \left(\frac{c}{\rho}\right)^{1/2} \tag{1.2.18}$$

in the long wavelength limit. The stiffness constant here is

$$c \equiv \frac{KK'a}{(K+K')} \tag{1.2.19}$$

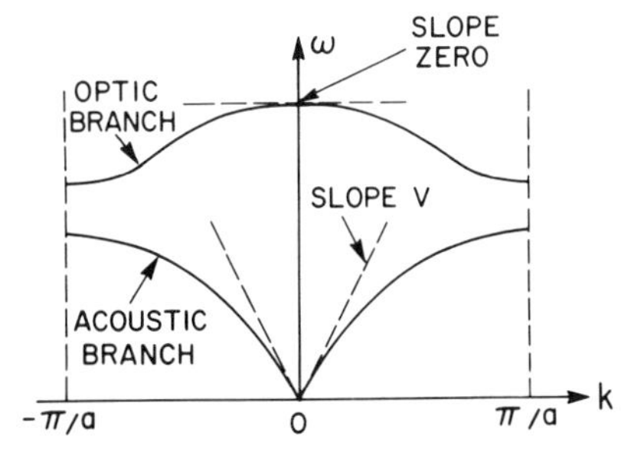

Fig. 1.4. The dispersion relation $\omega(k)$ for a one-dimensional diatomic lattice. The long wavelength velocities of the optic and acoustic modes are shown as tangents to the two branches at $k=0$.

and the mass density is

$$\rho \equiv \frac{M_1 + M_2}{a}. \tag{1.2.20}$$

The second mode has a zero group velocity in the long wavelength limit (and an infinite phase velocity). It is thus a nonpropagating vibration in this limit. It is called an *optic mode*.

The character of the acoustic and optic modes can be understood further by calculating the amplitude ratio $b^{(1)}/b^{(2)}$ for each type of mode from (1.2.14) with the use of their respective long wavelength dispersion relations (1.2.16) and (1.2.17). For the acoustic mode we find

$$\frac{b^{(1)}}{b^{(2)}} = 1 \tag{1.2.21}$$

showing that the acoustic mode is a center-of-mass motion with both atoms vibrating *in unison*. For the optic mode we find

$$\frac{b^{(1)}}{b^{(2)}} = -\frac{M_2}{M_1} \tag{1.2.22}$$

showing that the displacements of the two atoms are in the *opposite* directions and are inversely proportional to their masses. Thus the optic mode is a vibration between the two atoms.

For this simple system the most general motion is a superposition of the eigenmodes having the dispersion relation (1.2.15) and amplitude ratios $b^{(1)}/b^{(2)}$ given by (1.2.14) with the strength of each eigenmode determined by an initial condition. There is no coupling between the eigenmodes, since no nonlinearities or boundaries, which by breaking the periodicity could couple modes, are introduced here. Neither is any coupling to an electromagnetic field permitted. These and other effects are included in the chapters ahead.

PROBLEMS

1.1. Consider the primitive one-dimensional lattice of Fig. 1.1 with Hooke's law forces between all pairs of atoms. Find the Lagrangian, the equations of motion, and the dispersion relation.

1.2. Find the long wavelength limit of the dispersion relation of Prob. 1.1 and obtain an expression for the stiffness constant. *Hint:* Prove, using homogeneity, that $K^{mn}=K(|m-n|)$ where K^{mn} is the force constant between the m and n atoms.

1.3. Assume that the force constants of Prob. 1.1 obey $K(|p|)=Ke^{-c|p|}$ $(c>0)$. Find a closed form expression for the general dispersion relation. Find the long wavelength limit of it and determine the stiffness constant.

1.4. Consider the primitive one-dimensional lattice of Fig. 1.1 with Hooke's law forces between nearest neighbors only. A uniform tension T is applied along the length of the lattice and the particles are allowed to move in two dimensions (one longitudinal and one transverse to the chain). Find the discrete particle Lagrangian, retaining no potential energy terms higher than quadratic. Do linear terms occur? *Hint:* Take the unstrained position of the n particle as $\mathbf{X}^n=na\mathbf{i}$, the uniformly strained position as $[na+u_x^n(0)]\mathbf{i}$, and the deformed position as $\mathbf{x}^n=[na+u_x^n(0)]\mathbf{i}+\mathbf{u}^n$ where $\mathbf{i}$ is a unit vector in the x direction along the chain and $\mathbf{u}^n$ is the displacement vector which has x and y components. Eliminate $u_x^n(0)-u_x^{n-1}(0)$ from the potential energy in terms of T.

1.5. Find the Lagrange equations for u_x^n and u_y^n for Prob. 1.4. Are they coupled? What symmetry of the potential energy has led to this? Comment on the dependence of the stiffness constant for each mode on the tension T.

1.6. Find the dispersion relation for each mode of Prob. 1.5. Sketch them over the first Brillouin zone. Find the group velocity of each mode throughout the zone. What are the approximate dispersion relations and group velocities near $k=0$?

CHAPTER 2

Continuum Lagrange Mechanics

In the preceding chapter a long wavelength limit was taken in the solution for the vibrations of a discrete lattice in order to find the behavior of the system when viewed as a continuum. In the development of the general theory in later chapters, it is essential to take a continuum limit before the dynamical equations are solved if the degree of structural complexity at which we aim is to be included. In fact, it is most convenient to take the continuum limit in the Lagrangian itself before the dynamical equations are found from it.

For this reason we begin this chapter by reviewing the derivation of the dynamical Lagrange equations from a continuum Lagrangian. We then consider several relevant examples of continuum Lagrangians and the equations that follow from each. First we reconsider the one-dimensional monoatomic and diatomic lattices of Chapter 1 from the new viewpoint. These serve to illustrate in simple examples concepts whose generalizations are used in subsequent chapters. Among these are passage from a discrete to a continuum description, spatial and material coordinates of matter, internal coordinates of lattices, adiabatic elimination of internal coordinates for acoustic phenomena, and internal coordinate effects on the elastic stiffness. Then the continuum Lagrangian of the electromagnetic field is considered and the Maxwell-Lorentz equations are obtained from it. Finally, a simple example of a continuum Lagrangian for both the electromagnetic field and matter is considered.

2.1 Lagrange Equations for a Continuum

We begin from *Hamilton's principle* which states that the variation, defined below, of the time integral of the Lagrangian, called the *action integral*, vanishes,

$$\delta\int_{\Delta t} L\,dt = 0. \tag{2.1.1}$$

For a continuous system the Lagrangian L is a volume integral over the system,

$$L = \int_V \mathfrak{L}\, dV, \tag{2.1.2}$$

where $dV \equiv dX_1\, dX_2\, dX_3$ and $\mathfrak{L}$ is the *Lagrangian density* regarded as a function of the fields $\mathbf{x}$, $\dot{\mathbf{x}}$, $\partial\mathbf{x}/\partial X_I (I = 1,2,3)$ and the independent space and time variables $\mathbf{X}$ and t. The variation δ is a variation of the Lagrangian density with respect to the fields holding the independent variables $\mathbf{X}$ and t and the limits of their integration fixed. Also, the variation of the dependent field $\mathbf{x}$ evaluated on the surface of the volume or at the beginning or ending of the time interval is assumed to vanish. Thus we successively obtain

$$\begin{aligned} 0 &= \delta \int_{\Delta t} \int_V \mathfrak{L}\, dV\, dt \\ &= \int_{\Delta t} \int_V \delta\, \mathfrak{L}\, dV\, dt \\ &= \int_{\Delta t} \int_V \left[\frac{\partial \mathfrak{L}}{\partial x_j} \delta x_j + \frac{\partial \mathfrak{L}}{\partial \dot{x}_j} \delta \dot{x}_j + \frac{\partial \mathfrak{L}}{\partial(\partial x_j/\partial X_K)} \delta\left(\frac{\partial x_j}{\partial X_K} \right) \right] dV\, dt \end{aligned} \tag{2.1.3}$$

where, as throughout this book, summation over repeated subscripts is intended.

We now wish to manipulate the integrand of (2.1.3) into a form depending only on the three variations δx_j. The second term of the integrand can be integrated by parts with respect to time to yield

$$\begin{aligned} \int_{\Delta t} \frac{\partial \mathfrak{L}}{\partial \dot{x}_j} \delta \dot{x}_j\, dt &= \left[\frac{\partial \mathfrak{L}}{\partial \dot{x}_j} \delta x_j \right]_{t_1}^{t_2} - \int_{\Delta t} \frac{d}{dt} \left(\frac{\partial \mathfrak{L}}{\partial \dot{x}_j} \right) \delta x_j\, dt \\ &= - \int_{\Delta t} \frac{d}{dt} \left(\frac{\partial \mathfrak{L}}{\partial \dot{x}_j} \right) \delta x_j\, dt \end{aligned} \tag{2.1.4}$$

since the variations δx_j vanish at the end points of the time integration. Here d/dt is the total time derivative that acts on both explicit and implicit time dependence. In the third term in the integrand of (2.1.3) the operations δ and $\partial/\partial X_K$ can be interchanged and a partial integration performed,

$$\begin{aligned} \int_V \frac{\partial \mathfrak{L}}{\partial(\partial x_j/\partial X_K)} \delta\left(\frac{\partial x_j}{\partial X_K} \right) dV &= \int_V \frac{\partial \mathfrak{L}}{\partial(\partial x_j/\partial X_K)} \frac{\partial}{\partial X_K} (\delta x_j)\, dV \\ &= - \int_V \delta x_j \frac{d}{dX_K} \frac{\partial \mathfrak{L}}{\partial(\partial x_j/\partial X_K)} dV, \end{aligned} \tag{2.1.5}$$

with the integrated terms dropped, since the variations δx_j vanish on the surface of the volume. Here d/dX_K is the total derivative acting on both explicit and implicit space dependence. Inserting (2.1.4) and (2.1.5) into (2.1.3) results in

$$\int_{\Delta t}\int_V \delta x_j\left[\frac{\partial \mathcal{L}}{\partial x_j}-\frac{d}{dt}\frac{\partial \mathcal{L}}{\partial \dot{x}_j}-\frac{d}{dX_K}\frac{\partial \mathcal{L}}{\partial(\partial x_j/\partial X_K)}\right]dV\,dt=0. \tag{2.1.6}$$

Since each variation δx_j ($j=1,2,3$) is independent and arbitrary and the time interval and system volume are arbitrary, the integral of (2.1.6) can vanish only if the coefficients of δx_j in the integrand vanish. Therefore

$$\frac{d}{dt}\frac{\partial \mathcal{L}}{\partial \dot{x}_j}=\frac{\partial \mathcal{L}}{\partial x_j}-\frac{d}{dX_K}\frac{\partial \mathcal{L}}{\partial(\partial x_j/\partial X_K)}. \tag{2.1.7}$$

Thus we see that this *Lagrange equation* for the x_j field, or *generalized coordinate* as it is called, involves the *Lagrangian density* $\mathcal{L}$ rather than the Lagrangian L itself.

The form of the Lagrange equations (2.1.7) is dependent on the assumption made earlier that the functional dependence of the Lagrange density is $\mathcal{L}=\mathcal{L}\,(\mathbf{x},\dot{\mathbf{x}},\partial\mathbf{x}/\partial X_I,\mathbf{X},t)$. If the dependence on fields is extended to include $\ddot{\mathbf{x}}$, $\partial\dot{\mathbf{x}}/\partial X_I$, and $\partial^2\mathbf{x}/\partial X_I\partial X_J$, then the analogous derivation yields Lagrange equations of the form

$$\begin{aligned}\frac{d}{dt}\frac{\partial \mathcal{L}}{\partial \dot{x}_j}-\frac{d^2}{dt^2}\frac{\partial \mathcal{L}}{\partial \ddot{x}_j}=\frac{\partial \mathcal{L}}{\partial x_j}&-\frac{d}{dX_K}\frac{\partial \mathcal{L}}{\partial(\partial x_j/\partial X_K)}\\&+\frac{d^2}{dX_K dX_L}\frac{\partial \mathcal{L}}{\partial\left(\partial^2 x_j/\partial X_K\partial X_L\right)}\\&+\frac{d}{dt}\frac{d}{dX_K}\frac{\partial \mathcal{L}}{\partial(\partial \dot{x}_j/\partial X_K)}.\end{aligned} \tag{2.1.8}$$

We can compact this equation by introducing a four-dimensional coordinate vector $q_\sigma(\sigma=1,2,3,4)$ such that $q_1=X_1$, $q_2=X_2$, $q_3=X_3$, and $q_4=t$. Equation (2.1.8) then becomes

$$\frac{\partial \mathcal{L}}{\partial x_j}-\frac{d}{dq_\sigma}\frac{\partial \mathcal{L}}{\partial(\partial x_j/\partial q_\sigma)}+\frac{d^2}{dq_\sigma dq_\nu}\frac{\partial \mathcal{L}}{\partial\left(\partial^2 x_j/\partial q_\sigma\partial q_\nu\right)}=0. \tag{2.1.9}$$

The derivation of the last equation is left as an exercise.

The Lagrangian density $\mathfrak{L}$ that yields a particular set of Lagrange equations is not unique. For instance, it is simple to show that the Lagrange equations (2.1.7) are unaffected by either of the replacements

$$\mathfrak{L} \rightarrow \mathfrak{L} + \frac{d}{dt}\Omega_4(\mathbf{x}), \tag{2.1.10}$$

$$\mathfrak{L} \rightarrow \mathfrak{L} + \frac{d}{dX_K}\Omega_K(\mathbf{x}) \tag{2.1.11}$$

where the limited functional dependence of Ω_4 and Ω_K is essential. More generally it may be shown that the Lagrange equation (2.1.9) is form invariant to the inclusion of a four-divergence in the Lagrangian density,

$$\mathfrak{L} \rightarrow \mathfrak{L} + \Omega_{\sigma,\sigma}(\mathbf{x}, \partial\mathbf{x}/\partial q_\nu). \tag{2.1.12}$$

Demonstration of this is also left as an exercise.

2.2 Continuum Primitive One-Dimensional Lattice

Let us take a continuum limit of the primitive one-dimensional discrete lattice of Section 1.1 to illustrate the use of the continuum Lagrange equations. We must define a continuous field that describes the motion of the system and is a function of a continuous independent variable designating the material medium. A continuum limit of the Lagrangian then yields an integral over a continuous Lagrange density needed for the Lagrange equations.

The discrete Lagrangian from Section 1.1 is

$$L = \tfrac{1}{2}\sum_m \left[M(\dot{u}^m)^2 - K(u^m - u^{m-1})^2 \right]. \tag{2.2.1}$$

Clearly, we must convert the discrete displacements u^m into a single continuous displacement u. It must be a function of some independent variable characterizing the particular point of the continuous medium considered. Looking at Fig. 1.1 we see that the natural choice for such a variable is the continuum variable X corresponding to the equilibrium or undisturbed position X^m of the particles. We refer to this variable as the *material coordinate* of the medium. It designates or names the material point and, as such, moves with the point as its position changes. The continuum deformed position or *spatial coordinate* of the point is x corresponding to the discrete deformed positions x^m. Thus the continuum limit of (1.1.2) is

$$x(X,t) = X + u(X,t) \tag{2.2.2}$$

where u is the *displacement*.

The potential energy term in (2.2.1) depends on the elongation, that is, the difference of displacements of neighboring mass points. If the continuum limit of the displacement were taken simply as $u^m \to u(X)$, as was adequate in (2.2.2), then the continuum limit of the elongation would be zero. Thus it is clear that we must consider first order corrections by letting

$$u^m \to u(X^m) = u(X) + (X^m - X)\frac{\partial u}{\partial X}. \tag{2.2.3}$$

Thus we obtain to first order

$$u^m - u^{m-1} \to (X^m - X^{m-1})\frac{\partial u}{\partial X} = a\frac{\partial u}{\partial X} \tag{2.2.4}$$

where a is the cell size. To first order we also obtain

$$\dot{u}^m \to \dot{u}(X). \tag{2.2.5}$$

To make the transformation of the Lagrangian (2.2.1) to a continuum Lagrangian, we multiply and divide by a and define the mass per unit length ρ by

$$\rho \equiv \frac{M}{a} \tag{2.2.6}$$

and Young's modulus E, the force per elongation per unit length, by

$$E \equiv Ka. \tag{2.2.7}$$

Next we convert the sum to an integral by associating $a \to dX$ and setting

$$\tfrac{1}{2}\sum_m \left[\rho(\dot{u})^2 - E\left(\frac{\partial u}{\partial X}\right)^2\right] a \to \int \tfrac{1}{2}\left[\rho(\dot{u})^2 - E\left(\frac{\partial u}{\partial X}\right)^2\right] dX. \tag{2.2.8}$$

Thus the continuum Lagrangian density is

$$\mathfrak{L} = \tfrac{1}{2}\left[\rho(\dot{u})^2 - E\left(\frac{\partial u}{\partial X}\right)^2\right]. \tag{2.2.9}$$

Recognizing the displacement field u as the generalized coordinate and substituting the Lagrangian density (2.2.9) into (2.1.7) yields the continuum force equation for u,

$$\rho \ddot{u} = E \frac{\partial^2 u}{\partial X^2}. \tag{2.2.10}$$

A traveling wave,

$$u = b e^{i(kX - \omega t)}, \tag{2.2.11}$$

is obviously a solution provided that

$$\omega^2 = \frac{E}{\rho} k^2, \tag{2.2.12}$$

found from substitution of (2.2.11) into (2.2.10). This is the dispersion relation $\omega(k)$ in the continuum limit and implies that both the phase and group velocities are given by

$$v = v_g = \left(\frac{E}{\rho}\right)^{1/2}. \tag{2.2.13}$$

Note that these results are in agreement with those of (1.1.15) to (1.1.18) with the stiffness constant now called Young's modulus.

What we call the continuum limit here can now be seen to be identical with the long wavelength limit of Section 1.1. The reason why they are identical here is that in taking the continuum limit terms higher than first order in the Taylor series expansion of (2.2.3) are neglected. This implies that the change in the displacement from one cell to the next is small. This is true only if the wavelength of the disturbance is long compared to the cell size, that is, only if the long wavelength limit is considered.

2.3 Continuum Diatomic One-Dimensional Lattice

In this section we convert the diatomic one-dimensional discrete lattice to a continuum whose motion is described by two continuous fields. We begin, as in the preceding section, with the Lagrangian of the discrete system. Forces between neighbors of all separations are included. From Section 1.2 we have

$$L = \tfrac{1}{2} \sum_{m\alpha} M_\alpha (\dot{u}^{\alpha m})^2 - \tfrac{1}{4} \sum_{\alpha m, \beta n} K^{\alpha m \beta n} (u^{\alpha m} - u^{\beta n})^2. \tag{2.3.1}$$

In the continuum limit we expect that two continuous fields are needed to describe the possible motions of the system just as in the discrete problem two modes characterized by two sets of b^{γ} values are found. One of these modes, called the acoustic mode, is a center-of-mass motion while the other, the optic mode, is a vibration between the two types of atoms. The position of the center of mass of the p cell in the deformed state is

$$\begin{aligned} x^{p} &= \frac{\sum_{\alpha} M_{\alpha}(X^{\alpha p} + u^{\alpha p})}{\sum_{\alpha} M_{\alpha}} \\ &= X^{p} + \frac{\sum_{\alpha} M_{\alpha} u^{\alpha p}}{\sum_{\alpha} M_{\alpha}} \end{aligned} \tag{2.3.2}$$

where (1.2.9) is used. The displacement of the center of mass of the p cell is naturally defined by

$$\begin{aligned} u^{p} &= x^{p} - X^{p} \\ &= \frac{\sum_{\alpha} M_{\alpha} u^{\alpha p}}{\sum_{\alpha} M_{\alpha}}. \end{aligned} \tag{2.3.3}$$

In this case it becomes

$$u^{p} = \frac{1}{M}\left(M_{1} u^{1p} + M_{2} u^{2p}\right) \tag{2.3.4}$$

where

$$M \equiv M_{1} + M_{2}. \tag{2.3.5}$$

To describe the optic mode we introduce an *internal coordinate* y^{m} defined by

$$y^{p} \equiv u^{2p} - u^{1p}. \tag{2.3.6}$$

Since the center of mass carries all the momentum of the motion, an internal coordinate is chosen to be displacement invariant and so to carry none. If (2.3.4) and (2.3.6) are solved for u^{1p} and u^{2p}, we find

$$u^{1m} = u^{m} - \frac{M_{2}}{M} y^{m}, \tag{2.3.7}$$

$$u^{2m} = u^{m} + \frac{M_{1}}{M} y^{m}. \tag{2.3.8}$$

Note that the ratio of the coefficients of u^m in these equations agrees with that found for the acoustic mode in (1.2.21) and that the ratio of the coefficients of y^m agrees with that found for the optic mode in (1.2.22).

Substitution of (2.3.7) and (2.3.8) into the Lagrangian (2.3.1) yields

$$L=\tfrac{1}{2}\sum_n\left[M(\dot{u}^n)^2+m(\dot{y}^n)^2\right]-\tfrac{1}{4}\sum_{np}\left\{\left[u^n-u^p-\frac{M_2}{M}(y^n-y^p)\right]^2K^{1n1p}\right.$$
$$+2\left[u^n-u^p-\frac{1}{M}(M_2y^n+M_1y^p)\right]^2K^{1n2p}$$
$$\left.+\left[u^n-u^p+\frac{M_1}{M}(y^n-y^p)\right]^2K^{2n2p}\right\} \tag{2.3.9}$$

where

$$m\equiv\frac{M_1M_2}{M}. \tag{2.3.10}$$

Note that the choice of new variables retains the diagonality of the kinetic energy or, in other words, no terms in $\dot{u}^n\dot{y}^n$ appear. We see in the following chapters that this condition can always be imposed on internal coordinates.

We are now ready to obtain the continuum or long wavelength limit of the Lagrangian. We replace

$$u^n-u^p\rightarrow(X^n-X^p)\frac{\partial u}{\partial X}=(n-p)a\frac{\partial u}{\partial X} \tag{2.3.11}$$

using reasoning similar to that leading to (2.2.4). The internal coordinate y^n becomes

$$y^n\rightarrow y(X^n)=y(X)+(X^n-X)\frac{\partial y}{\partial X}. \tag{2.3.12}$$

Since the zero order term $y(X)$ in the Taylor series expansion remains in one of the three terms in the potential energy, it is correct in the long wavelength limit to ignore the first order terms.

Just as in the preceding section we multiply and divide by the cell size a. The a in the denominator is absorbed into the definitions of the two mass densities needed in this case,

$$\rho\equiv\frac{M}{a}, \tag{2.3.13}$$

$$\rho'\equiv\frac{m}{a}, \tag{2.3.14}$$

and into the continuum force constants,

$$H_0 \equiv \frac{1}{2a} \sum_{n-p} K^{12}(n-p), \tag{2.3.15}$$

$$H_1 \equiv -\sum_{n-p} (n-p) K^{12}(n-p), \tag{2.3.16}$$

$$H_2 \equiv \frac{a}{4} \sum_{n-p} (n-p)^2 \left[K^{11}(n-p) + 2K^{12}(n-p) + K^{22}(n-p) \right]. \tag{2.3.17}$$

The factor a in the numerator becomes dX and the remaining sum becomes the integral over dX. The continuum result is

$$L = \int \left[\frac{\rho}{2} (\dot{u})^2 + \frac{\rho'}{2} (\dot{y})^2 - H_0 y^2 - H_1 y \frac{\partial u}{\partial X} - H_2 \left(\frac{\partial u}{\partial X} \right)^2 \right] dX \tag{2.3.18}$$

where the integrand is the Lagrangian density $\mathfrak{L}$.

Two Lagrange equations can now be found from (2.1.7) for u and y. They are respectively

$$\rho \ddot{u} = 2H_2 \frac{\partial^2 u}{\partial X^2} + H_1 \frac{\partial y}{\partial X} \tag{2.3.19}$$

and

$$\rho' \ddot{y} = -2H_0 y - H_1 \frac{\partial u}{\partial X}, \tag{2.3.20}$$

which are coupled together.

Consider the second or internal motion equation first. From the previous discussion we know that y represents an optic mode. Such a mode's resonant frequency is typically in the infrared frequency range, well above frequencies at which acoustic or center-of-mass motion can respond. This can be seen from Fig. 1.4 in the region near $k=0$. For this reason the last term on the right side of (2.3.20) can be dropped. Since no space derivatives of y appear on the right side, we see that no traveling wave solution exists. Rather, only a time oscillating solution

$$y = de^{-i\omega t} \tag{2.3.21}$$

exists. Substitution of this into (2.3.20) yields

$$\omega = \left(\frac{2H_0}{\rho'} \right)^{1/2}, \tag{2.3.22}$$

which implies a zero group velocity,

$$v_g = \frac{\partial \omega}{\partial k} = 0, \tag{2.3.23}$$

consistent with the lack of a traveling wave solution. This is just what was found for the long wavelength optic mode behavior in Section 1.2. Specifically, (2.3.22) agrees with (1.2.17) when the nearest neighbor force conditions (1.2.13) are imposed on the definition of H_0 in (2.3.15).

Consider next the acoustic equation (2.3.19). The frequencies for which a traveling wave solution of this equation are of interest are much below the resonant frequency of the internal motion. Thus for these frequencies the internal motion follows the acoustic motion without inertia. This means that the inertial term $\rho'\ddot{y}$ in the internal motion equation can be dropped leading to

$$y = -\frac{H_1}{2H_0}\frac{\partial u}{\partial X}. \tag{2.3.24}$$

This can then be substituted for y in (2.3.19) to yield

$$\rho \ddot{u} = c\frac{\partial^2 u}{\partial X^2} \tag{2.3.25}$$

with

$$c = 2H_2 - \frac{(H_1)^2}{2H_0} \tag{2.3.26}$$

being the stiffness constant. This procedure is called *adiabatic elimination of the internal coordinate*. A traveling acoustic wave solution of (2.3.25),

$$u = be^{i(kX - \omega t)}, \tag{2.3.27}$$

has the dispersion relation

$$\omega^2 = \frac{c}{\rho}k^2, \tag{2.3.28}$$

which is in agreement with (1.2.16) when the conditions (1.2.13) for nearest neighbor forces are imposed on the definitions of H_0, H_1, H_2 in the expression (2.3.26) for c.

The interpretation of this expression for c is of some interest, since it illustrates a phenomenon that occurs quite generally. The first term is

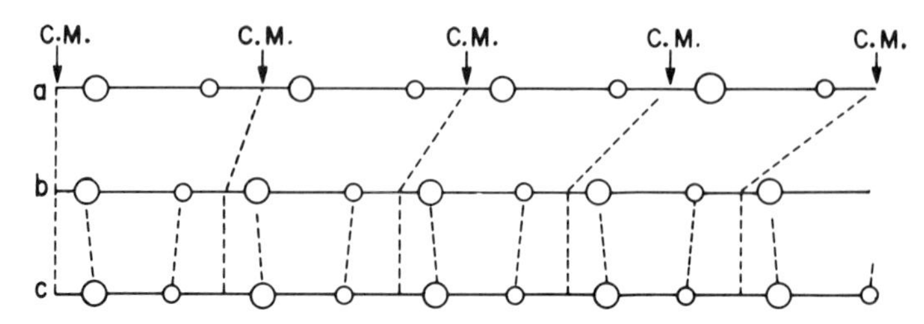

Fig. 2.1. A diatomic one-dimensional lattice in three conditions: (*a*) undeformed, (*b*) simply strained such that the interparticle spacing is altered proportionally to the overall strain, and (*c*) realistically strained where the interparticle spacing has adjusted for minimum deformation energy and has led to a reduced stiffness constant compared to (*b*). The center of mass (C.M.) of each cell is shown.

simply the stiffness determined by the direct response of the system to a strain. The second term is a *lowering* of the stiffness caused by a rearrangement of the particles in the unit cell by way of their internal mode of motion in response to the strain. This is illustrated in Fig. 2.1.

2.4 Lagrangian for Electromagnetic Fields

Since we aim at a study of crystals in interaction with the electromagnetic fields, it is necessary to have a Lagrangian that yields the electromagnetic equations in the Maxwell-Lorentz form. By *Maxwell–Lorentz form* we mean the vacuum form of the Maxwell equations in which the *electric field* **E** and the *magnetic induction* **B** are regarded as the fundamental fields and matter is represented by a charge density q and a current density **j**. These two densities can contain both free and bound charge. Since we are studying dielectrics, we assume that they contain only bound charge in succeeding chapters.

The Maxwell-Lorentz equations are

$$\frac{1}{\mu_0}\nabla\times\mathbf{B}-\epsilon_0\frac{\partial\mathbf{E}}{\partial t}=\mathbf{j}, \tag{2.4.1}$$

$$\epsilon_0\nabla\cdot\mathbf{E}=q, \tag{2.4.2}$$

$$\nabla\times\mathbf{E}+\frac{\partial\mathbf{B}}{\partial t}=0, \tag{2.4.3}$$

$$\nabla\cdot\mathbf{B}=0 \tag{2.4.4}$$

where ϵ_0 is the permittivity of free space and μ_0 is the permeability of free space. We use SI units throughout the treatment. It is well known that the

E and **B** fields can be expressed in terms of the *scalar potential* Φ and the *vector potential* **A** by

$$\mathbf{E} = -\nabla\Phi - \frac{\partial \mathbf{A}}{\partial t}, \tag{2.4.5}$$

$$\mathbf{B} = \nabla \times \mathbf{A}. \tag{2.4.6}$$

Since the **E** and **B** fields can exist at every point in space, the Lagrangian that yields the Maxwell-Lorentz equations must necessarily be a continuum Lagrangian. Of course, it cannot be constructed by forming the difference between kinetic and potential energies as done for particles of matter, since these quantities are not defined for the electromagnetic fields. Nevertheless, a Lagrangian that works has been found. It might be expected that it would be a function of the **E** and **B** fields. However, these fields clearly contain a redundancy, since their six components ($E_x, E_y, E_z, B_x, B_y, B_z$) can be expressed in (2.4.5) and (2.4.6) in terms of four quantities (Φ, A_x, A_y, A_z) making up the scalar and vector potentials. It is thus at least reasonable that the generalized coordinates of the electromagnetic fields in the Lagrangian density are these potentials.

We now assert that a Lagrangian density that will yield the Maxwell-Lorentz equations is

$$\mathcal{L}(\Phi(\mathbf{z}), \mathbf{A}(\mathbf{z})) = \frac{\epsilon_0}{2}\left[\left(-\nabla\Phi - \frac{\partial \mathbf{A}}{\partial t}\right)^2 - c^2(\nabla \times \mathbf{A})^2\right] - q\Phi + \mathbf{j}\cdot\mathbf{A}. \tag{2.4.7}$$

Here **z** is the coordinate vector of the point in space considered and c is the velocity of light in vacuum. The bracketed term, which is simply $(\epsilon_0/2)[E^2 - c^2B^2]$, is called the *electromagnetic field Lagrangian density* and the remaining terms the *interaction Lagrangian density*.

It is a simple matter to show that the Maxwell-Lorentz equations follow from this Lagrangian with the help of the ancillary conditions (2.4.5) and (2.4.6). First, consider the Lagrange equation for Φ,

$$\frac{d}{dt}\frac{\partial \mathcal{L}}{\partial(\partial\Phi/\partial t)} = \frac{\partial \mathcal{L}}{\partial \Phi} - \frac{d}{dz_j}\frac{\partial \mathcal{L}}{\partial \Phi_{,j}}, \tag{2.4.8}$$

where the comma notation means $\Phi_{,j} \equiv \partial\Phi/\partial z_j$. Since no $\partial\Phi/\partial t$ appears in the Lagrangian density (2.4.7), this becomes

$$0 = -q - \epsilon_0 \frac{d}{dz_j}\left(\Phi_{,j} + \frac{\partial A_j}{\partial t}\right) \tag{2.4.9}$$

which with the use of (2.4.5) yields

$$\epsilon_0 \nabla \cdot \mathbf{E} = q. \tag{2.4.10}$$

Next, consider the Lagrange equation for $\mathbf{A}$,

$$\frac{d}{dt}\frac{\partial \mathcal{L}}{\partial(\partial A_i/\partial t)} = \frac{\partial \mathcal{L}}{\partial A_i} - \frac{d}{dz_j}\frac{\partial \mathcal{L}}{\partial(A_{i,j})}. \tag{2.4.11}$$

Use of (2.4.7) leads to

$$\epsilon_0 \frac{d}{dt}\left(\Phi_{,i} + \frac{\partial A_i}{\partial t}\right) = j_i - \epsilon_0 c^2 \frac{d}{dz_j}(-\epsilon_{kji}\epsilon_{klm}A_{m,l}), \tag{2.4.12}$$

where the permutation symbol ϵ_{klm} is used to express

$$B_k = (\nabla \times \mathbf{A})_k = \epsilon_{klm}\frac{\partial A_m}{\partial z_l} = \epsilon_{klm}A_{m,l}. \tag{2.4.13}$$

With the use of $\mu_0\epsilon_0 \equiv c^{-2}$ and (2.4.5) and (2.4.6), (2.4.12) becomes

$$\frac{1}{\mu_0}\nabla \times \mathbf{B} - \epsilon_0 \frac{\partial \mathbf{E}}{\partial t} = \mathbf{j}. \tag{2.4.14}$$

The remaining equations (2.4.3) and (2.4.4) are not found from the Lagrangian but are automatic consequences of the ancillary conditions (2.4.5) and (2.4.6) which define the $\mathbf{E}$ and $\mathbf{B}$ fields in terms of Φ and $\mathbf{A}$.

2.5 Charged Particle in Electromagnetic Fields

We have considered a continuum Lagrangian for a material system and one for the electromagnetic fields. We now consider an example of a system consisting of both matter and the electromagnetic fields from a Lagrangian viewpoint. The matter is taken simply to be a point particle of mass m and charge e. Its position is taken as $\mathbf{x}$; its velocity is thus $\dot{\mathbf{x}}$ and its current is $e\dot{\mathbf{x}}$.

We write the total Lagrangian L as a sum of the *electromagnetic field Lagrangian* L_F, the *interaction Lagrangian* L_I, and the *matter Lagrangian* L_M

$$L = L_F + L_I + L_M. \tag{2.5.1}$$

From the preceding section the field Lagrangian L_F is

$$L_F \equiv \int \mathcal{L}_F \, dv = \int \frac{\epsilon_0}{2}(\mathbf{E}^2 - c^2\mathbf{B}^2)\, dv \tag{2.5.2}$$

where we intend $\mathbf{E}$ and $\mathbf{B}$ to be regarded as functions of Φ and $\mathbf{A}$ by way of (2.4.5) and (2.4.6) and $dv \equiv dz_1 dz_2 dz_3$. The interaction Lagrangian is

$$L_I = e\dot{\mathbf{x}}(t) \cdot \mathbf{A}(\mathbf{x}(t),t) - e\Phi(\mathbf{x}(t),t) \tag{2.5.3}$$

where the potential functions are evaluated at the position of the particle $\mathbf{z} = \mathbf{x}(t)$. The matter Lagrangian for this system is simply

$$L_M = \tfrac{1}{2} m(\dot{\mathbf{x}})^2. \tag{2.5.4}$$

Since the matter in this system is a discrete particle, the equation of motion for $\mathbf{x}$ is found from the total Lagrangian (not a Lagrangian density) by

$$\frac{d}{dt}\frac{\partial L}{\partial \dot{x}_i} = \frac{\partial L}{\partial x_i}. \tag{2.5.5}$$

Note that the field Lagrangian gives no contribution to this equation because it is an integral over all $\mathbf{z}$-space and is not evaluated at the position of the particle $\mathbf{x}(t)$. The interaction Lagrangian gives a contribution as, of course, the matter Lagrangian also does. Substituting (2.5.1) to (2.5.4) into the equation above leads to

$$\frac{d}{dt}(m\dot{x}_i + eA_i) = e\dot{x}_j A_{j,i} - e\Phi_{,i}. \tag{2.5.6}$$

It must be remembered here that $\mathbf{A}$ depends on the time both implicitly and explicitly, $\mathbf{A} = \mathbf{A}(\mathbf{x}(t),t)$, and so

$$\frac{d}{dt}A_i(\mathbf{x}(t),t) = A_{i,j}\dot{x}_j + \frac{\partial A_i}{\partial t}. \tag{2.5.7}$$

Therefore

$$m\ddot{x}_i = e\left[\dot{x}_j(A_{j,i} - A_{i,j}) - \Phi_{,i} - \frac{\partial A_i}{\partial t}\right]. \tag{2.5.8}$$

This combination of space derivatives of **A** can be recognized as

$$\begin{aligned} A_{j,i} - A_{i,j} &= (\delta_{im}\delta_{jn} - \delta_{in}\delta_{jm})A_{n,m} \\ &= \epsilon_{kij}\epsilon_{kmn}A_{n,m} \\ &= \epsilon_{ijk}B_k, \end{aligned} \tag{2.5.9}$$

where δ_{ij} is the Kronecker delta, the $\epsilon-\delta$ identity,

$$\epsilon_{kij}\epsilon_{kmn} = \delta_{im}\delta_{jn} - \delta_{in}\delta_{jm}, \tag{2.5.10}$$

is used, and the definition of the vector potential (2.4.13) is substituted. Equation (2.5.8) now becomes

$$m\ddot{x}_i = eE_i + e\epsilon_{ijk}\dot{x}_j B_k \tag{2.5.11}$$

or

$$m\ddot{\mathbf{x}} = e(\mathbf{E} + \dot{\mathbf{x}} \times \mathbf{B}). \tag{2.5.12}$$

Thus the charge is acted on by an electric force $e\mathbf{E}$ and the Lorentz magnetic force $e\dot{\mathbf{x}} \times \mathbf{B}$.

The total Lagrangian of (2.5.1) should also yield the electromagnetic equations just as in the preceding section. Before it can be used for this, however, it is necessary to reexpress the interaction Lagrangian in the spatial coordinate system as is done in (2.4.7). The charge density q and the current density $\mathbf{j}$ of this equation must be expressed for the single charged particle by

$$q(\mathbf{z},t) = e\delta(\mathbf{z} - \mathbf{x}(t)), \tag{2.5.13}$$

$$\mathbf{j}(\mathbf{z},t) = e\dot{\mathbf{x}}(t)\delta(\mathbf{z} - \mathbf{x}(t)), \tag{2.5.14}$$

where $\delta(\mathbf{z} - \mathbf{x}(t))$ is the three-dimensional Dirac delta function. It is defined by

$$\delta(\mathbf{z} - \mathbf{x}(t)) = 0 \qquad [\mathbf{z} \neq \mathbf{x}(t)] \tag{2.5.15}$$

and

$$\int_{\text{all space}} \delta(\mathbf{z} - \mathbf{x}(t))\, dv = 1. \tag{2.5.16}$$

Its usefulness derives from its properties

$$\int f(\mathbf{z})\delta(\mathbf{z}-\mathbf{x}(t))\,dv = f(\mathbf{x}(t)), \tag{2.5.17}$$

$$\int f(\mathbf{z})\delta'(\mathbf{z}-\mathbf{x}(t))\,dv = -f'(\mathbf{x}(t)) \tag{2.5.18}$$

where the prime indicates the derivative. Using the property (2.5.17), we can reexpress (2.5.3) as

$$\begin{aligned} L_I &\equiv \int \mathfrak{L}_I\,dv \\ &= \int e\delta(\mathbf{z}-\mathbf{x}(t))\left[\dot{\mathbf{x}}(t)\cdot\mathbf{A}(\mathbf{z},t) - \Phi(\mathbf{z},t)\right] dv \\ &= \int \left[\mathbf{j}(\mathbf{z},t)\cdot\mathbf{A}(\mathbf{z},t) - q(\mathbf{z},t)\Phi(\mathbf{z},t)\right] dv. \end{aligned} \tag{2.5.19}$$

With this form for the interaction Lagrangian density the total Lagrangian density will now give the Maxwell-Lorentz equations (2.4.10) and (2.4.14) with $q(\mathbf{z},t)$ and $\mathbf{j}(\mathbf{z},t)$ given by (2.5.13) and (2.5.14). The matter Lagrangian, of course, does not contribute to these equations.

PROBLEMS

2.1. Derive (2.1.9) from Hamilton's principle assuming the functional dependence

$$\mathfrak{L} = \mathfrak{L}\left(\mathbf{x}, \frac{\partial \mathbf{x}}{\partial q_\nu}, \frac{\partial^2 \mathbf{x}}{\partial q_\nu \partial q_\sigma}, q_\mu\right). \tag{2.P.1}$$

2.2. Show that (2.1.7) is form invariant to the replacements in (2.1.10) and (2.1.11). Also show that (2.1.9) is form invariant to the replacement in (2.1.12).

2.3. Find the continuum Lagrangian from the discrete Lagrangian of Prob. 1.4. Show that the dispersion relations and velocities of the longitudinal and transverse modes agree with the long wavelength limit of the results of the discrete problem.

CHAPTER 3

Kinematics of Elastic Deformation

In the simple examples of elastic motion studied in Chapter 2, we simply invoked a Hooke's law restoring force in order to construct the potential energy of the matter. To obtain the generality we wish in later chapters, it is necessary to construct a potential energy in a general form constrained only by the requirements of the conservation laws. In particular, angular momentum conservation requires that the potential energy be invariant to spatial rotations. We meet this requirement by expressing the potential energy as a function of rotationally invariant variables. Since we do not wish to restrict the theory to infinitesimal deformations, we develop some of the tools of finite deformation theory in this chapter to aid in constructing the rotational invariants in Chapter 6.

The key to understanding finite deformation theory is an appreciation of the dualism in the description of the *undeformed* and the *deformed* states of matter and an appreciation of the two sets of coordinates, *material* and *spatial*, that are used to describe those states respectively. We attempt to elucidate these concepts while presenting those portions of the theory of finite deformation that we need. To aid clarity we employ only rectangular Cartesian coordinate systems, rather than the general curvilinear coordinate systems often used in this field of study. This choice, of course, cannot affect the physics involved. We also expand many of the expressions we obtain for the case of small deformations in order to illustrate their meaning and to obtain relations useful to our later work. The first part of the chapter is based on Truesdell and Toupin [1960].

3.1 Deformation

Deformation of matter, as used here, refers to changing its size, shape, orientation, and location *without* causing breakage, cracking, or slippage, which destroy the continuity and reversibility of the process. Deformation is depicted in Fig. 3.1. Let $\mathbf{X}(P)$ denote the position vector of a point of

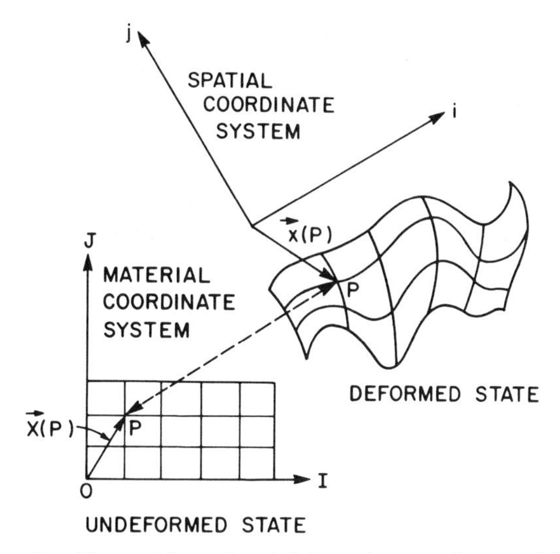

Fig. 3.1. Deformation. The undeformed and deformed states of a material body are shown schematically. A particular mass point P is shown in both states. The transformations in (3.1.1) and (3.1.2) are indicated by the double headed dashed arrow. The material and spatial coordinate systems in which the undeformed position $\mathbf{X}(P)$ and the deformed position $\mathbf{x}(P)$ are measured respectively are also shown.

matter P in the undeformed state and $\mathbf{x}(P)$ denote the position vector of the same point of matter in the deformed state. The deformation is thus characterized by the transformation,

$$\mathbf{x}=\mathbf{x}(\mathbf{X},t), \tag{3.1.1}$$

which carries the position $\mathbf{X}$ of each point in the undeformed region V into the position $\mathbf{x}$ in the deformed region v at the instant of time t. The inverse transformation is

$$\mathbf{X}=\mathbf{X}(\mathbf{x},t). \tag{3.1.2}$$

The continuity required of a deformation in the definition above implies that the transformation (3.1.1) and its inverse (3.1.2) are single valued and differentiable as many times as needed. It also implies that no region of finite volume is deformed into one of zero or infinite volume. This implies that the Jacobian of either the transformation (3.1.1) or (3.1.2) cannot be infinite or zero. We do, however, allow the Jacobian to be either positive or negative. In Section 3.8 we show that this continuity implies conservation of mass in a deformation as defined here.

The components X_1, X_2, X_3 of the position vector $\mathbf{X}$ in the undeformed state are called the *material coordinates* of the mass point. The material coordinates are the *name* of a mass point and as such remain with the mass point in the deformed state. An uppercase Latin letter is used as a subscript to denote these components as $X_K(K=1,2,3)$. The components x_1, x_2, x_3 of the position vector $\mathbf{x}$ in the deformed state are called the *spatial coordinates* of the mass point. A lowercase Latin letter is used as a subscript to denote these components as $x_k(k=1,2,3)$. The two coordinate systems with respect to which the components X_K and x_k respectively are measured are considered to be *different* rectangular Cartesian systems for the time being. The distinction between the two systems is useful to us when we come to considering the various invariances of the potential energy in Chapter 6. Later in the chapter when we introduce the displacement vector into the theory, we choose the two coordinate systems to be identical.

It should be understood that (3.1.1) and (3.1.2) represent a *physical* transformation, not a coordinate transformation, that is, they do not represent the relation between the two sets of coordinate axes. Neither does the physical transformation affect the coordinate axes; they remain fixed and rectangular. The function $\mathbf{x}(\mathbf{X},t)$ represents the deformed body as pictured in Fig. 3.1, that is, if the material coordinates X_1, X_2, X_3 of any point P of the undeformed body and the time t are given, the function $\mathbf{x}(\mathbf{X},t)$ determines the coordinates x_1, x_2, x_3 of that point in the deformed body relative to the spatial coordinate axes. Conversely, the function $\mathbf{X}(\mathbf{x},t)$ represents the undeformed body as pictured in Fig. 3.1, that is, if the spatial coordinates x_1, x_2, x_3 of any point P in the deformed body and the time t are given, the function $\mathbf{X}(\mathbf{x},t)$ determines the coordinates X_1, X_2, X_3 of that point in the undeformed body relative to the material coordinate axes.

When the material coordinates X_K and the time t are considered the independent variables, we say the theory is set in the *material description*; when the spatial coordinates x_k and the time t are considered the independent variables, we say the theory is set in the *spatial description*. This leads to a *dualism* in the theory whereby analogous formulas exist in the two descriptions. The formulas can be related to each other formally by replacing uppercase letters with lowercase letters and vice versa.

The *deformation gradients* $\partial x_k/\partial X_K$ and $\partial X_K/\partial x_k$, which we denote by

$$x_{k,K} \equiv \frac{\partial x_k}{\partial X_K}, \tag{3.1.3}$$

$$X_{K,k} \equiv \frac{\partial X_K}{\partial x_k}, \tag{3.1.4}$$

are the fundamental measures of local deformation. Clearly, they are inverses of each other,

$$x_{k,K}X_{K,l}=\delta_{kl}, \tag{3.1.5}$$

$$X_{K,k}x_{k,L}=\delta_{KL} \tag{3.1.6}$$

as seen by the chain rule of differentiation.

The Jacobian $J(\mathbf{x}/\mathbf{X})$ of the transformation (3.1.1) from the material coordinates X_K to the spatial coordinates x_k is defined in terms of the deformation gradients by

$$J(\mathbf{x}/\mathbf{X})\equiv\det\left(\frac{\partial\mathbf{x}}{\partial\mathbf{X}}\right) \tag{3.1.7a}$$

$$\equiv\begin{vmatrix} \dfrac{\partial x_1}{\partial X_1} & \dfrac{\partial x_1}{\partial X_2} & \dfrac{\partial x_1}{\partial X_3} \\ \dfrac{\partial x_2}{\partial X_1} & \dfrac{\partial x_2}{\partial X_2} & \dfrac{\partial x_2}{\partial X_3} \\ \dfrac{\partial x_3}{\partial X_1} & \dfrac{\partial x_3}{\partial X_2} & \dfrac{\partial x_3}{\partial X_3} \end{vmatrix} \tag{3.1.7b}$$

$$\equiv\epsilon_{IJK}x_{1,I}x_{2,J}x_{3,K} \tag{3.1.7c}$$

$$\equiv\epsilon_{ijk}x_{i,1}x_{j,2}x_{k,3} \tag{3.1.7d}$$

$$\equiv\tfrac{1}{6}\epsilon_{ijk}\epsilon_{IJK}x_{i,I,}x_{j,J}x_{k,K} \tag{3.1.7e}$$

where in (3.1.7c) and (3.1.7d) the numbers 1,2,3 can be replaced with any even permutation of them. The Jacobian $j(\mathbf{X}/\mathbf{x})$ of the inverse transformation (3.1.2) is the inverse Jacobian $J^{-1}(\mathbf{x}/\mathbf{X})$ of the transformation (3.1.1),

$$j\left(\frac{\mathbf{X}}{\mathbf{x}}\right)=J^{-1}\left(\frac{\mathbf{x}}{\mathbf{X}}\right). \tag{3.1.8}$$

For brevity expressions for $j(\mathbf{X}/\mathbf{x})$ that are dual to (3.1.7) are not recorded.

Explicit expressions for either of the deformation gradients can be found from either of the relations (3.1.5) or (3.1.6). Each of these represents nine linear equations in nine unknowns, $x_{i,J}$ or $X_{J,i}$. Cramer's rule for the solution of linear equations yields

$$X_{L,l}=\frac{\operatorname{cof}(x_{l,L})}{\det(\partial\mathbf{x}/\partial\mathbf{X})}=\frac{\operatorname{cof}(x_{l,L})}{J(\mathbf{x}/\mathbf{X})} \tag{3.1.9}$$

where the cofactor, $\mathrm{cof}(x_{l,L})$, of the $x_{l,L}$ element of the determinant $\det(\partial\mathbf{x}/\partial\mathbf{X})$ is given by

$$\begin{aligned}\mathrm{cof}(x_{l,L}) \equiv & \epsilon_{LJK}\delta_{1l}x_{2,J}x_{3,K}\\ & +\epsilon_{ILK}x_{1,I}\delta_{2l}x_{3,K}\\ & +\epsilon_{IJL}x_{1,I}x_{2,J}\delta_{3l}\\ = & \tfrac{1}{2}\epsilon_{ljk}\epsilon_{LJK}x_{j,J}x_{k,K}.\end{aligned} \tag{3.1.10}$$

Equations dual to (3.1.9) and (3.1.10) exist but need not be written down.

3.2 Deformation Identities

Several simple identities involving deformation gradients that are useful in succeeding chapters can be derived at this point. First, an expression for the derivative of the deformation gradient $x_{k,K}$ with respect to the dual deformation gradient $X_{J,j}$ can be found by differentiating (3.1.5),

$$\frac{\partial x_{k,L}}{\partial X_{J,j}}X_{L,l}+x_{k,L}\delta_{LJ}\delta_{lj}=0. \tag{3.2.1}$$

We then form the scalar product with $x_{l,K}$,

$$\frac{\partial x_{k,L}}{\partial X_{J,j}}X_{L,l}x_{l,K}+x_{k,J}x_{j,K}=0, \tag{3.2.2}$$

and use (3.1.6) to obtain the identity

$$\frac{\partial x_{k,K}}{\partial X_{J,j}}=-x_{k,J}x_{j,K}. \tag{3.2.3}$$

The derivative of the Jacobian with respect to a deformation gradient can be found by differentiating (3.1.7c),

$$\begin{aligned}\frac{\partial J(\mathbf{x}/\mathbf{X})}{\partial x_{l,L}} &= \frac{\partial}{\partial x_{l,L}}\left[\epsilon_{IJK}x_{1,I}x_{2,J}x_{3,K}\right]\\ &= \epsilon_{LJK}\delta_{1l}x_{2,J}x_{3,K}\\ &\quad +\epsilon_{ILK}x_{1,I}\delta_{2l}x_{3,K}\\ &\quad +\epsilon_{IJL}x_{1,I}x_{2,J}\delta_{3l}\\ &= \mathrm{cof}(x_{l,L})\\ &= X_{L,l}J\left(\frac{\mathbf{x}}{\mathbf{X}}\right),\end{aligned} \tag{3.2.4}$$

where use of (3.1.10) and (3.1.9) is made in the last two steps. Equation (3.2.4) is called *Jacobi's identity*. Since

$$\frac{\partial J^{-1}(\mathbf{x}/\mathbf{X})}{\partial x_{l,L}} = -J^{-2}\left(\frac{\mathbf{x}}{\mathbf{X}}\right)\frac{\partial J(\mathbf{x}/\mathbf{X})}{\partial x_{l,L}}, \tag{3.2.5}$$

we also find

$$\frac{\partial J^{-1}(\mathbf{x}/\mathbf{X})}{\partial x_{l,L}} = -X_{L,l}J^{-1}\left(\frac{\mathbf{x}}{\mathbf{X}}\right). \tag{3.2.6}$$

Next we derive the important identity of Euler, Piola, and Jacobi. Consider the quantity $[J^{-1}(\mathbf{x}/\mathbf{X})x_{j,A}]_{,j}$. Performing the differentiation with due regard to the functional dependences of the various quantities yields

$$\left[J^{-1}\left(\frac{\mathbf{x}}{\mathbf{X}}\right)x_{j,K}\right]_{,j} = \frac{\partial J^{-1}(\mathbf{x}/\mathbf{X})}{\partial x_{l,L}}x_{l,LM}X_{M,j}x_{j,K} + J^{-1}\left(\frac{\mathbf{x}}{\mathbf{X}}\right)x_{j,KL}X_{L,j}. \tag{3.2.7}$$

Modifying the first term on the right side with the use of (3.2.6) and (3.1.6) yields

$$\left[J^{-1}\left(\frac{\mathbf{x}}{\mathbf{X}}\right)x_{j,K}\right]_{,j} = -J^{-1}\left(\frac{\mathbf{x}}{\mathbf{X}}\right)X_{L,l}x_{l,LM}\delta_{MK} + J^{-1}\left(\frac{\mathbf{x}}{\mathbf{X}}\right)X_{L,j}x_{j,KL} \tag{3.2.8}$$

whose right side is seen to vanish resulting in the *Euler-Piola-Jacobi identity*

$$\left[J^{-1}\left(\frac{\mathbf{x}}{\mathbf{X}}\right)x_{j,K}\right]_{,j} = 0. \tag{3.2.9}$$

By the dualism between the material and spatial descriptions, we know that there are identities dual to those just derived. For later reference we record them here. Equations dual respectively to (3.2.3), (3.2.4), (3.2.6), and (3.2.9) are

$$\frac{\partial X_{J,j}}{\partial x_{k,K}} = -X_{J,k}X_{K,j}, \tag{3.2.10}$$

$$\frac{\partial J^{-1}(\mathbf{x}/\mathbf{X})}{\partial X_{L,l}} = x_{l,L}J^{-1}\left(\frac{\mathbf{x}}{\mathbf{X}}\right), \tag{3.2.11}$$

$$\frac{\partial J\left(\frac{\mathbf{x}}{\mathbf{X}}\right)}{\partial X_{L,l}} = -x_{l,L}J\left(\frac{\mathbf{x}}{\mathbf{X}}\right), \tag{3.2.12}$$

$$\left[J\left(\frac{\mathbf{x}}{\mathbf{X}}\right)X_{J,k}\right]_{,J} = 0 \tag{3.2.13}$$

where use of (3.1.8) is made.

3.3 Transformation of Elements of Arc, Area, and Volume

We next consider the transformation of elements of arc, area, and volume. Consider two mass points with coordinates X_K and $X_K + dX_K$ in the undeformed medium. Denote the coordinates of the same two mass points in the deformed medium by x_k and $x_k + dx_k$. The transformation (3.1.1) then implies a relation between the elements of arc, $d\mathbf{x}$ and $d\mathbf{X}$,

$$dx_k = x_{k,K} dX_K. \tag{3.3.1}$$

This relation can be inverted by forming the scalar product of it with $X_{J,k}$ and using (3.1.6). The result is

$$dX_J = X_{J,k} dx_k \tag{3.3.2}$$

which can also be looked upon as implied by the transformation (3.1.2).

Next let us consider how an element of area transforms. Consider a surface in the undeformed material parameterized by L and M. Its equation is thus

$$\mathbf{X} = \mathbf{X}(L, M), \tag{3.3.3}$$

which is shown in Fig. 3.2. From vector analysis, it is known that the vector product of two vectors (1) has a magnitude equal to the area formed by the parallelogram two of whose adjoining sides are the vectors and (2) has a direction normal to the plane of the vectors in the right-handed sense. Thus the two vector elements of arc $(\partial\mathbf{X}/\partial L)dL$ and $(\partial\mathbf{X}/\partial M)dM$ form an oriented element of area $d\mathbf{A}$ in the undeformed medium of

$$dA_I = \epsilon_{IJK} \frac{\partial X_J}{\partial L} dL \frac{\partial X_K}{\partial M} dM. \tag{3.3.4}$$

In the deformed medium the surface represented by (3.3.3) is transformed to a surface represented by

$$\mathbf{x} = \mathbf{x}(\mathbf{X}(L, M)). \tag{3.3.5}$$

The oriented element of area in the deformed medium corresponding to dA_I is

$$\begin{aligned} da_i &= \epsilon_{ijk} \frac{\partial x_j}{\partial L} dL \frac{\partial x_k}{\partial M} dM \\ &= \epsilon_{ijk} x_{j,J} \frac{\partial X_J}{\partial L} dL \, x_{k,K} \frac{\partial X_K}{\partial M} dM. \end{aligned} \tag{3.3.6}$$

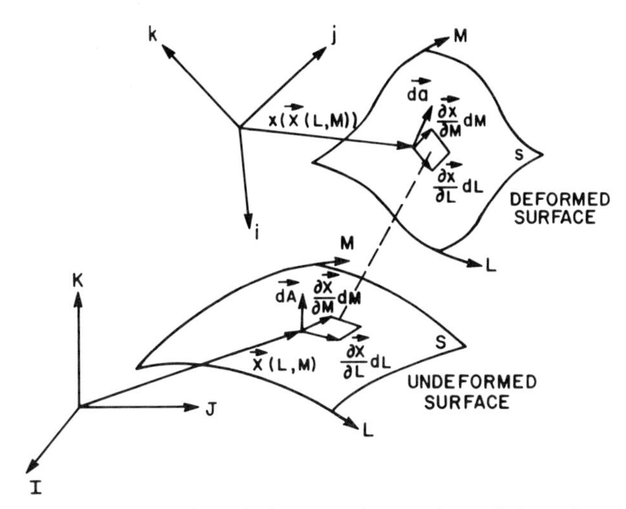

Fig. 3.2. Transformation of oriented elements of area. An undeformed surface S, parameterized by L, M, is transformed by the deformation into the deformed surface s. The undeformed oriented area element $d\mathbf{A}$ is transformed into the deformed oriented area element $d\mathbf{a}$. The material (I, J, K) and spatial (i, j, k) coordinate systems used for the measurement of the undeformed and deformed states respectively are shown.

We now need the transformation equation for the permutation symbol between the material and spatial coordinates. We can derive this from the definition of the Jacobian. Consider the dual of (3.1.7d),

$$j\left(\frac{\mathbf{X}}{\mathbf{x}}\right) = \epsilon_{PQR} X_{P,1} X_{Q,2} X_{R,3}. \tag{3.3.7}$$

It can be seen from simply expanding (3.3.7) that (1) if the indices 1,2,3 are replaced by an even permutation of them, the right side remains the same, (2) if they are replaced by an odd permutation, the right side changes sign, and (3) if they are replaced by a combination, two of which are equal, the right side vanishes. This we recognize as just the meaning of the permutation symbol ϵ_{ijk} except for normalization. If (3.3.7) is divided by $j(\mathbf{X}/\mathbf{x}) = J^{-1}(\mathbf{x}/\mathbf{X})$, the right side is properly normalized and equal to ϵ_{ijk},

$$\epsilon_{ijk} = J\left(\frac{\mathbf{x}}{\mathbf{X}}\right) \epsilon_{PQR} X_{P,i} X_{Q,j} X_{R,k}. \tag{3.3.8}$$

We note for later reference that a similar transformation equation,

$$\epsilon_{ijk} = J^{-1}\left(\frac{\mathbf{x}}{\mathbf{X}}\right) \epsilon_{ABC} x_{i,A} x_{j,B} x_{k,C}, \tag{3.3.9}$$

follows from (3.1.7c).

The result (3.3.8) can be substituted into (3.3.6) and (3.1.6) used to obtain

$$da_i = J\left(\frac{\mathbf{x}}{\mathbf{X}}\right) X_{P,i} \epsilon_{PJK} \frac{\partial X_J}{\partial L} dL \frac{\partial X_K}{\partial M} dM. \tag{3.3.10}$$

Equation (3.3.4) can now be substituted into this equation to get the final result,

$$da_i = J\left(\frac{\mathbf{x}}{\mathbf{X}}\right) X_{P,i} dA_P. \tag{3.3.11}$$

This is *Nanson's formula*, which relates oriented area elements in the undeformed and deformed configurations.

Separate transformation formulas between the magnitudes of the area elements, da and dA, of the deformed and undeformed media and between the associated unit normals, $\mathbf{n}$ and $\mathbf{N}$, are also useful. Thus we set

$$d\mathbf{a} = \mathbf{n}\, da, \tag{3.3.12}$$

$$d\mathbf{A} = \mathbf{N}\, dA. \tag{3.3.13}$$

Substituting these two equations into (3.3.11) gives

$$n_i\, da = J\left(\frac{\mathbf{x}}{\mathbf{X}}\right) X_{P,i} N_P\, dA. \tag{3.3.14}$$

By forming the scalar product with itself we obtain

$$da = J\left(\frac{\mathbf{x}}{\mathbf{X}}\right) \left[X_{P,i} N_P X_{Q,i} N_Q \right]^{1/2} dA. \tag{3.3.15}$$

Dividing this into the previous equation gives

$$n_i = X_{R,i} N_R \left[X_{P,j} N_P X_{Q,j} N_Q \right]^{-1/2}. \tag{3.3.16}$$

The transformation of a volume element from the undeformed medium to the deformed medium can be found by a procedure similar to that just used for the area element. It is based on the notion from vector analysis that the scalar triple product is the volume of a parallelopiped three of whose intersecting edges are the three vectors of the product. If the volume is parameterized in terms of L, M, N, then an element of volume $dV \equiv dX_1 dX_2 dX_3$ of the undeformed medium is the scalar triple product of $(\partial \mathbf{X}/\partial L)dL$, $(\partial \mathbf{X}/\partial M)dM$, and $(\partial \mathbf{X}/\partial N)dN$,

$$dV = \epsilon_{IJK} \frac{\partial X_I}{\partial L} \frac{\partial X_J}{\partial M} \frac{\partial X_K}{\partial N} dL\, dM\, dN, \tag{3.3.17}$$

while the deformed element of volume $dv = dx_1 dx_2 dx_3$ corresponding to dV is

$$dv = \epsilon_{ijk} \frac{\partial x_i}{\partial L} \frac{\partial x_j}{\partial M} \frac{\partial x_k}{\partial N} dL\, dM\, dN. \tag{3.3.18}$$

Substitution of (3.3.8) and (3.3.17) into the latter yields

$$dv = J\left(\frac{\mathbf{x}}{\mathbf{X}}\right) dV. \tag{3.3.19}$$

If $J(\mathbf{x}/\mathbf{X})$ is negative, the limits of integration are reversed in the deformed medium so that the volume remains positive.

3.4 Strain

Strain is the change in size or shape of a body resulting from changes in length and relative orientation of the various portions of the body. A number of measures of finite strain are in common use; we discuss several of them in this section. The infinitesimal strain tensor is introduced in Section 3.11.

Denote the length of the element of arc dx_k, given in (3.3.1), by ds. Thus

$$ds^2 = dx_k\, dx_k \tag{3.4.1a}$$

$$= x_{k,M} x_{k,N}\, dX_M\, dX_N \tag{3.4.1b}$$

$$= C_{MN}\, dX_M\, dX_N \tag{3.4.1c}$$

where

$$C_{MN} \equiv x_{k,M} x_{k,N}. \tag{3.4.2}$$

The quantity $\mathbf{C}$ is called the *Green deformation tensor*. Analogously, the length of the element of arc dX_K, given in (3.3.2), is denoted by dS. Thus

$$dS^2 = dX_K\, dX_K \tag{3.4.3a}$$

$$= X_{K,m} X_{K,n}\, dx_m\, dx_n \tag{3.4.3b}$$

$$= c_{mn}\, dx_m\, dx_n \tag{3.4.3c}$$

where

$$c_{mn} \equiv X_{K,m} X_{K,n}. \tag{3.4.4}$$

The quantity **c** is called the *Cauchy deformation tensor*. It is apparent from the forms (3.4.1a) and (3.4.1c) that a sphere about **x** of radius ds is transformed into an ellipsoid at **X**. Hence (3.4.1c) is called the *strain ellipsoid at* **X** or the *material strain ellipsoid*. In an analogous manner we see from the forms (3.4.3a) and (3.4.3c) that a sphere about **X** of radius dS is transformed into an ellipsoid at **x**. Thus (3.4.3c) is called the *strain ellipsoid at* **x** or the *spatial strain ellipsoid*. A deformation is said to be *rigid* if at every point

$$C_{MN} = \delta_{MN}, \qquad c_{mn} = \delta_{mn}, \tag{3.4.5}$$

that is, if the distance between every pair of material points remains unchanged.

Define **l** and **L** as unit vectors in the directions of $d\mathbf{x}$ and $d\mathbf{X}$, given in (3.3.1) and (3.3.2). Thus we set

$$d\mathbf{x} = \mathbf{l}\,ds, \tag{3.4.6}$$

$$d\mathbf{X} = \mathbf{L}\,dS. \tag{3.4.7}$$

Substitution of these into (3.4.3c) and (3.4.1c) respectively yields

$$ds = \left[C_{MN} L_M L_N \right]^{1/2} dS, \tag{3.4.8}$$

$$dS = \left[c_{mn} l_m l_n \right]^{1/2} ds. \tag{3.4.9}$$

The *stretch* λ is defined by

$$\lambda \equiv \frac{ds}{dS}. \tag{3.4.10}$$

Since ds/dS can be expressed as a function of **L** or **l**, the same is true of λ. Two expressions for the stretch are thus obtained,

$$\lambda = \lambda(\mathbf{L}) = \left[C_{MN} L_M L_N \right]^{1/2}, \tag{3.4.11}$$

$$\lambda = \lambda(\mathbf{l}) = \left[c_{mn} l_m l_n \right]^{-1/2}, \tag{3.4.12}$$

the first being spoken of as the stretch in the direction **L** and the second as the stretch in the direction **l**. They are, as indicated, numerically equal.

If (3.4.6) and (3.4.7) are combined with (3.3.1), we find

$$l_i = \frac{x_{i,I} L_I}{\lambda(\mathbf{L})} \tag{3.4.13}$$

with $\lambda(\mathbf{L})$ given by (3.4.11). This equation relates the directions of the elements of arc in the undeformed and deformed states. The dual of this equation is

$$L_I = \lambda(\mathbf{l}) X_{I,i} l_i \tag{3.4.14}$$

with $\lambda(\mathbf{l})$ now given by (3.4.12).

Before closing this section we wish to point out that, though the Cauchy and Green deformation tensors, $\mathbf{c}$ and $\mathbf{C}$, are the natural strain measures for a discussion of the strain ellipsoids, neither reduces to the infinitesimal strain tensor when the deformation gradients are small. The *Green finite strain tensor* $\mathbf{E}$,

$$E_{AB} \equiv \tfrac{1}{2}(C_{AB} - \delta_{AB}), \tag{3.4.15}$$

does, however, and for this reason we use this measure of strain most often in later chapters.

3.5 Principle Directions of Strain Ellipsoids

In this section we wish to find the principal axes of the two strain ellipsoids and explore their connection. Consider the spatial strain ellipsoid first. The principal axes are in the directions of three particular unit vectors $\mathbf{l}^\alpha$ $(\alpha = 1,2,3)$ among the class of unit vectors of (3.4.13). We call these the *principal vectors*. They are defined as being those vectors which diagonalize $\mathbf{c}$,

$$c_{mn} l_n^\alpha = c_\alpha l_m^\alpha. \tag{3.5.1}$$

These three equations have a solution for the three vectors $\mathbf{l}^\alpha$ only if the determinant of the coefficients vanishes,

$$\det(\mathbf{c} - c_\alpha \mathbf{1}) = 0, \tag{3.5.2}$$

which gives a cubic equation for the *principal values* c_α. By forming the scalar product of (3.5.1) with l_m^α and comparing with (3.4.12), it can be seen that the principal values c_α of the tensor $\mathbf{c}$ are the inverse squares of the *principal stretches*, $\lambda_\alpha \equiv \lambda(\mathbf{l}^\alpha)$, which are the stretches along the principal axes $\mathbf{l}^\alpha$,

$$c_\alpha = l_m^\alpha c_{mn} l_n^\alpha = \lambda_\alpha^{-2}. \tag{3.5.3}$$

The orthogonality of the principal vectors $\mathbf{l}^\alpha$ is shown by forming the scalar product of (3.5.1) with $\mathbf{l}^\beta$ and subtracting from it the same equation with α and β interchanged,

$$(c_\alpha - c_\beta) l_m^\alpha l_m^\beta = l_m^\beta c_{mn} l_n^\alpha - l_m^\alpha c_{mn} l_n^\beta = 0, \tag{3.5.4}$$

where the vanishing of the right side follows from the interchange symmetry of $\mathbf{c}$,

$$c_{mn} = c_{nm}, \tag{3.5.5}$$

apparent from (3.4.4). It now follows from (3.5.4) and the fact that the $\mathbf{l}^\alpha$ $(\alpha = 1,2,3)$ are unit vectors that

$$\mathbf{l}^\alpha \cdot \mathbf{l}^\beta = \delta^{\alpha\beta} \tag{3.5.6}$$

provided that $c_\alpha \neq c_\beta$. If the latter is not true, the three $\mathbf{l}^\alpha$ vectors may still be chosen to satisfy (3.5.6). Equation (3.5.6) is the *orthonormality condition* for the principal vectors.

The orthogonal triad of vectors $\mathbf{l}^\alpha$ $(\alpha = 1,2,3)$ can be used to express an arbitrary vector $\mathbf{a}$ in the spatial frame as

$$\mathbf{a} = \sum_{\alpha=1}^{3} a_\alpha \mathbf{l}^\alpha \tag{3.5.7}$$

where a_α are numbers found by forming the scalar product of this equation with $\mathbf{l}^\beta$ and using (3.5.6). Thus

$$a_\alpha = \mathbf{a} \cdot \mathbf{l}^\alpha. \tag{3.5.8}$$

Combining the last two equations yields

$$a_i = a_j \sum_\alpha l_j^\alpha l_i^\alpha. \tag{3.5.9}$$

Since $\mathbf{a}$ is an arbitrary vector, this implies

$$\sum_\alpha l_j^\alpha l_i^\alpha = \delta_{ji}, \tag{3.5.10}$$

which is the *completeness relation* for the set of vectors $\mathbf{l}^\alpha$ $(\alpha = 1,2,3)$.

We can now express the stretch in an arbitrary direction **l** in terms of the principal stretches λ_α. From (3.5.1) and (3.5.3) we can form

$$l_m c_{mn} l_n^\alpha = \lambda_\alpha^{-2} l_m l_m^\alpha = \lambda_\alpha^{-2} \cos\theta_\alpha \tag{3.5.11}$$

where θ_α is the angle between **l** and $\mathbf{l}^\alpha$. Multiplication by $\mathbf{l}\cdot\mathbf{l}^\alpha = \cos\theta_\alpha$, summation over α, and use of (3.5.9) give

$$l_m c_{mn} l_n = \sum_\alpha \lambda_\alpha^{-2} \cos^2\theta_\alpha. \tag{3.5.12}$$

Now using the expression (3.4.12) for the stretch in direction **l** we obtain the desired result,

$$\lambda(l) = \left[\sum_\alpha \frac{\cos^2\theta_\alpha}{\lambda_\alpha^2} \right]^{-1/2}. \tag{3.5.13}$$

The completeness relation can be used to obtain another expression of use to us later. Multiply (3.5.1) by l_p^α and sum over α,

$$c_{mn} \sum_\alpha l_n^\alpha l_p^\alpha = \sum_\alpha c_\alpha l_m^\alpha l_p^\alpha. \tag{3.5.14}$$

Use of the completeness relation (3.5.10) on the left side yields

$$c_{mp} = \sum_\alpha c_\alpha l_m^\alpha l_p^\alpha, \tag{3.5.15}$$

which is an expansion of the Cauchy deformation tensor in terms of its principal vectors and principal values. Since the square root of a tensor is defined as that tensor which, when multiplied by itself, yields the original tensor, it is immediately apparent from (3.5.15) that

$$\begin{aligned}(c^{1/2})_{mn} &= \sum_\alpha c_\alpha^{1/2} l_m^\alpha l_n^\alpha \\ &= \sum_\alpha \lambda_\alpha^{-1} l_m^\alpha l_n^\alpha \end{aligned} \tag{3.5.16}$$

where (3.5.3) is used in the latter form.

Let us next investigate the significance of the unit vectors $\mathbf{L}^\alpha$ in the undeformed state which are the transforms by (3.4.14) of $\mathbf{l}^\alpha$,

$$L_I^\alpha = \lambda(\mathbf{l}^\alpha) X_{I,i} l_i^\alpha. \tag{3.5.17}$$

Consider the quantity $C_{MN}L_N^\alpha$. Using successively (3.4.2), (3.5.17), (3.1.5), (3.5.1), (3.5.3), (3.4.4), (3.1.6), and (3.5.17), we obtain

$$\begin{aligned} C_{MN}L_N^\alpha &= x_{k,M}x_{k,N}\lambda(\mathbf{l}^\alpha)X_{N,i}l_i^\alpha \\ &= x_{k,M}\lambda_\alpha l_k^\alpha \\ &= x_{k,M}\lambda_\alpha^3 c_{kn}l_n^\alpha \\ &= \lambda_\alpha^3 x_{k,M}X_{K,k}X_{K,n}l_n^\alpha \\ &= \lambda_\alpha^3 X_{M,n}l_n^\alpha \\ &= C_\alpha L_M^\alpha \end{aligned} \tag{3.5.18}$$

where

$$C_\alpha = \lambda_\alpha^2 \equiv \lambda^2(\mathbf{L}^\alpha) = C_{MN}L_M^\alpha L_N^\alpha. \tag{3.5.19}$$

Thus we see that the three principal vectors $\mathbf{L}^\alpha$ diagonalize the tensor $\mathbf{C}$, that is, lie respectively along the three principal axes of the material strain ellipsoid and that the three principal values C_α are the squares of the principal stretches λ_α. To show that the $\mathbf{L}^\alpha$ vectors also form an orthonormal set of vectors we evaluate

$$\begin{aligned} L_I^\alpha L_I^\beta &= \lambda(\mathbf{l}^\alpha)\lambda(\mathbf{l}^\beta)X_{I,i}X_{I,j}l_i^\alpha l_j^\beta \\ &= \lambda_\alpha\lambda_\beta c_{ij}l_i^\alpha l_j^\beta \\ &= \lambda_\alpha\lambda_\beta^{-1}l_i^\alpha l_i^\beta \\ &= \delta^{\alpha\beta} \end{aligned} \tag{3.5.20}$$

with successive use of (3.5.17), (3.4.4), (3.5.1), (3.5.3), and (3.5.6). Since the three $\mathbf{L}^\alpha$ vectors can be used to express any vector in the material frame, a completeness relation,

$$\sum_\alpha L_I^\alpha L_J^\alpha = \delta_{IJ}, \tag{3.5.21}$$

analogous to (3.5.10), holds for them also.

Several other formulas analogous to ones found in the spatial frame are worth recording here. By a procedure similar to that leading to (3.5.13) it can be shown that the stretch in an arbitrary direction $\mathbf{L}$ in the material frame can be expressed in terms of the principal stretches by

$$\lambda(L) = \left[\sum_\alpha \lambda_\alpha^2 \cos^2\Theta_\alpha\right]^{1/2} \tag{3.5.22}$$

where Θ_α is the angle between $\mathbf{L}$ and $\mathbf{L}^\alpha$. The duals of (3.5.15) and (3.5.16),

$$C_{MP}=\sum_\alpha C_\alpha L_M^\alpha L_P^\alpha, \tag{3.5.23}$$

$$\begin{aligned}(C^{1/2})_{MN}&=\sum_\alpha C_\alpha^{1/2} L_M^\alpha L_N^\alpha\\ &=\sum_\alpha \lambda(\mathbf{L}^\alpha) L_M^\alpha L_N^\alpha,\end{aligned} \tag{3.5.24}$$

which are expansions of the Green deformation tensor and its square root in terms of the principal vectors and principal values of the former, are important relations that we need in succeeding chapters.

The major results of this section are illustrated in Fig. 3.3 and can be summarized as follows. First, the principal vectors $\mathbf{l}^\alpha$ of the spatial strain ellipsoid transform by way of (3.5.17) into the principal vectors $\mathbf{L}^\alpha$ of the material strain ellipsoid. Through the inverse transformation,

$$l_i^\alpha=\frac{x_{i,I}L_I^\alpha}{\lambda(\mathbf{L}^\alpha)}, \tag{3.5.25}$$

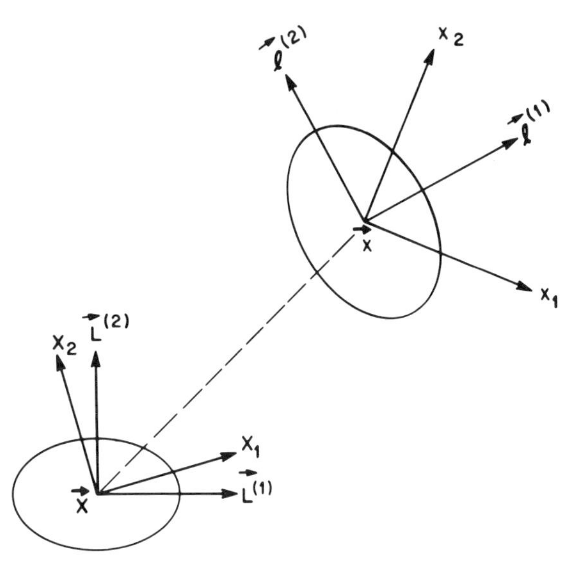

Fig. 3.3. Relation between strain ellipsoids. A two-dimensional projection of the material strain ellipsoid at $\mathbf{X}$ and the spatial strain ellipsoid at $\mathbf{x}$ are shown. The principal axes of one are rotated into those of the other. The ratio of the lengths of the principal axes of the spatial strain ellipsoid equals the ratio of the corresponding principal stretches; the ratio of the lengths of the principal axes of the material strain ellipsoid equals the ratio of the inverses of the corresponding principal stretches.

the inverse statement, of course, also holds true. Second, the principal values of the spatial strain ellipsoid are the inverse squares of the principal stretches and those of the material strain ellipsoid are the squares of the principal stretches.

3.6 Rotation

We now wish to define a measure of the part of a general deformation that is a *rotation*. Consider the two sets of principal vectors, that is, unit vectors along the principal axes, found in Section 3.5. The vectors $\mathbf{L}^\alpha$ $(\alpha = 1,2,3)$ define the principal axes of the material strain ellipsoid at $\mathbf{X}$ while the vectors $\mathbf{l}^\alpha$ $(\alpha = 1,2,3)$ define the principal axes of the spatial strain ellipsoid at $\mathbf{x}$. We call the object R_{kK}, which transforms the triad $\mathbf{L}^\alpha$ into the triad $\mathbf{l}^\alpha$ by

$$l_k^\alpha = R_{kK} L_K^\alpha, \tag{3.6.1}$$

the *finite rotation tensor*. Note that one tensor index is referred to the spatial coordinate system while the other index is referred to the material coordinate system. This does not affect the fact that it is a rotation tensor, since the two triads are, of course, both in ordinary space as seen in Fig. 3.3.

The finite rotation tensor can be decomposed into two parts according to

$$R_{kK} = R_{kl} g_{lK}. \tag{3.6.2}$$

Here R_{kl} is the tensor, referred entirely to the spatial coordinate system, that represents the physical rotation of the principal vectors of the material strain ellipsoid at $\mathbf{X}$ into those of the spatial strain ellipsoid at $\mathbf{x}$. This can occur only if the principal vectors at $\mathbf{X}$ are first displaced to $\mathbf{x}$ while remaining parallel to their original orientations. This parallel displacement of a vector is accomplished by the g_{lK} tensor, which also changes the referencing of the vector components from the material coordinate system to the spatial coordinate system. The tensor g_{lK} is called the *parallel displacement tensor* or the *shifter*. For rectangular Cartesian coordinate systems, which we use exclusively (see Fig. 3.3), the parallel displacement tensor g_{lK} is a rotation tensor that carries the material coordinate axes into the spatial coordinate axes. If the handedness of these two coordinate systems differs, the g_{lK} tensor represents an improper rotation. If the spatial and material coordinate systems have their corresponding axes aligned, then the parallel displacement tensor g_{lK} becomes simply the Kronecker delta δ_{lK},

$$g_{lK}=\delta_{lK}. \tag{3.6.3}$$

Later we will make this choice; for the present, however, we leave the tensor as a more general rotation tensor. In summary, R_{kK} is a product of two factors, one a physical or active rotation and the second a coordinate or passive rotation.

The transformation, inverse to (3.6.1), that carries the triad of principal vectors $\mathbf{l}^{\alpha}$ of the spatial strain ellipsoid into the triad $\mathbf{L}^{\alpha}$ of the material strain ellipsoid is thus

$$L_K^{\alpha}=(R^{-1})_{Kk}l_k^{\alpha}. \tag{3.6.4}$$

Combining this with (3.6.1) yields both

$$R_{kK}(R^{-1})_{Kl}=\delta_{kl}, \tag{3.6.5}$$

$$(R^{-1})_{Kk}R_{kL}=\delta_{KL}. \tag{3.6.6}$$

Since both R_{kK} and $(R^{-1})_{Kk}$ can be expanded as a linear combination of the l_k^{β} and L_K^{β}, it is apparent from (3.6.1) and (3.6.4) that

$$R_{kK}=(R^{-1})_{Kk}=\sum_{\beta} l_k^{\beta}L_K^{\beta}. \tag{3.6.7}$$

This equation allows (3.6.5) and (3.6.6) to be written as orthonormality relations,

$$R_{kK}R_{lK}=(R^{-1})_{Kk}(R^{-1})_{Kl}=\delta_{kl}, \tag{3.6.8}$$

$$R_{kK}R_{kL}=(R^{-1})_{Kk}(R^{-1})_{Lk}=\delta_{KL}. \tag{3.6.9}$$

Lastly, we remark that when there is no physical rotation, that is,

$$R_{kK}=g_{kK}, \tag{3.6.10}$$

the medium is said to be in a state of *pure strain*.

3.7 Fundamental Theorem of Deformation

We have now assembled the tools to prove the *fundamental theorem of deformation* which states: the deformation at any point may be regarded as resulting from a displacement, a rigid rotation of the principal axes of strain, and stretches along these axes. To prove this we reexpress the deformation gradient $x_{k,K}$ as

$$\begin{aligned} x_{k,K} &= x_{k,L}\delta_{LK} \\ &= \sum_\alpha x_{k,L} L_L^\alpha L_K^\alpha \\ &= \sum_\alpha \lambda(\mathbf{l}^\alpha) x_{k,L} X_{L,i} l_i^\alpha L_K^\alpha \\ &= \sum_\alpha \lambda(\mathbf{l}^\alpha) l_k^\alpha L_K^\alpha \\ &= R_{kL} \sum_\alpha \lambda(\mathbf{L}^\alpha) L_L^\alpha L_K^\alpha \\ &= R_{kL}(C^{1/2})_{LK} \end{aligned} \tag{3.7.1}$$

with the successive use of (3.5.21), (3.5.17), (3.1.5), (3.6.1), and (3.5.24). This equation is the mathematical statement of the fundamental theorem.

To see that (3.7.1) corresponds to the words of the theorem it should be recalled from the preceding section that R_{kL} is composed of two factors, one corresponding to a displacement from $\mathbf{x}$ to $\mathbf{X}$ and one a rigid rotation of the principal axes of strain. The factor $\mathbf{C}^{1/2}$ is seen from (3.5.24), used in the final step of (3.7.1) above, to correspond to stretches $\lambda(\mathbf{L}^{(\alpha)})$ along the principal axes $\mathbf{L}^{(\alpha)}$. These effects can occur singly, in pairs, or all three at once. For a pure strain ($\mathbf{R}=\mathbf{1}$), (3.7.1) becomes

$$x_{k,K} = g_{kL}(C^{1/2})_{LK}. \tag{3.7.2}$$

For a *simple displacement*, that is, a *rigid pure strain* ($\mathbf{R}=\mathbf{1}, \mathbf{C}=\mathbf{1}$, and so $\mathbf{C}^{1/2}=\mathbf{1}$), (3.7.1) becomes

$$x_{k,K} = g_{kK}. \tag{3.7.3}$$

That this can be a simple displacement is seen by integrating it,

$$x_k = g_{kK} X_K + d_k, \tag{3.7.4}$$

where $\mathbf{d}$ is a displacement vector independent of $\mathbf{X}$. If no displacement occurs, $\mathbf{d}=0$. For a *rigid displacement* ($\mathbf{C}^{1/2}=\mathbf{1}$), (3.7.1) becomes

$$x_{k,K} = R_{kl} g_{lK}. \tag{3.7.5}$$

The fundamental theorem gives us an alternate expression for the finite rotation tensor R_{kM} simply by forming the scalar product of (3.7.1) with $(C^{-1/2})_{KM}$,

$$R_{kM} = x_{k,K}(C^{-1/2})_{KM}. \tag{3.7.6}$$

This form is more useful in calculating R_{kM} to a given order of accuracy than the expression (3.6.7).

3.8 Time Derivatives

The characterization of deformation in this chapter so far has concerned only its spatial aspects at a given instant of time. Now we consider its evolution in time. We first assume that the transformations (3.1.1) and (3.1.2) possess continuous partial time derivatives of as many orders as needed. Though the theory can admit isolated singularities such as the singular surface associated with a shock wave, we are not concerned with such phenomena in this treatment. The meaning of a partial time derivative depends on the functional dependence of the quantity that is being differentiated and on whether the material or spatial description is being employed.

In the spatial description the partial time derivative, called the *spatial time derivative*, is taken holding the spatial point $\mathbf{z}$ constant,

$$\frac{\partial F(\mathbf{z},t)}{\partial t} \equiv \left.\frac{\partial F(\mathbf{z},t)}{\partial t}\right|_{\mathbf{z}=\text{const.}} \tag{3.8.1}$$

This derivative is the change in the function F with time at a particular point $\mathbf{z}$ in space. This point in space may be occupied by the material point $\mathbf{X}$ with position $\mathbf{x}(\mathbf{X},t)$; in this case the derivative is taken holding $\mathbf{z} = \mathbf{x}(\mathbf{X},t)$ constant. Note that in the spatial description

$$\frac{\partial \mathbf{x}}{\partial t} = \left.\frac{\partial \mathbf{x}}{\partial t}\right|_{\mathbf{x}=\text{const.}} = 0. \tag{3.8.2}$$

In the material description the partial time derivative, called the *material time derivative*, is taken holding the material point $\mathbf{X}$ constant and is denoted by

$$\dot{F}(\mathbf{X},t) \equiv \frac{dF(\mathbf{X},t)}{dt} \equiv \left.\frac{\partial F(\mathbf{X},t)}{\partial t}\right|_{\mathbf{X}=\text{const.}} \tag{3.8.3}$$

This derivative is the change in the function F with time as seen by an observer riding with the moving mass point $\mathbf{X}$. Note that two notations, a dot and d/dt, are used for the material time derivative. Note also that in the material description

$$\dot{\mathbf{X}} = \frac{d\mathbf{X}}{dt} = \left.\frac{\partial \mathbf{X}}{\partial t}\right|_{\mathbf{X}=\text{const.}} = 0. \tag{3.8.4}$$

Often in the material description the function being differentiated contains implicit time dependence through dependence on $\mathbf{x}(\mathbf{X},t)$ as well as explicit dependence on t. In this case the material time derivative is

$$\begin{aligned}\dot{F}(\mathbf{x}(\mathbf{X},t),t) &\equiv \frac{dF(\mathbf{x}(\mathbf{X},t),t)}{dt} \\ &= \left.\frac{\partial F(\mathbf{x}(\mathbf{X},t),t)}{\partial t}\right|_{\mathbf{x}=\text{const.}} + \dot{x}_i \left.\frac{\partial F(\mathbf{x}(\mathbf{X},t),t)}{\partial x_i}\right|_{\substack{t=\text{const.}\\ x_j=\text{const.}\\ (j\neq i)}} \\ &= \frac{\partial F}{\partial t} + F_{,i}\dot{x}_i \end{aligned} \tag{3.8.5}$$

where

$$\dot{\mathbf{x}} \equiv \left.\frac{\partial \mathbf{x}(\mathbf{X},t)}{\partial t}\right|_{\mathbf{X}=\text{const.}} \tag{3.8.6}$$

is the *velocity* of the mass point $\mathbf{X}$. Here we see that the material time derivative accounts for both the explicit change in time of the function F and for the effective change in time of the function caused by the motion of the mass point through the spatial variation of the function F.

Consider next the material time derivative of the Jacobian. We obtain

$$\begin{aligned}\frac{dJ(\mathbf{x}/\mathbf{X})}{dt} &= \frac{\partial J}{\partial x_{l,L}}\frac{dx_{l,L}}{dt} \\ &= JX_{L,l}\dot{x}_{l,L} \\ &= J\dot{x}_{l,l}\end{aligned} \tag{3.8.7}$$

where we use (3.2.4), the commutativity of d/dt ($\mathbf{X}$ fixed) and $\partial/\partial X_L$ (t fixed), and the chain rule of differentiation. The quantity $\dot{x}_{l,l}$ is called the *expansion* and (3.8.7) is called *Euler's expansion formula*.

3.9 Mass Conservation

We are now in a position to explore the consequence that the assumed continuity and single-valueness of the deformation have on physical properties of the matter. Let $\Gamma(\mathbf{X},t)$ represent some density in the material description and $\gamma(\mathbf{z}=\mathbf{x}, t)$ the corresponding density in the spatial description. These quantities can stand for a scalar density such as the mass density, a vector density such as the polarization, or a tensor density. Components, however, must be referred to a common rectangular frame. We thus must have

$$\gamma\, dv = \Gamma\, dV. \tag{3.9.1}$$

Use of (3.3.19) yields

$$\gamma(\mathbf{x}(\mathbf{X},t),t) = \frac{\Gamma(\mathbf{X},t)}{J(\mathbf{x}/\mathbf{X})}. \tag{3.9.2}$$

Taking the material time derivative of each side results in

$$\frac{d\gamma}{dt} = \frac{1}{J}\frac{d\Gamma}{dt} - \frac{\Gamma}{J^2}\frac{dJ}{dt}. \tag{3.9.3}$$

Use of (3.8.5) and (3.8.7) gives

$$\begin{aligned}\frac{\partial\gamma}{\partial t} + \gamma_{,l}\dot{x}_l &= \frac{1}{J}\frac{d\Gamma}{dt} - \frac{\Gamma \dot{x}_{l,l}}{J}\\ &= \frac{1}{J}\frac{d\Gamma}{dt} - \left(\frac{\Gamma}{J}\dot{x}_l\right)_{,l} + \left(\frac{\Gamma}{J}\right)_{,l}\dot{x}_l.\end{aligned} \tag{3.9.4}$$

Substituting (3.9.2) to eliminate Γ and canceling like terms, we find

$$\frac{\partial\gamma}{\partial t} + \frac{\partial}{\partial z_l}(\gamma\dot{x}_l) = \frac{1}{J}\frac{d}{dt}(J\gamma). \tag{3.9.5}$$

We recognize this as a continuity relation for γ in the spatial description with a source term on the right side. It is called the *spatial equation of continuity*.

Now let us consider the implications of (3.9.5) for the mass density. We replace Γ with ρ^0, the mass per unit material frame volume, and γ with ρ, the mass per unit spatial frame volume. Thus

$$\rho(\mathbf{z},t) = \frac{\rho^0}{J} \tag{3.9.6}$$

where J is regarded here as a function of $\mathbf{z}=\mathbf{x}$ and t. Since ρ^0 is a constant, this replacement in the spatial equation of continuity yields

$$\frac{\partial\rho}{\partial t} + \frac{\partial}{\partial z_l}(\rho\dot{x}_l) = 0 \tag{3.9.7}$$

because the source term vanishes. This is the statement of *conservation of mass* expressed in the spatial description. This physical statement results from the original continuity assumptions concerning a deformation, that is, that there is a one-to-one mapping of mass points between the material and spatial descriptions.

That (3.9.7) represents conservation of mass can be seen by integrating it over a fixed volume v in the spatial frame

$$\int_v \frac{\partial \rho}{\partial t}\, dv + \int_v \frac{\partial}{\partial z_l}(\rho \dot{x}_l)\, dv = 0. \tag{3.9.8}$$

The time derivative may be withdrawn from the first integral and the second one may be converted to an integral over the surface s of the volume by way of Gauss' theorem with the result

$$\frac{\partial}{\partial t}\int_v \rho\, dv + \int_s \rho \dot{x}_l n_l\, ds = 0, \tag{3.9.9}$$

where $\mathbf{n}$ is the outward unit normal of the surface. This equation states that the sum of the time rate of increase of mass in the volume plus the flow outward through the surface must vanish. Such a balance results from the absence of any source or sink term for mass on the right side. Thus mass is a conserved quantity. This can be seen even more clearly if the volume v considered encloses the *entire* system. Then there can be no mass flow through the surface s and so the surface integral vanishes. The volume integral is the total mass M of the system and (3.9.9) becomes

$$\frac{\partial M}{\partial t} = 0, \tag{3.9.10}$$

which states that M is a constant of the motion, that is, a conserved quantity.

3.10 Displacement Vector

A useful quantity in describing a deformation is the vector that is the change in position of a mass point arising from the deformation, namely, its position in the deformed state minus its position in the undeformed state. This vector is called the *displacement vector* or simply the displacement. The mathematical definition of its components must involve the relation of the material and spatial coordinate systems. In Section 3.6 we pointed out that this relation is specified by the parallel displacement tensor g_{lK} which becomes simply the Kronecker delta (3.6.3) when the material and spatial coordinate systems are taken as identical. We now

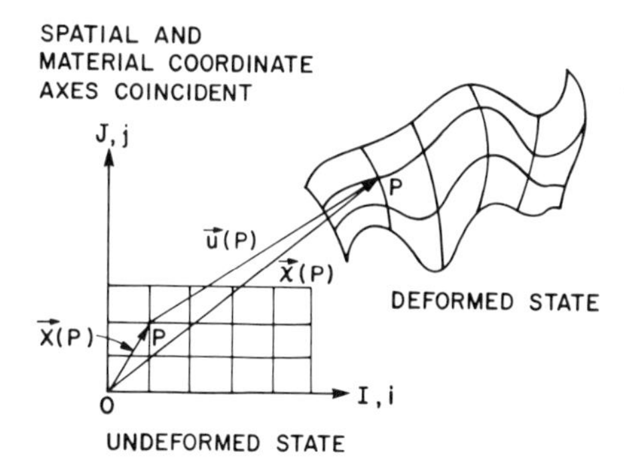

Fig. 3.4. Displacement vector. In comparison to Fig. 3.1, the spatial and material coordinate systems are chosen as identical and the displacement vector **u** is introduced as the difference of the deformed and undeformed positions, **x** − **X**.

make this choice and use it whenever the displacement is being used (see Fig. 3.4). Thus we define the displacement **u** by

$$u_i(\mathbf{X},t) \equiv x_i(\mathbf{X},t) - \delta_{iI} X_I. \tag{3.10.1}$$

It is written here as a spatial vector in the material description. Though a dual quantity can be defined, we do not find use for it in this treatment.

The *displacement gradient*,

$$u_{i,K} = x_{i,K} - \delta_{iK}, \tag{3.10.2}$$

is closely related to the deformation gradient $x_{i,K}$. By writing

$$u_{i,K}\, dX_K = dx_i - \delta_{iK}\, dX_K \tag{3.10.3}$$

we see that the displacement gradients $u_{i,K}$ are measures of the local changes in length and angle resulting from the deformation. When deformations are small, as typical in crystals but not in rubber, the displacement gradients are convenient expansion parameters because of their smallness. We also note that the material time derivative of the displacement,

$$\dot{\mathbf{u}} = \dot{\mathbf{x}}, \tag{3.10.4}$$

is the velocity of the mass point.

Next let us express the Green finite strain tensor E_{AB} in terms of displacement gradients. From (3.4.15), (3.4.2), and (3.10.2) we find

$$\begin{aligned} E_{AB} &= \tfrac{1}{2}(C_{AB} - \delta_{AB}) \\ &= \tfrac{1}{2}(x_{k,A} x_{k,B} - \delta_{AB}) \\ &= \tfrac{1}{2}\left[(u_{k,A} + \delta_{kA})(u_{k,B} + \delta_{kB}) - \delta_{AB}\right] \\ &= \tfrac{1}{2}(\delta_{kA} u_{k,B} + \delta_{kB} u_{k,A} + u_{k,A} u_{k,B}) \\ &= u_{(A,B)} + \tfrac{1}{2} u_{k,A} u_{k,B}. \end{aligned} \tag{3.10.5}$$

Here we use a notation that we frequently employ for its compactness. Parentheses enclosing two tensor subscripts denote the symmetric part of the tensor represented by those subscripts,

$$u_{(A,B)} \equiv \tfrac{1}{2}(u_{A,B} + u_{B,A}) = u_{(B,A)}. \tag{3.10.6}$$

Clearly, the notation can be used when the two subscripts involved are on two different tensors such as $A_{i(j}B_{k)l}$. Returning now to (3.10.5) we wish to point out that we can write

$$u_{A,B} = \delta_{kA} u_{k,B} \tag{3.10.7}$$

without confusion, since the material and spatial coordinate systems are now taken as identical and so components are measured identically in both systems. However, we must remember that

$$u_{A,k} \neq u_{A,B}\delta_{kB} \tag{3.10.8}$$

since $u_{A,k} = \partial u_A / \partial x_k$ and $u_{A,B} = \partial u_A / \partial X_B$, that is to say, the use of a lowercase or uppercase letter after a comma to indicate differentiation refers not only to the component designation of the variable used in differentiation but also to which quantity, $\mathbf{x}$ or $\mathbf{X}$, is used as that variable. Equation (3.10.5) is used in a later section to obtain an expression for the infinitesimal strain tensor.

The finite rotation tensor R_{iB} given in (3.7.6) can be expressed in terms of displacement gradients as

$$\begin{aligned} R_{iB} &= x_{i,A}(C^{-1/2})_{AB} \\ &= (\delta_{iA} + u_{i,A})\left[(\mathbf{1} + 2\mathbf{E})^{-1/2}\right]_{AB} \end{aligned} \tag{3.10.9}$$

where $\mathbf{1}$ stands for the Kronecker delta tensor and $\mathbf{E}$ is given by (3.10.5). An expansion of this expression in a later section for small displacement gradients yields the infinitesimal rotation tensor.

3.11 Derivative Transformations Between Spatial and Material Descriptions

We have frequent need in the subsequent chapters to transform space and time derivatives in the material description into corresponding derivatives in the spatial description. It is the purpose of this section to derive such connections. We have less need for connections of the inverse type because final equations are usually preferred to be expressed in spatial coordinates, since they correspond to a laboratory coordinate system; in other words, most (but not all!) measurements are made relative to the spatial positions in the laboratory rather than relative to the material positions fixed to and moving with the deforming body. We discuss the contrary situation in Chapter 17.

Consider a function $F(\mathbf{X},t)$ which is a function of the variables in the material description. Denote the same function as $\hat{F}(\mathbf{z},t)$ when its arguments have been transformed to the variables of the spatial description, $\mathbf{z}=\mathbf{x}(\mathbf{X},t)$ and t, that is,

$$\hat{F}(\mathbf{z},t)\equiv F(\mathbf{X}(\mathbf{z},t),t). \tag{3.11.1}$$

Consider first the material time derivative of F given by (3.8.5),

$$\dot{F}=\left.\frac{dF}{dt}\right|_{\mathbf{X}}=\left.\frac{\partial\hat{F}}{\partial t}\right|_{\mathbf{z}}+\left.\frac{\partial\hat{F}}{\partial z_i}\right|_{\substack{t,z_j\\(j\neq i)}}\cdot\left.\frac{dx_i}{dt}\right|_{\mathbf{X}}, \tag{3.11.2}$$

where the variables held fixed during differentiation are indicated at the lower right of each derivative. Note that the two derivatives of $\hat{F}$ on the right are spatial frame derivatives as desired, but the $\dot{x}_i$ derivative is another material time derivative.

In order to find an expression for $\dot{\mathbf{x}}$ replace F with $\mathbf{X}$ to obtain

$$\left.\frac{dX_A}{dt}\right|_{\mathbf{X}}=0=\left.\frac{\partial X_A}{\partial t}\right|_{\mathbf{z}}+\left.\frac{\partial X_A}{\partial z_i}\right|_{\substack{t,z_j\\(j\neq i)}}\cdot\left.\dot{x}_i\right|_{\mathbf{X}}. \tag{3.11.3}$$

By forming the scalar product with $x_{k,A}$,

$$0=x_{k,A}\frac{\partial X_A}{\partial t}+x_{k,A}X_{A,i}\dot{x}_i, \tag{3.11.4}$$

and using (3.1.5) we find

$$\dot{x}_k=-x_{k,A}\frac{\partial X_A}{\partial t}=-(\delta_{kA}+u_{k,A})\frac{\partial X_A}{\partial t}. \tag{3.11.5}$$

This relation between the *velocity* $\dot{\mathbf{x}}$ and the *flow* $\partial\mathbf{X}/\partial t$ of matter is illustrated in Fig. 3.5. The second factor on the right can be expressed in terms of the displacement by first expressing (3.10.1) in the spatial description,

$$X_A=\delta_{Ai}x_i-\delta_{Ai}\hat{u}_i, \tag{3.11.6}$$

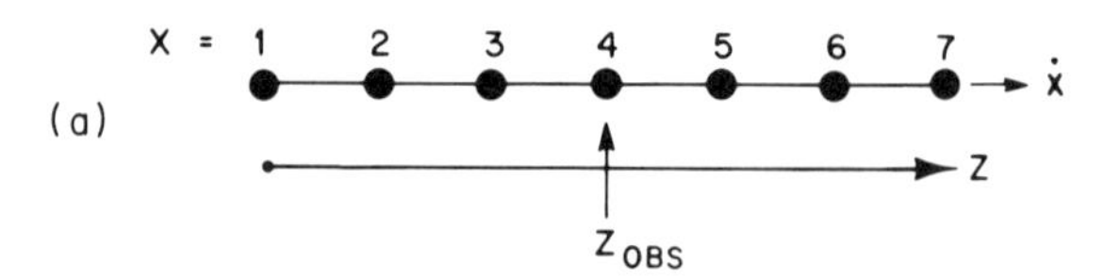

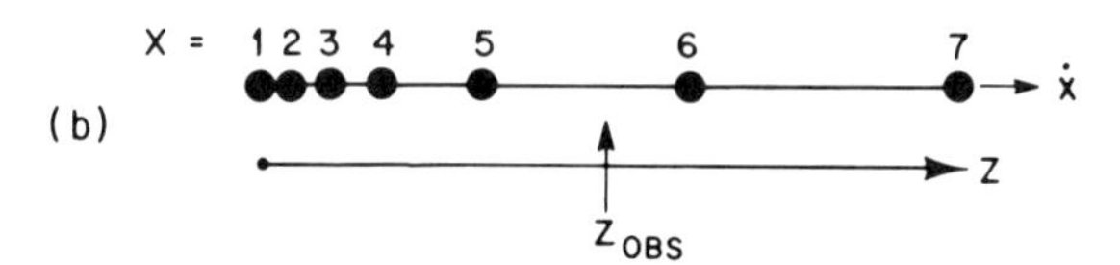

Fig. 3.5. Illustration of relation between velocity and flow (3.11.5). To aid visualization consider the material continuum to be a line of attached mass points labeled by integers in the material coordinate system and traveling to the right at velocity $\dot{x}$. The flow $\partial X/\partial t$ is observed at z_{OBS}. (*a*) In the absence of deformation, $\partial x/\partial X=1$, the flow is equal in magnitude to the velocity but is negative. (*b*) In the presence of deformation the deformation gradient $\partial x/\partial X$ becomes smaller at z_{OBS} as time passes while the flow $\partial X/\partial t$, which is negative, becomes larger in magnitude such that the product $-(\partial x/\partial X)(\partial X/\partial t)$ remains equal to the velocity.

and then by taking a spatial time derivative

$$\frac{\partial X_A}{\partial t} = -\delta_{Ai}\frac{\partial \hat{u}_i}{\partial t}\bigg|_{\mathbf{z}} = -\frac{\partial \hat{u}_A}{\partial t}\bigg|_{\mathbf{z}} \tag{3.11.7}$$

where the derivative of x_i vanishes by (3.8.2). Combining now (3.11.2), (3.11.5), and (3.11.7) results in

$$\dot{F} = \frac{\partial \hat{F}}{\partial t}\bigg|_{\mathbf{z}} + \frac{\partial \hat{F}}{\partial z_i}\bigg|_{\substack{t,z_j\\(j\neq i)}} \cdot(\delta_{iA}+u_{i,A})\frac{\partial \hat{u}_A}{\partial t}\bigg|_{\mathbf{z}}. \tag{3.11.8}$$

Now a material space derivative $u_{i,A}$ appears on the right side and must be eliminated.

Consider the transformation of a material space derivative of the arbitrary function F,

$$F_{,A} \equiv \frac{\partial F}{\partial X_A}\bigg|_t = \frac{\partial \hat{F}}{\partial z_j}\bigg|_{\substack{t,z_i\\(i\neq j)}} \cdot x_{j,A} = \hat{F}_{,j}(\delta_{jA}+u_{j,A}). \tag{3.11.9}$$

We see that we have only managed to shift the material derivative from F to $\mathbf{u}$. Hence apply (3.11.9) to $\mathbf{u}$ itself to get

$$u_{i,A} = \hat{u}_{i,j}(\delta_{jA}+u_{j,A}). \tag{3.11.10}$$

Transpose the last term to the left side,

$$(\delta_{ij} - \hat{u}_{i,j})u_{j,A} = \hat{u}_{i,j}\delta_{jA}, \tag{3.11.11}$$

and solve formally for $u_{k,A}$,

$$u_{k,A} = \left[(\mathbf{1} - \nabla_z \hat{\mathbf{u}})^{-1}\right]_{ki}\hat{u}_{i,j}\delta_{jA}. \tag{3.11.12}$$

This is finally the key that unlocks the puzzle. When all the displacement gradients are small compared to unity as is typical for crystal deformations within the elastic limit, the right side of this equation can be expanded in an infinite series. The terms up to third order are

$$u_{k,A} = \left[\delta_{ki} + \hat{u}_{k,i} + \hat{u}_{k,l}\hat{u}_{l,i} + \cdots\right]\hat{u}_{i,j}\delta_{jA}. \tag{3.11.13}$$

This formula gives a method of calculating a material space derivative of the displacement to any degree of accuracy from spatial space derivatives.

The result for $u_{k,A}$, (3.11.13), can now be substituted into (3.11.9) for $F_{,A}$ to obtain

$$F_{,A} = \hat{F}_{,k}\left[\delta_{kA} + (\delta_{ki} + \hat{u}_{k,i} + \hat{u}_{k,l}\hat{u}_{l,i} + \cdots)\hat{u}_{i,j}\delta_{jA}\right] \tag{3.11.14}$$

and into (3.11.8) for $\dot{F}$ to obtain

$$\dot{F} = \frac{\partial \hat{F}}{\partial t} + \hat{F}_{,k}\left[\delta_{kA} + (\delta_{ki} + \hat{u}_{k,i} + \hat{u}_{k,l}\hat{u}_{l,i} + \cdots)\hat{u}_{i,j}\delta_{jA}\right]\frac{\partial \hat{u}_A}{\partial t}. \tag{3.11.15}$$

These are now the final expressions, since only spatial frame derivatives appear on the right side. We also can obtain a comparable expression for $\dot{\mathbf{x}}$ by substituting (3.11.13) and (3.11.7) into (3.11.5) to obtain

$$\dot{x}_k = \frac{\partial \hat{u}_j}{\partial t}\left[\delta_{kj} + (\delta_{ki} + \hat{u}_{k,i} + \hat{u}_{k,l}\hat{u}_{l,i} + \cdots)\hat{u}_{i,j}\right]. \tag{3.11.16}$$

Carrying this procedure a step further by letting $\dot{\mathbf{x}}$ be F in (3.11.2) yields

$$\begin{aligned}
\ddot{x}_i &= \frac{\partial}{\partial t}\left(\frac{\partial \hat{u}_i}{\partial t} + \hat{u}_{i,j}\frac{\partial \hat{u}_j}{\partial t} + \cdots\right) + \dot{x}_k\frac{\partial}{\partial z_k}\left(\frac{\partial \hat{u}_i}{\partial t} + \hat{u}_{i,j}\frac{\partial \hat{u}_j}{\partial t} + \cdots\right) \\
&= \frac{\partial^2 \hat{u}_i}{\partial t^2} + \hat{u}_{i,j}\frac{\partial^2 \hat{u}_j}{\partial t^2} + 2\frac{\partial \hat{u}_{i,j}}{\partial t}\frac{\partial \hat{u}_j}{\partial t} + \cdots
\end{aligned} \tag{3.11.17}$$

with the use of (3.11.16).

3.12 Infinitesimal Strain and Rotation Tensors

The formula (3.11.13) for $u_{k,A}$ has other uses. For one thing, it can be substituted into the expression (3.10.5) for the Green finite strain tensor to calculate it to any order of accuracy desired. In particular, we note from (3.11.13) for $u_{k,A}$ that to linear order

$$\frac{\partial u_k}{\partial X_A} \cong \frac{\partial \hat{u}_k}{\partial z_A} = \frac{\partial \hat{u}_k}{\partial x_A}. \tag{3.12.1}$$

Thus to linear order the Green finite strain tensor becomes

$$E_{AB} \cong \frac{1}{2}\left(\frac{\partial \hat{u}_A}{\partial x_B} + \frac{\partial \hat{u}_B}{\partial x_A}\right) \equiv S_{AB} \tag{3.12.2}$$

where S_{AB} is the *infinitesimal strain tensor*. Since the linearized expression (3.12.2) contains components of $\mathbf{x}$ rather than $\mathbf{X}$, it is more in keeping with our previous convention on subscripts to write

$$S_{ij} \equiv \frac{1}{2}\left(\frac{\partial u_i}{\partial x_j} + \frac{\partial u_j}{\partial x_i}\right) = u_{(i,j)} \tag{3.12.3}$$

where the caret on the $\mathbf{u}$ is now dropped for simplicity. Clearly, however, *to the linear order* it makes no difference whether $\mathbf{x}$ or $\mathbf{X}$ is used for taking derivatives and hence whether uppercase or lowercase letters are used after the comma in subscripts to denote differentiation.

Next we expand the finite rotation tensor to linear order from (3.10.9) to obtain

$$\begin{aligned} R_{iB} &= (\delta_{iA} + u_{i,A})[\delta_{AB} - E_{AB} + \cdots] \\ &\cong (\delta_{iA} + u_{i,A})(\delta_{AB} - u_{(A,B)}) \\ &\cong \delta_{iB} + u_{[i,B]}. \end{aligned} \tag{3.12.4}$$

Here brackets enclosing two tensor subscripts denote the antisymmetric part, as

$$u_{[i,j]} \equiv \frac{1}{2}\left(\frac{\partial u_i}{\partial x_j} - \frac{\partial u_j}{\partial x_i}\right) = -u_{[j,i]}. \tag{3.12.5}$$

Just as pointed out for the parenthesis notation, the brackets can be used to enclose subscripts on different tensors such as $A_{i[j}B_{k]l}$ with an analogous meaning. Equation (3.12.4) is just the *infinitesimal rotation tensor*. If **a** is a unit vector in the direction of the axis of rotation and $\delta\theta$ is the infinitesimal angle of rotation, positive in the right-hand sense, then

$$u_{[i,j]} = \epsilon_{ijk} a_k \delta\theta, \tag{3.12.6}$$

$$R_{ij} = \delta_{ij} + \epsilon_{ijk} a_k \delta\theta. \tag{3.12.7}$$

Since any second rank tensor may be expressed as a sum of symmetric plus antisymmetric parts, we may write the linearized displacement gradient as

$$\begin{aligned} u_{i,j} &= u_{(i,j)} + u_{[i,j]} \\ &= S_{ij} + (R_{ij} - \delta_{ij}), \end{aligned} \tag{3.12.8}$$

which shows that the displacement gradient consists of a strain part and a rotation part. To visualize this consider the two-dimensional object of Fig. 3.6. Applied tractions deform the rectangle a into the parallelogram b. The bottom side of the object is constrained to lie on the x_2 axis as if resting on a rigid bench. The displacement gradients from the figure are

$$\begin{aligned} u_{1,1} &= -\frac{\delta h}{h}, & u_{1,2} &= 0, \\ u_{2,1} &= \frac{\delta s}{h} \cong 2\delta\theta, & u_{2,2} &= \frac{\delta w}{w}. \end{aligned} \tag{3.12.9}$$

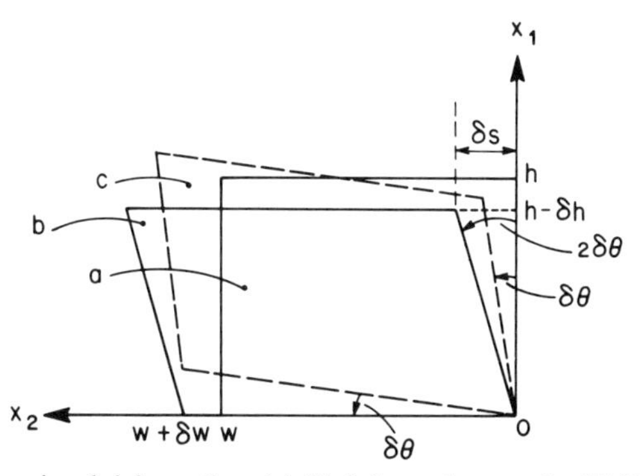

Fig. 3.6. Two-dimensional deformation. (a) Undeformed rectangle. (b) Deformed parallelogram which has both strain and rotation. (c) Deformed parallelogram which has strain but no rotation.

Thus the strain components are

$$S_{11} = -\frac{\delta h}{h}, \qquad S_{12} = \delta\theta,$$
$$S_{21} = \delta\theta, \qquad S_{22} = \frac{\delta w}{w} \tag{3.12.10}$$

and the rotation components are

$$R_{11} - \delta_{11} = 0\ , \qquad R_{12} - \delta_{12} = -\delta\theta$$
$$R_{21} - \delta_{21} = \delta\theta, \qquad R_{22} - \delta_{22} = 0. \tag{3.12.11}$$

The configuration c in the figure has the strain components of (3.12.10) but no rotation components.

3.13 Material Time Derivative of a Surface Integral

The manipulative tools developed in this chapter permit us to derive an identity for the material time derivative of a surface integral which has important uses in Chapters 7 and 16. Consider an integral over a surface S fixed in the material coordinate system. The integrand is a vector function $\mathbf{F}(\mathbf{x}, t)$ which is expressed in the spatial coordinate system and may be interpreted as a flux crossing the surface S. It may be written as

$$\begin{aligned} \frac{d}{dt}\int_S \mathbf{F}\cdot d\mathbf{a} &= \frac{d}{dt}\int_S F_j J X_{P,j}\, dA_P \\ &= \int_S \frac{d}{dt}(F_j J X_{P,j})\, dA_P \\ &= \int_S \frac{d}{dt}(F_j J X_{P,j}) J^{-1} x_{i,P}\, da_i. \end{aligned} \tag{3.13.1}$$

The first equality of (3.13.1) shows that $J X_{P,j} F_j$ is the material measure of the spatial frame flux F_j. The last integrand in (3.13.1) is called the *convected time derivative* and denoted by

$$\overset{*}{F}_i = J^{-1} x_{i,P} \frac{d}{dt}(F_j J X_{P,j}). \tag{3.13.2}$$

If we rewrite this equation as

$$\frac{d}{dt}(J X_{R,j} F_j) = J X_{R,i} \overset{*}{F}_i, \tag{3.13.3}$$

we see that the material time derivative of the material measure of **F** is the material measure of the convected time derivative of **F**. Equation (3.13.2) may be rearranged as

$$\overset{*}{F}_i = J^{-1} x_{i,P}\left[X_{P,j}\frac{d}{dt}(JF_j) + JF_j \frac{dX_{P,j}}{dt}\right]$$

$$= J^{-1}\frac{d}{dt}(JF_i) - F_j X_{P,j}\frac{dx_{i,P}}{dt}$$

$$= \frac{\partial F_i}{\partial t} + (F_i \dot{x}_k)_{,k} - F_j X_{P,j}\frac{\partial \dot{x}_i}{\partial X_P}$$

$$= \frac{\partial F_i}{\partial t} + (F_i \dot{x}_k)_{,k} - F_j \dot{x}_{i,j}$$

$$= \frac{\partial F_i}{\partial t} + (F_i \dot{x}_k - F_k \dot{x}_i)_{,k} + F_{j,j}\dot{x}_i \tag{3.13.4}$$

with the use of the spatial equation of continuity (3.9.5), the material time derivative of (3.1.5), the commutativity of material time and space derivatives, and the chain rule of differentiation. In vector form the preceding equation is

$$\overset{*}{\mathbf{F}} = \frac{\partial \mathbf{F}}{\partial t} + \nabla \times (\mathbf{F} \times \dot{\mathbf{x}}) + \dot{\mathbf{x}} \nabla \cdot \mathbf{F}. \tag{3.13.5}$$

Combining this equation with (3.13.1) and (3.13.2) we have

$$\frac{d}{dt}\int_S \mathbf{F}\cdot d\mathbf{a} = \int_S \left[\frac{\partial \mathbf{F}}{\partial t} + \nabla \times (\mathbf{F} \times \dot{\mathbf{x}}) + (\nabla \cdot \mathbf{F})\dot{\mathbf{x}}\right]\cdot d\mathbf{a} \tag{3.13.6}$$

for the material time derivative of the surface integral.

PROBLEMS

3.1. Prove (3.2.10) to (3.2.13).

3.2. Prove

$$\frac{\partial \dot{x}_i}{\partial X_{J,k}} = -x_{i,J}\dot{x}_k. \tag{3.P.1}$$

3.3. Prove

$$\hat{J}^{-1}\left(\frac{\mathbf{x}}{\mathbf{X}}\right) = 1 - \hat{u}_{l,l} + \tfrac{1}{2}(\hat{u}_{k,k}\hat{u}_{l,l} - \hat{u}_{l,k}\hat{u}_{k,l}) - \det(\hat{u}_{i,j}). \tag{3.P.2}$$

3.4. Prove that to linear order in $u_{j,k}$

$$\frac{\partial R_{jB}}{\partial x_{i,A}} \cong \delta_{i[j}\delta_{B]A} - \tfrac{1}{2}u_{i,B}\delta_{jA} - \tfrac{1}{2}u_{j,A}\delta_{iB} + \tfrac{1}{4}\left[3u_{(j,A)}\delta_{iB} - u_{(i,j)}\delta_{AB} + 3u_{(i,B)}\delta_{jA} - u_{(A,B)}\delta_{ij}\right]. \quad (3.P.3)$$

3.5. Prove both of the following:

$$(c^{-1})_{kl} = x_{k,A}x_{l,A}, \quad (3.P.4)$$

$$(C^{-1})_{KL} = X_{K,i}X_{L,i}. \quad (3.P.5)$$

3.6. Prove

$$2j\left(\frac{\mathbf{X}}{\mathbf{x}}\right)\delta_{KL} = \epsilon_{ijk}\epsilon_{IJL}X_{I,i}X_{J,j}X_{K,k}. \quad (3.P.6)$$

3.7. Derive an alternate statement of the fundamental theorem of deformation having the form

$$X_{K,k} = (R^{-1})_{Kl}(c^{1/2})_{lk}. \quad (3.P.7)$$

CHAPTER 4

Lagrangian of a Crystal and Electromagnetic Fields

This chapter begins the construction of the Lagrangian [Lax and Nelson, 1971] from which the results in the remainder of the book are derived. The Lagrangian represents a crystalline dielectric in interaction with the electromagnetic field. This is treated as a closed system, that is, one that obeys energy, momentum, and angular momentum conservation.

A microscopic, discrete particle viewpoint is employed to obtain the Lagrangian after which a continuum limit produces a Lagrangian for macroscopic long wavelength phenomena. The continuum limit introduces the material coordinate system in a natural way. We find it advantageous to transform from the simple position coordinates of the various particles (ions or electrons) to the center-of-mass coordinate (that gives rise to the acoustic modes) and a set of internal coordinates that describe the various internal motions of the crystal (the optic modes).

Inclusion of the internal coordinates allows this theory to describe all the resonances and interactions in the material medium up through visible light frequencies. Thus the constitutive relations for optical interactions and their relation to lower frequency interactions, which was not possible with continuum theories [Toupin, 1963; Dixon and Eringen, 1965] prior to this, is possible.

The interaction Lagrangian is expanded in a series of multipole moments. This is a useful procedure for two reasons. First, in dielectrics the monopole or free charge density term may be set equal to zero. Second, for most phenomena of interest the electric quadrupole, the magnetic dipole, and higher multipole terms are of negligible importance and can be dropped.

The construction of the most general potential or stored energy of the crystal consistent with the conservation laws is done in Chapter 6.

4.1 Total Discrete Particle Lagrangian

We regard a crystal as an array of point particles periodic in three dimensions. We allow the primitive cell, the smallest volume unit of structural repetition, to contain N particles, which are labeled by a lower-case Greek letter. These particles should include all of the ions and at least one electron.* Two atoms (ions) of the same element that are in different locations in the primitive cell are labeled differently, such as $\alpha=1$ and $\alpha=2$. In a general way we speak of the Greek letter index as labeling the type of particle. Each particle has a fixed charge e^{α} and a fixed mass m^{α}. Its position is $\mathbf{x}^{\alpha n}$ where n, which has three integer components, labels the primitive unit cell of which the particle is a constituent.

The particles reside in a vacuum and are subject to bonding forces between themselves. The bonding forces arise from *short range* electromagnetic fields (wavelengths$\lesssim$cell size) and quantum mechanical effects such as the exclusion principle. They are obtainable from a potential energy $V(\{\mathbf{x}^{\alpha n}\})$ which depends on the set of positions $\{\mathbf{x}^{\alpha n}\}$ of all the particles. Electric and magnetic forces on the particles caused by *long range* electric and magnetic fields (wavelengths$\gg$cell size) existing in the vacuum around the particles arise from the interaction Lagrangian. These electric and magnetic fields can be created by the medium itself or can be created externally.

The total Lagrangian L of this system of matter and electromagnetic fields consists of three parts,

$$L=L_M+L_F+L_I, \tag{4.1.1}$$

the matter Lagrangian L_M, the electromagnetic field Lagrangian L_F, and the field-matter interaction Lagrangian L_I. The matter Lagrangian for a collection of particles is the difference of the total kinetic energy and the total potential energy,

$$L_M=\tfrac{1}{2}\sum_{\alpha n} m^{\alpha}(\dot{\mathbf{x}}^{\alpha n})^2-V(\{\mathbf{x}^{\beta m}\}), \tag{4.1.2}$$

where the summation in the kinetic energy term extends over the N types

*Wemple and DiDomenico [1969, 1970] showed that a single classical electronic oscillator is sufficient to account for the strength and frequency dispersion of the dielectric and elastooptic tensors in a wide variety of crystals in their transparent regions; they showed that a second electronic oscillator is needed to account for the dispersion of the elastooptic tensor in crystals having a very strong exciton band at the edge of the transparent region.

of particles in a primitive unit cell and over $\mathfrak{N}$ such cells in the crystal. The matter Lagrangian has a nonrelativistic form and thus excludes matter velocities comparable to the velocity of light.

The field Lagrangian, given previously in (2.5.2), is

$$L_F = \int \mathcal{L}_F \, dv, \tag{4.1.3}$$

$$\mathcal{L}_F = \frac{\epsilon_0}{2}\left[\mathbf{E}^2 - c^2\mathbf{B}^2\right] \tag{4.1.4}$$

where $\mathbf{E}$ and $\mathbf{B}$ are *long range* electric and magnetic induction fields and are regarded as functions of Φ and $\mathbf{A}$ according to (2.4.5) and (2.4.6). The interaction Lagrangian has the form (2.5.3) for each of the particles labeled by n and α. Thus

$$L_I = \sum_{\alpha n} e^\alpha\left[\dot{\mathbf{x}}^{\alpha n}(t) \cdot \mathbf{A}(\mathbf{x}^{\alpha n}(t), t) - \Phi(\mathbf{x}^{\alpha n}(t), t)\right] \tag{4.1.5}$$

where it should be noted that the potential functions $\mathbf{A}$ and Φ are evaluated at the position of each particle. As discussed in Section 2.5 this form of the interaction Lagrangian is appropriate for the Lagrange equations of discrete particles but must be converted to an integral over a Lagrangian density in the spatial frame for use in obtaining the Maxwell-Lorentz electromagnetic equations. Using the procedure of that section we find

$$L_I = \int \mathcal{L}_{IS} \, dv, \tag{4.1.6}$$

$$\mathcal{L}_{IS} = \mathbf{j}(\mathbf{z}, \mathbf{t}) \cdot \mathbf{A}(\mathbf{z}, t) - q(\mathbf{z}, t)\Phi(\mathbf{z}, t) \tag{4.1.7}$$

where

$$q(\mathbf{z}, t) \equiv \sum_{\alpha n} e^\alpha \delta(\mathbf{z} - \mathbf{x}^{\alpha n}(t)), \tag{4.1.8}$$

$$\mathbf{j}(\mathbf{z}, t) \equiv \sum_{\alpha n} e^\alpha \dot{\mathbf{x}}^{\alpha n}(t) \delta(\mathbf{z} - \mathbf{x}^{\alpha n}(t)). \tag{4.1.9}$$

The interaction Lagrangian density is now denoted by $\mathcal{L}_{IS}$, the subscript S denoting that it is a density in the spatial frame. The charge and current densities $q(\mathbf{z}, t)$ and $\mathbf{j}(\mathbf{z}, t)$ are spatial frame densities because the Dirac delta function is a density in the space of its argument as is implied by (2.5.16).

4.2 Passage to Continuum Limit

We now wish to define a limiting process by which the discrete particle lattice of a crystal described in the preceding section can be converted to a material continuum useful for characterizing long wavelength phenomena. The material continuum so defined must retain all of the modes of motion of the discrete lattice and all the symmetry and anisotropy of the crystal. This can be done by replacing the discrete cell index n (which contains three components) by a continuous variable $\mathbf{X}$,

$$n \rightarrow \mathbf{X}, \tag{4.2.1}$$

which fulfills the same function of labeling the matter. It is just what we have called the material coordinate in Chapter 3. The particle index α is retained but is now called a sublattice index. Thus the $\mathfrak{N} \times N$ discrete position vectors $\mathbf{x}^{\alpha n}(t)$ are replaced by N continuum position vectors $\mathbf{x}^{\alpha}(\mathbf{X}, t)$,

$$\mathbf{x}^{\alpha n}(t) \rightarrow \mathbf{x}^{\alpha}(\mathbf{X}, t), \tag{4.2.2}$$

that are functions of the continuous material coordinate vector $\mathbf{X}$.

Because the continuum limit replacement (4.2.1) does not involve α, we have obtained a single material coordinate vector $\mathbf{X}$ to suffice for the argument of the N spatial position vectors $\mathbf{x}^{\alpha}$. Thus we are describing the matter of the crystal by a manifold of N spatial vector continua, not a single vector continuum $\mathbf{x}(\mathbf{X}, t)$ as done throughout Chapter 3. This is necessary in order to account for all the various internal motions (optic modes) of the crystal and the dispersion that these modes give to the various properties of the crystal. But there are a number of other effects that this expanded description of a material continuum has. Its effect on the mass density is discussed in the following section. Others are pointed out in later chapters.

In order to connect the material coordinate vector $\mathbf{X}$ introduced here firmly to that employed throughout Chapter 3 we must first determine what the spatial coordinates of a mass point in the sense of Chapter 3 are in the present context. Let us consider the center-of-mass position of the primitive unit cell labeled n. It is defined by

$$\mathbf{x}^{n}(t) \equiv \frac{\sum_{\alpha} m^{\alpha} \mathbf{x}^{\alpha n}(t)}{\sum_{\alpha} m^{\alpha}}. \tag{4.2.3}$$

In the continuum limit this becomes

$$\mathbf{x}(\mathbf{X},t) \equiv \frac{\sum_{\alpha} m^{\alpha} \mathbf{x}^{\alpha}(\mathbf{X},t)}{\sum_{\alpha} m^{\alpha}}. \tag{4.2.4}$$

It is clear that to lowest order the macroscopic deformation of a crystal is described by the center-of-mass position $\mathbf{x}(\mathbf{X},t)$ and that redistributions of the mass within a unit cell allowed by the internal motions can only lead to corrections to the simple deformation of the center-of-mass position. The diatomic lattice example of Sections 1.2 and 2.3 supports this view. Thus the center-of-mass position is what was called the spatial position of a mass point in Chapter 3. Hence we consider the material coordinates introduced here to be the *label of the center of mass.* It then has the same meaning as the material coordinates of a mass point used in Chapter 3. It is for these reasons that we have used the notation $\mathbf{x}$ and $\mathbf{X}$ here to denote the center of mass in spatial and material coordinates. When the coordinate systems used to measure the spatial and material coordinates are made identical and the displacement vector introduced by way of

$$\mathbf{x}(\mathbf{X},t) = \mathbf{X} + \mathbf{u}(\mathbf{X},t), \tag{4.2.5}$$

the material coordinate vector is seen to be the position vector of the center of mass in the absence of deformation ($\mathbf{u}(\mathbf{X},t)=0$).

In the continuum limit sums over the cell index n become integrals over the continuous material coordinate $\mathbf{X}$. The sum is multiplied and divided by the primitive unit cell volume Ω_0; the one in the numerator becomes an element of volume in the material coordinate system as the sum becomes an integral. Thus we have the prescription for changing sums to integrals

$$\sum_{n} F(\mathbf{x}^{\alpha n}(t)) \rightarrow \frac{1}{\Omega_0} \int F(\mathbf{x}^{\alpha}(\mathbf{X},t))\, dV. \tag{4.2.6}$$

Let us now take the continuum limit of the Lagrangian. The kinetic energy term becomes

$$\tfrac{1}{2} \sum_{\alpha n} m^{\alpha} \left[\dot{\mathbf{x}}^{\alpha n}(t) \right]^2 \rightarrow \tfrac{1}{2} \sum_{\alpha} \rho^{\alpha} \int \left[\dot{\mathbf{x}}^{\alpha}(\mathbf{X},t) \right]^2 dV \tag{4.2.7}$$

where

$$\rho^{\alpha} \equiv \frac{m^{\alpha}}{\Omega_0}, \tag{4.2.8}$$

$$\dot{\mathbf{x}}^{\alpha} = \left.\frac{\partial \mathbf{x}^{\alpha}(\mathbf{X},t)}{\partial t}\right|_{\mathbf{X}} \tag{4.2.9}$$

are the mass density and particle velocity of the α sublattice.

The continuum limit of the potential energy requires further thought. The potential energy is a function of the set of the positions of all particles $\{\mathbf{x}^{\alpha n}\}$. If we took the continuum limit of this set to be $\{\mathbf{x}^{\alpha}(\mathbf{X})\}$, we would be expressing the potential energy only as a function of the positions of the sublattices from one cell since the value of $\mathbf{X}$ here is the same for each $\mathbf{x}^{\alpha}$. The potential energy, however, arises from binding forces between *all* pairs of particles in the crystal, not just pairs within one cell. Since different cells correspond to different values of the material coordinate $\mathbf{X}$ in the continuum limit, the derivatives of $\mathbf{x}^{\alpha}$ with respect to $\mathbf{X}$ must be included in the argument of the potential energy in the continuum limit. We found in Section 2.2 for a simple Hooke's law restoring force that only the first derivative with respect to the material coordinate was needed. More generally one might consider including all orders of derivatives, $\mathbf{x}^{\alpha}_{,A}, \mathbf{x}^{\alpha}_{,AB}, \mathbf{x}^{\alpha}_{,ABC}, \ldots$, as arguments of the potential energy of the continuum. However, the relative change in the position of mass points Δu^{α} is a very small fraction of a unit cell dimension and the length in the crystal over which the change occurs is a wavelength λ of the disturbance. Hence the first derivative is of order $\Delta u^{\alpha}/\lambda$ and represents a small correction to the position. This clearly says that second and higher derivatives are negligible in the long wavelength or continuum limit. It also says that first derivatives are negligible if the coordinate itself is present. We show in Chapter 5 that the center-of-mass position cannot be an argument of the potential energy. For this reason its first derivatives must be kept as the only measures involving that variable. This agrees with the developments of Chapter 3, which show the deformation gradients as playing the fundamental role in the elasticity of solids. Since a portion of the gradient of the center of mass is present in each $\mathbf{x}^{\alpha}_{,A}$, as can be inferred from (4.2.4), the gradients $\mathbf{x}^{\alpha}_{,A}$ should be retained as arguments of the potential energy. Thus in the continuum limit we write

$$V(\{\mathbf{x}^{\beta n}(t)\}) \rightarrow \int \rho^0 \overline{\Sigma}(\mathbf{x}^{\beta}(\mathbf{X},t), \mathbf{x}^{\beta}_{,A}(\mathbf{X},t))\, dV \tag{4.2.10}$$

where $\overline{\Sigma}$ is the potential or stored energy per unit undeformed mass, the undeformed mass density ρ^0 is defined by

$$\rho^0 \equiv \sum_\alpha \rho^\alpha = \Omega_0^{-1} \sum_\alpha m^\alpha, \tag{4.2.11}$$

and the $\{\}$ notation to denote the set of quantities is implied but omitted from the continuum form.

The continuum limit of the interaction Lagrangian gives

$$\sum_{\alpha n} e^\alpha \left[\dot{\mathbf{x}}^{\alpha n}(t)\cdot\mathbf{A}(\mathbf{x}^{\alpha n}(t),t) - \Phi(\mathbf{x}^{\alpha n}(t),t)\right]$$

$$\to \sum_\alpha q^\alpha \int \left[\dot{\mathbf{x}}^\alpha(\mathbf{X},t)\cdot\mathbf{A}(\mathbf{x}^\alpha(\mathbf{X},t),t) - \Phi(\mathbf{x}^\alpha(\mathbf{X},t),t)\right] dV \tag{4.2.12}$$

where

$$q^\alpha \equiv \frac{e^\alpha}{\Omega_0}. \tag{4.2.13}$$

This form gives the interaction Lagrangian density in the material frame. Its form in the spatial frame remains (4.1.7) but now with the continuum charge and current densities of (4.1.8) and (4.1.9) given by

$$q(\mathbf{z},t) = \sum_\alpha q^\alpha \int \delta(\mathbf{z} - \mathbf{x}^\alpha(\mathbf{X},t))\, dV, \tag{4.2.14}$$

$$\mathbf{j}(\mathbf{z},t) = \sum_\alpha q^\alpha \int \dot{\mathbf{x}}^\alpha(\mathbf{X},t)\,\delta(\mathbf{z} - \mathbf{x}^\alpha(\mathbf{X},t))\, dV. \tag{4.2.15}$$

In summary, the matter Lagrangian in the material frame is given by

$$L_M = \int \mathcal{L}_{MM}\, dV, \tag{4.2.16}$$

$$\mathcal{L}_{MM} = \tfrac{1}{2} \sum_\alpha \rho^\alpha \left[\dot{\mathbf{x}}^\alpha(\mathbf{X},t)\right]^2 - \rho^0 \bar{\Sigma}\left(\mathbf{x}^\beta(\mathbf{X},t), \mathbf{x}^\beta_{,A}(\mathbf{X},t), \mathbf{X}, t\right) \tag{4.2.17}$$

where we have inserted $\mathbf{X}, t$ as explicit arguments of $\bar{\Sigma}$ to allow it possibly to vary with location and time. The interaction Lagrangian in the material frame is given by

$$L_I = \int \mathcal{L}_{IM}\, dV, \tag{4.2.18}$$

$$\mathcal{L}_{IM} = \sum_\alpha q^\alpha \left[\dot{\mathbf{x}}^\alpha(\mathbf{X},t)\cdot\mathbf{A}(\mathbf{x}^\alpha(\mathbf{X},t),t) - \Phi(\mathbf{x}^\alpha(\mathbf{X},t),t)\right]. \tag{4.2.19}$$

4.3 Mass Density

We consider in this section one of the simpler implications of representing the matter in a crystal by N vector continua rather than just the center-of-mass position vector continuum. The mass density in the spatial frame $\rho(\mathbf{z},t)$ in the discrete particle view is

$$\rho(\mathbf{z},t)=\sum_{\alpha n} m^{\alpha}\delta(\mathbf{z}-\mathbf{x}^{\alpha n}(t)). \tag{4.3.1}$$

In the continuum limit this becomes

$$\rho(\mathbf{z},t)=\sum_{\alpha}\rho^{\alpha}\int\delta(\mathbf{z}-\mathbf{x}^{\alpha}(\mathbf{X},t))\,dV. \tag{4.3.2}$$

Note that contributions to the mass density arise from each of the N continua $\mathbf{x}^{\alpha}(\mathbf{X},t)$.

We expand the Dirac delta function in a Taylor series about the center of mass

$$\begin{aligned}\rho(\mathbf{z},t)&=\sum_{\alpha}\rho^{\alpha}\int\delta(\mathbf{z}-\mathbf{x}(\mathbf{X},t)-\mathbf{x}^{\alpha}(\mathbf{X},t)+\mathbf{x}(\mathbf{X},t))\,dV\\&=\rho^{M}(\mathbf{z},t)+\rho^{D}(\mathbf{z},t)+\rho^{Q}(\mathbf{z},t)+\cdots\end{aligned} \tag{4.3.3}$$

where

$$\rho^{M}(\mathbf{z},t)=\sum_{\alpha}\rho^{\alpha}\int\delta(\mathbf{z}-\mathbf{x}(\mathbf{X},t))\,dV, \tag{4.3.4}$$

$$\rho^{D}(\mathbf{z},t)=-\sum_{\alpha}\rho^{\alpha}\int\left[x_i^{\alpha}(\mathbf{X},t)-x_i(\mathbf{X},t)\right]\frac{\partial}{\partial z_i}\delta(\mathbf{z}-\mathbf{x}(\mathbf{X},t))\,dV, \tag{4.3.5}$$

$$\begin{aligned}\rho^{Q}(\mathbf{z},t)=\tfrac{1}{2}\sum_{\alpha}\rho^{\alpha}\int&\left[x_i^{\alpha}(\mathbf{X},t)-x_i(\mathbf{X},t)\right]\\&\times\left[x_j^{\alpha}(\mathbf{X},t)-x_j(\mathbf{X},t)\right]\frac{\partial^2}{\partial z_i\,\partial z_j}\delta(\mathbf{z}-\mathbf{x}(\mathbf{X},t))\,dV.\end{aligned} \tag{4.3.6}$$

The superscripts on the mass densities refer to *m*onopole, *d*ipole, and *q*uadrupole. The monopole mass density can be reexpressed as

$$\begin{aligned}\rho^M(\mathbf{z},t)&=\rho^0\int\delta(\mathbf{z}-\mathbf{x})\,dV\\&=\rho^0\int\delta(\mathbf{z}-\mathbf{x})\frac{dv}{J(\mathbf{x}/\mathbf{X})}\\&=\frac{\rho^0}{\hat{J}(\mathbf{z},t)}\end{aligned}\tag{4.3.7}$$

where

$$\hat{J}(\mathbf{z},t)\equiv\left[J\left(\frac{\mathbf{x}}{\mathbf{X}}\right)\right]_{\mathbf{x}(\mathbf{X},t)=\mathbf{z}}\tag{4.3.8}$$

with the successive use of (4.2.11), (3.3.19), and (2.5.17).

Next consider the dipole mass density. It can be reexpressed as

$$\begin{aligned}\rho^D(\mathbf{z},t)&=-\sum_\alpha\rho^\alpha\frac{\partial}{\partial z_i}\int[x_i^\alpha-x_i]\delta(\mathbf{z}-\mathbf{x})\,dV\\&=-\sum_\alpha\rho^\alpha\frac{\partial}{\partial z_i}\int[x_i^\alpha-x_i]\frac{\delta(\mathbf{z}-\mathbf{x})\,dv}{J(\mathbf{x}/\mathbf{X})}\\&=-\frac{\partial}{\partial z_i}\left[\frac{\sum_\alpha\rho^\alpha(x_i^\alpha-x_i)}{J(\mathbf{x}/\mathbf{X})}\right]_{\mathbf{x}(\mathbf{X},t)=\mathbf{z}}\\&=-\frac{\partial}{\partial z_i}\left[\frac{\rho^0x_i-\rho^0x_i}{J(\mathbf{x}/\mathbf{X})}\right]_{\mathbf{x}(\mathbf{X},t)=\mathbf{z}}\\&=0\end{aligned}\tag{4.3.9}$$

where we remove the derivative with respect to z_i from the integral, since its limits are independent of z_i, and then use (3.3.19), (2.5.17), (4.2.4), (4.2.8), and (4.2.11). Thus, not surprisingly, we find that there is no dipole mass density. The quadrupole term can be manipulated similarly, obtaining successively

$$\rho^Q(\mathbf{z},t)=\tfrac{1}{2}\sum_\alpha \rho^\alpha \frac{\partial^2}{\partial z_i \partial z_j}\int [x_i^\alpha - x_i][x_j^\alpha - x_j]\frac{\delta(\mathbf{z}-\mathbf{x})\,dv}{J(\mathbf{x}/\mathbf{X})}$$

$$=\tfrac{1}{2}\frac{\partial^2}{\partial z_i\,\partial z_j}\left[J^{-1}\left(\frac{\mathbf{x}}{\mathbf{X}}\right)\sum_\alpha \rho^\alpha(x_i^\alpha x_j^\alpha - x_i^\alpha x_j - x_j^\alpha x_i + x_i x_j)\right]_{\mathbf{x}=\mathbf{z}}$$

$$=\tfrac{1}{2}\frac{\partial^2}{\partial z_i\,\partial z_j}\left[J^{-1}\left(\frac{\mathbf{x}}{\mathbf{X}}\right)\left(\sum_\alpha \rho^\alpha x_i^\alpha x_j^\alpha - \rho^0 x_i x_j - \rho^0 x_j x_i + \rho^0 x_i x_j\right)\right]_{\mathbf{x}=\mathbf{z}}$$

$$=\frac{\partial^2 \rho_{ij}^Q(\mathbf{z},t)}{\partial z_i\,\partial z_j} \tag{4.3.10}$$

where

$$\rho_{ij}^Q(\mathbf{z},t)\equiv\tfrac{1}{2}\left[J^{-1}\left(\frac{\mathbf{x}}{\mathbf{X}}\right)\left(\sum_\alpha \rho^\alpha x_i^\alpha x_j^\alpha - \rho^0 x_i x_j\right)\right]_{\mathbf{x}(\mathbf{X},t)=\mathbf{z}}. \tag{4.3.11}$$

Collecting results we find

$$\rho(\mathbf{z},t)=\frac{\rho^0}{\hat{J}(\mathbf{z},t)}+\frac{\partial^2 \rho_{ij}^Q(\mathbf{z},t)}{\partial z_i\,\partial z_j}+\cdots. \tag{4.3.12}$$

Comparison of this with (3.9.6), which holds for a single mass continuum, the center-of-mass continuum, shows that the necessary representation of the matter by N continua leads to a quadrupole (and higher order) contribution to the mass density. The monopole term here is the entire result (3.9.6) when only the center-of-mass position represented the matter.

The quadrupole term is a small effect and we ignore it when using the mass density later. Its smallness results from the presence of two space derivatives. A space derivative brings in a factor of the wavevector, which is inversely proportional to the wavelength of the disturbance, and so indicates the term is a correction to the long wavelength limit of the phenomenon considered. Such corrections are called *wavevector dispersion* corrections, or *spatial dispersion* corrections. A physical view of the effect is the following. For a wave to be affected by the distribution of the mass in the unit cell its phase must be at least somewhat different at different points in the cell. It then is affected by the changes in mass density across the cell, that is, the various multipole moments of the distribution. The effect is thus seen to be numerically large only when the wavelength of the disturbance is comparable to cell dimensions and negligible in the long wavelength limit.

4.4 Multipole Expansion of Charge Density

In contrast to the multipole expansion of the mass density, the multipole expansion of the charge and current densities is very useful. One reason is that dielectric media have no free charge distribution. Thus the monopole term may be set equal to zero. The electric dipole interaction is thus the lowest order interaction. A second reason is that the electric quadrupole and the associated (same order of smallness) magnetic dipole contributions are usually numerically negligible for long wavelength phenomena in dielectrics for the same reason as discussed for the quadrupole contribution to the mass density. This means that only the electric dipole contribution need be retained for most long wavelength phenomena in dielectrics.

The mathematical procedure here is very similar to that used for the mass density of the last section and so can be shortened somewhat here. We begin with the charge density expression (4.2.14) and expand the Dirac delta function about the center of mass. We obtain

$$q(\mathbf{z},t)=\sum_{\alpha} q^{\alpha}\int\left[\delta(\mathbf{z}-\mathbf{x})-(x_i^{\alpha}-x_i)\frac{\partial}{\partial z_i}\delta(\mathbf{z}-\mathbf{x})\right.$$
$$\left.+\tfrac{1}{2}(x_i^{\alpha}-x_i)(x_j^{\alpha}-x_j)\frac{\partial^2}{\partial z_i\,\partial z_j}\delta(\mathbf{z}-\mathbf{x})+\cdots\right]dV. \qquad (4.4.1)$$

We now invoke the *dielectric assumption*,

$$\sum_{\alpha} q^{\alpha}=0, \qquad (4.4.2)$$

that is, electrical neutrality of the unit cell under all conditions. Thus the first or monopole term in (4.4.1) vanishes. By withdrawing the derivatives from the integral, transforming the integration variable to the spatial frame, and using the delta function property we obtain

$$q(\mathbf{z},t)=-\sum_{\alpha} q^{\alpha}\frac{\partial}{\partial z_i}\int\frac{(x_i^{\alpha}-x_i)\delta(\mathbf{z}-\mathbf{x})}{J(\mathbf{x}/\mathbf{X})}\,dv$$
$$+\tfrac{1}{2}\sum_{\alpha} q^{\alpha}\frac{\partial^2}{\partial z_i\,\partial z_j}\int\frac{(x_i^{\alpha}-x_i)(x_j^{\alpha}-x_j)\delta(\mathbf{z}-\mathbf{x})}{J(\mathbf{x}/\mathbf{X})}\,dv$$
$$=-P_{i,i}(\mathbf{z},t)+Q_{ij,ij}(\mathbf{z},t) \qquad (4.4.3)$$

where the *polarization* is defined by

$$P_i(\mathbf{z},t) \equiv \sum_\alpha q^\alpha \left[\frac{(x_i^\alpha - x_i)}{J(\mathbf{x}/\mathbf{X})} \right]_{\mathbf{x}(\mathbf{X},t)=\mathbf{z}} = \sum_\alpha q^\alpha \frac{\hat{u}_i^\alpha(\mathbf{z},t)}{\hat{J}(\mathbf{z},t)} \tag{4.4.4}$$

and the *quadrupolarization* is defined by

$$Q_{ij}(\mathbf{z},t) \equiv \tfrac{1}{2} \sum_\alpha q^\alpha \left[\frac{(x_i^\alpha - x_i)(x_j^\alpha - x_j)}{J(\mathbf{x}/\mathbf{X})} \right]_{\mathbf{x}(\mathbf{X},t)=\mathbf{z}} = \tfrac{1}{2} \sum_\alpha q^\alpha \frac{\hat{u}_i^\alpha(\mathbf{z},t)\hat{u}_j^\alpha(\mathbf{z},t)}{\hat{J}(\mathbf{z},t)}, \tag{4.4.5}$$

the latter obviously being a symmetric tensor. Here we introduce the *internal displacement* $\mathbf{u}^\alpha$ of the α sublattice from the center of mass by

$$\mathbf{u}^\alpha \equiv \mathbf{x}^\alpha - \mathbf{x} \tag{4.4.6}$$

and use the caret to indicate dependence on $\mathbf{z}, t$. Note that $\mathbf{u}^\alpha$ in both the polarization and the quadrupolarization is the moment arm of the charge distribution relative to the center of mass. An alternative expression for (4.4.3) in vector notation is

$$q = -\nabla \cdot \mathbf{P} + \nabla\nabla : \mathbf{Q}. \tag{4.4.7}$$

4.5 Multipole Expansion of Current Density

The multipole expansion procedure is now applied to the current density $\mathbf{j}(\mathbf{z},t)$ of (4.2.15). Following the procedure of the preceding two sections, we obtain

$$\begin{aligned}
j_k(\mathbf{z},t) = {} & \sum_\alpha q^\alpha \int [\dot{x}_k + \dot{x}_k^\alpha - \dot{x}_k] \Bigg[\delta(\mathbf{z}-\mathbf{x}) - (x_i^\alpha - x_i) \frac{\partial}{\partial z_i} \delta(\mathbf{z}-\mathbf{x}) \\
& + \tfrac{1}{2}(x_i^\alpha - x_i)(x_j^\alpha - x_j) \frac{\partial^2}{\partial z_i \partial z_j} \delta(\mathbf{z}-\mathbf{x}) + \cdots \Bigg] \frac{dv}{J(\mathbf{x}/\mathbf{X})} \\
= {} & \sum_\alpha q^\alpha \frac{\hat{\dot{u}}_k^\alpha(\mathbf{z},t)}{\hat{J}(\mathbf{z},t)} - \sum_\alpha q^\alpha \frac{\partial}{\partial z_i} \left[\frac{\hat{\dot{x}}_k(\mathbf{z},t)\hat{u}_i^\alpha(\mathbf{z},t)}{\hat{J}(\mathbf{z},t)} \right] \\
& - \sum_\alpha q^\alpha \frac{\partial}{\partial z_i} \left[\frac{\hat{\dot{u}}_k^\alpha(\mathbf{z},t)\hat{u}_i^\alpha(\mathbf{z},t)}{\hat{J}(\mathbf{z},t)} \right] \\
& + \tfrac{1}{2} \sum_\alpha q^\alpha \frac{\partial^2}{\partial z_i \partial z_j} \left[\frac{\hat{\dot{x}}_k(\mathbf{z},t)\hat{u}_i^\alpha(\mathbf{z},t)\hat{u}_j^\alpha(\mathbf{z},t)}{\hat{J}(\mathbf{z},t)} \right] + \cdots.
\end{aligned} \tag{4.5.1}$$

We wish to rearrange this expression in terms of the polarization, the quadrupolarization, and the magnetization. If the polarization of (4.4.4) is substituted for $\gamma(\mathbf{z},t)$ in the spatial equation of continuity (3.9.5), the result is

$$\sum_\alpha q^\alpha \frac{\hat{u}_k^\alpha}{\hat{J}} = \frac{1}{\hat{J}}\frac{d}{dt}(\hat{J}P_k) = \frac{\partial P_k}{\partial t} + \frac{\partial}{\partial z_i}(\dot{x}_i P_k), \tag{4.5.2}$$

which reexpresses the first term on the right of (4.5.1). The second term on the right in that equation is

$$-\sum_\alpha q^\alpha \frac{\partial}{\partial z_i}\left[\frac{\hat{\dot{x}}_k \hat{u}_i^\alpha}{\hat{J}}\right] = -\frac{\partial}{\partial z_i}(\dot{x}_k P_i) \tag{4.5.3}$$

and can be combined with (4.5.2) to give

$$\begin{aligned}\frac{1}{\hat{J}}\frac{d}{dt}(\hat{J}P_k) - \frac{\partial}{\partial z_i}(\dot{x}_k P_i) &= \frac{\partial P_k}{\partial t} + \frac{\partial}{\partial z_i}(\dot{x}_i P_k - \dot{x}_k P_i)\\ &= \frac{\partial P_k}{\partial t} + \epsilon_{kip}\frac{\partial}{\partial z_i}\epsilon_{plm}P_l \dot{x}_m.\end{aligned} \tag{4.5.4}$$

In vector notation this can be written

$$\frac{1}{\hat{J}}\frac{d}{dt}(\hat{J}\mathbf{P}) - \frac{\partial}{\partial z_i}(\dot{\mathbf{x}}P_i) = \frac{\partial \mathbf{P}}{\partial t} + \nabla\times(\mathbf{P}\times\dot{\mathbf{x}}). \tag{4.5.5}$$

The third term on the right of (4.5.1) can be divided into symmetric and antisymmetric parts according to

$$\begin{aligned}-\sum_\alpha q^\alpha \frac{\partial}{\partial z_i}\left[\frac{\hat{u}_k^\alpha \hat{u}_i^\alpha}{\hat{J}}\right] = &-\frac{\partial}{\partial z_i}\left[\hat{J}^{-1}\sum_\alpha q^\alpha \hat{u}^\alpha_{(k}\, \hat{u}^\alpha_{i)}\right]\\ &-\frac{\partial}{\partial z_i}\left[\hat{J}^{-1}\sum_\alpha q^\alpha \hat{u}^\alpha_{[k}\, \hat{u}^\alpha_{i]}\right].\end{aligned} \tag{4.5.6}$$

The antisymmetric part is the negative of the *magnetization tensor* M_{ik},

$$M_{ik}(\mathbf{z},t) \equiv \hat{J}^{-1}(\mathbf{z},t)\sum_\alpha q^\alpha \hat{u}^\alpha_{[i}(\mathbf{z},t)\hat{u}^\alpha_{k]}(\mathbf{z},t). \tag{4.5.7}$$

Since any antisymmetric second rank tensor can be related to an axial vector, we define the *magnetization vector* **M** by

$$M_i(\mathbf{z},t) \equiv \tfrac{1}{2}\epsilon_{ijk}M_{jk}(\mathbf{z},t). \tag{4.5.8}$$

This can be inverted with the use of the $\epsilon-\delta$ identity (2.5.10) to give

$$M_{ij}(\mathbf{z},t)=\epsilon_{ijk}M_k(\mathbf{z},t). \tag{4.5.9}$$

The current contribution from the magnetization can now be written in two forms,

$$\frac{\partial M_{ki}}{\partial z_i}=\epsilon_{kij}\frac{\partial M_j}{\partial z_i}. \tag{4.5.10}$$

We have two remaining current terms, the last one in (4.5.1) and the first one on the right side of (4.5.6), which need to be reexpressed. With successive use of the spatial equation of continuity (3.9.5) with $\gamma(\mathbf{z},t)$ replaced by Q_{ki} and the $\epsilon-\delta$ identity we get

$$\begin{aligned}
&\tfrac{1}{2}\sum_\alpha q^\alpha\frac{\partial^2}{\partial z_i\,\partial z_j}\left[\frac{\hat{\dot{x}}_k\hat{u}_i^\alpha\hat{u}_j^\alpha}{\hat{J}}\right]-\sum_\alpha q^\alpha\frac{\partial}{\partial z_i}\left[\frac{\hat{\dot{u}}^\alpha_{(k}\,\hat{u}^\alpha{}_{i)}}{\hat{J}}\right]\\
&=\frac{\partial^2}{\partial z_i\,\partial z_j}\left[\dot{x}_k Q_{ij}\right]-\frac{\partial}{\partial z_i}\left[\frac{1}{\hat{J}}\frac{d}{dt}(\hat{J}Q_{ki})\right]\\
&=\frac{\partial}{\partial z_i}\left[\frac{\partial}{\partial z_j}(\dot{x}_k Q_{ij})-\frac{\partial}{\partial z_j}(\dot{x}_j Q_{ki})-\frac{\partial Q_{ki}}{\partial t}\right]\\
&=-\frac{\partial^2 Q_{ki}}{\partial t\,\partial z_i}-\epsilon_{kjl}\frac{\partial}{\partial z_j}\epsilon_{lmn}\frac{\partial}{\partial z_i}(Q_{im}\dot{x}_n).
\end{aligned} \tag{4.5.11}$$

Substitution of the reexpressed current terms into (4.5.1) yields

$$\begin{aligned}
j_k=&\frac{\partial P_k}{\partial t}+\epsilon_{kip}\frac{\partial}{\partial z_i}\epsilon_{plm}P_l\dot{x}_m-\frac{\partial^2 Q_{ki}}{\partial t\,\partial z_i}\\
&-\epsilon_{kjl}\frac{\partial}{\partial z_j}\epsilon_{lmn}\frac{\partial}{\partial z_i}(Q_{im}\dot{x}_n)+\epsilon_{kij}\frac{\partial M_j}{\partial z_i}.
\end{aligned} \tag{4.5.12}$$

This expression in tensor notation is complicated but unambiguous. In vector notation it becomes

$$\begin{aligned}
\mathbf{j}=&\frac{\partial \mathbf{P}}{\partial t}+\nabla\times(\mathbf{P}\times\dot{\mathbf{x}})-\frac{\partial}{\partial t}(\nabla\cdot\mathbf{Q})\\
&-\nabla\times\left[(\nabla\cdot\mathbf{Q}+\mathbf{Q}\cdot\nabla)\times\dot{\mathbf{x}}\right]+\nabla\times\mathbf{M},
\end{aligned} \tag{4.5.13}$$

which is more compact but somewhat difficult to interpret. The expression for $\mathbf{j}$ and that for q from the preceding section can now be used to express the interaction Lagrangian to a given multipole order. Before doing this we wish to introduce a more convenient set of spatial coordinates.

4.6 Internal Coordinates

The discussion of the continuum limit above brought out the fact that gradients of the center-of-mass position correspond to the deformation gradients of Chapter 3 in determining the elastic response of the crystalline medium. For this reason it is advantageous to have the center-of-mass position as one of the independent set of spatial coordinates. Since the center of mass carries all the momentum of the matter, the remaining $N-1$ vector coordinates can be chosen so that they individually do not carry any momentum. This choice can be guaranteed by requiring the $N-1$ vector coordinates to be *invariant to displacements* of the entire crystal in space. These coordinates are called *internal coordinates*.

The new set of coordinates, denoted by $\mathbf{y}^{T\mu}(\mathbf{X},t)$, are defined in terms of the simple continuum position coordinates by

$$\mathbf{y}^{T\mu}(\mathbf{X},t) \equiv \sum_{\alpha=1}^{N} U^{\mu\alpha}\mathbf{x}^{\alpha}(\mathbf{X},t) \qquad \mu=0,1,2,\ldots,N-1 \tag{4.6.1}$$

where $\mathbf{y}^{T0}(\mathbf{X},t)$ is taken as the center of mass

$$\mathbf{y}^{T0}(\mathbf{X},t) \equiv \mathbf{x}(\mathbf{X},t) = \frac{1}{\rho^0}\sum_{\alpha}\rho^{\alpha}\mathbf{x}^{\alpha}(\mathbf{X},t). \tag{4.6.2}$$

Only the $\mathbf{y}^{T\nu}$ coordinates for $\nu=1,2,\ldots,N-1$ are called internal coordinates. Note that the numbering of the N vector coordinates is shifted from 1 to N for the $\mathbf{x}^{\alpha}$ coordinates to 0 to $N-1$ for the $\mathbf{y}^{T\mu}$ coordinates. Note also that we use Greek letters late in the alphabet as superscripts for the $\mathbf{y}^{T\mu}$ while we use Greek letters early in the alphabet for the $\mathbf{x}^{\alpha}$. The superscript T, standing for total, indicates that the internal coordinates contain a constant part as well as a varying part. Comparison of (4.6.1) for $\mu=0$ with (4.6.2) leads to the identification

$$U^{0\alpha} = \frac{\rho^{\alpha}}{\rho^0}. \tag{4.6.3}$$

We denote the inverse transformation by

$$\mathbf{x}^{\alpha}(\mathbf{X},t)=\sum_{\mu=0}^{N-1} V^{\alpha\mu}\mathbf{y}^{T\mu}(\mathbf{X},t) \qquad \alpha=1,2,\ldots,N. \tag{4.6.4}$$

Combining this with (4.6.1) yields both

$$\sum_{\alpha=1}^{N} U^{\mu\alpha}V^{\alpha\nu}=\delta^{\mu\nu} \tag{4.6.5}$$

and

$$\sum_{\nu=0}^{N-1} V^{\alpha\nu}U^{\nu\beta}=\delta^{\alpha\beta}. \tag{4.6.6}$$

The internal coordinates are displacement invariant provided that

$$\mathbf{y}^{T\mu}=\sum_{\alpha} U^{\mu\alpha}\mathbf{x}^{\alpha}=\sum_{\alpha} U^{\mu\alpha}(\mathbf{x}^{\alpha}+\mathbf{d}) \qquad \mu\neq 0 \tag{4.6.7}$$

where $\mathbf{d}$ is any constant vector. This requires that

$$\sum_{\alpha=1}^{N} U^{\mu\alpha}=0 \qquad \mu\neq 0. \tag{4.6.8}$$

Since (4.6.3) yields

$$\sum_{\alpha=1}^{N} U^{0\alpha}=1 \tag{4.6.9}$$

by the definition of ρ^0 (4.2.11), (4.6.8) can be written for any value of μ as

$$\sum_{\alpha=1}^{N} U^{\mu\alpha}=\delta^{\mu 0}. \tag{4.6.10}$$

The implication of this for the $V^{\alpha\nu}$ matrix can be found by summing (4.6.6) over β,

$$\sum_{\nu=0}^{N-1} V^{\alpha\nu}\sum_{\beta} U^{\nu\beta}=1, \tag{4.6.11}$$

and substituting (4.6.10) into this to obtain

$$V^{\alpha 0}=1. \tag{4.6.12}$$

Both (4.6.8) and (4.6.12) are mathematical statements of the displacement invariance of the internal coordinates. This invariance can be understood qualitatively by thinking of an internal coordinate as a difference coordinate, that is, as the difference between the position of a particle and the center of mass or between the positions of two particles.

As a further convenience the new coordinates can be chosen to retain the diagonality of the kinetic energy, that is, the kinetic energy need not contain terms like $\dot{\mathbf{y}}^{T\mu}\cdot\dot{\mathbf{y}}^{T\nu}$. This is accomplished by requiring that

$$\begin{aligned}\sum_{\alpha}\rho^{\alpha}(\dot{\mathbf{x}}^{\alpha})^2 &= \sum_{\nu=0}^{N-1} m^{\nu}(\dot{\mathbf{y}}^{T\nu})^2\\ &= \sum_{\alpha\beta}\sum_{\nu=0}^{N-1} m^{\nu}U^{\nu\alpha}U^{\nu\beta}\dot{\mathbf{x}}^{\alpha}\cdot\dot{\mathbf{x}}^{\beta}\end{aligned} \tag{4.6.13}$$

and

$$\sum_{\mu=0}^{N-1} m^{\mu}(\dot{\mathbf{y}}^{T\mu})^2=\sum_{\alpha}\sum_{\mu\nu=0}^{N-1}\rho^{\alpha}V^{\alpha\mu}V^{\alpha\nu}\dot{\mathbf{y}}^{T\mu}\cdot\dot{\mathbf{y}}^{T\nu}, \tag{4.6.14}$$

which imply

$$\sum_{\nu=0}^{N-1} m^{\nu}U^{\nu\alpha}U^{\nu\beta}=\rho^{\alpha}\delta^{\alpha\beta}, \tag{4.6.15}$$

$$\sum_{\alpha}\rho^{\alpha}V^{\alpha\mu}V^{\alpha\nu}=m^{\mu}\delta^{\mu\nu}. \tag{4.6.16}$$

These are satisfied if

$$m^{\nu}U^{\nu\alpha}=\rho^{\alpha}V^{\alpha\nu}. \tag{4.6.17}$$

This equation can be regarded as determining m^{ν}, the mass density associated with the $\mathbf{y}^{T\nu}$ coordinate.

The internal displacements $\mathbf{u}^{\alpha}$, defined in (4.4.6), are related to the internal coordinates by

$$\mathbf{u}^{\alpha}(\mathbf{X},t)=\sum_{\nu=1}^{N-1} V^{\alpha\nu}\mathbf{y}^{T\nu}(\mathbf{X},t). \tag{4.6.18}$$

This allows the polarization, quadrupolarization, and magnetization to be expressed as

$$\mathbf{P}(\mathbf{z},t)=\sum_{\nu}\frac{q^{\nu}\hat{\mathbf{y}}^{T\nu}}{\hat{J}},\tag{4.6.19}$$

$$\mathbf{Q}(\mathbf{z},t)=\frac{1}{2}\sum_{\mu\nu}\frac{q^{\mu\nu}\hat{\mathbf{y}}^{T\mu}\hat{\mathbf{y}}^{T\nu}}{\hat{J}},\tag{4.6.20}$$

$$\mathbf{M}(\mathbf{z},t)=\frac{1}{2}\sum_{\mu\nu}\frac{q^{\mu\nu}\hat{\mathbf{y}}^{T\mu}\times\dot{\hat{\mathbf{y}}}^{T\nu}}{\hat{J}}\tag{4.6.21}$$

where

$$q^{\nu}\equiv\sum_{\alpha}q^{\alpha}V^{\alpha\nu},\tag{4.6.22}$$

$$q^{\mu\nu}\equiv\sum_{\alpha}q^{\alpha}V^{\alpha\mu}V^{\alpha\nu},\tag{4.6.23}$$

and a sum over the internal coordinate designations that does not explicitly show its limits is one which omits 0 from the summation variable. The importance of the convention for Greek letter superscripts being early or late in the alphabet is seen from (4.6.22). Note that (4.6.12) and (4.4.2) imply that

$$q^{\nu}=0 \text{ for } \nu=0,\qquad q^{\mu\nu}=0 \text{ for } \mu=\nu=0,\qquad q^{0\nu}=q^{\nu 0}=q^{\nu}.\tag{4.6.24}$$

4.7 Lagrangian Density in Internal Coordinates and Electric Dipole Approximation

The Lagrangian density must now be expressed in terms of the new spatial coordinates, the center of mass and the internal coordinates. For simplicity we henceforth consider the interaction Lagrangian to be truncated at the electric dipole level. Electric quadrupole and magnetic dipole contributions in dielectric phenomena are not only small but are experimentally distinguishable in principle from electric dipole effects. Thus a study of higher multipole effects can be done separately.

To the electric dipole level the charge and current densities in the spatial frame are given by (4.4.7) and (4.5.13),

$$q^{D}=-\nabla\cdot\mathbf{P},\tag{4.7.1}$$

$$\mathbf{j}^{D}=\frac{\partial\mathbf{P}}{\partial t}+\nabla\times(\mathbf{P}\times\dot{\mathbf{x}})\tag{4.7.2}$$

where q^D and $\mathbf{j}^D$ are called the *dielectric charge density* and the *dielectric current density*. To this level of approximation the interaction Lagrangian in the spatial frame is

$$L_I = \int \left\{ \mathbf{A} \cdot \left[\frac{\partial \mathbf{P}}{\partial t} + \nabla \times (\mathbf{P} \times \dot{\mathbf{x}}) \right] + \Phi \nabla \cdot \mathbf{P} \right\} dv. \tag{4.7.3}$$

A more convenient form can be obtained for this by rearranging the derivatives as

$$\mathbf{A} \cdot \frac{\partial \mathbf{P}}{\partial t} = \frac{\partial}{\partial t}(\mathbf{A} \cdot \mathbf{P}) - \mathbf{P} \cdot \frac{\partial \mathbf{A}}{\partial t}, \tag{4.7.4}$$

$$\Phi \nabla \cdot \mathbf{P} = \nabla \cdot (\Phi \mathbf{P}) - \mathbf{P} \cdot \nabla \Phi, \tag{4.7.5}$$

$$\begin{aligned} \mathbf{A} \cdot [\nabla \times (\mathbf{P} \times \dot{\mathbf{x}})] &= \nabla \cdot [(\mathbf{P} \times \dot{\mathbf{x}}) \times \mathbf{A}] + [(\mathbf{P} \times \dot{\mathbf{x}}) \times \nabla] \cdot \mathbf{A} \\ &= \nabla \cdot [(\mathbf{P} \times \dot{\mathbf{x}}) \times \mathbf{A}] + (\mathbf{P} \times \dot{\mathbf{x}}) \cdot (\nabla \times \mathbf{A}) \\ &= \nabla \cdot [(\mathbf{P} \times \dot{\mathbf{x}}) \times \mathbf{A}] + \mathbf{P} \cdot [\dot{\mathbf{x}} \times (\nabla \times \mathbf{A})]. \end{aligned} \tag{4.7.6}$$

Each of these derivative rearrangements produces a perfect derivative as the first term on the right side. These terms can be simply discarded because a perfect derivative in the Lagrangian density cannot affect the equations of motion as shown in Section 2.1. The presence of $\dot{\mathbf{x}}$ within the perfect derivative in (4.7.6) means that the equations of motion (2.1.8) must be considered for the dropping of the perfect derivative to be exact. However, if the equations of motion (2.1.7) are used the error in dropping the perfect derivative is no larger than the higher order derivative terms that are ignored.

The interaction Lagrangian in the electric dipole approximation thus becomes

$$\begin{aligned} L_I &= \int \mathbf{P} \cdot \left[-\nabla \Phi - \frac{\partial \mathbf{A}}{\partial t} + \dot{\mathbf{x}} \times (\nabla \times \mathbf{A}) \right] dv \\ &= \int \mathfrak{L}_{IS} \, dv \end{aligned} \tag{4.7.7}$$

where

$$\begin{aligned} \mathfrak{L}_{IS} &= \sum_{\nu} \frac{q^{\nu} \mathbf{y}^{T\nu}}{\hat{J}} \cdot \left[-\nabla \Phi - \frac{\partial \mathbf{A}}{\partial t} + \dot{\mathbf{x}} \times (\nabla \times \mathbf{A}) \right] \\ &= \mathbf{P} \cdot [\mathbf{E} + \mathbf{x} \times \mathbf{B}], \end{aligned} \tag{4.7.8}$$

The last form has mnemonic value but we must remember that Φ and $\mathbf{A}$

are the Lagrangian coordinates of the electromagnetic fields and the internal coordinates $\mathbf{y}^{T\nu}$ (as well as the center of mass $\mathbf{x}$) are the Lagrangian coordinates of the matter. In the spatial description all fields are functions of the independent variables $\mathbf{z}$ and t and $\dot{\mathbf{x}}$ should be regarded as expressed by (3.11.16).

The interaction Lagrangian density in the electric dipole approximation in the material description can be obtained from that in (4.7.7) and (4.7.8) by transforming the integration variables by (3.3.19) and evaluating the electromagnetic potentials at $\mathbf{z}=\mathbf{x}(\mathbf{X},t)$. The result is

$$L_I=\int\sum_\nu q^\nu\mathbf{y}^{T\nu}\cdot\left[-\nabla\Phi(\mathbf{x},t)-\frac{\partial\mathbf{A}(\mathbf{x},t)}{\partial t}+\dot{\mathbf{x}}\times(\nabla\times\mathbf{A}(\mathbf{x},t))\right]dV$$

$$=\int\mathcal{L}_{IM}\,dV \tag{4.7.9}$$

where

$$\mathcal{L}_{IM}=\sum_\nu q^\nu\mathbf{y}^{T\nu}\cdot\left[-\nabla\Phi(\mathbf{x},t)-\frac{\partial\mathbf{A}(\mathbf{x},t)}{\partial t}+\dot{\mathbf{x}}\times(\nabla\times\mathbf{A}(\mathbf{x},t))\right]$$

$$=\sum_\nu q^\nu\mathbf{y}^{T\nu}\cdot\left[\mathbf{E}(\mathbf{x},t)+\dot{\mathbf{x}}\times\mathbf{B}(\mathbf{x},t)\right]. \tag{4.7.10}$$

Here $\mathbf{y}^{T\nu}$, $\mathbf{x}$, and $\dot{\mathbf{x}}$ are regarded as functions of $\mathbf{X}$ and t.

The matter Lagrangian density is easily transformed to dependence on the internal coordinates, since they have retained the diagonality of the kinetic energy. The dependence of the stored energy $\bar{\Sigma}$ is now shifted to $\mathbf{x},\mathbf{y}^{T\nu}(\nu=1,2,\ldots,N-1)$, and $\mathbf{x}_{,A},\mathbf{y}^{T\nu}_{,B}(\nu=1,2,\ldots,N-1)$. We denote this new dependence by a tilde rather than a bar over Σ. Thus

$$L_M=\int\mathcal{L}_{MM}\,dV, \tag{4.7.11}$$

$$\mathcal{L}_{MM}=\tfrac{1}{2}\rho^0(\dot{\mathbf{x}})^2+\tfrac{1}{2}\sum_\nu m^\nu(\dot{\mathbf{y}}^{T\nu})^2-\rho^0\tilde{\Sigma}\left(x_i,y_i^{T\nu},x_{i,A},y_{i,A}^{T\nu},X_A,t\right) \tag{4.7.12}$$

where all the fields are functions of $\mathbf{X}$ and t.

It is instructive to transform the matter Lagrangian density to the spatial description. Since $\mathbf{z}$ and t are the independent variables in this description, derivatives appearing must be taken with respect to one of $\mathbf{z}$ and t while holding the others fixed. Thus the material time derivatives in the center-of-mass kinetic energy must be transformed by

$$\begin{aligned}\tfrac{1}{2}\rho^0\dot{x}_i\dot{x}_i &= \frac{1}{2}\rho^0\left(-x_{i,K}\frac{\partial X_K}{\partial t}\right)\left(-x_{i,L}\frac{\partial X_L}{\partial t}\right)\\ &= \frac{1}{2}\rho^0 C_{KL}\frac{\partial X_K}{\partial t}\frac{\partial X_L}{\partial t}\end{aligned} \tag{4.7.13}$$

where use of (3.11.5) and (3.4.2) is made. The kinetic energy of the internal coordinates must be reexpressed in terms of spatial description derivatives. Using successively (3.8.5) and (3.11.5) and the dual of (3.1.9) we obtain

$$\begin{aligned}\dot{y}_i^{T\nu} &= \frac{\partial y_i^{T\nu}}{\partial t} + \dot{x}_j y_{i,j}^{T\nu}\\ &= \frac{\partial y_i^{T\nu}}{\partial t} - x_{j,K}\frac{\partial X_K}{\partial t}y_{i,j}^{T\nu}\\ &= \frac{\partial y_i^{T\nu}}{\partial t} - \frac{\operatorname{cof}(X_{K,j})}{j(\mathbf{X}/\mathbf{x})}\frac{\partial X_K}{\partial t}y_{i,j}^{T\nu}.\end{aligned} \tag{4.7.14}$$

The matter Lagrangian now becomes

$$L_M = \int \mathcal{L}_{MS}\,dv, \tag{4.7.15}$$

$$\begin{aligned}\mathcal{L}_{MS} = \tfrac{1}{2}\rho C_{KL}\frac{\partial X_K}{\partial t}\frac{\partial X_L}{\partial t} + \frac{1}{2}j(\mathbf{X}/\mathbf{x})\sum_{\nu} m^{\nu}\left(\frac{\partial y_i^{T\nu}}{\partial t} - \frac{\operatorname{cof}(X_{K,j})}{j(\mathbf{X}/\mathbf{x})}\frac{\partial X_K}{\partial t}y_{i,j}^{T\nu}\right)^2\\ -\rho\tilde{\tilde{\Sigma}}\left(X_A, y_i^{T\nu}, X_{A,i}, y_{i,j}^{T\nu}, z_i, t\right)\end{aligned} \tag{4.7.16}$$

where the double tilde over Σ indicates its new functional dependence and the transformation of volume elements (3.3.19) and the definition of the deformed state mass density (3.9.6) are used.

We conclude by noting that the field Lagrangian,

$$L_F = \int \mathcal{L}_{FS}\,dv, \tag{4.7.17}$$

$$\mathcal{L}_{FS} = \frac{\epsilon_0}{2}\left[\left(-\nabla\Phi - \frac{\partial \mathbf{A}}{\partial t}\right)^2 - c^2(\nabla\times\mathbf{A})^2\right], \tag{4.7.18}$$

where $\mathbf{z}$ and $\mathbf{t}$ are the independent variables, is unaffected by taking the electric dipole approximation and introducing the center of mass and internal coordinates. We postpone the transformation of the electromagnetic field Lagrangian density to the material description until Chapter 16.

PROBLEMS

4.1. Show that the Lagrangian densities in (4.2.17) and (4.2.19) lead to

$$\rho^\alpha \ddot{\mathbf{x}}^\alpha(\mathbf{X},t) = q^\alpha\big[\mathbf{E}(\mathbf{x}^\alpha(\mathbf{X},t),t) + \dot{\mathbf{x}}^\alpha(\mathbf{X},t)\times\mathbf{B}(\mathbf{x}^\alpha(\mathbf{X},t),t)\big] + \mathbf{f}^\alpha, \tag{4.P.1}$$

$$f_i^\alpha \equiv -\rho^0 \frac{\partial\bar{\Sigma}}{\partial x_i^\alpha} + \rho^0 \frac{d}{dX_A}\frac{\partial\bar{\Sigma}}{\partial x_{i,A}^\alpha} \tag{4.P.2}$$

for the dynamical equation for $\mathbf{x}^\alpha$ in the material description.

4.2. Show from the result of Prob. 4.1 that the dynamical equation for the center of mass in the material description is

$$\rho^0\ddot{\mathbf{x}}(\mathbf{X},t) = \sum_\alpha q^\alpha\big[\mathbf{E}(\mathbf{x}^\alpha(\mathbf{X},t),t) + \dot{\mathbf{x}}^\alpha(\mathbf{X},t)\times\mathbf{B}(\mathbf{x}^\alpha(\mathbf{X},t),t)\big] + \mathbf{f} \tag{4.P.3}$$

with

$$f_i = -\rho^0\frac{\partial\bar{\Sigma}}{\partial x_i} + \rho^0\frac{d}{dX_A}\frac{\partial\bar{\Sigma}}{\partial x_{i,A}}. \tag{4.P.4}$$

4.3. Show by a Taylor series expansion (about the center of mass) of the $\mathbf{E}$ and $\mathbf{B}$ fields in the electromagnetic force terms of (4.P.3) to quadratic order in $\mathbf{u}^\alpha = \mathbf{x}^\alpha - \mathbf{x}$ that the center-of-mass equation of Prob. 4.2 in the material description can be expressed as

$$\begin{aligned}\rho^0\ddot{x}_i = {} & f_i + p_j E_{i,j} + \epsilon_{ijk}\dot{p}_j B_k + \epsilon_{ijk}\dot{x}_j B_{k,l} p_l \\ & + q_{kl}E_{i,kl} + \epsilon_{ijk}\dot{x}_j B_{k,lm} q_{lm} \\ & + \epsilon_{ijk}\dot{q}_{jl}B_{k,l} + m_k B_{k,i}\end{aligned} \tag{4.P.5}$$

where

$$\mathbf{p} \equiv J\mathbf{P}, \qquad \mathbf{q} \equiv J\mathbf{Q}, \qquad \mathbf{m} \equiv J\mathbf{M} \tag{4.P.6}$$

and the $\mathbf{E}$ and $\mathbf{B}$ fields are now functions of $\mathbf{x}(\mathbf{X},t)$ and t.

4.4. Show that the dynamical equation (4.P.5) for the center of mass may be expressed in the spatial description to electric dipole order as

$$\rho\ddot{x}_i = \hat{J}^{-1}f_i + \left[P_l E_i + P_l \epsilon_{ijk} \dot{x}_j B_k \right]_{,l} + q^D E_i + \epsilon_{ijk} j_j^D B_k \qquad (4.P.7)$$

where the fields are now functions of $\mathbf{z}, t$.

CHAPTER 5

Lagrangian Invariance

The dynamical laws of physics are unchanged in form by several transformations, called *symmetry transformations*, in time and space. A symmetry transformation that consists of a simple displacement in time, $t \rightarrow t+\tau$, leads to no change in the basic physical laws, since time is homogeneous. Similarly, the homogeneity of space leads to the laws of physics being invariant in form to displacements in space, $\mathbf{x} \rightarrow \mathbf{x}+\mathbf{d}$, while the isotropy of space leads to the physical laws being invariant in form to rotations in space, $\mathbf{x} \rightarrow \mathbf{R} \cdot \mathbf{x}$.

The purposes of this chapter are twofold. One purpose is to show that each invariance of the basic dynamical equations leads to a conservation law. For instance, invariance to displacements in time, displacements in space, and rotations in space yield the energy, momentum, and angular momentum conservation laws respectively. The second purpose is to show that a conservation law will result from a Lagrangian formulation only if the functional dependence of the Lagrangian meets a certain requirement resulting from the invariance. This fact allows us in the next chapter to construct the remaining portion of the Lagrangian, the stored energy, in a form as general as possible while being consistent with the basic conservation laws. The discussion is patterned after that of Hill [1951], which is based on the work of Noether [1918].

5.1 Functional Variation of Action Integral

In Section 2.1 we show that Hamilton's principle, that is, the vanishing of a certain functional variation of the action integral I, leads to the Lagrange equations of motion. For the study of invariances in this chapter we need to develop a more general type of functional variation of the action integral. It is also helpful to use a somewhat more general notation. We write the action integral as

$$I \equiv \int \mathcal{L}(q_k, \psi^\alpha, \psi^\alpha_{,k})\, dQ. \tag{5.1.1}$$

Here q_k are the independent variables, which for our purposes include q_1, q_2, q_3 as the space coordinates (either material or spatial) and $q_4 = t$ as the time coordinate. The volume element is $dQ \equiv dq_1\, dq_2\, dq_3\, dq_4$. The ψ^α $(\alpha = 1, 2, \ldots, N)$ are the N dependent fields and $\psi^\alpha_{,k} \equiv \partial\psi^\alpha / \partial q_k$ their first derivatives. We have no need to consider higher derivatives in the Lagrangian density $\mathfrak{L}$.

The functional variation is denoted by δ' and involves both variations of the independent variables

$$\delta' q_k \equiv q_k' - q_k \tag{5.1.2}$$

and variations of the dependent fields and their derivatives

$$\delta' \psi^\alpha(q) \equiv \psi^{\alpha'}(q') - \psi^\alpha(q), \tag{5.1.3}$$

$$\delta' \psi^\alpha_{,k}(q) \equiv \psi^{\alpha'}_{,k}(q') - \psi^\alpha_{,k}(q). \tag{5.1.4}$$

Here q in a functional dependence denotes the set of all $q_k (k = 1, 2, 3, 4)$. The variations of the dependent fields may at this point be regarded as independent of the variations of the independent variables. The variation δ' is more general than the variation δ used in Hamilton's principle, since the latter did not involve a variation of the independent variables.

The functional variation of I is thus

$$\delta' I \equiv \int_{R'} \mathfrak{L}(q_k', \psi^{\alpha'}, \psi^{\alpha'}_{,k})\, dQ' - \int_R \mathfrak{L}(q_k, \psi^\alpha, \psi^\alpha_{,k})\, dQ \tag{5.1.5a}$$

$$= \int_{R'} \mathfrak{L}(q_k + \delta' q_k, \psi^\alpha + \delta'\psi^\alpha, \psi^\alpha_{,k} + \delta'\psi^\alpha_{,k})\, dQ' - \int_R \mathfrak{L}(q_k, \psi^\alpha, \psi^\alpha_{,k})\, dQ \tag{5.1.5b}$$

$$= \int_{R'} \left[\mathfrak{L}(q_k, \psi^\alpha, \psi^\alpha_{,k}) + \frac{\partial \mathfrak{L}}{\partial q_k} \delta' q_k + \frac{\partial \mathfrak{L}}{\partial \psi^\alpha} \delta'\psi^\alpha + \frac{\partial \mathfrak{L}}{\partial \psi^\alpha_{,k}} \delta'\psi^\alpha_{,k} \right] dQ' - \int_R \mathfrak{L}(q_k, \psi^\alpha, \psi^\alpha_{,k})\, dQ \tag{5.1.5c}$$

where summation over repeated superscripts α as well as subscripts k is implied and the partial derivatives are taken holding the remaining explicit arguments fixed. The volume element dQ' can be transformed by way of the Jacobian to dQ by

$$dQ' = J\left(\frac{q'}{q}\right) dQ = \left(1 + \frac{\partial \delta' q_k}{\partial q_k}\right) dQ \tag{5.1.6}$$

to first order by using (3.P.2). Thus $\delta' I$ becomes to first order

$$\delta' I = \int_R \left[\mathcal{L} \frac{\partial \delta' q_k}{\partial q_k} + \frac{\partial \mathcal{L}}{\partial q_k} \delta' q_k + \frac{\partial \mathcal{L}}{\partial \psi^\alpha} \delta' \psi^\alpha + \frac{\partial \mathcal{L}}{\partial \psi^\alpha_{,k}} \delta' \psi^\alpha_{,k} \right] dQ. \quad (5.1.7)$$

We now wish to express as much of the integrand of this functional variation as possible in terms of a total derivative with respect to q_k. Manipulation of the integrand of (5.1.7), however, is made difficult by the fact that the variation δ' and differentiation $\partial/\partial q_k$ do not commute, since the variation involves evaluation of a function at two points in the space of the independent variables q_k. For this reason we introduce the variation δ, which varies only the dependent fields at a fixed point in the space of the independent variables, by

$$\delta\psi^\alpha(q') \equiv \psi^{\alpha'}(q') - \psi^\alpha(q'), \quad (5.1.8)$$

$$\delta\psi^\alpha_{,k}(q') \equiv \psi^{\alpha'}_{,k}(q') - \psi^\alpha_{,k}(q'). \quad (5.1.9)$$

The operations δ and $\partial/\partial q_k$ do commute. The variation δ' can be expressed in terms of δ by rewriting (5.1.3) and (5.1.4) as

$$\begin{aligned} \delta'\psi^\alpha(q) &= \left[\psi^{\alpha'}(q') - \psi^\alpha(q')\right] + \psi^\alpha(q') - \psi^\alpha(q) \\ &= \delta\psi^\alpha(q') + \psi^\alpha_{,k}(q)\delta' q_k, \end{aligned} \quad (5.1.10)$$

$$\begin{aligned} \delta'\psi^\alpha_{,k}(q) &= \left[\psi^{\alpha'}_{,k}(q') - \psi^\alpha_{,k}(q')\right] + \psi^\alpha_{,k}(q') - \psi^\alpha_{,k}(q) \\ &= \delta\psi^\alpha_{,k}(q') + \psi^\alpha_{,kl}(q)\delta' q_l. \end{aligned} \quad (5.1.11)$$

The fact that the variations on the two sides of (5.1.10) and (5.1.11) are evaluated at different points, q' and q, is insignificant when considering variations to first order, since

$$\delta\psi^\alpha(q') = \delta\psi^\alpha(q) + \text{higher order terms} \quad (5.1.12)$$

and similarly for the other variations.

Equation (5.1.7) can now be reexpressed as

$$\begin{aligned} \delta' I &= \int_R \left[\mathcal{L} \frac{\partial \delta' q_k}{\partial q_k} + \frac{\partial \mathcal{L}}{\partial q_k} \delta' q_k + \frac{\partial \mathcal{L}}{\partial \psi^\alpha} (\delta\psi^\alpha + \psi^\alpha_{,k} \delta' q_k) \right. \\ &\quad \left. + \frac{\partial \mathcal{L}}{\partial \psi^\alpha_{,l}} (\delta\psi^\alpha_{,l} + \psi^\alpha_{,lk} \delta' q_k) \right] dQ \\ &= \int_R \left[\frac{d}{dq_k} (\mathcal{L}\delta' q_k) + \frac{\partial \mathcal{L}}{\partial \psi^\alpha} \delta\psi^\alpha + \frac{\partial \mathcal{L}}{\partial \psi^\alpha_{,l}} \delta\psi^\alpha_{,l} \right] dQ \end{aligned} \quad (5.1.13)$$

where d/dq_k is the total derivative with respect to q_k, that is, the derivative which acts on both explicit and implicit dependence on q_k. The last term in the integrand can be rewritten as

$$\begin{aligned}\frac{\partial \mathcal{L}}{\partial \psi^{\alpha}_{,l}}\delta\psi^{\alpha}_{,l} &= \frac{\partial \mathcal{L}}{\partial \psi^{\alpha}_{,l}}\frac{\partial \delta\psi^{\alpha}}{\partial q_l} \\ &= \frac{\partial \mathcal{L}}{\partial \psi^{\alpha}_{,l}}\frac{d\delta\psi^{\alpha}}{dq_l} \\ &= \frac{d}{dq_l}\left(\frac{\partial \mathcal{L}}{\partial \psi^{\alpha}_{,l}}\delta\psi^{\alpha}\right) - \frac{d}{dq_l}\left(\frac{\partial \mathcal{L}}{\partial \psi^{\alpha}_{,l}}\right)\delta\psi^{\alpha} \end{aligned} \tag{5.1.14}$$

where we use successively the commutativity of the δ and $\partial/\partial q_l$ operations, the identity of meanings of $\partial/\partial q_l$ and d/dq_l when applied to ψ^{α} which is a function only of q_l ($l=1,2,3,4$), and the rule of differentiating a product. If we now define the *variational* or *Lagrangian derivative* $\delta/\delta\psi^{\alpha}$ by

$$\frac{\delta \mathcal{L}}{\delta \psi^{\alpha}} \equiv \frac{\partial \mathcal{L}}{\partial \psi^{\alpha}} - \frac{d}{dq_l}\frac{\partial \mathcal{L}}{\partial \psi^{\alpha}_{,l}}, \tag{5.1.15}$$

then (5.1.13) becomes after substitution of (5.1.14)

$$\delta' I = \int_R \left[\frac{d}{dq_k}\left(\mathcal{L}\delta' q_k + \frac{\partial \mathcal{L}}{\partial \psi^{\alpha}_{,k}}\delta\psi^{\alpha}\right) + \frac{\delta \mathcal{L}}{\delta \psi^{\alpha}}\delta\psi^{\alpha}\right] dQ \tag{5.1.16}$$

or, with the use of (5.1.10),

$$\begin{aligned}\delta' I = \int_R \Bigg[& \frac{d}{dq_k}\left(\mathcal{L}\delta' q_k - \frac{\partial \mathcal{L}}{\partial \psi^{\alpha}_{,k}}\psi^{\alpha}_{,l}\delta' q_l + \frac{\partial \mathcal{L}}{\partial \psi^{\alpha}_{,k}}\delta'\psi^{\alpha}\right) \\ & + \frac{\delta \mathcal{L}}{\delta \psi^{\alpha}}(\delta'\psi^{\alpha} - \psi^{\alpha}_{,k}\delta' q_k)\Bigg] dQ. \end{aligned} \tag{5.1.17}$$

Either (5.1.16) or (5.1.17) can be regarded as the final expression for the *functional variation of the action integral*.

Hamilton's principle can be applied to the form (5.1.16). It states that for the special type of variation defined by

$$\delta' q_k = 0 \text{ everywhere,} \tag{5.1.18a}$$

$$\delta\psi^{\alpha} = 0 \text{ on boundary of } R, \tag{5.1.18b}$$

$$R \text{ constant} \tag{5.1.18c}$$

the variation of the action integral vanishes,

$$\delta' I = \delta I = \int_R \left[\frac{d}{dq_k} \left(\frac{\partial \mathfrak{L}}{\partial \psi^\alpha_{,k}} \delta\psi^\alpha \right) + \frac{\delta \mathfrak{L}}{\delta \psi^\alpha} \delta\psi^\alpha \right] dQ = 0. \qquad (5.1.19)$$

Since the first term can be converted by Gauss' theorem to a surface integral which vanishes by (5.1.18b), this gives

$$\int_R \frac{\delta \mathfrak{L}}{\delta \psi^\alpha} \delta\psi^\alpha dQ = 0. \qquad (5.1.20)$$

Since the $\delta\psi^\alpha$ $(\alpha = 1, 2, \ldots, N)$ are independent and arbitrary, this integral can vanish only if

$$\frac{\delta \mathfrak{L}}{\delta \psi^\alpha} = 0 \qquad (\alpha = 1, 2, \ldots, N), \qquad (5.1.21)$$

which from (5.1.15) is seen to be the Lagrange equation of motion for the field ψ^α.

5.2 Symmetry Transformations

A *symmetry transformation* is a transformation which leaves the equations of motion of a physical system unchanged in form, that is, form invariant. Since most finite symmetry transformations are successions of infinitesimal transformations (spatial inversion is an exception), we consider infinitesimal symmetry transformations. One may be represented in the form of (5.1.2) to (5.1.4) by

$$q_k' = q_k + \delta^S q_k, \qquad (5.2.1)$$

$$\psi^{\alpha'}(q') = \psi^\alpha(q) + \delta^S \psi^\alpha(q), \qquad (5.2.2)$$

$$\psi^{\alpha'}_{,k}(q') = \psi^\alpha_{,k}(q) + \delta^S \psi^\alpha_{,k}(q) \qquad (5.2.3)$$

where the variation is now represented by δ^S to indicate it corresponds to a symmetry transformation. In a symmetry transformation the variations $\delta^S q_k$ in the independent variables are the imposed variations and the variations in the dependent fields $\delta^S \psi^\alpha(q)$ and their derivatives $\delta^S \psi^\alpha_{,k}(q)$ result from them.

The variations $\delta^S q_k$ arise from a change in the relative position or orientation of a body and the coordinate axes used in specifying its location. Thus the transformation may be regarded as a change in the coordinate axes holding the body fixed or a change in the location of the body holding the coordinate axes fixed. We prefer the latter as it makes the physical nature of the transformation more apparent.

Clearly, the equations of motion are form invariant if the Lagrangian density itself is form invariant under the symmetry transformation (5.2.1) to (5.2.3). However, the equations of motion can remain form invariant under a more general condition on the Lagrangian density. We saw in Section 2.1 that a four-divergence could be added to the Lagrangian density without effect on the equations of motion. Thus under a symmetry transformation the Lagrangian density need obey only a modified form invariance condition

$$\mathcal{L}'\left(q',\psi^{\alpha'},\psi^{\alpha'}_{,k}\right)=\mathcal{L}\left(q',\psi^{\alpha'},\psi^{\alpha'}_{,k}\right)+\frac{d\delta^S\Omega_k}{dq'_k}. \tag{5.2.4}$$

This states that the Lagrangian density $\mathcal{L}'$ resulting from the symmetry transformation must be the *same function* as the Lagrangian density $\mathcal{L}$ existing before the transformation to within the addition of a four-divergence vector. Since the symmetry transformation is infinitesimal, this vector can be taken as an infinitesimal.

The transformation of the Lagrangian under a symmetry (or other) transformation is also restricted by the Lagrangian being a scalar density. When multiplied by the volume element, it becomes a scalar and so must transform as

$$\mathcal{L}'\left(q',\psi^{\alpha'},\psi^{\alpha'}_{,k}\right)dQ'=\mathcal{L}\left(q,\psi^{\alpha},\psi^{\alpha}_{,k}\right)dQ, \tag{5.2.5}$$

or in terms of the Jacobian $J(q/q')$,

$$\mathcal{L}'\left(q',\psi^{\alpha'},\psi^{\alpha'}_{,k}\right)=\mathcal{L}\left(q,\psi^{\alpha},\psi^{\alpha}_{,k}\right)J\left(\frac{q}{q'}\right). \tag{5.2.6}$$

The combination of (5.2.4) and (5.2.6) can be used to obtain a condition that can be used to test whether a particular transformation is in fact a symmetry transformation for the dynamical equations being studied. Eliminating $\mathcal{L}'$ between these equations yields

$$\mathcal{L}\left(q,\psi^{\alpha},\psi^{\alpha}_{,k}\right)J\left(\frac{q}{q'}\right)=\mathcal{L}\left(q',\psi^{\alpha'},\psi^{\alpha'}_{,k}\right)+\frac{d\delta^S\Omega_k}{dq'_k}. \tag{5.2.7}$$

We now expand all quantities to first order. The first order expansion of the Jacobian can be obtained from (3.P.2). Since $\delta^S\Omega_k$ is already a first order quantity, the divergence with respect to q_k differs from that with respect to q_k' only by a second order quantity which may be dropped. Thus we get

$$\mathcal{L}\left(1-\frac{\partial \delta^S q_k}{\partial q_k}\right)=\mathcal{L}+\frac{\partial \mathcal{L}}{\partial q_k}\delta^S q_k+\frac{\partial \mathcal{L}}{\partial \psi^\alpha}\delta^S\psi^\alpha + \frac{\partial \mathcal{L}}{\partial \psi^\alpha_{,k}}\delta^S\psi^\alpha_{,k}+\frac{d\delta^S\Omega_k}{dq_k}, \tag{5.2.8}$$

which after rearranging is

$$-\frac{d\delta^S\Omega_k}{dq_k}=\frac{\partial \mathcal{L}}{\partial q_k}\delta^S q_k+\frac{\partial \mathcal{L}}{\partial \psi^\alpha}\delta^S\psi^\alpha+\frac{\partial \mathcal{L}}{\partial \psi^\alpha_{,k}}\delta^S\psi^\alpha_{,k}+\mathcal{L}\frac{\partial \delta^S q_k}{\partial q_k}. \tag{5.2.9}$$

This states that the combination of terms on the right side for a particular Lagrangian density and a particular transformation (5.2.1) to (5.2.3) must be a four-divergence (or zero). If not, the transformation is not a symmetry transformation for the physical system represented by the particular Lagrangian density used. If the right side of (5.2.9) is zero, the Lagrangian density (as well as the dynamical equations) is then form invariant according to (5.2.4) for the symmetry transformation involved.

5.3 Conservation Laws

We now wish to show that a symmetry transformation implies a conservation law. Since (5.2.7) contains the effect that the symmetry transformation has on the Lagrangian density, it can be the starting point. Multiply that equation by dQ' and integrate over R',

$$\int_{R'}\mathcal{L}(q,\psi^\alpha,\psi^\alpha_{,k})J\left(\frac{q}{q'}\right)dQ'=\int_{R'}\left[\mathcal{L}(q',\psi^{\alpha'},\psi^{\alpha'}_{,k})+\frac{d\delta^S\Omega_k}{dq_k'}\right]dQ'. \tag{5.3.1}$$

The integral on the left can be converted to an integral over R from the definition of the Jacobian. The last term on the right side is already a first order term and so to first order accuracy may be converted to an integral over R. Thus we get successively

$$
\begin{aligned}
0 &= \int_{R'} \mathfrak{L}(q',\psi^{\alpha'},\psi^{\alpha'}_{,k})\,dQ' - \int_R \mathfrak{L}(q,\psi^{\alpha},\psi^{\alpha}_{,k})\,dQ + \int_R \frac{d\delta^S\Omega_k}{dq_k}\,dQ \\
&= \delta' I + \int_R \frac{d\delta^S\Omega_k}{dq_k}\,dQ \\
&= \int_R \left[\frac{d}{dq_k}\left(\mathfrak{L}\delta^S q_k - \frac{\partial \mathfrak{L}}{\partial \psi^{\alpha}_{,k}} \psi^{\alpha}_{,l}\delta^S q_l + \frac{\partial \mathfrak{L}}{\partial \psi^{\alpha}_{,k}} \delta^S\psi^{\alpha} + \delta^S\Omega_k \right) \right. \\
&\qquad \left. + \frac{\delta \mathfrak{L}}{\delta \psi^{\alpha}} (\delta^S\psi^{\alpha} - \psi^{\alpha}_{,k}\delta^S q_k) \right] dQ \\
&= \int_R \frac{d}{dq_k}\left(\mathfrak{L}\delta^S q_k - \frac{\partial \mathfrak{L}}{\partial \psi^{\alpha}_{,k}} \psi^{\alpha}_{,l}\delta^S q_l + \frac{\partial \mathfrak{L}}{\partial \psi^{\alpha}_{,k}} \delta^S\psi^{\alpha} + \delta^S\Omega_k \right) dQ
\end{aligned}
\tag{5.3.2}
$$

where we use the definition of $\delta' I$ in (5.1.5a), the final derived expression for $\delta' I$ in (5.1.17), and the fact that the Lagrangian density $\mathfrak{L}$ represents a physical system for which the Lagrange equations (5.1.21) hold. Now, since the region of integration R is arbitrary in (5.3.2), the integrand must vanish,

$$
\frac{d}{dq_k}\left[\left(\mathfrak{L}\delta_{kl} - \frac{\partial \mathfrak{L}}{\partial \psi^{\alpha}_{,k}} \psi^{\alpha}_{,l} \right) \delta^S q_l + \frac{\partial \mathfrak{L}}{\partial \psi^{\alpha}_{,k}} \delta^S\psi^{\alpha} + \delta^S\Omega_k \right] = 0. \tag{5.3.3}
$$

This is the *conservation law* resulting from the symmetry transformation (5.2.1) to (5.2.3). The statement that the conservation law (5.3.3) results from the invariance of the action integral (5.1.1) to within a divergence under the symmetry transformation (5.2.1) to (5.2.3) for a physical system obeying the Lagrange equations (5.1.21) is *Noether's theorem* [Noether, 1918].

5.4 Spatial Displacement Invariance

In this section the general results of the last two sections are applied to an infinitesimal *spatial displacement* defined by

$$
\mathbf{z}' = \mathbf{z} + \delta\mathbf{z} \qquad (\delta\mathbf{z} \text{ constant}), \tag{5.4.1a}
$$

$$
t' = t \tag{5.4.1b}
$$

where, as is evident, the four-dimensional notation has been dropped. Since space is homogeneous, we know that (5.4.1) is a symmetry transformation. We wish to find out what functional restriction the Lagrangian must satisfy so that it can produce dynamical equations that are form invariant to this transformation. We also wish to find the conservation law implied by the existence of this symmetry transformation.

Since the transformation we consider here is a spatial one, the Lagrangian should be expressed in the spatial description, that is, with $\mathbf{z}$, t as independent variables. The dependent fields and their derivatives can be seen by examining the Lagrangians in (4.7.8), (4.7.16), and (4.7.18). Thus the dependence of the Lagrangian is

$$\mathfrak{L}=\mathfrak{L}\left(z_i,t,y_i^{T\nu}(\nu\neq 0),X_A,y_{i,j}^{T\nu},X_{A,i},\frac{\partial y_i^{T\nu}}{\partial t},\frac{\partial X_A}{\partial t},A_i,A_{i,j},\frac{\partial A_i}{\partial t},\Phi,\Phi_{,i}\right). \tag{5.4.2}$$

First the variations of the dependent fields and their derivatives induced by the symmetry transformation (5.4.1) must be found. Consider the material coordinate $\mathbf{X}(\mathbf{z},t)$. Since its components are referred to the material coordinate system, it is a scalar in the spatial coordinate system in which the displacement (5.4.1a) is made. Thus the numerical value of $\mathbf{X}$ must not change under the transformation $\mathbf{z}\rightarrow\mathbf{z}'$. This requires that the function $\mathbf{X}$ be changed into the function $\mathbf{X}'$ such that

$$X_A'(\mathbf{z}',t')=X_A(\mathbf{z},t). \tag{5.4.3}$$

From (5.2.2) we then obtain

$$\delta^S X_A\equiv X_A'(\mathbf{z}',t')-X_A(\mathbf{z},t)=0. \tag{5.4.4}$$

When a vector in the spatial coordinate system such as $\mathbf{y}^{T\nu}$ $(\nu\neq 0)$ is considered, the above-mentioned reasoning must be modified to account for referring the components to new coordinates. From the transformation law of a vector field we have

$$y_i^{T\nu'}(\mathbf{z}',t')=\frac{\partial z_i'}{\partial z_j}y_j^{T\nu}(\mathbf{z},t). \tag{5.4.5}$$

From (5.4.1a) we find

$$\frac{\partial z_i'}{\partial z_j}=\delta_{ij}+\frac{\partial \delta z_i}{\partial z_j}=\delta_{ij} \tag{5.4.6}$$

since $\delta\mathbf{z}$ is a constant. Hence

$$\delta^S y_i^{T\nu} \equiv y_i^{T\nu'}(\mathbf{z}',t') - y_i^{T\nu}(\mathbf{z},t) = 0 \qquad (\nu \neq 0). \tag{5.4.7}$$

Combining (5.4.5) with the transformation of a derivative we find for the transformation of $y_{i,j}^{T\nu}$

$$y_{i,j}^{T\nu'}(\mathbf{z}',t') = \frac{\partial z_i'}{\partial z_k}\frac{\partial z_l}{\partial z_j'} y_{k,l}^{T\nu}(\mathbf{z},t). \tag{5.4.8}$$

In addition to (5.4.6) we have

$$\frac{\partial z_l}{\partial z_j'} = \delta_{lj} - \frac{\partial \delta z_l}{\partial z_j'} = \delta_{lj} \tag{5.4.9}$$

since again $\delta\mathbf{z}$ is a constant. Thus

$$\delta^S y_{i,j}^{T\nu} \equiv y_{i,j}^{T\nu'}(\mathbf{z}',t') - y_{i,j}^{T\nu}(\mathbf{z},t) = 0 \qquad (\nu \neq 0). \tag{5.4.10}$$

Similar reasoning also yields

$$\begin{aligned}\delta^S\left(\frac{\partial y_i^{T\nu}}{\partial t}\right) &= \delta^S X_{A,i} = \delta^S\left(\frac{\partial X_A}{\partial t}\right) = \delta^S A_i = \delta^S A_{i,j} \\ &= \delta^S\left(\frac{\partial A_i}{\partial t}\right) = \delta^S \Phi = \delta^S \Phi_{,i} = 0.\end{aligned} \tag{5.4.11}$$

Equation (5.2.9) can now be used to test whether the Lagrangian density (5.4.2) is consistent with (5.4.1) being a symmetry transformation. Substitution of (5.4.5), (5.4.8), (5.4.10), and (5.4.11) into (5.2.9) results in

$$-\frac{d\delta^S \Omega_i}{dz_i} - \frac{d\delta^S \Omega_4}{dt} = \frac{\partial \mathfrak{L}}{\partial z_i}\delta z_i. \tag{5.4.12}$$

Since the right side is not a perfect derivative, it is necessary to require that

$$\frac{\partial \mathfrak{L}}{\partial z_i} = 0. \tag{5.4.13}$$

Since within a material body $\mathbf{z} = \mathbf{x}$, this also implies that

$$\frac{\partial \mathfrak{L}}{\partial x_i} = 0. \tag{5.4.14}$$

Conditions (5.4.13) and (5.4.14) are important, since they prohibit any *explicit* dependence of $\mathfrak{L}$ on $\mathbf{z}$ or $\mathbf{x}$. With the right side of (5.4.12) now zero we may take

$$\delta^S \Omega_i = 0, \qquad \delta^S \Omega_4 = 0. \tag{5.4.15}$$

This means by (5.2.4) that the Lagrangian is form invariant under the symmetry transformation of spatial displacement.

Substitution of the various quantities into the general conservation law (5.3.3) yields

$$\frac{\partial g_i^C}{\partial t} - \frac{\partial t_{ij}^C}{\partial z_j} = 0 \tag{5.4.16}$$

where

$$g_i^C \equiv -\sum_\nu \frac{\partial \mathfrak{L}}{\partial(\partial y_j^{T\nu}/\partial t)} y_{j,i}^{T\nu} - \frac{\partial \mathfrak{L}}{\partial(\partial X_A/\partial t)} X_{A,i} - \frac{\partial \mathfrak{L}}{\partial(\partial A_j/\partial t)} A_{j,i}, \tag{5.4.17}$$

$$t_{ij}^C \equiv \sum_\nu \frac{\partial \mathfrak{L}}{\partial y_{k,j}^{T\nu}} y_{k,i}^{T\nu} + \frac{\partial \mathfrak{L}}{\partial X_{A,j}} X_{A,i} + \frac{\partial \mathfrak{L}}{\partial A_{k,j}} A_{k,i} + \frac{\partial \mathfrak{L}}{\partial \Phi_{,j}} \Phi_{,i} - \mathfrak{L}\delta_{ij}. \tag{5.4.18}$$

The minus sign in (5.4.16) is introduced to give the canonical stress tensor the conventional sign. In obtaining (5.4.16) we have used the fact that the components of $\delta\mathbf{z}$ are independent and arbitrary and so their individual coefficients must vanish. In this way a conservation law is obtained without δz_j variations appearing. The derivatives $\partial/\partial t$ and $\partial/\partial z_i$ appearing in (5.4.16) are total derivatives in the sense that they differentiate both explicit and implicit dependence on the respective variables.

Equation (5.4.16) is the *law of conservation of momentum* in the spatial frame. The quantity g_i^C is the *canonical momentum density* while the quantity t_{ij}^C is the *canonical stress tensor*. The names indicate that these quantities are not the most interesting physical quantities for the system involved having the dimensions of momentum density and stress. The lack of uniqueness of such quantities stems from the possibility of mathematically rearranging terms between the time and space derivatives or adding time independent or divergenceless quantities and still retaining the conservation law form. We return to this question in Chapter 8 and there decide what should be called the total stress tensor and why. In so doing the momentum density is also defined. Until then the importance of

(5.4.16) is simply to say that a physical system whose Lagrangian density in the spatial frame satisfies (5.4.13) and (5.4.14) conserves momentum.

5.5 Temporal Displacement Invariance

We show two things in this section: first, that a displacement in time is a symmetry transformation for a physical system provided its Lagrangian contains no explicit dependence on time and, second, that the presence of this symmetry transformation leads to energy conservation for the system. An infinitesimal *temporal displacement* is defined by

$$\mathbf{z}' = \mathbf{z}, \tag{5.5.1a}$$

$$t' = t + \delta t \qquad (\delta t \text{ constant}). \tag{5.5.1b}$$

Since time is homogeneous, we know that this transformation must be a symmetry transformation for any system studied. We use the Lagrangian density (5.4.2) in the spatial description.

The variations of the dependent fields and their derivatives are easily found by the reasoning of the preceding section along with

$$\frac{\partial z_i}{\partial z'_j} = \frac{\partial z'_i}{\partial z_j} = \delta_{ij}, \qquad \frac{\partial t}{\partial t'} = 1 \tag{5.5.2}$$

where the constancy of δt is used. It follows that the variations of all the dependent fields and their derivatives vanish for this transformation,

$$\begin{aligned} \delta^S X_A &= \delta^S X_{A,j} = \delta^S\left(\frac{\partial X_A}{\partial t}\right) = \delta^S y_i^{Tv} = \delta^S y_{i,j}^{Tv} = \delta^S\left(\frac{\partial y_i^{Tv}}{\partial t}\right) = \delta^S A_i \\ &= \delta^S A_{i,j} = \delta^S\left(\frac{\partial A_i}{\partial t}\right) = \delta^S \Phi = \delta^S \Phi_{,j} = 0. \end{aligned} \tag{5.5.3}$$

Substitution of these variations into the consistency condition (5.2.9) results in

$$-\frac{d\delta^S \Omega_k}{dz_k} - \frac{d\delta^S \Omega_4}{dt} = \frac{\partial \mathfrak{L}}{\partial t}\delta t. \tag{5.5.4}$$

Since the right side is not a perfect derivative, we must require

$$\frac{\partial \mathfrak{L}}{\partial t} = 0, \tag{5.5.5}$$

which states that the Lagrangian density cannot have any *explicit* dependence on time if a temporal displacement is to be a symmetry transformation of the system represented by $\mathfrak{L}$. Since the right side of (5.5.4) is now zero, we may take

$$\delta^S\Omega_i = 0, \qquad \delta^S\Omega_4 = 0. \tag{5.5.6}$$

By (5.2.4) this means that the Lagrangian is form invariant to a temporal displacement transformation.

With use of the conditions in (5.5.3) and (5.5.6) the general conservation law (5.3.3) becomes

$$\frac{\partial H}{\partial t} + \frac{\partial S_j}{\partial z_j} = 0 \tag{5.5.7}$$

with

$$H \equiv \frac{\partial \mathfrak{L}}{\partial(\partial X_A/\partial t)}\frac{\partial X_A}{\partial t} + \sum_\nu \frac{\partial \mathfrak{L}}{\partial(\partial y_i^{T\nu}/\partial t)}\frac{\partial y_i^{T\nu}}{\partial t} + \frac{\partial \mathfrak{L}}{\partial(\partial A_i/\partial t)}\frac{\partial A_i}{\partial t} - \mathfrak{L}, \tag{5.5.8}$$

$$S_j \equiv \frac{\partial \mathfrak{L}}{\partial X_{A,j}}\frac{\partial X_A}{\partial t} + \sum_\nu \frac{\partial \mathfrak{L}}{\partial y_{i,j}^{T\nu}}\frac{\partial y_i^{T\nu}}{\partial t} + \frac{\partial \mathfrak{L}}{\partial \Phi_{,j}}\frac{\partial \Phi}{\partial t} + \frac{\partial \mathfrak{L}}{\partial A_{i,j}}\frac{\partial A_i}{\partial t}. \tag{5.5.9}$$

Equation (5.5.7) is the *energy conservation law* which results from temporal displacement invariance, H is the *energy density* or Hamiltonian, and $\mathbf{S}$ is the *flux of energy*. We return to a consideration of this equation in Chapter 8.

5.6 Spatial Rotation Invariance

Since space is isotropic, a spatial rotation transformation should be a symmetry transformation for any closed system. We develop the condition the Lagrangian must meet for it to represent such a system and we exhibit the conservation law demanded by the presence of this symmetry transformation.

Once again we use the Lagrangian density (5.4.2) in the spatial description. The form of an infinitesimal rotation is found in Section 3.12. In the present notation an infinitesimal *spatial rotation* is expressed as

$$z_i' = z_i + \delta z_{[i,m]} z_m, \tag{5.6.1a}$$

$$t' = t \tag{5.6.1b}$$

where $\delta z_{[i,m]}$ are constants expressed as $\epsilon_{imk} a_k \delta\theta$ in (3.12.6). To first order we have

$$\frac{\partial z_i'}{\partial z_j} = \delta_{ij} + \delta z_{[i,j]}, \tag{5.6.2}$$

$$\begin{aligned}\frac{\partial z_j}{\partial z_i'} &= \delta_{ij} - \delta z_{[j,i]} \\ &= \delta_{ij} + \delta z_{[i,j]},\end{aligned} \tag{5.6.3}$$

$$\frac{\partial t'}{\partial t} = 1. \tag{5.6.4}$$

The variations of the independent variables by (5.6.1a) are

$$\delta^S z_i = \delta z_{[i,m]} z_m. \tag{5.6.5}$$

The variations in the dependent fields and their derivatives induced by this variation are found from the transformation laws of vectors, scalars, and derivatives used in Section 5.4. For example, we find here that

$$\begin{aligned} y_{k,l}^{T\nu\prime}(\mathbf{z}',t') &= \frac{\partial z_k'}{\partial z_j}\frac{\partial z_i}{\partial z_l'} y_{j,i}^{T\nu}(\mathbf{z},t) \\ &= (\delta_{kj} + \delta z_{[k,j]})(\delta_{il} - \delta z_{[i,l]}) y_{j,i}^{T\nu}(\mathbf{z},t) \\ &= y_{k,l}^{T\nu} + \delta z_{[k,j]} y_{j,l}^{T\nu} - \delta z_{[i,l]} y_{k,i}^{T\nu} \end{aligned} \tag{5.6.6}$$

and so the variation is

$$\delta^S y_{k,l}^{T\nu} = \delta z_{[i,j]}\left[\delta_{ik} y_{j,l}^{T\nu} - \delta_{jl} y_{k,i}^{T\nu}\right]. \tag{5.6.7}$$

Similar procedures lead to

$$\delta^S y_i^{T\nu} = \delta z_{[i,j]} y_j^{T\nu}, \qquad \delta^S\left(\frac{\partial y_i^{T\nu}}{\partial t}\right) = \delta z_{[i,j]}\frac{\partial y_j^{T\nu}}{\partial t}, \tag{5.6.8}$$

$$\delta^S X_A = 0, \qquad \delta^S\left(\frac{\partial X_A}{\partial t}\right) = 0, \qquad \delta^S X_{A,i} = \delta z_{[i,j]} X_{A,j}, \tag{5.6.9}$$

$$\delta^S A_i = \delta z_{[i,j]} A_j, \qquad \delta^S\left(\frac{\partial A_i}{\partial t}\right) = \delta z_{[i,j]}\frac{\partial A_j}{\partial t}, \tag{5.6.10}$$

$$\delta^S A_{k,l} = \delta z_{[i,j]}(\delta_{ik} A_{j,l} - \delta_{jl} A_{k,i}), \tag{5.6.11}$$

$$\delta^S \Phi = 0, \qquad \delta^S \Phi_{,i} = \delta z_{[i,j]} \Phi_{,j}. \tag{5.6.12}$$

Substitution of these variations into the test condition (5.2.9) leads to

$$-\frac{d\delta^S\Omega_k}{dz_k}-\frac{d\delta^S\Omega_4}{dt}=\left[\frac{\partial\mathcal{L}}{\partial z_i}z_j+\sum_\nu\frac{\partial\mathcal{L}}{\partial y_i^{T\nu}}y_j^{T\nu}+\frac{\partial\mathcal{L}}{\partial A_i}A_j+\frac{\partial\mathcal{L}}{\partial X_{A,i}}X_{A,j}\right.$$
$$+\sum_\nu\frac{\partial\mathcal{L}}{\partial y_{k,l}^{T\nu}}\left(\delta_{ki}y_{j,l}^{T\nu}-\delta_{jl}y_{k,i}^{T\nu}\right)+\frac{\partial\mathcal{L}}{\partial\Phi_{,i}}\Phi_{,j}$$
$$+\frac{\partial\mathcal{L}}{\partial A_{k,l}}\left(\delta_{ki}A_{j,l}-\delta_{jl}A_{k,i}\right)+\sum_\nu\frac{\partial\mathcal{L}}{\partial\left(\partial y_i^{T\nu}/\partial t\right)}\frac{\partial y_j^{T\nu}}{\partial t}$$
$$\left.+\frac{\partial\mathcal{L}}{\partial(\partial A_i/\partial t)}\frac{\partial A_j}{\partial t}\right]\delta z_{[i,j]}. \tag{5.6.13}$$

The right side of this equation cannot be expressed as a perfect derivative and so must be required to vanish. Since $\delta z_{[i,j]}$ contains three independent arbitrary components, the coefficients of these must separately vanish, leading to

$$0=\frac{\partial\mathcal{L}}{\partial z_{[i}}z_{j]}+\sum_\nu\frac{\partial\mathcal{L}}{\partial y_{[i}^{T\nu}}y_{j]}^{T\nu}+\frac{\partial\mathcal{L}}{\partial A_{[i}}A_{j]}+\frac{\partial\mathcal{L}}{\partial X_{A,k}}X_{A,[j}\delta_{i]k}$$
$$+\sum_\nu\frac{\partial\mathcal{L}}{\partial y_{k,l}^{T\nu}}\left(\delta_{k[i}y_{j],l}^{T\nu}+y_{k,[j}^{T\nu}\delta_{i]l}\right)+\frac{\partial\mathcal{L}}{\partial\Phi_{,[i}}\Phi_{,j]}$$
$$+\frac{\partial\mathcal{L}}{\partial A_{k,l}}\left(\delta_{k[i}A_{j],l}+A_{k,[j}\delta_{i]l}\right)+\sum_\nu\frac{\partial\mathcal{L}}{\partial\left(\partial y_{[i}^{T\nu}/\partial t\right)}\frac{\partial y_{j]}^{T\nu}}{\partial t}$$
$$+\frac{\partial\mathcal{L}}{\partial(\partial A_{[i}/\partial t)}\frac{\partial A_{j]}}{\partial t} \tag{5.6.14}$$

where the bracket notation has the meaning given in Section 3.12. After the imposition of this condition on $\mathcal{L}$ we are justified by (5.6.13) in setting

$$\delta^S\Omega_k=0,\qquad\delta^S\Omega_4=0. \tag{5.6.15}$$

Thus we find by (5.2.4) that the Lagrangian is form invariant to a spatial rotation transformation.

The conservation law implied by the existence of this symmetry transformation is found by substitution of (5.6.5) to (5.6.12) and (5.6.15) into (5.3.3). We could once again set the antisymmetric part of the individual coefficients of $\delta z_{[i,j]}$ equal to zero in the resulting equation. However, to obtain an axial vector quantity rather than an antisymmetric tensor quantity [which are closely related; see (4.5.8) and (4.5.9)] we express $\delta z_{[i,j]}$ by (3.12.6)

$$\delta z_{[i,j]} = \epsilon_{ijk} a_k \,\delta\theta \tag{5.6.16}$$

and set the coefficients of $a_k\,\delta\theta$ equal to zero. The result may be expressed as

$$\frac{\partial \omega_k^C}{\partial t} + \frac{\partial f_{km}^C}{\partial z_m} = 0 \tag{5.6.17}$$

where

$$\omega_k^C \equiv \epsilon_{kij} z_i g_j^C + \sum_\nu \frac{\partial \mathfrak{L}}{\partial(\partial y_i^{T\nu}/\partial t)} \epsilon_{ijk} y_j^{T\nu} + \frac{\partial \mathfrak{L}}{\partial(\partial A_i/\partial t)} \epsilon_{ijk} A_j, \tag{5.6.18}$$

$$f_{km}^C \equiv \epsilon_{kij} z_i t_{jm}^C + \sum_\nu \frac{\partial \mathfrak{L}}{\partial y_{i,m}^{T\nu}} \epsilon_{ijk} y_j^{T\nu} + \frac{\partial \mathfrak{L}}{\partial A_{i,m}} \epsilon_{ijk} A_j. \tag{5.6.19}$$

Equation (5.6.17) is the *angular momentum conservation law* in the spatial frame, ω_k^C is the *canonical angular momentum density*, and f_{km}^C is the *flux of canonical angular momentum*. We return to a discussion of these quantities in Chapter 8.

5.7 Material Translation Invariance

If a body is homogeneous, an arbitrary translation of the body in the material coordinate system should leave the equations of motion form invariant and so should be a symmetry transformation in the material description. It is obvious that this symmetry transformation does not possess the fundamental significance that those of the last three sections have. Nevertheless, for homogeneous bodies, which we wish to consider for simplicity, this symmetry transformation produces a restriction on the functional form of the Lagrangian density useful in its construction and also produces a conservation law, that of crystal momentum.

As before, we need consider only an infinitesimal transformation. An infinitesimal *material translation* is given by

$$\mathbf{X}' = \mathbf{X} + \delta\mathbf{X} \qquad (\delta\mathbf{X}\text{ constant}), \tag{5.7.1a}$$

$$t' = t. \tag{5.7.1b}$$

Since $\delta\mathbf{X}$ can be taken as constant, we have

$$\frac{\partial X_A}{\partial X'_B}=\frac{\partial X'_A}{\partial X_B}=\delta_{AB}. \tag{5.7.2}$$

Such a transformation does not affect components of vectors referred to the spatial coordinate system nor time derivatives. We work in the material description, since we want $\mathbf{X}$ to be an independent variable. Thus the functional dependence of the Lagrangian density is

$$\mathcal{L}=\mathcal{L}\left(X_A,t,x_i,y_i^{T\nu},x_{i,A},y_{i,A}^{T\nu},\dot{x}_i,\dot{y}_i^{T\nu},A_i,A_{i,A},\dot{A}_i,\Phi,\Phi_{,A}\right). \tag{5.7.3}$$

The procedure now used is entirely dual to that used in studying spatial displacement invariance in Section 5.4. We find that the variation of every field induced by the transformation (5.7.1) vanishes. The consistency condition (5.2.9) then yields

$$-\frac{d\delta^S\Omega_A}{dX_A}-\frac{d\delta^S\Omega_4}{dt}=\frac{\partial\mathcal{L}}{\partial X_A}\delta X_A. \tag{5.7.4}$$

Once again, since the right side is not a perfect derivative, we require it to vanish,

$$\frac{\partial\mathcal{L}}{\partial X_A}=0. \tag{5.7.5}$$

This condition prohibits the Lagrangian density of a homogeneous body from being an *explicit* function of the material coordinate $\mathbf{X}$. We now may set

$$\delta^S\Omega_A=0,\qquad \delta^S\Omega_4=0, \tag{5.7.6}$$

which implies that the Lagrangian density is form invariant to a material translation.

The conservation law found from (5.3.3) for this symmetry transformation is

$$\begin{aligned}0=\frac{d}{dt}&\left[-\frac{\partial\mathcal{L}}{\partial\dot{x}_i}\dot{x}_{i,A}-\sum_\nu\frac{\partial\mathcal{L}}{\partial\dot{y}_i^{T\nu}}y_{i,A}^{T\nu}-\frac{\partial\mathcal{L}}{\partial\dot{A}_i}A_{i,A}\right]+\frac{d}{dX_B}\left[-\frac{\partial\mathcal{L}}{\partial x_{i,B}}x_{i,A}\right.\\&\left.-\sum_\nu\frac{\partial\mathcal{L}}{\partial y_{i,B}^{T\nu}}y_{i,A}^{T\nu}-\frac{\partial\mathcal{L}}{\partial\Phi_{,B}}\Phi_{,A}-\frac{\partial\mathcal{L}}{\partial A_{i,B}}A_{i,A}+\mathcal{L}\delta_{AB}\right].\end{aligned} \tag{5.7.7}$$

The quantity acted on by d/dt is the *canonical crystal momentum density*. The quantity acted on by d/dX_B is the flux of canonical crystal momentum or the *canonical material stress tensor*. Equation (5.7.7) is the *law of conservation of crystal momentum.*

PROBLEMS

5.1. Show that the *gauge transformation*,

$$\mathbf{A}' = \mathbf{A} - \nabla(\delta G), \tag{5.P.1a}$$

$$\Phi' = \Phi + \frac{\partial \delta G}{\partial t}, \tag{5.P.1b}$$

is a symmetry transformation and that it leads to *charge conservation*,

$$\nabla \cdot \mathbf{j} + \frac{\partial q}{\partial t} = 0. \tag{5.P.2}$$

Use the general expression (4.1.7) for the interaction Lagrangian. Show that the test condition (5.2.9) is satisfied provided the Lagrangian transforms under the gauge transformation as

$$\mathcal{L}'\left(\mathbf{A}', \nabla\mathbf{A}', \frac{\partial \mathbf{A}'}{\partial t}, \Phi', \nabla\Phi'\right) = \mathcal{L}\left(\mathbf{A}', \nabla\mathbf{A}', \frac{\partial \mathbf{A}'}{\partial t}, \Phi', \nabla\Phi'\right) + \nabla \cdot (\mathbf{j}\,\delta G) + \frac{\partial}{\partial t}(q\,\delta G). \tag{5.P.3}$$

5.2. Show that the Lagrangian of Chapter 4 gives the canonical momentum density g_i^C of (5.4.17) the form

$$g_i^C = \rho \dot{x}_i - (\mathbf{P}\times\mathbf{B})_i + (\epsilon_0 E_j + P_j) A_{j,i} \tag{5.P.4}$$

and the canonical stress tensor t_{ij}^C of (5.4.18) the form

$$t_{ij}^C = t_{ij}^y - \rho \dot{x}_i \dot{x}_j - \frac{\epsilon_0}{2} E_k E_k \delta_{ij} + \frac{1}{2\mu_0} B_k B_k \delta_{ij} + (\mathbf{P}\times\mathbf{B})_i \dot{x}_j + \left(E_i + \frac{\partial A_i}{\partial t}\right)(\epsilon_0 E_j + P_j) - \left[\frac{1}{\mu_0} B_l - (\mathbf{P}\times\dot{\mathbf{x}})_l\right] \epsilon_{ljk} A_{k,i} \tag{5.P.5}$$

where

$$t_{ij}^{y} \equiv \rho x_{j,B} \frac{\partial \tilde{\Sigma}}{\partial x_{i,B}} . \tag{5.P.6}$$

Note that neither of these expressions are gauge invariant since some terms involving the vector potential cannot be expressed entirely in terms of the electric field **E** and the magnetic induction **B**. Thus neither g_i^C nor t_{ij}^C can have a direct physical interpretation.

5.3 Prove that a conservation law expressed in the spatial frame,

$$\frac{\partial d}{\partial t} + \frac{\partial f_i}{\partial z_i} = 0, \tag{5.P.7}$$

where d is a density of the conserved quantity and **f** the flux of that quantity, can be expressed in the material frame as

$$\frac{dD}{dt} + \frac{\partial F_J}{\partial X_J} = 0 \tag{5.P.8}$$

where

$$D \equiv Jd, \tag{5.P.9}$$

$$F_J \equiv J X_{J,i}(f_i - \dot{x}_i d) \tag{5.P.10}$$

are the material frame measures of the density and flux.

5.4 Beginning from the Lagrangian density (5.7.4) in the material description find the energy conservation law

$$\frac{\partial H^M}{\partial t} + \frac{\partial S_A^M}{\partial X_A} = 0 \tag{5.P.11}$$

along with expressions for the material frame energy density H^M and the material frame flux of energy $\mathbf{S}^M$. By transforming the Lagrangian density by $\mathfrak{L}_M = J\mathfrak{L}_S$ show that H^M and $\mathbf{S}^M$ are related to H and $\mathbf{S}$ of (5.5.8) and (5.5.9) by (5.P.9) and (5.P.10).

CHAPTER 6

Stored Energy of a Crystal

The only task remaining in the construction of the Lagrangian for a crystal in interaction with the electromagnetic field is determining the form of the stored or potential energy of the crystal. We wish this to be as general as possible while still being consistent with the laws of conservation of energy, momentum, and angular momentum. As seen in Chapter 5, these require, respectively, the Lagrangian to be invariant to displacements in time, displacements in space, and rotations in space. These imply, in turn, that the Lagrangian cannot be an explicit function of time, cannot be an explicit function of the center-of-mass position, and must be a rotational invariant.

Other invariances are also considered. Even though parity conservation is not a fundamental law of nature, no violation of it in atomic interactions has so far been detected and so its obeyance in bonding interactions in crystals can reasonably be assumed. This implies invariance of the Lagrangian to spatial inversions which, in turn, prohibits the Lagrangian from having any pseudoscalar part. We also consider, for simplicity, only homogeneous crystals in their natural state. As seen in the preceding chapter, material homogeneity requires invariance of the Lagrangian to translations in the material coordinate system and leads to conservation of crystal momentum. This invariance prohibits the Lagrangian from being an explicit function of the material coordinate $\mathbf{X}$.

After these invariances are applied to the stored energy, we expand it in a power series. Application of crystal group invariance can then give relations between the coefficients in the power series corresponding to the space group symmetry of the crystal considered.

Toupin [1956] gave the first careful discussion of the construction of a general stored energy subject only to the conservation laws. The treatment of this chapter [Lax and Nelson, 1971] generalizes his work by using a complete set of internal coordinates in place of Toupin's single inertialess polarization variable.

6.1 Invariance to Displacements in Time and Space

The stored energy per unit volume in the spatial frame is found to have the form

$$\rho\tilde{\tilde{\Sigma}}=\rho\tilde{\tilde{\Sigma}}\left(X_A,y_i^{T\nu},X_{A,i},y_{i,j}^{T\nu},z_i,t\right) \tag{6.1.1}$$

in Section 4.7. In Section 5.5 the dynamical equations of a system represented by a total Lagrangian density $\mathfrak{L}$ are found to be form invariant to displacements in time, $t\to t+\delta t$, and so possess energy conservation, provided that

$$\frac{\partial\mathfrak{L}}{\partial t}=0. \tag{6.1.2}$$

Since all parts of $\mathfrak{L}$ except the stored energy part (see Section 4.7) obey this equation, it reduces to

$$\frac{\partial\tilde{\tilde{\Sigma}}}{\partial t}=0. \tag{6.1.3}$$

Since the partial derivative denotes differentiation with respect to explicit time dependence only, this condition prohibits such dependence. Thus t must be removed from the list of arguments in (6.1.1).

In Section 5.4 the dynamical equations of a system represented by a total Lagrangian density $\mathfrak{L}$ are found to be form invariant to displacements in space, $\mathbf{z}\to\mathbf{z}+\delta\mathbf{z}$, and so possess momentum conservation, provided that

$$\frac{\partial\mathfrak{L}}{\partial z_i}=\frac{\partial\mathfrak{L}}{\partial x_i}=0. \tag{6.1.4}$$

Since all parts of $\mathfrak{L}$ considered except $\tilde{\tilde{\Sigma}}$ already obey this equation, it reduces to

$$\frac{\partial\tilde{\tilde{\Sigma}}}{\partial z_i}=\frac{\partial\tilde{\tilde{\Sigma}}}{\partial x_i}=0, \tag{6.1.5}$$

which thus prohibits explicit dependence of $\tilde{\tilde{\Sigma}}$ on $\mathbf{z}$ and $\mathbf{x}$. Thus $\mathbf{z}$ must be dropped from the list of arguments in (6.1.1). No requirement similar to (6.1.5) exists for the internal coordinates $\mathbf{y}^{T\nu}$, since they are displacement invariant coordinates by definition.

6.2 Homogeneity

A body is homogeneous if a translation in the material coordinate system, $\mathbf{X} \to \mathbf{X} + \delta\mathbf{X}$, leaves the body unchanged. It is shown in Section 5.7 that a homogeneous body must be described by a Lagrangian density such that

$$\frac{\partial \mathcal{L}}{\partial X_A} = 0 \tag{6.2.1}$$

and that this invariance corresponds to conservation of what is called crystal momentum. Examination of the various parts of the Lagrangian density in Section 4.7 shows that all parts except the stored energy already obey this equation and hence that equation becomes

$$\frac{\partial \tilde{\tilde{\Sigma}}}{\partial X_A} = 0. \tag{6.2.2}$$

This requires that $\tilde{\tilde{\Sigma}}$ not be an explicit function of $\mathbf{X}$. Hence $\mathbf{X}$ is deleted from the list of arguments of $\tilde{\tilde{\Sigma}}$ in (6.1.1). It should be remembered that this requirement is made for simplicity, and not for fundamental reasons.

6.3 Invariance to Spatial Rotations

In Section 5.6 it is shown that form invariance of the dynamical equations of a physical system under spatial rotation transformations, demanded by the isotropy of space, leads to angular momentum conservation for the system. Furthermore, it is shown that a Lagrangian density for such a system must obey (5.6.14). It is straightforward to show that all parts of the Lagrangian density (4.7.8), (4.7.16), and (4.7.18) except the stored energy already satisfy (5.6.14) and so we leave those demonstrations as exercises. Thus we need only require the stored energy to satisfy it in order to guarantee angular momentum conservation.

Since X_A, $z_i = x_i$, and t are no longer present in the list of arguments for $\tilde{\tilde{\Sigma}}$ in (6.1.1), only $y_i^{T\nu}$, $X_{A,i}$, and $y_{i,j}^{T\nu}$ remain. Before proceeding further we wish to drop the third of these, $y_{i,j}^{T\nu}$, since it represents a correction to the long wavelength limit or, in other words, a wavevector dispersion correction. This is simply because it involves a spatial derivative of a quantity, $y_i^{T\nu}$, which also appears in the stored energy. The effect of space derivatives giving corrections to the long wavelength limit is discussed in Sections 4.2 and 4.3. With this deletion the stored energy is now simply

$$\rho\tilde{\tilde{\Sigma}} = j\left(\frac{\mathbf{X}}{\mathbf{x}}\right)\rho^0\tilde{\tilde{\Sigma}}\left(y_i^{T\nu}, X_{A,i}\right) \tag{6.3.1}$$

where $j(\mathbf{X}/\mathbf{x})$ is the Jacobian defined in (3.1.8). With this functional dependence (5.6.14) becomes

$$\sum_\nu \frac{\partial\left(j\rho^0\tilde{\tilde{\Sigma}}\right)}{\partial y_{[i}^{T\nu}} y_{j]}^{T\nu} + \frac{\partial\left(j\rho^0\tilde{\tilde{\Sigma}}\right)}{\partial X_{A,k}} X_{A,[j}\delta_{i]k} = 0. \tag{6.3.2}$$

Since it is easy to show that

$$\frac{\partial j(\mathbf{X}/\mathbf{x})}{\partial X_{A,k}} X_{A,[j}\delta_{i]k} = 0 \tag{6.3.3}$$

with the use of (3.1.8), (3.2.11), and (3.1.5), the expression (6.3.2) becomes

$$\sum_\nu \frac{\partial\tilde{\tilde{\Sigma}}}{\partial y_{[i}^{T\nu}} y_{j]}^{T\nu} + \frac{\partial\tilde{\tilde{\Sigma}}}{\partial X_{A,k}} X_{A,[j}\delta_{i]k} = 0. \tag{6.3.4}$$

The meaning of this restriction, which rotational invariance places on the stored energy, becomes apparent if we substitute for $\tilde{\tilde{\Sigma}}$ various quantities invariant to rotations of the body in the spatial coordinate system, that is, scalar products of spatial coordinate system vectors. Since $X_{A,i}$ is a spatial system vector for each value of A, we can form a number of rotational invariants, for example, $y_i^{T\nu}y_i^{T\mu}$, $y_i^{T\nu}X_{A,i}$, and $X_{A,i}X_{B,i}$, from the $(N-1)$ spatial vectors $y_{\underset{\sim}{i}}^{T\nu}$ and the three spatial vectors $X_{A,i}$. Substitution of any one of these for $\tilde{\tilde{\Sigma}}$ in (6.3.4) satisfies the equation. Consider, for example, $y_i^{T\mu}X_{A,i}$. Equation (6.3.4) becomes

$$\sum_\nu \delta^{\mu\nu} X_{A,[i}y_{j]}^{T\nu} + y_k^{T\mu}X_{A,[j}\delta_{i]k} = 0, \tag{6.3.5}$$

or, in expanded form,

$$X_{A,i}y_j^{T\mu} - X_{A,j}y_i^{T\mu} + y_i^{T\mu}X_{A,j} - y_j^{T\mu}X_{A,i} = 0, \tag{6.3.6}$$

which is satisfied identically. It is clear from this exercise that any scalar product of the spatial system vectors considered satisfies (6.3.4).

This makes the following theorem on invariant functions of several vectors, originally proved by Cauchy, quite reasonable. It states: if $F(\mathbf{V}^\alpha)$ is a single-valued function of the components of n vectors, $\mathbf{V}^\alpha$ $(\alpha=1,2,\dots,n)$, and is invariant to an arbitrary rigid (proper) rotation of the system of vectors, F must reduce to a function of their lengths $\mathbf{V}^\alpha\cdot\mathbf{V}^\alpha$, scalar products $\mathbf{V}^\alpha\cdot\mathbf{V}^\beta$, and the determinants of their components taken three at a time, $\epsilon_{ijk}V_i^\alpha V_j^\beta V_k^\gamma=\mathbf{V}^\alpha\cdot\mathbf{V}^\beta\times\mathbf{V}^\gamma$. From this theorem we conclude that $\tilde{\tilde{\Sigma}}(y_i^{T\nu},X_{A,i})$ can always be expressed as a function of

$$(C^{-1})_{AB}\equiv X_{A,i}X_{B,i}, \tag{6.3.7}$$

$$L^{\nu\mu}\equiv y_i^{T\nu}y_i^{T\mu} \qquad (\nu,\mu=1,2,\dots,N-1), \tag{6.3.8}$$

$$\Pi_A^{T\nu}\equiv y_i^{T\nu}X_{A,i} \qquad (\nu=1,2,\dots,N-1), \tag{6.3.9}$$

$$j\left(\frac{\mathbf{X}}{\mathbf{x}}\right)\equiv\tfrac{1}{6}\epsilon_{ijk}\epsilon_{IJK}X_{I,i}X_{J,j}X_{K,k}, \tag{6.3.10}$$

$$D_K^\nu\equiv\tfrac{1}{2}\epsilon_{ijk}\epsilon_{IJK}X_{I,i}X_{J,j}y_k^{T\nu} \qquad (\nu=1,2,\dots,N-1), \tag{6.3.11}$$

$$D_{JK}^{\nu\mu}\equiv\tfrac{1}{2}\epsilon_{ijk}\epsilon_{IJK}X_{I,i}y_j^{T\nu}y_k^{T\mu} \qquad (\nu,\mu=1,2,\dots,N-1;\nu\neq\mu), \tag{6.3.12}$$

$$D^{\nu\mu\lambda}\equiv\epsilon_{ijk}y_i^{T\nu}y_j^{T\mu}y_k^{T\lambda} \qquad (\nu,\mu,\lambda=1,2,\dots,N-1;\nu\neq\mu\neq\lambda) \tag{6.3.13}$$

where the designation $(C^{-1})_{AB}$ in (6.3.7) follows from (3.P.5).

Cauchy's theorem does not say that this set of rotational invariants is the smallest set of such invariants that can be used. Thus we must examine them to see if any of them can be expressed in terms of the others and so may be eliminated from the list. The invariant (6.3.8) may be reexpressed as

$$\begin{aligned}L^{\nu\mu}&=y_j^{T\nu}\delta_{jk}\delta_{kl}y_l^{T\mu}\\&=y_j^{T\nu}X_{A,j}x_{k,A}x_{k,B}X_{B,l}y_l^{T\mu}\\&=\Pi_A^{T\nu}C_{AB}\Pi_B^{T\mu},\end{aligned} \tag{6.3.14}$$

which shows that it may be expressed in terms of $\Pi_A^{T\nu}$ and $(C^{-1})_{AB}$ and so can be eliminated from the list of invariants necessary for the expression of $\tilde{\tilde{\Sigma}}$.

Next we consider D_K^ν. From (3.P.6) we have

$$\epsilon_{IJK}\epsilon_{ijk}X_{I,i}X_{J,j}X_{L,k}=2\delta_{KL}\,j(\mathbf{X}/\mathbf{x}). \tag{6.3.15}$$

Forming the scalar product with $(\frac{1}{2})x_{l,L}y_l^{T\nu}$ yields

$$\tfrac{1}{2}\epsilon_{IJK}\epsilon_{ijk}X_{I,i}X_{J,j}X_{L,k}x_{l,L}y_l^{T\nu}=\delta_{KL}\, j(\mathbf{X}/\mathbf{x})x_{l,L}y_l^{T\nu}. \tag{6.3.16}$$

The left side becomes D_K^ν and so we obtain

$$\begin{aligned} D_K^\nu &= j(\mathbf{X}/\mathbf{x})x_{m,K}\delta_{ml}y_l^{T\nu} \\ &= j(\mathbf{X}/\mathbf{x})x_{m,K}x_{m,N}X_{N,l}y_l^{T\nu} \\ &= j(\mathbf{X}/\mathbf{x})C_{KN}\Pi_N^{T\nu}. \end{aligned} \tag{6.3.17}$$

Thus D_K^ν is expressible in terms of the other invariants and so may be dropped from the list. By similar manipulations both $D_{JK}^{\nu\mu}$ and $D^{\nu\mu\lambda}$ can be shown to be expressible in terms of $j(\mathbf{X}/\mathbf{x})$, C_{IJ}, and $\Pi_K^{T\nu}$. We leave these demonstrations for the problems.

Lastly the Jacobian $j(\mathbf{X}/\mathbf{x})$ must be considered. From the rule that a determinant of a product is the product of the determinants we find

$$\begin{aligned} \det\left[(C^{-1})_{AB}\right] &= \det(X_{A,i}X_{B,i}) \\ &= \det(X_{A,i})\det(X_{B,i}) \\ &= j^2\left(\frac{\mathbf{X}}{\mathbf{x}}\right). \end{aligned} \tag{6.3.18}$$

Thus

$$j\left(\frac{\mathbf{X}}{\mathbf{x}}\right) = \pm\left\{\det\left[(C^{-1})_{AB}\right]\right\}^{1/2}. \tag{6.3.19}$$

This states that the magnitude of the Jacobian is a redundant invariant. If only *proper rotations* are considered, as is the case up to this point, the sign of the Jacobian can only be positive. In that case, (6.3.19) shows that the Jacobian (magnitude and sign) is redundant and may be dropped from the list of needed rotational invariants. However, if at this point we also consider *improper rotations*, that is, *spatial inversions*, then the sign of the Jacobian can be either positive or negative. This sign, denoted by sgn(j), cannot be expressed in terms of the other two remaining rotational invariants, $(C^{-1})_{AB}$ and $\Pi_C^{T\nu}$, and so must be retained in the stored energy.

We thus conclude at this point that the most general stored energy per unit mass which permits energy, momentum, and angular momentum conservation is expressible as

$$\hat{\Sigma} = \hat{\Sigma}\left((C^{-1})_{AB}, \Pi_C^{T\nu}\right) + \operatorname{sgn}(j)\hat{\Sigma}'\left((C^{-1})_{AB}, \Pi_C^{T\nu}\right) \tag{6.3.20}$$

for a homogeneous crystal in the long wavelength limit. The caret over the function denotes its new functional dependence.

The restriction of this expression to the long wavelength limit results from neglecting the first (and higher) derivatives of the internal coordinates and the second (and higher) derivatives of the center-of-mass coordinate. If these derivatives were retained, the theory would be capable of accounting for wavevector dispersion effects such as optical activity and the analogous acoustic activity [Toupin, 1962; Pine, 1970; Joffrin and Levelut, 1970]. They would also force us to include a considerably greater number of rotational invariants as stored energy arguments in (6.3.20). We do not feel that the inclusion of the above two phenomena in this treatment justifies the much more complicated formulation that would be needed.

6.4 Invariance to Spatial Inversions

It is obvious that the transformation of *spatial inversion*,

$$\mathbf{z}' = -\mathbf{z}, \tag{6.4.1a}$$

$$t' = t, \tag{6.4.1b}$$

does not represent a physically possible transformation in the sense that spatial rotations and displacements do. It thus should be regarded as questionable whether it represents a symmetry transformation.

Under the spatial inversion operation a function that does not undergo any change of sign is called a *scalar* and is said to have *even parity*; one that does undergo a change of sign is called a *pseudoscalar* and is said to have *odd parity*. A physical system represented by a Lagrangian that has a definite (even or odd) parity is said to possess *parity conservation*. If the Lagrangian were a sum of scalar and pseudoscalar parts, then it would not have a definite parity and parity would not be conserved in the system represented.

The belief that parity conservation was a basic law of nature grew gradually over a long period of time. The possible violation of this law in the weak interactions that govern neutrino processes was first suggested by Lee and Yang [1956] and soon thereafter demonstrated experimentally [Wu et al., 1957; Garwin et al., 1957]. Since then parity conservation has been tested for the electromagnetic interaction in molecules and a very small upper limit on the level of its possible violation has been found [Feinberg, 1969]. No tests of parity conservation in crystals have been performed. Since the electromagnetic interaction governs the bonding forces of crystals as well as those of molecules, it seems likely that parity violation in crystals is equally small. Though tests of this should be made before final acceptance of this, for the remainder of the book we make the tentative, and perhaps only approximate, assumption of parity conservation.

This assumption means that the transformation (6.4.1) is a symmetry transformation. Note, however, that there is no way to construct it from a succession of infinitesimal transformations. Consideration of the spatial inversion transformation thus does not fit into the treatment of symmetry transformations of Chapter 5. The consequences of it, however, can be handled simply.

The spatial inversion transformation reverses the sign of all parts of the Lagrangian density except the term in the stored energy $\operatorname{sgn}(j)\hat{\Sigma}'$ because the Jacobian is involved in the transformation of the Lagrangian density (5.2.6). This sign reversal does not affect the Lagrangian itself because it is compensated by a reversal of the orientation of the region of integration of the Lagrangian density. Thus this portion of the Lagrangian is a true scalar.

The sign reversal of all parts of the Lagrangian density does not affect the equations of motion, since they are homogeneous in the Lagrangian density. However, since the stored energy term $\operatorname{sgn}(j)\hat{\Sigma}'$ reverses its sign relative to the remainder of the Lagrangian density, the Lagrange equations of motion cannot be form invariant to spatial inversions if this term remains. Thus we conclude that $\hat{\Sigma}'$ must vanish. The stored energy per unit mass (6.3.20) then becomes

$$\hat{\Sigma} = \hat{\Sigma}\big((C^{-1})_{AB}, \Pi_C^{T\nu}\big). \tag{6.4.2}$$

6.5 Series Expansion of Stored Energy

Now that the functional dependence of the stored energy is greatly restricted by the invariance requirements, the functional form can be considered. A series expansion of the stored energy in terms of the rotational invariants is the most convenient form provided that higher order terms correspond to higher order (smaller) effects. The latter is possible only if the rotational invariants represent deviations from the *natural state* of the crystal, that is, from a stable state in which there is no deformation, no applied fields, and no change of properties in time or space. This means that the rotational invariants used in the series expansion must vanish in the natural state. We know, however, from the definition of $\mathbf{C}^{-1}$ that it becomes $\mathbf{1}$ in the natural state and we point out in

Section 4.6 that $\mathbf{y}^{T\mu}$ contains a constant part also. Thus we must introduce new rotational invariants related to $\mathbf{C}^{-1}$ and $\mathbf{\Pi}^{T\nu}$ that vanish in the natural state.

One of the needed invariants is defined in Section 3.4; it is the Green finite strain tensor,

$$E_{AB} \equiv \tfrac{1}{2}(C_{AB} - \delta_{AB}), \tag{6.5.1}$$

which vanishes in the natural state and becomes the conventional infinitesimal strain tensor for small deformations. In order to introduce a rotationally invariant measure of each internal coordinate that vanishes in the natural state we must define the following notation. Let Y_B^ν be the natural state value of the internal coordinate $\mathbf{y}^{T\nu}$ referred to the material coordinate system and $\mathbf{y}^\nu$ be the varying part of the internal coordinate referred to the spatial coordinate system. Thus

$$y_i^{T\nu} \equiv \delta_{iB} Y_B^\nu + y_i^\nu \qquad (\nu = 0, 1, 2, \ldots, N-1) \tag{6.5.2}$$

where the use of the Kronecker delta for the shifter indicates that the material and spatial coordinate system axes are taken identical. For $\nu = 0$ this equation is intended to be identical to (3.10.1) which defines the displacement vector. In the new notation we have

$$Y_B^0 \equiv X_B, \tag{6.5.3}$$

$$y_i^0 \equiv u_i \tag{6.5.4}$$

since $\mathbf{y}^{T0}$ is defined as the center-of-mass position in (4.6.2).

In the natural state the displacement vector is zero and so $X_{A,i} = \delta_{Ai}$. Also, $\mathbf{y}^\nu$ vanishes. Thus the value of the rotationally invariant measure of the internal coordinate in the natural state is

$$\Pi_A^{S\nu} \equiv \delta_{Ai}\delta_{iB} Y_B^\nu = Y_A^\nu \qquad (\nu = 1, 2, \ldots, N-1). \tag{6.5.5}$$

The rotationally invariant measure of the internal coordinate that vanishes in the natural state can now be defined by

$$\begin{aligned} \Pi_A^\nu &\equiv \Pi_A^{T\nu} - \Pi_A^{S\nu} \qquad (\nu = 1, 2, \ldots, N-1) \\ &= X_{A,i}(\delta_{iB} Y_B^\nu + y_i^\nu) - Y_A^\nu. \end{aligned} \tag{6.5.6}$$

Thus our final choice of functional dependence for the stored energy is

$$\rho^0 \Sigma = \rho^0 \Sigma(E_{AB}, \Pi_C^\nu) \tag{6.5.7}$$

where no marking above Σ is now used to indicate this functional dependence. The simplicity of this important result conceals the amount of physics that is used to obtain it from the stored energy form (6.1.1).

Now that the stored energy is a function of rotational invariants that vanish in the natural state of the crystal, a series expansion of the stored energy may be written. The last requirement on the form of the stored energy, that it be invariant to the choice of the material coordinate system, requires each term of the series expansion to be a scalar with respect to material coordinate system transformations. Thus the final form of the stored energy is

$$\begin{aligned}\rho^0\Sigma(E_{AB},\Pi_C^\nu) = {}& {}^{01}M_{AB}E_{AB} + {}^{02}M_{ABCD}E_{AB}E_{CD} \\ & + {}^{03}M_{ABCDEF}E_{AB}E_{CD}E_{EF} + \sum_\nu {}^{10}M_A^\nu\Pi_A^\nu \\ & + \sum_{\nu\mu} {}^{20}M_{AB}^{\nu\mu}\Pi_A^\nu\Pi_B^\mu + \sum_{\nu\mu\lambda} {}^{30}M_{ABC}^{\nu\mu\lambda}\Pi_A^\nu\Pi_B^\mu\Pi_C^\lambda \\ & + \sum_\nu {}^{11}M_{ABC}^\nu\Pi_A^\nu E_{BC} + \sum_\nu {}^{12}M_{ABCDE}^\nu\Pi_A^\nu E_{BC}E_{DE} \\ & + \sum_{\nu\mu} {}^{21}M_{ABCD}^{\nu\mu}\Pi_A^\nu\Pi_B^\mu E_{CD} + \cdots . \end{aligned} \tag{6.5.8}$$

The constant term that represents the stored energy in the natural state is omitted, since it cannot affect the equations of motion. The ${}^{mn}M$ coefficients, which we call *material descriptors*, have tensor character indicated by their subscripts and matrix character indicated by the postsuperscripts. The presuperscripts on the material descriptors label the number m of $\Pi_A^{T\nu}$ factors and the number n of E_{BC} factors occurring in the term in question and serve simply as convenient designations of the coefficients.

We wish to emphasize the fact, apparent from the origin of this stored energy, that the various material descriptors are *numerical constants*. They do not depend on frequency or wavevector; they depend only on the configuration of the particles in the primitive unit cell in the natural state and on the bonding forces between the particles. The values of the material descriptors cannot be derived from a classical mechanics theory such as this; their values can be derived only with the use of quantum mechanics. We regard their values as determined by comparison of predictions of this theory with experiment. The material descriptors are the much generalized analogs of the Hooke's law spring constants introduced in the lattice dynamics examples of Chapters 1 and 2.

Note also that the stored energy function depends only on mechanical variables because of its origin as the potential energy of the configuration of particles in the crystal. In particular, the stored energy is not an explicit function of the electric field. However, we see in Chapter 17 when we treat nonlinear electroacoustics, which may be regarded as low frequency phenomena even in the ultrasonic and hypersonic regions, that the internal coordinates are algebraically related to the displacement gradient and the electric field. If this relation is used to eliminate the internal coordinates from the stored energy, the latter becomes a function of the electric field as well as the displacement gradient. This elimination cannot be done for frequencies near internal motion (optic mode) resonances in the infrared or visible parts of the electromagnetic spectrum, since there a dynamical equation governs each internal coordinate.

The choice of expanding the stored energy about the natural state is the most convenient choice for the great majority of uses. However, there are situations where another choice has advantages. If the crystal has two natural states (this is possible in a ferroelectric or a ferroelastic crystal) and if excitations strong enough to carry the material from one natural state to the other are considered, it is useful to expand the stored energy about a configuration that is midway between the two natural states. We do not, however, treat this case.

6.6 Crystal Symmetry Implications for Stored Energy

The space group symmetry of a particular crystal restricts the form that the material descriptors may take. The resultant conditions are quite complicated for several reasons. The crystal symmetry operations act on both the tensor subscripts of the material descriptors through rotation operators and on the superscripts that label the internal coordinates through relabeling operators. The latter operators can be understood only if the internal coordinate labels are first transformed to the sublattice labels associated with the original position coordinates $\mathbf{x}^{\alpha}$. In short, the series expansion (6.5.8) and consequently the coefficients in it are chosen for convenience in the dynamical equations, not in the application of crystal group operations.

Not only are the crystal symmetry restrictions on the material descriptors complicated, but we have no need to use them. This is because the form of all the measurable material tensors, such as the dielectric tensor or the stiffness tensor, that we encounter in later chapters are determined by the much simpler point group symmetry of the crystal. For these reasons we omit a derivation of crystal symmetry restrictions on the material descriptors.

6.7 Local Electric Field Contributions

It is well known that the applied electric field is not the entire field that is active in producing a force on an electric charge within a crystal. There is in addition a short range *local electric field* which contributes. It is induced by the applied electric field through polarization of the surrounding atoms. In general there is a contribution from each mode of polarization allowed by the crystal lattice, that is, from each internal coordinate. The additional electric force thus must be as general a function of the internal coordinates as allowed by the structure and symmetry of the lattice and the constraints implied by the conservation laws. Since the short range forces arising from the stored energy have just such a form, it is clear that local field contributions are already included in the stored energy of the Lagrangian. It is also clear that they are not separable from other stored energy contributions by measurements involving long wavelength interactions in perfect crystals.

PROBLEMS

6.1. Show that the field Lagrangian in the spatial description (4.7.18) satisfies the rotational invariance condition (5.6.14).

6.2. Show that the interaction Lagrangian in the spatial description (4.7.8) satisfies the rotational invariance condition (5.6.14).

6.3. Show that the kinetic energy portion of the matter Lagrangian in the spatial description (4.7.16) satisfies the rotational invariance condition (5.6.14).

6.4. Show that the quantity $D_{JK}^{\nu\mu}$ defined in (6.3.12) is a rotational invariant.

6.5. Show that $D_{JK}^{\nu\mu}$ and $D^{\nu\mu\lambda}$ can be expressed in terms of $j(\mathbf{X}/\mathbf{x})$, C_{IJ}, and $\Pi_K^{T\nu}$ and so are redundant as arguments of the stored energy.

CHAPTER 7

General Equations of Motion

The construction of the Lagrangian of a dielectric crystal in interaction with the electromagnetic field is complete and we can now begin to consider deductions from it. To find the electromagnetic field equations the Lagrange equations for the vector and scalar potentials are formed using the Lagrangian density in the spatial description for the field and for the field-matter interaction. The matter Lagrangian density makes no contribution. To find the matter equations of motion the Lagrange equations for the center of mass and each internal coordinate are formed using the Lagrangian density in the material description for the matter and for the field-matter interaction. The field Lagrangian density makes no contribution. An alternate procedure is to work from the Lagrangian density in the spatial description.

The equations that we get by these procedures in this chapter apply to crystals of any symmetry and structural complexity. All long wavelength (nondestructive) modes of motion of the crystal are included. Nonlinear effects of all orders in the interactions among these modes and between these modes and the electromagnetic fields are included with one exception. As a matter of convenience, not necessity, we use the charge and current densities in the electric dipole approximation. The equations obtained in this chapter form a complete set for the study of long wavelength phenomena in dielectrics.

7.1 Maxwell Equations in Electric Dipole Approximation

In this section we obtain the Maxwell electromagnetic equations from the total Lagrangian we now have. The procedure is the same as used in Section 2.4 except that we now use the interaction Lagrangian expanded and truncated at the electric dipole order.

The Maxwell charge equation follows from the Lagrange equation for the scalar potential Φ,

$$\frac{d}{dt}\frac{\partial \mathcal{L}_S}{\partial(\partial\Phi/\partial t)}=\frac{\partial \mathcal{L}_S}{\partial\Phi}-\frac{\partial}{\partial z_j}\frac{\partial \mathcal{L}_S}{\partial\Phi_{,j}} \tag{7.1.1}$$

where $\mathcal{L}_S$ is the total Lagrangian density in the spatial description. Since the matter Lagrangian does not contribute to this equation, we need consider only the interaction Lagrangian (4.7.8) and the field Lagrangian (4.7.18). Neither of these expressions contain Φ or $\partial\Phi/\partial t$ and so only the last term of (7.1.1) contributes. Thus it becomes

$$0=\epsilon_0\frac{\partial E_j}{\partial z_j}+\frac{\partial P_j}{\partial z_j} \tag{7.1.2}$$

or

$$\epsilon_0\nabla\cdot\mathbf{E}=-\nabla\cdot\mathbf{P}\equiv q^D \tag{7.1.3}$$

where the right side is the dielectric charge density (4.7.1) and the polarization $\mathbf{P}$ is given by (4.6.19).

The Maxwell current equation follows from the Lagrange equation for the vector potential $\mathbf{A}$,

$$\frac{d}{dt}\frac{\partial \mathcal{L}_S}{\partial(\partial A_i/\partial t)}=\frac{\partial \mathcal{L}_S}{\partial A_i}-\frac{\partial}{\partial z_j}\frac{\partial \mathcal{L}_S}{\partial A_{i,j}}. \tag{7.1.4}$$

Since no dependence of $\mathcal{L}_S$ on $\mathbf{A}$ occurs, this becomes

$$\frac{\partial}{\partial t}(-\epsilon_0 E_i-P_i)=-\frac{\partial}{\partial z_j}\left(-\epsilon_0 c^2 B_k\epsilon_{kji}+P_k\epsilon_{klm}\dot{x}_l\epsilon_{mji}\right) \tag{7.1.5}$$

or

$$\mu_0^{-1}\nabla\times\mathbf{B}-\epsilon_0\frac{\partial\mathbf{E}}{\partial t}=\frac{\partial\mathbf{P}}{\partial t}+\nabla\times(\mathbf{P}\times\dot{\mathbf{x}})\equiv\mathbf{j}^D \tag{7.1.6}$$

where the right side is the dielectric current (4.7.2).

From the forms of (7.1.3) and (7.1.6) we are led to define the *electric displacement vector* $\mathbf{D}$ by

$$\mathbf{D}\equiv\epsilon_0\mathbf{E}+\mathbf{P} \tag{7.1.7}$$

and the *magnetic intensity vector* **H** by

$$\mathbf{H} \equiv \mu_0^{-1}\mathbf{B} - \mathbf{P} \times \dot{\mathbf{x}}. \tag{7.1.8}$$

Introducing these fields into (7.1.3) and (7.1.6) results in

$$\nabla \cdot \mathbf{D} = 0, \tag{7.1.9}$$

$$\nabla \times \mathbf{H} - \frac{\partial \mathbf{D}}{\partial t} = 0, \tag{7.1.10}$$

the familiar forms of the Maxwell equations in the absence of free charges and free currents. As pointed out in Section 2.4, the remaining two Maxwell equations,

$$\nabla \times \mathbf{E} + \frac{\partial \mathbf{B}}{\partial t} = 0, \tag{7.1.11}$$

$$\nabla \cdot \mathbf{B} = 0, \tag{7.1.12}$$

are consequences of the definitions (2.4.5) and (2.4.6) of the electric and magnetic induction fields in terms of the scalar and vector potentials and so should be regarded as ancillary conditions to the Lagrangian approach to the electromagnetic equations.

7.2 Boundary Conditions for Electromagnetic Fields

We wish to develop from the differential equations of the preceding section the boundary conditions that **B**, **D**, **E**, and **H** must satisfy at a surface that is moving and deforming. Differentiability of these fields is implicitly assumed in the derivation of the differential equations. At a material boundary, where material characteristics undergo an abrupt change, it is reasonable that some components of these fields could possess discontinuities and lose differentiability.

Consider first the magnetic induction field **B**. Its differential equation is

$$\nabla \cdot \mathbf{B} = 0. \tag{7.2.1}$$

Let us regard the material surface not as a mathematically abrupt discontinuity but rather as a rapidly varying and continuous transition layer of thickness Δl in which all the material properties drop continuously to zero. In such a layer **B** also varies continuously. Now consider a small "pill box"

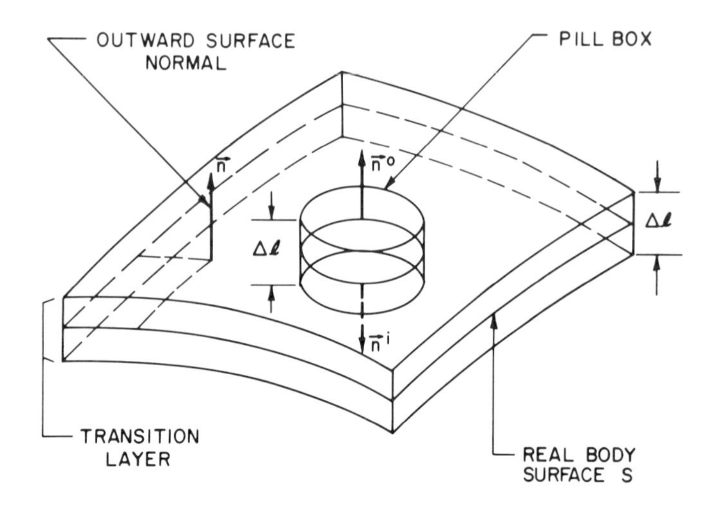

Fig. 7.1. Pill box geometry used for finding normal component boundary conditions.

whose flat and parallel ends are coincident with the (possibly moving) boundaries of this transition layer at a given instant of time as shown in Fig. 7.1. We can apply Gauss' theorem to a volume integral of (7.2.1) over the pill box obtaining

$$\int_A \mathbf{B}\cdot\mathbf{n}^P \, da^P = 0 \tag{7.2.2}$$

where $\mathbf{n}^P$ is the unit vector normal to an element of surface of the pill box whose area is da^P. The lateral extent of the flat surfaces of the pill box is small enough that the **B** field can be taken as constant over each one, $\mathbf{B}^o$ on the "outside" flat surface and $\mathbf{B}^i$ on the "inside" flat surface. The corresponding normals are $\mathbf{n}^o$ and $\mathbf{n}^i$. Equation (7.2.2) can then be rewritten as

$$\mathbf{B}^o\cdot\mathbf{n}^o\,\Delta a + \mathbf{B}^i\cdot\mathbf{n}^i\Delta a + \text{side wall contributions} = 0. \tag{7.2.3}$$

We now shrink the transition layer thickness Δl and the pill box height to zero simultaneously. Since the side wall contributions are proportional to Δl, they approach zero also. The field $\mathbf{B}^o$ becomes the magnetic induction field at the surface as approached from the outside and $\mathbf{B}^i$ that as approached from the inside. Letting $\mathbf{n} \equiv \mathbf{n}^o = -\mathbf{n}^i$ be the outward normal of the material surface, we obtain

$$(\mathbf{B}^o - \mathbf{B}^i)\cdot\mathbf{n} = 0 \tag{7.2.4}$$

as the boundary condition on the normal component of the magnetic induction at a moving deforming material surface.

Since the differential equation for the electric displacement **D** has the same form as that for **B**, a boundary condition analogous to (7.2.4) results for a true dielectric, a crystal with no free charge. We, however, have occasion later to consider dielectric crystals that by virtue of either a metal film electrode or an immobile extrinsic charge layer have a surface charge density σ. If the monopole or free charge density q^f were not set equal to zero in the multipole expansion of Section 4.4, it would appear in (7.1.9) as

$$\nabla \cdot \mathbf{D} = q^f. \tag{7.2.5}$$

If only the surface charge density σ^f (including both extrinsic mobile and immobile charge) is nonzero, then

$$q^f = \sigma^f \delta(s) \tag{7.2.6}$$

where

$$s \equiv \mathbf{n} \cdot (\mathbf{x} - \mathbf{x}^S) \tag{7.2.7}$$

is a coordinate measured normal to the surface at the point $\mathbf{x}^S$ on the surface. The equation of the surface is $s=0$ and $\delta(s)$ is the Dirac delta function. If a volume integral of (7.2.5) over the pill box of Fig. 7.1 is formed, Gauss' theorem is applied to the left side, and the Dirac delta function is used to simplify the right side, we obtain

$$\int_A \mathbf{D} \cdot \mathbf{n}^P \, da^P = \int_S \sigma^f \, da^S = \sigma^f \Delta a \tag{7.2.8}$$

where the integral on the right side is over the portion Δa of the material surface contained within the pill box. If the thickness of the pill box and the transition layer are shrunk to zero as before, this gives

$$(\mathbf{D}^o - \mathbf{D}^i) \cdot \mathbf{n} = \sigma^f \tag{7.2.9}$$

as the boundary conditions on the normal component of the electric displacement at a moving deforming material surface.

Consider next the electric field **E** which obeys

$$\nabla \times \mathbf{E} + \frac{\partial \mathbf{B}}{\partial t} = 0. \tag{7.2.10}$$

This equation can be integrated over the area of a rectangular loop, shown

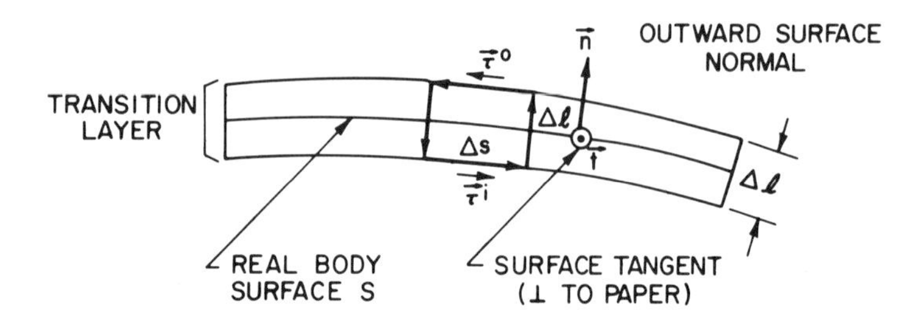

Fig. 7.2. Geometry used for finding tangential component boundary conditions.

in Fig. 7.2, whose sides of length Δs are coincident with the boundary layer edges at a material surface and whose ends of length Δl pass through the boundary layer. Because the area of the loop moves with the body, we must use the identity (3.13.6) for the time derivative of an area integral. The result is

$$\int_A (\nabla\times\mathbf{E})\cdot\mathbf{da}+\frac{d}{dt}\int_A \mathbf{B}\cdot\mathbf{da}-\int_A \left[\nabla\times(\mathbf{B}\times\dot{\mathbf{x}})\right]\cdot\mathbf{da}-\int_A(\nabla\cdot\mathbf{B})\dot{\mathbf{x}}\cdot\mathbf{da}=0 \tag{7.2.11}$$

where A denotes the region of the rectangle. The last integral vanishes because $\nabla\cdot\mathbf{B}=0$. Application of Stokes' theorem to the first and third integrals leads to

$$\int_C (\mathbf{E}+\dot{\mathbf{x}}\times\mathbf{B})\cdot\mathbf{ds}+\frac{d}{dt}\int_A \mathbf{B}\cdot\mathbf{t}\,da=0 \tag{7.2.12}$$

where $\mathbf{ds}$ is an element of arc around the perimeter C of the rectangle taken in the right-hand sense and $\mathbf{t}$ is a unit vector in the direction of $\mathbf{da}$, that is, normal to the plane of the rectangle. For a rectangle sufficiently small that the fields can be taken as constant on each of the two long sides, the last equation becomes

$$(\mathbf{E}+\dot{\mathbf{x}}\times\mathbf{B})^o\cdot\boldsymbol{\tau}^o\,\Delta s+(\mathbf{E}+\dot{\mathbf{x}}\times\mathbf{B})^i\cdot\boldsymbol{\tau}^i\,\Delta s+\text{end contributions}+\frac{d}{dt}(\mathbf{B}\cdot\mathbf{t}\,\Delta s\,\Delta l)=0. \tag{7.2.13}$$

The transition layer thickness and the length of the rectangle ends Δl are now shrunk to zero simultaneously. The end contributions, being proportional to Δl, vanish and the last term in (7.2.13) also vanishes for the same reason. If $\mathbf{n}$ is the outward normal of the material surface, then $\boldsymbol{\tau}^o = -\boldsymbol{\tau}^i = \mathbf{t}\times\mathbf{n}$ and the last equation becomes

$$(\mathbf{t}\times\mathbf{n})\cdot\left[(\mathbf{E}+\dot{\mathbf{x}}\times\mathbf{B})^o-(\mathbf{E}+\dot{\mathbf{x}}\times\mathbf{B})^i\right]$$
$$=\mathbf{t}\cdot\left\{\mathbf{n}\times\left[(\mathbf{E}+\dot{\mathbf{x}}\times\mathbf{B})^o-(\mathbf{E}+\dot{\mathbf{x}}\times\mathbf{B})^i\right]\right\}=0. \tag{7.2.14}$$

Since the orientation of the loop can be changed, $\mathbf{t}$ is an arbitrary material surface tangent vector. Hence its coefficients must vanish separately,

$$\mathbf{n}\times\left[(\mathbf{E}+\dot{\mathbf{x}}\times\mathbf{B})^o-(\mathbf{E}+\dot{\mathbf{x}}\times\mathbf{B})^i\right]=0, \tag{7.2.15}$$

which states that the tangential components of $\mathbf{E}+\dot{\mathbf{x}}\times\mathbf{B}$ are continuous at a moving deforming material surface.

Lastly, we turn our attention to the magnetic intensity $\mathbf{H}$. Since its differential equation is very similar to that for $\mathbf{E}$, a boundary condition completely analogous to (7.2.15) can be expected for a dielectric. However, once again we wish to relax the dielectric assumption on just the surface to allow for surface currents. In particular, we wish to include the possibilities of a thin, high conductivity metal film electrode on the surface or a moving surface having attached to it an extrinsic charge distribution.

If the free charge density and free current density were not set equal to zero in the current density expansion of Section 4.5, (7.1.10) would have the form

$$\nabla\times\mathbf{H}-\frac{\partial\mathbf{D}}{\partial t}=q^f\dot{\mathbf{x}}+\mathbf{j}^c \tag{7.2.16}$$

where the first term on the right arises from the motion of the medium having attached to it a free charge density and the second term arises from charge moving through a medium because of its conductivity. Since we are here considering the possibility of these currents only at a material surface whose equation, $s=0$, is expressed in terms of s of (7.2.7), we have both the surface charge condition (7.2.6) and

$$\mathbf{j}^c=\mathbf{k}^c\delta(s) \tag{7.2.17}$$

where $\mathbf{k}^c$ is the surface conduction current.

We now integrate (7.2.16) over the area of the rectangle of Fig. 7.2 and introduce the identity (3.13.6) for the integral of the time derivative $\partial \mathbf{D}/\partial t$,

$$\int_A (\nabla\times\mathbf{H})\cdot\mathbf{da} - \frac{d}{dt}\int_A \mathbf{D}\cdot\mathbf{da} + \int_A \left[\nabla\times(\mathbf{D}\times\dot{\mathbf{x}})\right]\cdot\mathbf{da} + \int_A (\nabla\cdot\mathbf{D})\dot{\mathbf{x}}\cdot\mathbf{da} = \int_A \left[q^f\dot{\mathbf{x}}+\mathbf{j}^c\right]\cdot\mathbf{da}. \tag{7.2.18}$$

Next we transform the first and third integrals on the left side by Stokes' theorem and cancel the fourth integral with the first on the right side. The result is

$$\int_C (\mathbf{H}-\dot{\mathbf{x}}\times\mathbf{D})\cdot\mathbf{ds} - \frac{d}{dt}\int_A \mathbf{D}\cdot\mathbf{t}\,da = \int_A \mathbf{j}^c\cdot\mathbf{t}\,da \tag{7.2.19}$$

where $\mathbf{ds}$ is the element of arc around the perimeter C of the rectangle and $\mathbf{t}$ is a unit vector in the direction of $\mathbf{da}$. For a sufficiently small rectangle this becomes

$$(\mathbf{H}-\dot{\mathbf{x}}\times\mathbf{D})^o\cdot\tau^o\Delta s + (\mathbf{H}-\dot{\mathbf{x}}\times\mathbf{D})^i\cdot\tau^i\Delta s + \text{end contributions} - \frac{d}{dt}(\mathbf{D}\cdot\mathbf{t}\Delta s\,\Delta l) = \mathbf{k}^c\cdot\mathbf{t}\,\Delta s \tag{7.2.20}$$

where (7.2.17) is used to obtain the right side. Now letting Δl approach zero as done in reaching (7.2.14), we find that

$$(\mathbf{t}\times\mathbf{n})\cdot\left[(\mathbf{H}-\dot{\mathbf{x}}\times\mathbf{D})^o - (\mathbf{H}-\dot{\mathbf{x}}\times\mathbf{D})^i\right] = \mathbf{t}\cdot\mathbf{k}^c. \tag{7.2.21}$$

Since $\mathbf{t}$ is an arbitrary surface tangent, we find that

$$\mathbf{n}\times\left[(\mathbf{H}-\dot{\mathbf{x}}\times\mathbf{D})^o - (\mathbf{H}-\dot{\mathbf{x}}\times\mathbf{D})^i\right] = \mathbf{k}^c \tag{7.2.22}$$

as the boundary condition on the tangential components of the $\mathbf{H}-\dot{\mathbf{x}}\times\mathbf{D}$ at a moving deforming material surface.

7.3 Internal Motion Equations

For a crystal with N particles per unit cell there are $N-1$ vector degrees of freedom for internal motion. We express these in terms of the internal coordinates whose Lagrange equations are

$$\frac{d}{dt}\frac{\partial \mathfrak{L}_M}{\partial \dot{y}_i^{T\nu}} = \frac{\partial \mathfrak{L}_M}{\partial y_i^{T\nu}} - \frac{d}{dX_A}\frac{\partial \mathfrak{L}_M}{\partial y_{i,A}^{T\nu}} \qquad (\nu = 1,2,\ldots,N-1) \tag{7.3.1}$$

where $\mathfrak{L}_M$ is the total Lagrangian density in the material description. The field Lagrangian contributes nothing to this equation. Also, since no $y_{i,A}^{T\nu}$ occurs in the material frame Lagrangian density that we use, the last term may be dropped. The matter Lagrangian density is given by (4.7.12) with the stored energy now expressed as the series in (6.5.8). The field-matter interaction Lagrangian density is given in (4.7.10). Performing the indicated differentiations leads to

$$\frac{d}{dt}\left(m^\nu \dot{y}_i^{T\nu}\right) = q^\nu\left[E_i + (\dot{\mathbf{x}}\times\mathbf{B})_i\right] - \rho^0 \frac{\partial \Sigma}{\partial y_i^{T\nu}} \tag{7.3.2}$$

or

$$m^\nu \ddot{y}_i^\nu = q^\nu \mathscr{E}^i - \rho^0 \frac{\partial \Sigma}{\partial \Pi_C^\nu} X_{C,i} \tag{7.3.3}$$

where

$$\mathscr{E} \equiv \mathbf{E} + \dot{\mathbf{x}}\times\mathbf{B}. \tag{7.3.4}$$

It should be remembered that the electric and magnetic induction fields are evaluated at $\mathbf{z} = \mathbf{x}(\mathbf{X}, t)$, the center of mass, because of the multipole expansion we use, and that the independent variables in (7.3.3) are $\mathbf{X}$ and t.

Let us consider the effect of the natural state definition of Section 6.5 on the internal motion equation (7.3.3). For an "ordinary" dielectric, by which we mean a nonpyroelectric dielectric, we have y_i^ν, $\mathscr{E}_i$, E_{AB}, and Π_C^ν all zero and $X_{C,i} = \delta_{Ci}$. Using the series expansion (6.5.8) for the stored energy, we then find (7.3.3) in the natural state for a dielectric to be simply the requirement

$$0 = {}^{10}M_C^\nu \delta_{Ci}. \tag{7.3.5}$$

Thus the ${}^{10}\mathbf{M}^\nu$ term can be dropped from (6.5.8) for an ordinary dielectric.

A pyroelectric crystal is one whose crystal symmetry allows the existence of a natural state polarization called the *spontaneous polarization*,

$$P_i^S \equiv \delta_{iA} \sum_\nu q^\nu Y_A^\nu. \tag{7.3.6}$$

The constant or spontaneous part Y_A^{ν} of the internal coordinate is introduced in Section 6.5. In its intrinsic state this spontaneous polarization creates a spontaneous electric field $\mathbf{E}^S$. Under normal conditions this spontaneous electric field, which exists inside and outside the pyroelectric crystal, attracts extrinsic charge to the crystal surface until it is completely canceled. The compensating surface charge may also arise from a small conductivity of the crystal. Under this extrinsic condition $\mathbf{E}^S=0$ while $\mathbf{P}^S$ remains but is opposed by an equal and opposite dipole moment on the crystal surfaces.

We now see that y_i^{ν}, B_i, E_{AB}, and Π_C^{ν} are zero, $X_{C,i}=\delta_{Ci}$, and $\mathbf{E}=\mathbf{E}^S$ for the *intrinsic natural state* (*NS*) of a pyroelectric. Equation (7.3.3) then becomes

$$q^{\nu}E_i^S=\left(\frac{\partial\rho^0\Sigma}{\partial\Pi_A^{\nu}}\right)^{NS}\delta_{iA}={}^{10}M_A^{\nu}\delta_{iA}. \tag{7.3.7}$$

This condition removes the constant terms from (7.3.3). If we wish to consider the *extrinsic natural state* in which $\mathbf{E}^S$ is canceled, we need merely to set $\mathbf{E}^S=0$ which then requires ${}^{10}\mathbf{M}^{\nu}=0$ also.

7.4 Center-of-Mass Equation in Material Frame

The center-of-mass equation plays a special role among the matter equations of motion because it is the dynamic equation governing the deformation of the crystal. Its Lagrange equation is

$$\frac{d}{dt}\frac{\partial\mathcal{L}_M}{\partial\dot{x}_i}=\frac{\partial\mathcal{L}_M}{\partial x_i}-\frac{d}{dX_A}\frac{\partial\mathcal{L}_M}{\partial x_{i,A}} \tag{7.4.1}$$

where $\mathcal{L}_M$ is the Lagrangian density in the material frame consisting of (4.7.10), (4.7.12), and (6.5.8). The electromagnetic field Lagrangian density does not contribute to the matter equations.

Before proceeding we note that the field-matter interaction Lagrangian density in the *material* description (4.7.10) does depend on the center-of-mass position $\mathbf{x}$. This is not inconsistent with the displacement invariance condition (5.4.14) since that condition was on the Lagrangian density in the *spatial* description. The restriction of displacement invariance on the stored energy (6.1.5) remains during the transformation of it from spatial to material description. However, the displacement invariance condition does not prevent the introduction of $\mathbf{x}$ dependence into the material frame interaction Lagrangian density through the transformation from (4.7.8) to (4.7.10).

We now substitute the Lagrangian density into (7.4.1) and carry out the differentiations,

$$\rho^0\ddot{x}_i + \frac{d}{dt}\sum_\nu q^\nu y_j^{T\nu}\epsilon_{jik}B_k = \sum_\nu q^\nu y_j^{T\nu}(E_{j,i} + \epsilon_{jkl}\dot{x}_k B_{l,i}) + \frac{d}{dX_A}\frac{\partial\rho^0\Sigma}{\partial x_{i,A}}. \tag{7.4.2}$$

It is convenient to introduce the induced electric dipole moment per unit volume in the material frame $\mathbf{p}$ by

$$\mathbf{p} \equiv \sum_\nu q^\nu \mathbf{y}^{T\nu} \tag{7.4.3}$$

and a mixed frame stress tensor T_{iA} called the *Piola-Kirchoff stress tensor* by

$$T_{iA} \equiv \rho^0 \frac{\partial\Sigma}{\partial x_{i,A}}\bigg|_{\mathbf{y}^{T\nu}}. \tag{7.4.4}$$

The latter represents a force acting in a direction in the spatial frame determined by the subscript i on an element of area whose normal in the material frame has a direction determined by the subscript A. Equation (7.4.2) then becomes

$$\rho^0\ddot{x}_i = (\mathbf{p}\times\dot{\mathbf{B}})_i + (\dot{\mathbf{p}}\times\mathbf{B})_i + \mathbf{p}\cdot\mathbf{E}_{,i} + \mathbf{p}\cdot(\dot{\mathbf{x}}\times\mathbf{B}_{,i}) + T_{iA,A}, \tag{7.4.5}$$

which is the center-of-mass equation in the material frame.

The electromagnetic force terms in (7.4.5) can be arranged in an alternate and useful form. First, we expand the material time derivative $\dot{\mathbf{B}}$ by

$$\begin{aligned}\dot{\mathbf{B}} &= \frac{\partial\mathbf{B}}{\partial t} + (\dot{\mathbf{x}}\cdot\nabla)\mathbf{B}\\ &= -\nabla\times\mathbf{E} + (\dot{\mathbf{x}}\cdot\nabla)\mathbf{B}\end{aligned} \tag{7.4.6}$$

where one of the Maxwell equations is used. This creates a second force term in (7.4.5) depending on $\mathbf{E}$. Together they may be expressed as

$$-\mathbf{p}\times(\nabla\times\mathbf{E}) + (\nabla\mathbf{E})\cdot\mathbf{p} = (\mathbf{p}\cdot\nabla)\mathbf{E}, \tag{7.4.7}$$

which follows simply from expanding the vector triple product.

The force term in (7.4.5) arising from the second term in (7.4.6) may be combined with the $\mathbf{p}\cdot(\dot{\mathbf{x}}\times\mathbf{B}_{,i})$ term to yield

$$[\mathbf{p}\times(\dot{\mathbf{x}}\cdot\nabla)\mathbf{B}]_i+\mathbf{p}\cdot(\dot{\mathbf{x}}\times\mathbf{B}_{,i})=[\dot{\mathbf{x}}\times(\mathbf{p}\cdot\nabla)\mathbf{B}]_i \tag{7.4.8}$$

or in unambiguous tensor notation

$$\epsilon_{ijk}p_j B_{k,l}\dot{x}_l+\epsilon_{jlk}p_j\dot{x}_l B_{k,i}=\epsilon_{ilk}\dot{x}_l B_{k,j}p_j. \tag{7.4.9}$$

To prove this relation we use the Maxwell equation (7.1.12) and the $\epsilon-\delta$ identity twice,

$$\begin{aligned}\epsilon_{jlk}p_j\dot{x}_l B_{k,i}&=-\epsilon_{ijl}p_j B_{k,k}\dot{x}_l+\epsilon_{kjl}p_j B_{k,i}\dot{x}_l\\&=-(\delta_{kn}\delta_{ip}-\delta_{kp}\delta_{in})\epsilon_{pjl}p_j B_{k,n}\dot{x}_l\\&=-\epsilon_{imk}\epsilon_{pmn}\epsilon_{pjl}p_j B_{k,n}\dot{x}_l\\&=-\epsilon_{imk}p_j B_{k,n}\dot{x}_l(\delta_{mj}\delta_{nl}-\delta_{ml}\delta_{nj})\\&=\epsilon_{ilk}\dot{x}_l B_{k,j}p_j-\epsilon_{ijk}p_j B_{k,l}\dot{x}_l\end{aligned} \tag{7.4.10}$$

Note that this is not a mathematical identity but a physical identity because of the use of $\nabla\cdot\mathbf{B}=0$.

With the use of (7.4.6), (7.4.7), and (7.4.8) the center-of-mass equation becomes

$$\rho^0\ddot{\mathbf{x}}=(\mathbf{p}\cdot\nabla)\mathbf{E}+\dot{\mathbf{p}}\times\mathbf{B}+\dot{\mathbf{x}}\times(\mathbf{p}\cdot\nabla)\mathbf{B}+\mathbf{f} \tag{7.4.11}$$

where

$$f_i\equiv T_{iA,A}. \tag{7.4.12}$$

This equation is still in the material description, that is, with $\mathbf{X}$ and t as the independent variables. The derivatives acting on the electromagnetic fields, however, are spatial frame space derivatives $\partial/\partial x_i$. This is very natural because the electromagnetic fields are evaluated at $\mathbf{z}=\mathbf{x}(\mathbf{X},t)$ for the interaction. Since the field-matter interaction forces can be given different expressions, as in (7.4.5) and (7.4.11), we do not dwell on their interpretation at this point. Their interpretation is more obvious when the center-of-mass equation is transformed to the spatial description.

7.5 Center-of-Mass Equation in Spatial Frame

The center-of-mass equation expressed in the spatial frame is more convenient for some uses than its expression in the material frame. Rather than obtain it from a spatial frame Lagrangian density we transform the equation of motion.

First, consider the force term arising from the Piola-Kirchoff stress tensor. We multiply the entire force equation by J^{-1} and so we write

$$\begin{aligned}
J^{-1}f_i &= J^{-1}\left(\frac{\partial \rho^0\Sigma}{\partial x_{i,A}}\right)_{,A} \\
&= J^{-1}\left(\frac{\partial \rho^0\Sigma}{\partial x_{i,A}}\right)_{,j} x_{j,A} \\
&= \left(J^{-1}\frac{\partial \rho^0\Sigma}{\partial x_{i,A}} x_{j,A}\right)_{,j} - \frac{\partial \rho^0\Sigma}{\partial x_{i,A}}\left(J^{-1}x_{j,A}\right)_{,j} \\
&= \left(J^{-1}\frac{\partial \rho^0\Sigma}{\partial x_{i,A}} x_{j,A}\right)_{,j} \\
&= t^y_{ij,j}.
\end{aligned} \tag{7.5.1}$$

Here we use the Euler-Piola-Jacobi identity (3.2.9) and the *local stress tensor* definition,

$$\begin{aligned}
t^y_{ij} &\equiv J^{-1}\left.\frac{\partial \rho^0\Sigma}{\partial x_{i,A}}\right|_{\mathbf{y}^{T\nu}} x_{j,A} \\
&= J^{-1}T_{iA}x_{j,A}.
\end{aligned} \tag{7.5.2}$$

The superscript y on the local stress tensor indicates that each $\mathbf{y}^{T\nu}$ is held constant during the differentiation. The converse of the last equation is

$$T_{i,A} = Jt^y_{ij}X_{A,j}. \tag{7.5.3}$$

Next consider the $J^{-1}\dot{\mathbf{p}}$ term in (7.4.11). With the use of

$$\mathbf{P} = J^{-1}\mathbf{p}, \tag{7.5.4}$$

apparent from (4.6.19) and (7.4.3), and the spatial equation of continuity

(3.9.5), we obtain

$$\frac{1}{J}\frac{d}{dt}\mathbf{p}=\frac{\partial \mathbf{P}}{\partial t}+\nabla\times(\mathbf{P}\times\dot{\mathbf{x}})+\nabla\cdot(\mathbf{P}\dot{\mathbf{x}})$$
$$=\mathbf{j}^D+\nabla\cdot(\mathbf{P}\dot{\mathbf{x}}). \tag{7.5.5}$$

Here the definition of the dielectric current $\mathbf{j}^D$ (4.7.2) is used and the notation of the last term means $(P_j\dot{x}_i)_{,j}$ for the i component.

The last term in (7.5.5) leads to a force term in the center-of-mass equation $\nabla\cdot(\mathbf{P}\dot{\mathbf{x}})\times\mathbf{B}$ which can be combined with the last interaction force term in (7.4.11) to yield

$$\nabla\cdot(\mathbf{P}\dot{\mathbf{x}})\times\mathbf{B}+\dot{\mathbf{x}}\times(\mathbf{P}\cdot\nabla)\mathbf{B}=\nabla\cdot[\mathbf{P}(\dot{\mathbf{x}}\times\mathbf{B})]. \tag{7.5.6}$$

The electric field force term in (7.4.11) becomes

$$J^{-1}(\mathbf{p}\cdot\nabla)\mathbf{E}=\nabla\cdot(\mathbf{PE})-\mathbf{E}(\nabla\cdot\mathbf{P})$$
$$=\nabla\cdot(\mathbf{PE})+q^D\mathbf{E} \tag{7.5.7}$$

where the definition of the dielectric charge (4.7.1) is used. The result of (7.5.6) and the divergence term on the right side of (7.5.7) now can be combined into

$$\nabla\cdot(\mathbf{PE})+\nabla\cdot[\mathbf{P}(\dot{\mathbf{x}}\times\mathbf{B})]=\nabla\cdot(\mathbf{P}\mathscr{E}) \tag{7.5.8}$$

where $\mathscr{E}$ from (7.3.4) is introduced.

Collecting all these results now yields

$$J^{-1}\rho^0\ddot{x}_i=(t_{ij}^y+\mathscr{E}_iP_j)_{,j}+q^DE_i+(\mathbf{j}^D\times\mathbf{B})_i. \tag{7.5.9}$$

As discussed in Section 4.3, $J^{-1}\rho^0=\rho$ in the long wavelength limit. We also introduce for notational convenience the *elastic stress tensor* t_{ij}^E by

$$t_{ij}^E\equiv t_{ij}^y+\mathscr{E}_iP_j. \tag{7.5.10}$$

The name simply indicates that this is the stress tensor appearing in the center-of-mass or elasticity equation in the spatial frame when the body force terms are arranged as in (7.5.9). We now have

$$\rho\ddot{x}_i=t_{ij,j}^E+q^DE_i+(\mathbf{j}^D\times\mathbf{B})_i \tag{7.5.11}$$

as the final form of the center-of-mass equation. This is basically a spatial frame form but we retain the inertial force term expressed in material time derivatives for compactness of notation.

The force terms in this equation are of two general types, *body* or *volume forces* and *surface* or *stress forces*. The body force terms are seen to have the electric force plus Lorentz force form, $q^D\mathbf{E}+\mathbf{j}^D\times\mathbf{B}$, a satisfying result. These are long range forces acting throughout the volume on the dielectric charge and dielectric current, both of which arise from *bound* charge in a dielectric. The surface force, which has the form of a divergence of a stress, is a short range force acting only at the surface of a volume element through "mechanical" interactions with the neighboring volume elements.

An attempt to obtain a boundary condition on a stress tensor from (7.5.11) alone by integrating it over the volume of a pill box at a body surface causes the body forces in the equation to produce surface tractions (infinite forces in a zero volume). Not surprisingly, the way to handle this difficulty is to introduce the electromagnetic equations into the considerations. We defer this question to the next chapter so that it may be discussed in connection with momentum conservation.

7.6 Asymmetry of Elastic Stress Tensor

In this section the elastic stress tensor,

$$t_{ij}^E \equiv t_{ij}^y + \mathscr{E}_i P_j, \tag{7.6.1}$$

which enters the elasticity equation (7.5.11), is shown to be asymmetric in general and symmetric only under special conditions.

Consider the local stress tensor first. Its derivative may be reexpressed as

$$\begin{aligned} t_{ij}^y &= J^{-1}\frac{\partial\rho^0\Sigma}{\partial x_{i,A}}\bigg|_{\mathbf{y}^{T\nu}} x_{j,A} \\ &= J^{-1}x_{j,A}\left[\frac{\partial\rho^0\Sigma}{\partial E_{BC}}\frac{\partial E_{BC}}{\partial x_{i,A}} + \sum_\nu \frac{\partial\rho^0\Sigma}{\partial \Pi_B^\nu}\frac{\partial \Pi_B^\nu}{\partial x_{i,A}}\right] \\ &= J^{-1}x_{j,A}\left[\frac{\partial\rho^0\Sigma}{\partial E_{BC}} x_{i,B}\delta_{CA} - \sum_\nu \frac{\partial\rho^0\Sigma}{\partial \Pi_B^\nu} y_k^{T\nu} X_{B,i} X_{A,k}\right] \\ &= t_{ij}^\Pi - J^{-1}X_{B,i}\sum_\nu y_j^{T\nu}\frac{\partial\rho^0\Sigma}{\partial \Pi_B^\nu} \end{aligned} \tag{7.6.2}$$

where

$$t_{ij}^{\Pi} \equiv \rho \frac{\partial \Sigma}{\partial E_{BC}} \bigg|_{\Pi_A^\nu} x_{i,B} x_{j,C} = t_{ji}^{\Pi} \tag{7.6.3}$$

is a symmetric tensor.

The last term in (7.6.2) can be reexpressed using the internal motion equation (7.3.3). By forming the scalar product of that equation with $x_{i,B}$ in order to solve for the stored energy derivative, we obtain

$$\frac{\partial \rho^0 \Sigma}{\partial \Pi_B^\nu} = q^\nu \, \mathscr{E}_i x_{i,B} - m^\nu \ddot{y}_i^\nu x_{i,B}. \tag{7.6.4}$$

Substitution of this into (7.6.2) now yields

$$\begin{aligned} t_{ij}^{y} &= t_{ij}^{\Pi} - J^{-1} x_{k,B} X_{B,i} \sum_\nu y_j^{T\nu} (q^\nu \mathscr{E}_k - m^\nu \ddot{y}_k^\nu) \\ &= t_{ij}^{\Pi} - P_j \mathscr{E}_i + \sum_\nu \rho^\nu y_j^{T\nu} \ddot{y}_i^\nu \end{aligned} \tag{7.6.5}$$

where the polarization (4.6.19) is introduced and

$$\rho^\nu \equiv J^{-1} m^\nu. \tag{7.6.6}$$

The elastic stress tensor now is given by

$$t_{ij}^{E} = t_{(ij)}^{\Pi} + \sum_\nu \rho^\nu \ddot{y}_i^\nu y_j^{T\nu}. \tag{7.6.7}$$

It is apparent from this form that the last term on the right possesses an antisymmetric part thus making the elastic stress tensor *asymmetric* under general conditions. However, for frequencies far enough below the resonant frequencies of the internal motions (optic modes) this last term, which arises from inertial effects of these motions, is negligible. In this low frequency regime, which normally will include ultrasonic and hypersonic frequencies, the elastic stress tensor becomes *symmetric*. Perhaps the asymmetric stresses will be observable in a crystal that has an optic mode whose frequency drops into the hypersonic region at a temperature near its phase transition (a "soft optic mode").

PROBLEMS

7.1. Show that the canonical momentum density g_i^C (5.P.4) and the canonical stress tensor t_{ij}^C (5.P.5) can be expressed as

$$g_i^C = \rho \dot{x}_i + \epsilon_0 (\mathbf{E} \times \mathbf{B})_i + A_{i,j} D_j, \tag{7.P.1}$$

$$t_{ij}^C = t_{ij}^y + m_{ij} + \mathscr{E}_i P_j - \rho \dot{x}_i \dot{x}_j + \frac{\partial A_i}{\partial t} D_j + A_{i,k} \epsilon_{kjl} H_l \tag{7.P.2}$$

where

$$m_{ij} \equiv \epsilon_0 E_i E_j + \frac{1}{\mu_0} B_i B_j - \frac{\epsilon_0}{2} E_k E_k \delta_{ij} - \frac{1}{2\mu_0} B_k B_k \delta_{ij} \tag{7.P.3}$$

is the Maxwell vacuum-field stress tensor.

7.2. Show that the terms in (7.P.1) and (7.P.2) that are not gauge invariant drop from the momentum conservation law,

$$\frac{\partial}{\partial t}(A_{i,j} D_j) - \frac{\partial}{\partial z_j}\left(\frac{\partial A_i}{\partial t} D_j + A_{i,k} \epsilon_{kjl} H_l\right) = 0. \tag{7.P.4}$$

This *suggests* that the remaining terms in g_i^C and t_{ij}^C represent the true momentum density and true total stress tensor in the spatial frame. We show in Chapter 8 that this suggestion is correct.

CHAPTER 8

Conservation Laws

We saw in Chapter 5 that placing certain invariance conditions on the Lagrangian density guaranteed the conservation of energy, momentum, and angular momentum in the system represented. However, the canonical forms of the conservation laws that result from the invariance arguments are sometimes difficult to interpret as illustrated by the problems in Chapters 5 and 7 on the canonical momentum conservation law. Thus rather than work from the canonical forms obtained before, we obtain the conservation laws in this chapter from the equations of motion. This brings us to a consideration of the proper definition of such quantities as the momentum density and the total stress tensor which enter the momentum conservation law. We show that the proper definitions of these quantities are closely tied to boundary conditions at body surfaces. Consideration of the energy conservation law leads us to a very general proof that the group velocity and the energy propagation velocity are equal. Much of this chapter is based on Lax and Nelson [1976a] and Nelson and Lax [1976a].

8.1 Maxwell Stress Tensor

In forming the momentum conservation equation it is necessary to eliminate the body force terms in the center-of-mass equation (7.5.11). We are led to search for a combination of the electromagnetic equations that has this combination of body forces.

Since the body forces involve the dielectric charge q^D and the dielectric current $\mathbf{j}^D$, we use the Maxwell-Lorentz form of the electromagnetic equations (7.1.3), (7.1.6), (7.1.11), and (7.1.12), which involve only the vacuum fields $\mathbf{E}$ and $\mathbf{B}$. We form the vector product of (7.1.6) with $-\mathbf{B}$ and of (7.1.11) with $-\epsilon_0\mathbf{E}$,

$$-\frac{1}{\mu_0}(\nabla\times\mathbf{B})\times\mathbf{B}+\epsilon_0\frac{\partial\mathbf{E}}{\partial t}\times\mathbf{B}=-\mathbf{j}^D\times\mathbf{B}, \tag{8.1.1}$$

$$-\epsilon_0(\nabla\times\mathbf{E})\times\mathbf{E}-\epsilon_0\frac{\partial\mathbf{B}}{\partial t}\times\mathbf{E}=0. \tag{8.1.2}$$

With the vector triple products expanded these equations in tensor notation are

$$-\frac{1}{\mu_0}B_jB_{i,j}+\frac{1}{\mu_0}B_{j,i}B_j+\epsilon_0\left(\frac{\partial\mathbf{E}}{\partial t}\times\mathbf{B}\right)_i=-(\mathbf{j}^D\times\mathbf{B})_i, \qquad (8.1.3)$$

$$-\epsilon_0E_jE_{i,j}+\epsilon_0E_{j,i}E_j+\epsilon_0\left(\mathbf{E}\times\frac{\partial\mathbf{B}}{\partial t}\right)_i=0. \qquad (8.1.4)$$

Next we multiply (7.1.3) by $-E_i$ and (7.1.12) by $-B_i/\mu_0$,

$$-\epsilon_0E_iE_{j,j}=-q^DE_i, \qquad (8.1.5)$$

$$-\frac{1}{\mu_0}B_iB_{j,j}=0. \qquad (8.1.6)$$

The addition of the last four equations can be written as

$$\frac{\partial}{\partial t}(\epsilon_0\mathbf{E}\times\mathbf{B})_i-m_{ij,j}=-q^DE_i-(\mathbf{j}^D\times\mathbf{B})_i, \qquad (8.1.7)$$

where

$$m_{ij}\equiv\epsilon_0E_iE_j+\frac{1}{\mu_0}B_iB_j-\tfrac{1}{2}\left(\epsilon_0E_kE_k+\frac{1}{\mu_0}B_kB_k\right)\delta_{ij}=m_{ji} \qquad (8.1.8)$$

is the symmetric *Maxwell stress tensor* for the vacuum fields **E** and **B**. Equation (8.1.7) has the form of an electromagnetic momentum continuity equation with source (or sink) terms on the right side representing the momentum transfer between the electromagnetic fields and the dielectric crystal.

8.2 Momentum Conservation

Since the internal coordinates are displacement invariant fields, which consequently can carry no real momentum, it is sufficient to combine just the center-of-mass equation and the electromagnetic momentum continuity equation (8.1.7). The inertial force term in the center-of-mass equation can be written as

$$\rho\ddot{x}_i=\frac{1}{J}\frac{d}{dt}(J\rho\dot{x}_i)=\frac{\partial}{\partial t}(\rho\dot{x}_i)+\frac{\partial}{\partial z_j}(\rho\dot{x}_i\dot{x}_j) \qquad (8.2.1)$$

with the use of the spatial equation of continuity (3.9.5). The center-of-mass equation (7.5.11) then becomes

$$\frac{\partial}{\partial t}(\rho\dot{x}_i)+\frac{\partial}{\partial z_j}\left(\rho\dot{x}_i\dot{x}_j-t_{ij}^E\right)=q^D E_i+(\mathbf{j}^D\times\mathbf{B})_i. \tag{8.2.2}$$

Adding (8.1.7) to this yields

$$\frac{\partial}{\partial t}(\rho\dot{x}_i+\epsilon_0(\mathbf{E}\times\mathbf{B})_i)+\frac{\partial}{\partial z_j}\left(-t_{ij}^E-m_{ij}+\rho\dot{x}_i\dot{x}_j\right)=0, \tag{8.2.3}$$

which expresses *conservation of momentum* in the spatial frame. This suggests that the quantity

$$t_{ij}^E+m_{ij}-\rho\dot{x}_i\dot{x}_j \tag{8.2.4}$$

is the total stress tensor.

8.3 Total Stress Tensor

In this section we wish to consider the question: what is the total stress tensor of the system consisting of a dielectric crystal in interaction with the electromagnetic fields? The question arises because of an apparent lack of uniqueness of the stress tensor caused by its occurrence within a divergence in the momentum conservation law.

The stress tensor (8.2.4) may be altered in many ways without affecting the validity of the momentum conservation equation (8.2.3). One way is simply to add an arbitrary function which is independent of $\mathbf{z}$ to the form (8.2.4). Another way is to add a curl-like quantity $\epsilon_{jkl}f_{il,k}$ to the stress tensor, since the divergence of a curl vanishes. A more complicated way is to add the spatial time derivative of an arbitrary tensor function $\partial h_{ij}/\partial t$ to the stress tensor while at the same time subtracting the spatial frame divergence $h_{ij,j}$ of the same function from the quantity within the time derivative in (8.2.3). Still other ways involve the use of the electromagnetic equations, as in (7.P.4), or of the internal coordinate equations.

If the form (8.2.4) may be altered, a further question arises: should the altered form of the stress tensor be symmetric? Usually the total stress tensor must be symmetric in order to have angular momentum conservation. But it is shown in Sections 5.6 and 6.3 without reference to the symmetry of any stress tensor that the system considered here possesses angular momentum conservation. Hence there is no reason to impose symmetry on the stress tensor.

A boundary condition on the elastic stress tensor $\mathbf{t}^E$ is not presented in Section 7.5 because the body forces present in the center-of-mass equation (7.5.11) become infinite within a zero volume at a body surface, where material properties drop abruptly to zero, and so produce surface traction terms. These terms must then be evaluated through the use of the electromagnetic equations. If a boundary condition is found from the momentum conservation law, there are no body forces to create surface traction terms. However, surface traction terms could arise from this procedure if space derivatives of material properties were to appear in either the total stress tensor or the momentum density. Since no such derivatives appear in the forms of these quantities in (8.2.3), it is clear that surface tractions can and should be avoided. In particular, we must avoid any of the transformations discussed above which produce such space derivatives of material quantities. We conclude from this discussion that the stress boundary condition *at a body surface* plays a key role in properly identifying the total stress tensor.

This is reasonable on other grounds. First, the boundary condition has the form of a scalar product of a surface normal and the stress tensor. Since the stress tensor is no longer within a derivative, the resultant arbitrariness is gone. Because of the scalar product one might suspect that a different arbitrariness now appears—the addition of an arbitrary tensor whose scalar product with the surface normal vanishes. However, this can be ruled out because the stress tensor must hold throughout the volume of the body and so cannot depend *ab initio* on the surface orientation. Second, it is natural that the proper definition of a total stress tensor, which is equivalently a flux of momentum density, should involve a body surface, since it is there that a contacting test system, whose properties are regarded as understood, measures the transfer of momentum to the body by way of the boundary condition.

Although appearance in the stress boundary condition is a more stringent requirement on the total stress tensor than its appearance in the momentum conservation law, it is clear that some of the stress tensors produced by the transformations discussed earlier in this section are not ruled out by a boundary condition requirement. Any null stress, which is a stress tensor having no divergence and possessing a continuous scalar product with the unit normal across every surface, could be added to the total stress tensor definition. A constant stress tensor is an example of a null stress. It is clear, however, that null stresses produce no observable effects and so should be excluded from the definition. This can be done by defining the total stress tensor in some *reference state* in which there is agreement on the proper form. The vacuum is obviously the appropriate reference state. For the vacuum the total stress tensor t_{ij}^T (T stands for total) and momentum density are

$$t_{ij}^{T} \equiv m_{ij} \qquad \text{(vacuum)}, \tag{8.3.1}$$

$$g_i \equiv \epsilon_0(\mathbf{E}\times\mathbf{B})_i \qquad \text{(vacuum)}. \tag{8.3.2}$$

The discussion above leads us to require the total stress tensor (1) to satisfy the momentum conservation law,

$$\frac{\partial g_i}{\partial t} - \frac{\partial t_{ij}^{T}}{\partial z_j} = 0, \tag{8.3.3}$$

(2) to possess a continuous scalar product with the outward unit normal $\mathbf{n}$ across any surface *fixed* in the *spatial frame*,

$$\left(t_{kl}^{To} - t_{kl}^{Ti}\right) n_l = 0, \tag{8.3.4}$$

where o and i stand for outside and inside the surface, and (3) to reduce to (8.3.1) when the medium is a vacuum.

We now find the stress boundary condition by using a pill box argument on the momentum conservation law (8.2.3). A volume integral of this law is taken over a pill box which encompasses a portion of a surface fixed in the spatial frame. The surface may lie within the volume of the crystal or at a particular instant of time coincide with the center of a surface transition layer of the body, as shown in Fig. 7.1. The volume integral of the divergence term is then converted to a surface integral over the pill box with the use of Gauss' theorem and the spatial frame time derivative is removed from the volume integral, since the pill box is stationary in this frame,

$$\frac{\partial}{\partial t}\int_V (\rho\dot{x}_i + \epsilon_0(\mathbf{E}\times\mathbf{B})_i)\, dv^P + \int_A \left(-t_{ij}^{E} - m_{ij} + \rho\dot{x}_i\dot{x}_j\right) n_j^P\, da^P = 0. \tag{8.3.5}$$

Here V is the volume of the pill box, A its area, $\mathbf{n}^P$ a unit outward normal to its surface, dv^P an element of its volume, and da^P an element of its surface area. For an infinitesimal pill box (8.3.5) becomes

$$\frac{\partial}{\partial t}\left[(\rho\dot{x}_i + \epsilon_0(\mathbf{E}\times\mathbf{B})_i)\Delta a\,\Delta l\right] + \left(-t_{ij}^{E} - m_{ij} + \rho\dot{x}_i\dot{x}_j\right)^{o} n_j^{Po}\,\Delta a$$
$$-\left(-t_{ij}^{E} - m_{ij} + \rho\dot{x}_i\dot{x}_j\right)^{i} n_j^{Pi}\,\Delta a + \text{side wall contributions} = 0. \tag{8.3.6}$$

Here Δa is the cross-sectional area of the pill box, Δl is the height of the pill box and the thickness of the transition layer, o stands for outside, and i stands for inside. We now shrink the transition layer thickness and the pill box height to zero ($\Delta l \to 0$). The first term in (8.3.6) and the side wall contributions, which are also proportional to Δl, vanish in the limit. Letting $\mathbf{n} = \mathbf{n}^{Po} = -\mathbf{n}^{Pi}$, we then obtain

$$\left[\left(-t_{ij}^{E} - m_{ij} + \rho \dot{x}_i \dot{x}_j\right)^{o} - \left(-t_{ij}^{E} - m_{ij} + \rho \dot{x}_i \dot{x}_j\right)^{i}\right] n_j = 0 \tag{8.3.7}$$

as the stress boundary condition for surfaces fixed in the spatial frame.

We may now identify the *total stress tensor* as

$$t_{ij}^{T} \equiv t_{ij}^{E} + m_{ij} - \rho \dot{x}_i \dot{x}_j \tag{8.3.8}$$

since it satisfies the momentum conservation law (8.3.3) [see (8.2.3)], the boundary condition (8.3.4) [see (8.3.7)], and reduces to (8.3.1) when the medium is a vacuum. The choice of sign in (8.3.8) is made so that the linearized form of t_{ij}^{T} has the conventionally chosen sign of the linear stress tensor. Note that the identification of (8.3.8) agrees with the result suggested by Probs. 7.1 and 7.2. Note also that the total stress tensor t_{ij}^{T} is *asymmetric* since t_{ij}^{E} is asymmetric (see Section 7.6) while the other two terms are symmetric. The asymmetry has significant size only for frequencies near the resonances of internal motions (optic modes). We interpret the origin of the asymmetry in Section 8.6. Since t_{ij}^{T} is the stress tensor appearing in the momentum conservation law (8.2.3), we are justified in identifying the quantity contained in the time derivative in that equation [which reduces to (8.3.2) for a vacuum] as the *total momentum density* **g** in the spatial frame,

$$\mathbf{g} \equiv \rho \dot{\mathbf{x}} + \epsilon_0 \mathbf{E} \times \mathbf{B}. \tag{8.3.9}$$

The simplicity of the derivation of the boundary condition (8.3.7) follows from the vanishing of the volume integral of the time derivative term and of the pill box side wall contributions to the surface integral. These conditions should not be taken for granted. As an example of a contrary situation consider another form of the momentum conservation equation,

$$\frac{\partial g_i'}{\partial t} + \frac{\partial t_{ij}'}{\partial z_j} = 0, \tag{8.3.10}$$

where

$$g_i' \equiv \rho \dot{x}_i + \epsilon_0 (\mathbf{E} \times \mathbf{B})_i - \left[\nabla \times \sum_\nu \frac{\rho^\nu}{2} \mathbf{y}^{T\nu} \times \dot{\mathbf{y}}^{T\nu} \right]_i, \tag{8.3.11}$$

$$t_{ij}' \equiv \rho \dot{x}_i \dot{x}_j - t_{ij}^{\Pi} - m_{ij} + \sum_\nu \rho^\nu y_{(i}^{T\nu} \ddot{y}_{j)}^{T\nu}$$

$$- \frac{\partial}{\partial z_p} \left[\dot{x}_{(i} \epsilon_{j)pk} \epsilon_{knm} \sum_\nu \frac{\rho^\nu}{2} y_n^{T\nu} \dot{y}_m^{T\nu} \right] = t_{ji}'. \tag{8.3.12}$$

This form is obtained from (8.2.3) by mathematical manipulation, which we do not present. Because of the space derivative in g_i' acting on the material quantity ρ^ν which becomes discontinuous in the limit as the pill box volume and transition layer shrink to zero, the volume integral of $\partial g_i'/\partial t$ does not vanish. Also, the last term in t_{ij}', which is a curl-like quantity, has components that contribute to the pill box side wall terms of the surface integral and, being derivatives of discontinuous quantities in the limit, do not vanish. If these troublesome terms are handled properly, the result is simply (8.3.7) and so the same total stress tensor (8.3.8) is again identified. Many other forms of quantities analogous to those of (8.3.11) and (8.3.12) may be found that present similar problems and lead to the same conclusion.

8.4 Stress Boundary Condition

In determining the total stress tensor relative to the spatial coordinate system in the preceding section it is necessary to consider the stress boundary condition across a surface *fixed* in that coordinate system. However, if momentum transfer from an electromagnetic field to a material body or from one material body to another is being considered, the surface of most interest for the transfer of momentum is the body surface which is, in general, a *moving deforming* surface when viewed from the spatial (laboratory) frame of reference. Thus it is of considerable importance to generalize the stress boundary condition (8.3.7) to the case of a moving material surface.

We begin by integrating the momentum conservation law over the volume of a pill box fixed to the transition layer at a moving body surface as done in (8.3.5). The spatial frame time derivative, however, must be left inside the volume integral, since the limits of integration (the body surface) depend on time in that frame of reference. Thus we have

$$\int_V \frac{\partial g_i}{\partial t} dv^P - \int_A t_{ij}^T n_j^P da^P = 0 \tag{8.4.1}$$

where the momentum density g_i of (8.3.9) and the total stress tensor of (8.3.8) are introduced. We now eliminate the spatial frame time derivative with the use of the spatial equation of continuity (3.9.5) and obtain

$$\int_V \left[\frac{1}{J} \frac{d(Jg_i)}{dt} - \frac{\partial}{\partial z_j}(\dot{x}_j g_i) \right] dv^P - \int_A t_{ij}^T n_j^P da^P = 0. \tag{8.4.2}$$

Changing the volume integration variable to the material frame in the first integral and using Gauss' theorem on the second yields

$$\int_V \frac{d(Jg_i)}{dt} dV^P - \int_A \left(t_{ij}^T + g_i \dot{x}_j\right) n_j^P da^P = 0. \tag{8.4.3}$$

Since the material frame time derivative holds the material coordinates $\mathbf{X}$ fixed, it may be removed from the volume integral,

$$\frac{d}{dt} \int_V J g_i dV^P - \int_A \left(t_{ij}^T + g_i \dot{x}_j\right) n_j^P da^P = 0. \tag{8.4.4}$$

We may now shrink the surface transition layer thickness and the pill box height to zero simultaneously. In this limit the volume integral and the pill box side wall contributions to the surface integral vanish. Replacing the pill box normal $\mathbf{n}^P$ by $\mathbf{n}$ on the outside surface and by $-\mathbf{n}$ on the inside surface, we obtain

$$\left[\left(t_{ij}^T + g_i \dot{x}_j\right)^o - \left(t_{ij}^T + g_i \dot{x}_j\right)^i \right] n_j = 0. \tag{8.4.5}$$

Here o refers to outside and i to inside. This stress boundary condition holds for any material surface whether it is moving at a velocity $\dot{\mathbf{x}}$ or is stationary ($\dot{\mathbf{x}} = 0$) and so generalizes the boundary condition (8.3.7). A further generalization [Eringen, 1967] to a singular surface traveling at a velocity v, such as a shock front, is possible but is omitted in this treatment.

8.5 Natural-State Stress

Just as it was important to consider the forces on the internal coordinates in the natural state (Section 7.3), it is important to consider the total stress in the natural state. The considerations divide naturally according to whether the crystal is an ordinary dielectric or a pyroelectric and whether the crystal is infinite or finite.

The natural state of an infinite homogeneous crystal may be taken as a stress-free state, that is, it may be presumed that the boundaries are playing no essential role and so the surfaces at infinity may be regarded as free. Since free surfaces can relax, no stress remains in a homogeneous crystal whether infinite or finite. For a dielectric setting the total stress tensor (8.3.8) in the natural state to zero leads simply to

$$\left(t_{ij}^{T}\right)^{NS}=\left(t_{ij}^{\Pi}\right)^{NS}={}^{01}M_{AB}\delta_{iA}\delta_{iB}=0 \tag{8.5.1}$$

with the use of (7.6.7), (7.6.3), and (6.5.8). Thus the ${}^{01}\mathbf{M}$ term in the stored energy expansion can be dropped for infinite homogeneous dielectrics.

For an infinite pyroelectric crystal the natural-state total stress tensor is

$$\left(t_{ij}^{T}\right)^{NS}={}^{01}M_{AB}\delta_{iA}\delta_{iB}+\epsilon_0 E_i^S E_j^S-\frac{\epsilon_0}{2}E_k^S E_k^S\delta_{ij} \tag{8.5.2}$$

where $\mathbf{E}^S$ is the spontaneous electric field. Setting the total stress tensor in the natural state to zero leads to

$$^{01}M_{CD}=\frac{\epsilon_0}{2}E_k^S E_k^S\delta_{CD}-\epsilon_0\delta_{jC}\delta_{iD}E_i^S E_j^S \tag{8.5.3}$$

for infinite homogeneous pyroelectric crystals. The relation of $\mathbf{E}^S$ to the spontaneous polarization $\mathbf{P}^S$ is determined by a boundary value problem and so depends on the shape of the crystal prior to letting all dimensions approach infinity.

Consider next a finite-sized nonpyroelectric dielectric crystal. Its natural-state stress is determined by the boundary condition

$$N_k\left(t_{jk}^{Ti}\right)^{NS}=N_k\left(t_{jk}^{To}\right)^{NS} \tag{8.5.4}$$

where i and o refer to inside and outside the crystal and the normal to the surface $\mathbf{N}$ is a vector function of position. Since the right side is zero, this leads immediately to

$$N_k {}^{01}M_{AB}\delta_{jA}\delta_{kB}=0. \tag{8.5.5}$$

Since **N** is a function of the position on the surface, the only solution to this is

$${}^{01}M_{AB}=0. \tag{8.5.6}$$

We conclude that a finite-sized homogeneous dielectric crystal must be stress-free in the natural state.

Use of (8.5.4) for a finite-sized pyroelectric crystal is more complicated, since the total stress is not zero outside the crystal because of the Maxwell stress from the spontaneous electric field. That equation now becomes

$$N_l\left(\delta_{kA}\delta_{lB}{}^{01}M_{AB}+\epsilon_0E_k^SE_l^S-\frac{\epsilon_0}{2}E_m^SE_m^S\delta_{kl}\right)^i=N_l\left(\epsilon_0E_k^SE_l^S-\frac{\epsilon_0}{2}E_m^SE_m^S\delta_{kl}\right)^o. \tag{8.5.7}$$

Since it is well known that there is no electric field that is separately homogeneous inside and outside a finite homogeneously polarized body, we can hope to find at best a natural-state solution that is homogeneous only inside the crystal.

We thus are led to express the outside electric field in terms of the inside field and then eliminate the outside field from (8.5.7). From the continuity of the normal component of the electric displacement,

$$\mathbf{N}\cdot\left(\epsilon_0\mathbf{E}^S+\mathbf{P}^S\right)^i=\mathbf{N}\cdot\left(\epsilon_0\mathbf{E}^S\right)^o, \tag{8.5.8}$$

we have at the surface

$$(\mathbf{E}^S)^o_{\text{norm}}=\mathbf{NN}\cdot\left(\mathbf{E}^S+\frac{\mathbf{P}^S}{\epsilon_0}\right)^i, \tag{8.5.9}$$

while from the continuity of tangential electric field,

$$\mathbf{N}\times(\mathbf{E}^S)^i=\mathbf{N}\times(\mathbf{E}^S)^o, \tag{8.5.10}$$

we have at the surface

$$(\mathbf{E}^S)^o_{\text{tang}}=(\mathbf{E}^S)^i-\mathbf{NN}\cdot(\mathbf{E}^S)^i. \tag{8.5.11}$$

Thus

$$(\mathbf{E}^S)^o = (\mathbf{E}^S)^o_{\text{norm}} + (\mathbf{E}^S)^o_{\text{tang}}$$
$$= (\mathbf{E}^S)^i + \frac{1}{\epsilon_0}\mathbf{N}(\mathbf{N}\cdot\mathbf{P}^S) \tag{8.5.12}$$

at the surface. Substitution of this into (8.5.7) yields

$$N_l \delta_{kA} \delta_{lB}{}^{01}M_{AB} = E_k^S(\mathbf{N}\cdot\mathbf{P}^S) + \frac{N_k}{2\epsilon_0}(\mathbf{N}\cdot\mathbf{P}^S)^2 \tag{8.5.13}$$

where the superscript i is dropped since all fields are inside the crystal.

The most generally shaped isotropic body that possesses both a homogeneous electric field and polarization interior to the body is an ellipsoid [Mason and Weaver, 1929]. Because of the cubic dependence on $\mathbf{N}$, which is a function of the position on the surface, in one of the terms and of the linear dependence on $\mathbf{N}$ in the other term, (8.5.13) cannot be satisfied for such a shape. The addition of anisotropy to the considerations does not help. We conclude that a pyroelectric body having a spontaneous electric field and having all of its dimensions finite cannot have a homogeneous natural state.

Consider an infinite cylinder of circular cross section with the spontaneous polarization parallel to the sides of the cylinder. The Maxwell charge equation when applied to the vacuum region around the cylinder gives

$$\nabla\cdot\mathbf{D} = -\epsilon_0\nabla^2\Phi = -\frac{\epsilon_0}{r}\frac{\partial}{\partial r}\left(r\frac{\partial\Phi}{\partial r}\right) = 0 \tag{8.5.14}$$

where r is a coordinate measured radially from the cylinder axis. Its solution is simply

$$\Phi = c_1 \log r + c_2. \tag{8.5.15}$$

Requiring finiteness of Φ at $r=\infty$ gives $c_1=0$ and so $\mathbf{E}^S = -\nabla\Phi = 0$ outside the cylinder. Equation (8.5.12) then gives $\mathbf{E}^S=0$ inside the cylinder and (8.5.13) gives ${}^{01}M_{AB}=0$. Hence a homogeneous natural state exists for this semiinfinite shape. By superposition of many such solutions, it is apparent that the same solution holds for an arbitrary cross section cylinder.

Consider next an infinite plate, that is, a body with one finite and constant dimension. Outside the plate the Maxwell charge equation becomes

$$\nabla\cdot\mathbf{D} = -\epsilon_0\nabla^2\Phi = -\epsilon_0\frac{\partial^2\Phi}{\partial z^2} = 0 \tag{8.5.16}$$

where z is measured normal to the plate. The solution,

$$\Phi = c_1 z + c_2, \tag{8.5.17}$$

can be finite at $z=\infty$ only if $c_1=0$, which leads to

$$\mathbf{E}^S = 0 \tag{8.5.18}$$

outside the plate. Equation (8.5.12) then gives

$$\mathbf{E}^S = -\frac{1}{\epsilon_0}\mathbf{N}(\mathbf{N}\cdot\mathbf{P}^S) \tag{8.5.19}$$

inside the plate. Substitution of (8.5.19) into (8.5.13) gives

$$N_l \delta_{kA}\delta_{lB}{}^{01}M_{AB} = -\frac{1}{2\epsilon_0}N_k(\mathbf{N}\cdot\mathbf{P}^S)^2. \tag{8.5.20}$$

In general, $\delta_{kA}\delta_{lB}{}^{01}M_{AB}$ must be expressible as

$$\delta_{kA}\delta_{lB}{}^{01}M_{AB} = AP_k^S P_l^S + BN_k P_l^S + CP_k^S N_l + DN_k N_l \tag{8.5.21}$$

where A, B, C, and D are arbitrary constants. Symmetry between k and l makes $B=C$. Substitution into (8.5.20) gives two conditions, the separate vanishing of the coefficients of **P** and **N**, because these two vectors are independent. These conditions give

$$B = -(\mathbf{N}\cdot\mathbf{P}^S)A, \tag{8.5.22}$$

$$D = -(\mathbf{N}\cdot\mathbf{P}^S)^2\left(\frac{1}{2\epsilon_0} - A\right). \tag{8.5.23}$$

Requiring the solution to become that of the infinite cylinder above, ${}^{01}M_{AB}=0$, when $\mathbf{N}\cdot\mathbf{P}^S=0$ gives $A=0$. The solution for the infinite plate is thus

$$\delta_{kA}\delta_{lB}{}^{01}M_{AB} = -\frac{N_k N_l(\mathbf{N}\cdot\mathbf{P}^S)^2}{2\epsilon_0} \tag{8.5.24}$$

showing that a homogeneous natural state exists for this semiinfinite shape also.

8.6 Angular Momentum Conservation

In this section we wish to find the angular momentum conservation law from the equations of motion rather than from the canonical form (5.6.17). It is found by combining contributions from the center-of-mass motion, each of the internal motions, and the electromagnetic fields. Using arguments similar to those applied to the momentum conservation law, we identify the flux of total angular momentum and the total angular momentum density. The latter includes an internal angular momentum density associated with the internal motions. Examination of the dynamical equation governing it leads to an interpretation of the antisymmetric part of the total stress tensor.

The center-of-mass contribution is found by forming the vector product of the spatial frame equation of motion (7.5.11) with $\mathbf{x}$ and reexpressing the inertial term by way of (8.2.1),

$$\epsilon_{ijk}x_j\left[\frac{\partial}{\partial t}(\rho\dot{x}_k)+\frac{\partial}{\partial z_l}(\rho\dot{x}_k\dot{x}_l)\right]=\epsilon_{ijk}x_j t^E_{kl,l}+q^D(\mathbf{x}\times\mathbf{E})_i+\left[\mathbf{x}\times(\mathbf{j}^D\times\mathbf{B})\right]_i. \tag{8.6.1}$$

Since the spatial frame time derivative of $\mathbf{x}$ is identically zero, x_j may be brought inside the time derivative. It may also be brought inside the divergence on the left side because differentiating it produces $\epsilon_{ilk}\rho\dot{x}_k\dot{x}_l$ which is identically zero. Thus (8.6.1) may be rewritten as

$$\begin{aligned}\frac{\partial}{\partial t}\left[\rho(\mathbf{x}\times\dot{\mathbf{x}})_i\right]+\frac{\partial}{\partial z_l}\left[\rho(\mathbf{x}\times\dot{\mathbf{x}})_i\dot{x}_l-\epsilon_{ijk}x_j t^E_{kl}\right]&\\=-\epsilon_{ilk}t^E_{kl}+q^D(\mathbf{x}\times\mathbf{E})_i+\left[\mathbf{x}\times(\mathbf{j}^D\times\mathbf{B})\right]_i.&\end{aligned} \tag{8.6.2}$$

The internal motion contribution is found by forming the vector product of (7.3.3) with $\mathbf{y}^{T\nu}$ and multiplying it by J^{-1} to obtain

$$\frac{1}{J}\frac{d}{dt}\left[J\rho^\nu\mathbf{y}^{T\nu}\times\dot{\mathbf{y}}^{T\nu}\right]_i=\frac{1}{J}q^\nu(\mathbf{y}^{T\nu}\times\mathscr{E})_i-\frac{1}{J}\epsilon_{ijk}y_j^{T\nu}\frac{\partial\rho^0\Sigma}{\partial y_k^{T\nu}} \tag{8.6.3}$$

where ρ^ν is defined by (7.6.6). If we now define the *internal angular momentum density* by

$$\mathbf{l}\equiv\sum_\nu\rho^\nu\mathbf{y}^{T\nu}\times\dot{\mathbf{y}}^{T\nu}, \tag{8.6.4}$$

sum (8.6.3) over ν from 1 to $N-1$, apply the spatial equation of continuity to the left side, and substitute the definition of the polarization (4.6.19) we obtain

$$\frac{1}{J}\frac{d}{dt}(Jl_i)=\frac{\partial l_i}{\partial t}+\frac{\partial}{\partial z_l}(l_i\dot{x}_l)=(\mathbf{P}\times\mathscr{E})_i-\frac{1}{J}\epsilon_{ijk}\sum_\nu y_j^{T\nu}\frac{\partial\rho^0\Sigma}{\partial y_k^{T\nu}}. \quad (8.6.5)$$

It should be remembered that internal angular momentum is distinct from intrinsic particle spin which is not included in the formulation.

The electromagnetic contribution is easily found by forming the vector product of the electromagnetic momentum equation (8.1.7) with $\mathbf{x}$

$$\left[\epsilon_0\mathbf{x}\times\frac{\partial}{\partial t}(\mathbf{E}\times\mathbf{B})\right]_i-\epsilon_{ijk}x_j m_{kl,l}=-q^D(\mathbf{x}\times\mathbf{E})_i-\left[\mathbf{x}\times(\mathbf{j}^D\times\mathbf{B})\right]_i. \quad (8.6.6)$$

The terms on the left side may be rearranged into

$$\frac{\partial}{\partial t}\left[\epsilon_0\mathbf{x}\times(\mathbf{E}\times\mathbf{B})\right]_i-\frac{\partial}{\partial z_l}\left[\epsilon_{ijk}x_j m_{kl}\right]=-q^D(\mathbf{x}\times\mathbf{E})_i-\left[\mathbf{x}\times(\mathbf{j}^D\times\mathbf{B})\right]_i \quad (8.6.7)$$

by using $\partial\mathbf{x}/\partial t=0$ and the interchange symmetry of m_{kl}.

We now add (8.6.2), (8.6.5), and (8.6.7) to obtain

$$\frac{\partial}{\partial t}\left[\mathbf{x}\times\mathbf{g}+\mathbf{l}\right]_i+\frac{\partial}{\partial z_l}\left[l_i\dot{x}_l-\epsilon_{ijk}x_j t_{kl}^T\right]=-\epsilon_{ijk}\left[\frac{1}{J}\sum_\nu y_j^{T\nu}\frac{\partial\rho^0\Sigma}{\partial y_k^{T\nu}}+t_{kj}^y\right]. \quad (8.6.8)$$

This equation has the conservation law form if we can show the right side vanishes.

This can be done with the aid of the rotational invariance condition (6.3.4). Forming the double scalar product of that equation with $\rho^0 J^{-1}\epsilon_{ijk}$, we find

$$\begin{aligned}\epsilon_{ijk}\frac{1}{J}\sum_\nu y_j^{T\nu}\frac{\partial\rho^0\Sigma}{\partial y_k^{T\nu}}&=\epsilon_{ijk}\frac{1}{J}\frac{\partial\rho^0\Sigma}{\partial X_{A,j}}X_{A,k}=\epsilon_{ijk}\frac{1}{J}\frac{\partial\rho^0\Sigma}{\partial x_{l,B}}\frac{\partial x_{l,B}}{\partial X_{A,j}}X_{A,k}\\&=-\epsilon_{ijk}\frac{1}{J}\frac{\partial\rho^0\Sigma}{\partial x_{l,B}}x_{j,B}x_{l,A}X_{A,k}=-\epsilon_{ijk}\frac{1}{J}\frac{\partial\rho^0\Sigma}{\partial x_{k,B}}x_{j,B}\\&=-\epsilon_{ijk}t_{kj}^y\end{aligned} \quad (8.6.9)$$

by regarding the dependence of Σ changed from $X_{A,j}$ to $x_{l,B}$, using the identities (3.2.3) and (3.1.5), and substituting the definition of t^{y}_{kj}, (7.5.2). Thus the right side of (8.6.8) vanishes and we are left with

$$\frac{\partial}{\partial t}\left[\mathbf{x}\times\mathbf{g}+\mathbf{l}\right]_i+\frac{\partial}{\partial z_l}\left[l_i\dot{x}_l-\epsilon_{ijk}x_jt^T_{kl}\right]=0, \tag{8.6.10}$$

the law of *angular momentum conservation* expressed in the spatial frame.

We now identify

$$k_{il}\equiv l_i\dot{x}_l-\epsilon_{ijk}x_jt^T_{kl} \tag{8.6.11}$$

as the *flux of total angular momentum* and

$$\boldsymbol{\omega}\equiv\mathbf{x}\times\mathbf{g}+\mathbf{l} \tag{8.6.12}$$

as the *total angular momentum density*. The identification of the flux of total angular momentum follows from arguments analogous to those used to identify the total stress tensor from the momentum conservation law in Section 8.3. These are: (1) the total angular momentum flux $\mathbf{k}$ appears in the divergence term of the angular momentum conservation law,

$$\frac{\partial\omega_i}{\partial t}+\frac{\partial k_{ij}}{\partial z_j}=0; \tag{8.6.13}$$

(2) it appears in the boundary condition for total angular momentum flux for surfaces *fixed* in the spatial frame,

$$\left(k^o_{jl}-k^i_{jl}\right)n_l=0, \tag{8.6.14}$$

a condition found by arguments analogous to those leading to the total stress boundary condition (8.3.7); (3) for a vacuum $\mathbf{k}$ reduces to

$$k_{il}=-\epsilon_{ijk}z_jm_{kl}, \tag{8.6.15}$$

which is the expected form and so is acceptable as a reference value.

It is important to realize that this closed system, consisting of a dielectric crystal in interaction with the electromagnetic field, possesses total angular momentum conservation even though the total stress tensor is *not* symmetric. We can understand this by considering the equation of motion of the internal angular momentum $\mathbf{l}$ (8.6.5). If the rotational invariance expression (8.6.9) is substituted into (8.6.5) and the definitions of the total stress tensor (8.3.8) and the elastic stress tensor (7.6.1) are inserted, we find that

$$\frac{1}{J}\frac{d}{dt}(Jl_i)=\epsilon_{ijk}t^T_{kj}. \tag{8.6.16}$$

This states that the torque created by the antisymmetric part of the total stress tensor is balanced by, that is, causes a change in, the internal angular momentum density. Note that, if the material medium were represented only by a center-of-mass coordinate **x** (and thus all internal coordinates were set equal to zero), the left side of (8.6.16) would be zero and the equation would then become a condition requiring the interchange symmetry of the total stress tensor. We thus see that the antisymmetric part of the total stress tensor results from representing the crystal at each point of the material by a manifold of N vector matter continua, $N-1$ of them representing internal motions. Though these motions cannot carry any real linear momentum (see Section 8.2), they can possess angular momentum as evidenced by the presence of the internal angular momentum density in the angular momentum conservation law.

It is apparent from the stress boundary condition derivation in Section 8.4 that a boundary condition can always be found from an equation having the conservation law form. Thus we can expect a boundary condition on the flux of total angular momentum for a moving deforming *material surface*. By a procedure completely analogous to that of Section 8.4 this is found to be

$$\left[(k_{jl}-\omega_j\dot{x}_l)^o-(k_{jl}-\omega_j\dot{x}_l)^i\right]n_l=0. \tag{8.6.17}$$

Though this equation is an important physical statement, it does not usually play an important role in problem solving, since it does not give a new condition on the fields that is independent of the stress boundary condition and the displacement field boundary condition (which we consider in Chapters 10 and 16).

8.7 Energy Conservation

Next we find the energy conservation law from the equations of motion. First, form the scalar product of the center-of-mass equation (7.5.11) with $\dot{x}_i$. The inertial term can be reexpressed by the spatial equation of continuity (3.9.5) to be

$$\rho \dot{x}_i \ddot{x}_i = \frac{1}{J}\frac{d}{dt}\left(J\frac{\rho \dot{\mathbf{x}}^2}{2}\right) = \frac{\partial}{\partial t}\left(\frac{\rho \dot{\mathbf{x}}^2}{2}\right) + \frac{\partial}{\partial z_j}\left(\frac{\rho \dot{\mathbf{x}}^2 \dot{x}_j}{2}\right). \tag{8.7.1}$$

Thus we obtain

$$\frac{\partial}{\partial t}\left(\frac{\rho \dot{\mathbf{x}}^2}{2}\right) + \frac{\partial}{\partial z_j}\left(\frac{\rho \dot{\mathbf{x}}^2 \dot{x}_j}{2} - t_{ij}^E \dot{x}_i\right) + t_{ij}^y \dot{x}_{i,j} = -\mathcal{E}_i P_j \dot{x}_{i,j} + \left[q^D E_i + (\mathbf{j}^D \times \mathbf{B})_i\right]\dot{x}_i. \tag{8.7.2}$$

Forming the scalar product of the internal motion equations with $\dot{y}_i^\nu / J$ and using the spatial equation of continuity on both the inertial term and the electromagnetic interaction term yields

$$\frac{\partial}{\partial t}\left(\frac{\rho^\nu \dot{\mathbf{y}}^{\nu 2}}{2}\right) + \frac{\partial}{\partial z_j}\left(\frac{\rho^\nu \dot{\mathbf{y}}^{\nu 2} \dot{x}_j}{2}\right) + \frac{1}{J}\frac{\partial \rho^0 \Sigma}{\partial y_i^{T\nu}} \dot{y}_i^\nu = q^\nu \left(\frac{\partial (y_i^{T\nu}/J)}{\partial t} + \frac{\partial (y_i^{T\nu} \dot{x}_j / J)}{\partial z_j}\right)\mathcal{E}_i. \tag{8.7.3}$$

The equation of continuity of electromagnetic energy can be found by first forming the scalar product of (7.1.6) with $-\mathbf{E}$ and of (7.1.11) with $\mathbf{B}/\mu_0$,

$$-\frac{\mathbf{E}\cdot(\nabla \times \mathbf{B})}{\mu_0} + \epsilon_0 \mathbf{E}\cdot\frac{\partial \mathbf{E}}{\partial t} = -\mathbf{j}^D\cdot\mathbf{E}, \tag{8.7.4}$$

$$\frac{\mathbf{B}\cdot(\nabla \times \mathbf{E})}{\mu_0} + \frac{1}{\mu_0}\mathbf{B}\cdot\frac{\partial \mathbf{B}}{\partial t} = 0. \tag{8.7.5}$$

Adding these equations and rearranging the result we find *Poynting's theorem*,

$$\frac{\partial}{\partial t}\left(\frac{\epsilon_0 \mathbf{E}^2}{2} + \frac{\mathbf{B}^2}{2\mu_0}\right) + \nabla\cdot\left(\frac{\mathbf{E}\times\mathbf{B}}{\mu_0}\right) = -\mathbf{j}^D\cdot\mathbf{E}. \tag{8.7.6}$$

The quantity in the time derivative is the *vacuum electromagnetic field energy density*, the quantity in the divergence is the energy flux or *Poynting vector* for the vacuum fields **E** and **B**, and the quantity on the right is the work done on the matter.

Equations (8.7.2), (8.7.3) (for each $\nu=1,2,\ldots,N-1$), and (8.7.6) are now added to obtain

$$\frac{\partial}{\partial t}\left[\frac{\rho\dot{\mathbf{x}}^2}{2}+\sum_{\nu}\frac{\rho^{\nu}\dot{\mathbf{y}}^{\nu 2}}{2}+\frac{\epsilon_0\mathbf{E}^2}{2}+\frac{\mathbf{B}^2}{2\mu_0}\right]+\frac{\partial}{\partial z_j}\left[\left(\frac{\rho\dot{\mathbf{x}}^2}{2}\delta_{ij}+\sum_{\nu}\frac{\rho^{\nu}\dot{\mathbf{y}}^{\nu 2}}{2}\delta_{ij}-t_{ij}^{E}\right)\dot{x}_i\right.$$
$$\left.+\frac{(\mathbf{E}\times\mathbf{B})_j}{\mu_0}\right]+t_{ij}^{y}\dot{x}_{i,j}+\frac{1}{J}\sum_{\nu}\frac{\partial\rho^0\Sigma}{\partial y_i^{T\nu}}\dot{y}_i^{\nu}=W \tag{8.7.7}$$

where the quantity W with the use of the definition of q^D, (4.7.1), and of $\mathbf{j}^D$, (4.7.2), can be reexpressed as

$$W=-\mathscr{E}_iP_j\dot{x}_{i,j}-(\mathbf{E}\cdot\dot{\mathbf{x}})(\nabla\cdot\mathbf{P})+\left(\frac{\partial\mathbf{P}}{\partial t}\times\mathbf{B}\right)\cdot\dot{\mathbf{x}}+\{[\nabla\times(\mathbf{P}\times\dot{\mathbf{x}})]\times\mathbf{B}\}\cdot\dot{\mathbf{x}}$$
$$+\frac{\partial\mathbf{P}}{\partial t}\cdot\mathscr{E}+[\nabla\times(\mathbf{P}\times\dot{\mathbf{x}})]\cdot\mathscr{E}+(P_j\dot{x}_i)_{,j}\mathscr{E}_i-\frac{\partial\mathbf{P}}{\partial t}\cdot\mathbf{E}-\mathbf{E}\cdot[\nabla\times(\mathbf{P}\times\dot{\mathbf{x}})]$$
$$=(\nabla\cdot\mathbf{P})\dot{\mathbf{x}}\cdot(\dot{\mathbf{x}}\times\mathbf{B})+\frac{\partial\mathbf{P}}{\partial t}\cdot(\dot{\mathbf{x}}\times\mathbf{B})-\left(\frac{\partial\mathbf{P}}{\partial t}\times\dot{\mathbf{x}}\right)\cdot\mathbf{B}$$
$$+[\nabla\times(\mathbf{P}\times\dot{\mathbf{x}})]\cdot(\dot{\mathbf{x}}\times\mathbf{B})+[\nabla\times(\mathbf{P}\times\dot{\mathbf{x}})]\cdot(\mathbf{B}\times\dot{\mathbf{x}})$$
$$=0. \tag{8.7.8}$$

It is also possible to reexpress

$$t_{ij}^{y}\dot{x}_{i,j}+\frac{1}{J}\sum_{\nu}\frac{\partial\rho^0\Sigma}{\partial y_i^{T\nu}}\dot{y}_i^{\nu}=\frac{1}{J}\left[\frac{\partial\rho^0\Sigma}{\partial x_{i,A}}\dot{x}_{i,j}x_{j,A}+\sum_{\nu}\frac{\partial\rho^0\Sigma}{\partial y_i^{T\nu}}\dot{y}_i^{\nu}\right]$$
$$=\frac{1}{J}\left[\frac{\partial\rho^0\Sigma}{\partial x_{i,A}}\dot{x}_{i,A}+\sum_{\nu}\frac{\partial\rho^0\Sigma}{\partial y_i^{T\nu}}\dot{y}_i^{\nu}\right]$$
$$=\frac{1}{J}\frac{d}{dt}(\rho^0\Sigma)$$
$$=\frac{1}{J}\frac{d}{dt}(J\rho\Sigma)$$
$$=\frac{\partial}{\partial t}(\rho\Sigma)+\frac{\partial}{\partial z_j}(\rho\Sigma\dot{x}_j) \tag{8.7.9}$$

where in the last step the spatial equation of continuity is used once again.

With this result (8.7.7) becomes

$$\frac{\partial}{\partial t}\left[\frac{\rho\dot{\mathbf{x}}^2}{2}+\sum_\nu\frac{\rho^\nu\dot{\mathbf{y}}^{\nu 2}}{2}+\rho\Sigma+\frac{\epsilon_0\mathbf{E}^2}{2}+\frac{\mathbf{B}^2}{2\mu_0}\right]+\frac{\partial}{\partial z_j}\left[\left(\frac{\rho\dot{\mathbf{x}}^2}{2}\delta_{ij}+\sum_\nu\frac{\rho^\nu\dot{\mathbf{y}}^{\nu 2}}{2}\delta_{ij}+\rho\Sigma\delta_{ij}-t_{ij}^E\right)\dot{x}_i+\frac{(\mathbf{E}\times\mathbf{B})_j}{\mu_0}\right]=0, \tag{8.7.10}$$

which is *conservation of energy* expressed in the spatial frame.

It appears at first glance that the flux term in this equation is sensitive to a constant term added to Σ. However, by noting that the two terms involving Σ in (8.7.10) can be expressed in terms of a total time derivative of $\rho^0\Sigma$, as done in (8.7.9), it is seen that a constant term in Σ has no effect on the energy conservation law.

We identify the total *energy density* H as

$$H\equiv\frac{\rho\dot{\mathbf{x}}^2}{2}+\sum_\nu\frac{\rho^\nu\dot{\mathbf{y}}^{\nu 2}}{2}+\rho\Sigma+\frac{\epsilon_0\mathbf{E}^2}{2}+\frac{\mathbf{B}^2}{2\mu_0} \tag{8.7.11}$$

and the energy flux $\mathbf{S}$ as

$$S_j\equiv\left(\frac{\rho\dot{\mathbf{x}}^2}{2}\delta_{ij}+\sum_\nu\frac{\rho^\nu\dot{\mathbf{y}}^{\nu 2}}{2}\delta_{ij}+\rho\Sigma\delta_{ij}-t_{ij}^E\right)\dot{x}_i+\frac{(\mathbf{E}\times\mathbf{B})_j}{\mu_0}. \tag{8.7.12}$$

The identification of the latter as the energy flux uses arguments analogous to those used to identify the total stress tensor in Section 8.3. In brief they are: (1) the energy flux $\mathbf{S}$ appears in the divergence term of the energy conservation law,

$$\frac{\partial H}{\partial t}+\frac{\partial S_j}{\partial z_j}=0; \tag{8.7.13}$$

(2) it appears in the boundary condition for the energy flux for surfaces *fixed* in the spatial frame,

$$(\mathbf{S}^o-\mathbf{S}^i)\cdot\mathbf{n}=0, \tag{8.7.14}$$

a condition found by arguments analogous to those leading to the total stress boundary condition (8.3.7); (3) for a vacuum **S** reduces to

$$\mathbf{S}=\frac{1}{\mu_0}(\mathbf{E}\times\mathbf{B}), \tag{8.7.15}$$

the well known vacuum form of the Poynting vector, which is thus an acceptable reference-state value. The energy density consists of the kinetic, potential, and vacuum electromagnetic field energy densities. Note that the interaction energy that entered the Lagrangian does not enter the energy density. The energy flux consists of the transport of kinetic and potential energies, the transport of stress, and electromagnetic energy flux.

The boundary condition on the energy flux for a moving deforming *material surface* follows from the energy conservation law (8.7.13) by arguments analogous to those of Section 8.4. The result is

$$\left[(\mathbf{S}-H\dot{\mathbf{x}})^{o}-(\mathbf{S}-H\dot{\mathbf{x}})^{i}\right]\cdot\mathbf{n}=0. \tag{8.7.16}$$

Just as in the case of the boundary condition on angular momentum flux, this is an important physical statement but one that usually gives no aid in problem solving.

8.8 Group Velocity

We prove in this section that the *energy propagation velocity* (or *ray velocity*) of a wave $\mathbf{v}_r$ in a lossless medium is equal to the *group velocity* of the wave $\mathbf{v}_g$ defined by

$$\mathbf{v}_g\equiv\nabla_k\omega \tag{8.8.1}$$

where ω is the angular frequency and the gradient is taken with respect to the components of the wavevector **k**. In component form the definition is

$$(v_g)_i\equiv\frac{\partial\omega}{\partial k_i}. \tag{8.8.2}$$

The term group velocity stems from the fact that it is the velocity of planes of constant amplitude in a group of waves of slightly differing frequencies. The definition of the group velocity has a simple geometrical meaning since the gradient of a surface is a vector normal to the surface. Figure 8.1 illustrates this. The velocity equality proof of this section applies to any

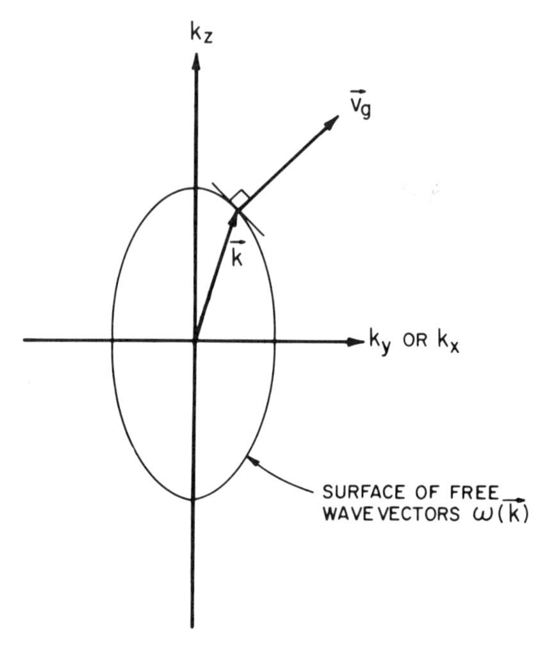

Fig. 8.1. Geometrical interpretation of group velocity. The group velocity $\mathbf{v}_g$ is normal to the $\omega(\mathbf{k})=\Omega$ surface for a specified frequency Ω.

plane periodic wave in a system that conserves energy, momentum, and crystal momentum. The system may be nonlinear and the wave may have finite amplitude. Our proof is patterned after Lighthill [1965].

In order to indicate the generality of the argument we denote the fields on which the Lagrangian density depends as $\psi^\alpha(\alpha=1,2,\ldots M)$ as is done in Section 5.1. We do not, however, use the four-dimensional notation used there. The first derivatives of ψ^α, on which the Lagrangian density also depends, are denoted by $\partial\psi^\alpha/\partial t$ and $\psi^\alpha_{,k}\equiv\partial\psi^\alpha/\partial z_k$. Thus we have

$$\mathcal{L}=\mathcal{L}\left(\psi^\alpha,\frac{\partial\psi^\alpha}{\partial t},\psi^\alpha_{,k}\right), \tag{8.8.3}$$

there being no dependence on t because the system conserves energy (see Section 5.5) or on $\mathbf{z}$ because the system conserves momentum (see Section 5.4) or on $\mathbf{X}$ because the system conserves crystal momentum (see Section 5.7). The following proof proceeds analogously if the material coordinate is taken as the independent space coordinate.

We found in Section 5.5 that the energy conservation law for such a system is expressed as

$$\frac{\partial H}{\partial t}+\frac{\partial S_j}{\partial z_j}=0 \tag{8.8.4}$$

where in the present notation the energy density H and the energy flow vector $\mathbf{S}$ are

$$H=\frac{\partial \mathfrak{L}}{\partial(\partial\psi^\alpha/\partial t)}\frac{\partial\psi^\alpha}{\partial t}-\mathfrak{L}, \tag{8.8.5}$$

$$S_j=\frac{\partial \mathfrak{L}}{\partial\psi^\alpha_{,j}}\frac{\partial\psi^\alpha}{\partial t}, \tag{8.8.6}$$

and summation over α is implied (see Prob. 5.4 for analogous quantities in the material description). The *energy propagation velocity* (*ray velocity*) $\mathbf{v}_r$ for periodic plane waves is defined as

$$\mathbf{v}_r\equiv\frac{\langle\mathbf{S}\rangle}{\langle H\rangle} \tag{8.8.7}$$

where $\langle\ \rangle$ denotes an average taken over one period T in time (or one wavelength in space),

$$\langle S_j\rangle\equiv\int_{t-T/2}^{t+T/2}S_j\,dt'. \tag{8.8.8}$$

Clearly, time integration and space differentiation commute,

$$\left\langle\frac{\partial S_j}{\partial z_j}\right\rangle=\frac{\partial}{\partial z_j}\langle S_j\rangle. \tag{8.8.9}$$

We also find that

$$\begin{aligned}\left\langle\frac{\partial H}{\partial t}\right\rangle&=\int_{t-T/2}^{t+T/2}\frac{\partial H}{\partial t}\,dt'\\&=H\left(t+\frac{T}{2}\right)-H\left(t-\frac{T}{2}\right)\\&=\frac{\partial}{\partial t}\int_{t-T/2}^{t+T/2}H\,dt'\\&=\frac{\partial}{\partial t}\langle H\rangle.\end{aligned} \tag{8.8.10}$$

Thus we can average the energy conservation law and obtain

$$\frac{\partial}{\partial t}\langle H\rangle+\frac{\partial}{\partial z_j}\langle S_j\rangle=0 \tag{8.8.11}$$

and with the use of the ray velocity definition obtain

$$\frac{\partial}{\partial t}\langle H\rangle+\nabla\cdot(\mathbf{v}_r\langle H\rangle)=0. \tag{8.8.12}$$

This equation shows clearly that the mean energy density $\langle H\rangle$ (a localized quantity) is transported by the energy propagation velocity $\mathbf{v}_r$ in a manner analogous to the transport of mass density ρ by the material velocity $\dot{\mathbf{x}}$ in the mass conservation law (3.9.7).

Next we need a special form of Hamilton's principle applicable to plane periodic waves for use in the proof below. In this special form,

$$\delta\int_{t_1}^{t_2}\int_v \mathcal{L}\left(\psi^\alpha,\frac{\partial\psi^\alpha}{\partial t},\partial\psi^\alpha_{,k}\right)dv\,dt=0, \tag{8.8.13}$$

the interval of time integration t_2-t_1, is taken as an integral multiple (one, for instance) of the wave period T and the volume v is taken as a rectangular parallelepiped with four surfaces perpendicular to the wave fronts and two surfaces parallel to the wave fronts, the latter pair being separated by an integral number of wavelengths. Each field ψ^α, its derivatives $\partial\psi^\alpha/\partial t$ and $\psi^\alpha_{,k}$, and consequently $\mathcal{L}$ are functions only of $\mathbf{k}\cdot\mathbf{z}-\omega t$. The variation δ is taken holding the frequency and wavevector of the wave fixed. Hamilton's principle for this case then gives

$$\begin{aligned}
\delta\int_{t-T/2}^{t+T/2}\int_v \mathcal{L}\,dv\,dt
&=\int_{t-T/2}^{t+T/2}\int_v\left[\frac{\partial\mathcal{L}}{\partial\psi^\alpha}\delta\psi^\alpha+\frac{\partial\mathcal{L}}{\partial(\partial\psi^\alpha/\partial t)}\delta\left(\frac{\partial\psi^\alpha}{\partial t}\right)+\frac{\partial\mathcal{L}}{\partial\psi^\alpha_{,k}}\delta\psi^\alpha_{,k}\right]dv\,dt\\
&=\int_{t-T/2}^{t+T/2}\int_v\left[\frac{\partial\mathcal{L}}{\partial\psi^\alpha}-\frac{d}{dt}\left(\frac{\partial\mathcal{L}}{\partial(\partial\psi^\alpha/\partial t)}\right)-\frac{d}{dz_k}\left(\frac{\partial\mathcal{L}}{\partial\psi^\alpha_{,k}}\right)\right]\delta\psi^\alpha\,dv\,dt\\
&\quad+\int_v\int_{t-T/2}^{t+T/2}\frac{d}{dt}\left(\frac{\partial\mathcal{L}}{\partial(\partial\psi^\alpha/\partial t)}\delta\psi^\alpha\right)dt\,dv\\
&\quad+\int_{t-T/2}^{t+T/2}\int_v\frac{d}{dz_k}\left(\frac{\partial\mathcal{L}}{\partial\psi^\alpha_{,k}}\delta\psi^\alpha\right)dv\,dt=0.
\end{aligned} \tag{8.8.14}$$

The first integral vanishes because the bracketed terms are just the Lagrange equations (2.1.7); the second integral vanishes because the time integral is over a period; the third integral vanishes because it can be converted by Gauss' theorem to a surface integral whose contributions from opposite faces cancel. This cancellation follows from the shape and positioning of the volume v and the dependence of the integrand on only $\mathbf{k}\cdot\mathbf{z}-\omega t$.

Now consider a variation δ which, in addition to varying the fields as in (8.8.14), also varies the frequency ω and wavevector $\mathbf{k}$ of the wave. For a plane periodic wave the fields ψ^α must be given by

$$\psi^\alpha = f^\alpha(\mathbf{k}\cdot\mathbf{z}-\omega t). \tag{8.8.15}$$

Their derivatives are thus

$$\frac{\partial \psi^\alpha}{\partial t} = -\omega f^{\alpha'}, \qquad \psi^\alpha_{,j} = k_j f^{\alpha'} \tag{8.8.16}$$

where the prime denotes differentiation. The variation of the averaged Lagrangian density now becomes

$$\begin{aligned}
\delta\langle \mathfrak{L}\rangle &= \delta\int_{t-T/2}^{t+T/2} \mathfrak{L}\,dt \\
&= \int_{t-T/2}^{t+T/2}\left[\frac{\partial \mathfrak{L}}{\partial \psi^\alpha}\,\delta f^\alpha - \frac{\partial \mathfrak{L}}{\partial(\partial\psi^\alpha/\partial t)}\,\omega\,\delta f^{\alpha'} + \frac{\partial \mathfrak{L}}{\partial \psi^\alpha_{,j}}\,k_j\,\delta f^{\alpha'}\right]dt \\
&\quad - \delta\omega\int_{t-T/2}^{t+T/2}\frac{\partial \mathfrak{L}}{\partial(\partial\psi^\alpha/\partial t)} f^{\alpha'}\,dt + \delta k_j\int_{t-T/2}^{t+T/2}\frac{\partial \mathfrak{L}}{\partial \psi^\alpha_{,j}} f^{\alpha'}\,dt
\end{aligned} \tag{8.8.17}$$

where we assume that the variations δf^α retain the period T of the functions f^α. Since the first integral on the right side is just the variation holding ω and $\mathbf{k}$ fixed, it can be put into the form (8.8.14) which vanishes by Hamilton's principle. Note, however, that we do not assume that the more general variation $\delta\langle \mathfrak{L}\rangle$ of the last equation vanishes. Multiplying (8.8.17) by ω now gives

$$\omega\,\delta\langle \mathfrak{L}\rangle = \left\langle \frac{\partial \mathfrak{L}}{\partial(\partial\psi^\alpha/\partial t)}\,\frac{\partial \psi^\alpha}{\partial t}\right\rangle\delta\omega - \left\langle \frac{\partial \mathfrak{L}}{\partial \psi^\alpha_{,j}}\,\frac{\partial \psi^\alpha}{\partial t}\right\rangle \delta k_j. \tag{8.8.18}$$

We now observe that, if $\langle \mathfrak{L}\rangle/\omega$ is held constant during the variation, we have

$$\omega\delta\langle \mathfrak{L} \rangle - \langle \mathfrak{L} \rangle \delta\omega = 0. \tag{8.8.19}$$

When combined with the previous equation, this yields

$$\langle H \rangle \delta\omega = \langle \mathbf{S} \rangle \cdot \delta\mathbf{k} \tag{8.8.20}$$

with the use of the definitions of $\langle H \rangle$ and $\langle \mathbf{S} \rangle$. We now only need to introduce the definitions of the group velocity $\mathbf{v}_g$ and the ray velocity $\mathbf{v}_r$ to obtain

$$\mathbf{v}_g \equiv \nabla_k \omega \Big|_{\frac{\langle \mathfrak{L} \rangle}{\omega}} = \frac{\langle \mathbf{S} \rangle}{\langle H \rangle} \equiv \mathbf{v}_r, \tag{8.8.21}$$

the theorem we wished to prove. The subscript $\langle \mathfrak{L} \rangle / \omega$ on the group velocity reminds us that this quantity (as well as the other components of $\mathbf{k}$) must be held constant during the differentiation.

Holding $\langle \mathfrak{L} \rangle / \omega$ constant during differentiation in (8.8.21) does not make that definition inconsistent with the usual definition of group velocity for linear waves, where no such condition is included, since for linear waves

$$\langle \mathfrak{L} \rangle = 0. \tag{8.8.22}$$

This is well known to apply to purely mechanical waves for which it results from the equality of the average kinetic and potential energies. It is also well known to be obeyed by purely electromagnetic waves for which it is a consequence of the equality of the average electric and magnetic energies. The condition (8.8.22) may be proved for any linear wave as follows. Waves having no nonlinearity must arise from a Lagrangian density *homogeneous* of the *second degree* in all of its dependent fields. Thus if the variation of each of the fields ψ^α produces the field $(1+\epsilon)\psi^\alpha$, then the variation of the averaged Lagrangian becomes

$$\begin{aligned} \delta\langle \mathfrak{L} \rangle &= (1+\epsilon)^2 \langle \mathfrak{L} \rangle - \langle \mathfrak{L} \rangle \\ &\cong 2\epsilon \langle \mathfrak{L} \rangle. \end{aligned} \tag{8.8.23}$$

Since (8.8.14) implies

$$\delta\langle \mathfrak{L} \rangle = 0, \tag{8.8.24}$$

and ϵ is arbitrary, we conclude that (8.8.22) must hold for all linear waves.

The importance of holding $\langle \mathfrak{L} \rangle / \omega$ fixed during differentiation in the group velocity definition lies in considering *nonlinear periodic waves*. The relation of ω and $\mathbf{k}$ for such waves must also involve some measure of the strength of the wave, such as its amplitude. The derivation above shows that the measure of strength that should be held constant during differentiation is $\langle \mathfrak{L} \rangle / \omega$.

Since the variations in frequency and wavevector in the derivation above are small quantities, the proof applies only to wave forms where the spread in these quantities is small. For a large spread more complicated formulas would be necessary to account for energy propagation. It should also be noted that there are other velocities, called *characteristic velocities*, that play important roles in nonlinear wave propagation and have also been regarded as the nonlinear generalization of the linear group velocity [Whitham, 1974].

PROBLEMS

8.1. Derive the internal motion equations in the spatial description,

$$\frac{\partial}{\partial t}\left[\rho^\nu\left(\frac{\partial y_i^\nu}{\partial t} - y_{i,j}^\nu x_{j,L}\frac{\partial X_L}{\partial t}\right)\right] + \frac{\partial}{\partial z_j}\left[\rho^\nu\left(\frac{\partial y_i^\nu}{\partial t} - y_{i,k}^\nu x_{k,L}\frac{\partial X_L}{\partial t}\right)\right.$$
$$\left.\cdot\left(-x_{j,K}\frac{\partial X_K}{\partial t}\right)\right] = \frac{q^\nu}{J}\left(E_i - \epsilon_{ijk}x_{j,L}\frac{\partial X_L}{\partial t}B_k\right) - \frac{1}{J}\frac{\partial \rho^0\Sigma}{\partial y_i^{T\nu}}, \qquad (8.P.1)$$

from the Lagrangian given by (4.7.8), (4.7.16), (4.7.18), and (6.5.7). Regard $x_{j,L}$ in (8.P.1) as shorthand for the inverse of $X_{L,j}$.

8.2. Transform the equation of Prob. 8.1 to the material description and so obtain (7.3.3). Note the more compact form of the equation in the material description.

8.3. Find the center-of-mass equation from the Lagrangian in the spatial description (see Prob. 8.1). *Hints:* Put all terms on the same side of the equation and show (*a*) the terms arising from the center-of-mass kinetic energy and the stored energy are

$$\frac{\partial}{\partial t}\left(\rho C_{JL}\frac{\partial X_L}{\partial t}\right) + \frac{\partial}{\partial z_i}\left[\left(\frac{\rho}{2}C_{KL}\frac{\partial X_K}{\partial t}\frac{\partial X_L}{\partial t} - \rho\Sigma\right)x_{i,J}\right.$$
$$\left. - \rho\frac{\partial X_K}{\partial t}\frac{\partial X_L}{\partial t}C_{JL}x_{i,K} - \rho\frac{\partial \Sigma}{\partial X_{J,i}}\right], \qquad (8.P.2)$$

(b) the terms from the internal motion kinetic energy are

$$-\frac{\partial}{\partial t}\left[\sum_{\nu}\rho^{\nu}\left(\frac{\partial y_l^{\nu}}{\partial t}-y_{l,j}^{\nu}x_{j,L}\frac{\partial X_L}{\partial t}\right)y_{l,k}^{\nu}x_{k,J}\right]$$

$$-\frac{\partial}{\partial z_i}\left[\sum_{\nu}\rho^{\nu}\left(\frac{\partial y_l^{\nu}}{\partial t}-y_{l,j}^{\nu}x_{j,L}\frac{\partial X_L}{\partial t}\right)y_{l,k}^{\nu}x_{k,J}x_{i,M}\frac{\partial X_M}{\partial t}\right.$$

$$\left.+\sum_{\nu}\frac{\rho^{\nu}}{2}\left(\frac{\partial y_l^{\nu}}{\partial t}-y_{l,j}^{\nu}x_{j,L}\frac{\partial X_L}{\partial t}\right)^2 x_{i,J}\right], \tag{8.P.3}$$

and (c) the terms from the interaction Lagrangian are

$$-\frac{\partial P_i}{\partial t}\epsilon_{ijk}x_{j,J}B_k+P_i\epsilon_{ijk}x_{j,J}\epsilon_{klm}E_{m,l}+x_{m,Jm}P_i\,\mathcal{E}_i$$

$$+x_{m,J}P_{i,m}\mathcal{E}_i+x_{m,J}P_i\left(E_{i,m}-\epsilon_{ijk}x_{j,K}\frac{\partial X_K}{\partial t}B_{k,m}\right)$$

$$+P_{i,m}\epsilon_{ijk}x_{j,J}B_k x_{m,K}\frac{\partial X_K}{\partial t}$$

$$+P_i\epsilon_{ijk}x_{j,J}B_{k,m}x_{m,K}\frac{\partial X_K}{\partial t}-P_i\epsilon_{ijk}x_{j,J}B_k\dot{x}_{m,m}. \tag{8.P.4}$$

8.4. Show that the sum of terms in (8.P.2) to (8.P.4) can be reexpressed in the form of (7.5.11). *Hint:* Express time derivatives as material time derivatives and show that (a) the terms in (8.P.2) may be converted to

$$\rho\ddot{x}_l-t_{li,i}^{y}+\rho\sum_{\nu}\frac{\partial\Sigma}{\partial y_j^{T\nu}}y_{j,l}^{\nu}, \tag{8.P.5}$$

(b) the terms in (8.P.3) may be converted to

$$-P_i\mathcal{E}_i x_{k,K}X_{K,lk}+P_{i,l}\mathcal{E}_i-\rho\sum_{\nu}\frac{\partial\Sigma}{\partial y_i^{T\nu}}y_{i,l}^{\nu}, \tag{8.P.6}$$

(c) the terms in (8.P.4) may be converted to

$$-(P_j\mathcal{E}_l)_{,j}-q^D E_l-\epsilon_{ljk}j_j^D B_k-P_{i,l}\mathcal{E}_i+P_i\mathcal{E}_i x_{m,J}X_{J,lm}, \tag{8.P.7}$$

and (d) the sum of (8.P.5) to (8.P.7) yields (7.5.11).

CHAPTER 9

Linear Crystal Optics

The general theory is now complete (except for some reformulation in Chapter 16) and we begin in this chapter the application of it to various long wavelength phenomena. Linear optical propagation in anisotropic media is the first application. It is the simplest application because the center-of-mass variable cannot respond at optical frequencies and so may be dropped and because only a simple constitutive relation, that relating polarization to electric field, need be derived.

Our treatment [Lax and Nelson, 1971] of crystal optics differs from the standard approach by being formulated in terms of *three* eigenvectors of the wave equation. This seemingly obvious approach has not been used before, we believe, because one of the three eigenmodes is not a freely propagating wave. Their use, even when considering just the two types of *free waves*, is convenient but their use in later chapters for characterizing the three types of *driven waves* such as occur in piezoelectricity and nonlinear optics is necessary and their form is particularly convenient.

One of the essential results of this treatment is the *derivation* of all the constitutive relations, linear and nonlinear, along with the dynamical equations. The first example of this, which appears in this chapter, is the linear electric susceptibility and the closely associated dielectric tensor. It is found to depend on the basic constants of the stored energy expansion and on the charge and mass of the various particles in the unit cell. The frequency dependence is derived from the dynamical equations of the internal motions. The interaction of light with the resonances of these internal motions or optic modes [Huang, 1951] produces what is known as *polariton propagation*. This is treated in detail in a section of this chapter.

A relation, originally found by Lyddane, Sachs, and Teller [1941] for cubic crystals, equates the ratio of the values of the dielectric tensor well below an ionic resonance and that well above to a ratio of the squares of the longitudinal and transverse optic mode frequencies. We derive this relation in its simplest form and then find a generalization of it for crystals having lower symmetry and any number of ionic resonances [Lax and Nelson, 1971].

The chapter concludes with a treatment of group velocity in crystal optics [Nelson, 1977]. The separate effects that anisotropy and frequency dispersion give to the group velocity are discussed and illustrated.

9.1 Linear Electric Susceptibility

An enormous simplification of the equations of motion of the matter and the electromagnetic field found in Chapter 7 can be effected by realizing that at optical frequencies in the infrared or visible portions of the spectrum the center-of-mass variable $\mathbf{x}$ can be ignored (see Prob. 10.8). By this we mean that the displacement $\mathbf{u}=\mathbf{x}-\mathbf{X}$ is so small that it is completely negligible. This results from the inertia of the center of mass being so large and thus the acoustic resonant frequencies so low that the center of mass cannot respond to optical frequencies.

With this simplification we need only consider the four electromagnetic equations, (7.1.3), (7.1.6), (7.1.11), and (7.1.12), and the internal motion equations (7.3.3). The latter equations must be considered because the resonant frequencies of these modes of matter motion have resonances in the infrared, called *ionic resonances*, and in the ultraviolet, called *electronic resonances*.

We note that the only material property that enters the electromagnetic equations in the electric dipole approximation is the polarization, which is here given by

$$\mathbf{P}=\sum_{\nu} q^{\nu}\mathbf{y}^{\nu}. \tag{9.1.1}$$

The Jacobian in the more general definition (4.6.19) is set equal to unity, since there is no dynamic deformation occurring and we are not considering any applied static deformation. Also, since the polarization enters the electromagnetic equations only in the form of space or time derivatives, the constant or spontaneous part of $\mathbf{P}$ (which exists in pyroelectrics) can be dropped, that is, $\mathbf{y}^{\nu}$ not $\mathbf{y}^{T\nu}$ appears in (9.1.1).

The appearance of $\mathbf{y}^{\nu}$ in (9.1.1) leads us to consider the internal motion equations

$$m^{\nu}\ddot{y}_i^{\nu}=q^{\nu}\mathscr{E}_i-\frac{\partial\rho^0\Sigma}{\partial\Pi_C^{\nu}}X_{C,i} \tag{9.1.2}$$

at this point. We can simplify them for application to linear optical propagation by dropping center-of-mass dependence and by considering only linear terms. From the formulas of Section 3.11 the second material time derivative $\ddot{y}_i^\nu$ can be replaced to linear accuracy by the second spatial time derivative $\partial^2 y_i^\nu/\partial t^2$. Also, to linear accuracy $\mathscr{E}_i$ becomes $E_i^S+E_i^V$ where $\mathbf{E}^S$ is the spontaneous electric field that may be present in a pyroelectric and $\mathbf{E}^V$ is the varying part of the electric field. The displacement gradient $X_{C,i}$ becomes simply a Kronecker delta δ_{iC}. Because the center-of-mass dependence and terms that yield nonlinear effects are now gone, the stored energy becomes simply

$$\rho^0\Sigma = \sum_\nu {}^{10}M_A^\nu \Pi_A^\nu + \sum_{\nu\mu} {}^{20}M_{AB}^{\nu\mu} \Pi_A^\nu \Pi_B^\mu. \tag{9.1.3}$$

Gathering these simplifications together yields

$$m^\nu \frac{\partial^2 y_i^\nu}{\partial t^2} = q^\nu\left(E_i^S+E_i^V\right) - \left[{}^{10}M_C^\nu + 2\sum_\mu {}^{20}M_{CB}^{\nu\mu}\Pi_B^\mu\right]\delta_{iC} \tag{9.1.4}$$

for the internal motion equations. The natural-state condition (7.3.7) eliminates the two constant terms from this equation. We can then for notational simplicity drop the superscript V from the varying electric field. The rotationally invariant measure of the internal coordinates Π_B^μ can be replaced here simply by the internal coordinate $y_i^\mu\delta_{iB}$ itself. Lastly, we point out that, once a certain level of expansion between material frame and spatial frame derivatives is made, the distinction between lowercase and uppercase subscripts no longer serves any purpose and so may be dropped. Thus we get

$$m^\nu \frac{\partial^2 y_i^\nu}{\partial t^2} = q^\nu E_i - 2\sum_\mu {}^{20}M_{ib}^{\nu\mu} y_b^\mu \tag{9.1.5}$$

for the linearized internal motion equations.

Several things should be noted about (9.1.5). Most important is the lack of any space derivative appearing on an internal coordinate. This means that propagating solutions do not exist for the internal coordinates in the long wavelength limit. Time oscillating solutions obviously do exist. This makes it clear that the internal coordinates represent optic modes of atomic motion or electronic modes of the bonding electrons. Equation (9.1.5) is seen to be analogous to, but a much generalized version of, (2.3.20) (with the $\partial u/\partial y$ term dropped) of the continuum diatomic lattice model discussed in Chapter 2. Note that an internal coordinate by (9.1.5) is coupled, in general, to every other internal coordinate by way of the $^{20}\mathbf{M}$ term as well as to the long wavelength electric field by way of the charge q^ν associated with the ν internal coordinate.

The lack of any space derivatives in the internal motion equation leads to a great simplification. If a single frequency variation, $\exp - i\omega t$, of all the fields is considered, the internal motion equations,

$$-m^\nu\omega^2 y_i^\nu = q^\nu E_i - 2\sum_\mu {}^{20}M_{ib}^{\nu\mu} y_b^\mu, \tag{9.1.6}$$

may be solved for $\mathbf{y}^\nu$ and these fields then eliminated from the polarization and consequently from the electromagnetic equations with only a frequency dependence left to denote their previous presence. To perform this elimination we gather all the terms in $\mathbf{y}^\mu$ on one side of the equation,

$$\sum_\mu \left[2\,{}^{20}M_{ib}^{\nu\mu} - m^\nu\omega^2\delta^{\nu\mu}\delta_{ib}\right] y_b^\mu = q^\nu E_i. \tag{9.1.7}$$

A *mechanical admittance* $\Upsilon_{ji}^{\lambda\nu}(\omega)$ is now defined as the inverse of the coefficient of y_b^μ by

$$\sum_\nu \Upsilon_{ji}^{\lambda\nu}(\omega)\left[2\,{}^{20}M_{ib}^{\nu\mu} - m^\nu\omega^2\delta^{\nu\mu}\delta_{ib}\right] \equiv \delta^{\lambda\mu}\delta_{jb} \qquad (\lambda,\mu \neq 0). \tag{9.1.8}$$

Equation (9.1.7) can now be solved for y_j^λ with the result

$$y_j^\lambda = \sum_\nu \Upsilon_{ji}^{\lambda\nu}(\omega) q^\nu E_i. \tag{9.1.9}$$

Because the bracketed quantity in (9.1.8) is symmetric upon simultaneous interchange of i and b along with ν and μ, its inverse $\Upsilon_{ji}^{\lambda\nu}(\omega)$ also has such symmetry,

$$\Upsilon_{ji}^{\lambda\nu}(\omega) = \Upsilon_{ij}^{\nu\lambda}(\omega). \tag{9.1.10}$$

Note from the definition (9.1.8) that each $\boldsymbol{\Upsilon}$ has a frequency dependence of the form of a sum of terms like

$$\boldsymbol{\Upsilon} \sim \frac{C}{\omega_R^2 - \omega^2}, \tag{9.1.11}$$

thus giving to any quantity containing $\boldsymbol{\Upsilon}$ a resonant character at the frequencies ω_R of the internal motions (see also Prob. 9.10). Damping of the resonances which keeps $\boldsymbol{\Upsilon}$ finite is ignored throughout this treatment.

It can be seen from (9.1.6) that the internal coordinates are not normal coordinates since all of the internal coordinates enter the equation through the restoring force term. *Normal coordinates* are those coordinates whose force equations are decoupled from one another. Transformation to normal coordinates is useful when studying the resonant excitation of a particular optic mode as occurs in the Raman effect, for instance. For the great majority of applications that we study internal coordinates suffice. For other applications we introduce normal coordinates in Prob. 9.9.

Substitution of (9.1.9) into the polarization (9.1.1) now yields

$$P_i = \epsilon_0 \chi_{ij}(\omega) E_j \tag{9.1.12}$$

where

$$\chi_{ij}(\omega) \equiv \frac{1}{\epsilon_0} \sum_{\nu\mu} q^\nu \Upsilon_{ij}^{\nu\mu}(\omega) q^\mu = \chi_{ji}(\omega) \tag{9.1.13}$$

is the *linear electric susceptibility*. It represents the induced dipolar response of the internal motions of the crystal to an applied electric field. Without the internal coordinates we would have no linear electric susceptibility. It contains the resonant frequency response of the $\Upsilon(\omega)$ terms of which it is a sum. All of the constants contained in its definition come from ${}^{20}\mathbf{M}$ coefficients of the stored energy expansion and the charges and masses of the constituent particles of the unit cell. We regard these constants as determined by a comparison of the theoretical constitutive expressions (9.1.12) and (9.1.13) with an adequate number of experimental observations of the linear electric susceptibility. Note that the *interchange symmetry* (9.1.13) of the linear electric susceptibility follows from the derivation. In terms of the susceptibility the dielectric charge density and current density are given by

$$q^D = -\epsilon_0 \chi_{ij}(\omega) E_{j,i}, \tag{9.1.14}$$

$$j_i^D = \epsilon_0 \chi_{ij}(\omega) \frac{\partial E_j}{\partial t} \tag{9.1.15}$$

for linear optical phenomena at frequency ω.

9.2 Electric Field Wave Equation

The electric field wave equation is a simple consequence of the two electromagnetic equations containing curls. Form the curl of (7.1.11),

$$\nabla \times (\nabla \times \mathbf{E}) + \nabla \times \frac{\partial \mathbf{B}}{\partial t} = 0, \tag{9.2.1}$$

and the time derivative of (7.1.6),

$$\frac{\partial}{\partial t}(\nabla \times \mathbf{B}) - \mu_0 \epsilon_0 \frac{\partial^2 \mathbf{E}}{\partial t^2} = \mu_0 \frac{\partial \mathbf{j}^D}{\partial t}. \tag{9.2.2}$$

Since the time and space derivatives are both spatial frame derivatives, they commute and these two equations can be combined into

$$\nabla \times (\nabla \times \mathbf{E}) + \frac{1}{c^2}\frac{\partial^2 \mathbf{E}}{\partial t^2} = -\mu_0 \frac{\partial \mathbf{j}^D}{\partial t}, \tag{9.2.3}$$

the *electric field wave equation*. Substitution of the linear optical form of $\mathbf{j}^D$ into this equation yields

$$\nabla \times (\nabla \times \mathbf{E}) + \frac{1}{c^2}\frac{\partial^2 \boldsymbol{\kappa} \cdot \mathbf{E}}{\partial t^2} = 0 \tag{9.2.4}$$

where the *dielectric tensor*,

$$\kappa_{ij}(\omega) \equiv \delta_{ij} + \chi_{ij}(\omega) = \kappa_{ji}(\omega), \tag{9.2.5}$$

is introduced.

Consider a plane wave electric field which thus contains the factor

$$e^{i(k\mathbf{s}\cdot\mathbf{r} - \omega t)} \tag{9.2.6}$$

where $\mathbf{k} = k\mathbf{s}$ is the propagation vector or *wavevector*, k being its magnitude and $\mathbf{s}$ being a unit vector. Substitution of such an electric field into the wave equation yields

$$-k^2 \mathbf{s} \times (\mathbf{s} \times \mathbf{E}) - \frac{\omega^2}{c^2} \boldsymbol{\kappa} \cdot \mathbf{E} = 0. \tag{9.2.7}$$

We now define the *index of refraction* by

$$n \equiv \frac{kc}{\omega} = \frac{c}{v} \tag{9.2.8}$$

where v is the *phase velocity* of the wave defined by

$$v \equiv \frac{\omega}{k}. \tag{9.2.9}$$

The phase velocity is the velocity of a plane of constant phase of a sinusoidally varying wave in the direction normal to the wave front. Using this definition and expanding the vector triple product, we find

$$(\mathbf{1}-\mathbf{ss})\cdot\mathbf{E}=\frac{1}{n^2}\boldsymbol{\kappa}(\omega)\cdot\mathbf{E}, \tag{9.2.10a}$$

or in index notation,

$$(\delta_{ij}-s_is_j)E_j=\frac{1}{n^2}\kappa_{ij}(\omega)E_j. \tag{9.2.10b}$$

The index of refraction, as is evident from this equation, should be regarded as a function of $\mathbf{s}$ and ω,

$$n=n(\mathbf{s},\omega). \tag{9.2.11}$$

9.3 Eigenvector Treatment of Crystal Optics

The form of the wave equation (9.2.10) for plane waves suggests treating the equation as an eigenvector problem with $(1/n^\alpha)^2$ being the eigenvalue, $\mathfrak{E}^\alpha$ being the *electric field eigenvector*, and α presumably running over three values,

$$(\mathbf{1}-\mathbf{ss})\cdot\mathfrak{E}^\alpha=\frac{1}{(n^\alpha)^2}\boldsymbol{\kappa}(\omega)\cdot\mathfrak{E}^\alpha. \tag{9.3.1}$$

We explore this connection in this section.

First we search for an orthogonality condition. Form the scalar product of (9.3.1) with $\mathfrak{E}^\beta$,

$$\mathfrak{E}^\beta\cdot(\mathbf{1}-\mathbf{ss})\cdot\mathfrak{E}^\alpha=\frac{1}{(n^\alpha)^2}\mathfrak{E}^\beta\cdot\boldsymbol{\kappa}\cdot\mathfrak{E}^\alpha. \tag{9.3.2}$$

With α and β interchanged we also have

$$\mathfrak{E}^\alpha\cdot(\mathbf{1}-\mathbf{ss})\cdot\mathfrak{E}^\beta=\frac{1}{(n^\beta)^2}\mathfrak{E}^\alpha\cdot\boldsymbol{\kappa}\cdot\mathfrak{E}^\beta. \tag{9.3.3}$$

Subtracting these two equations and using the interchange symmetry of $\boldsymbol{\kappa}$, we obtain

$$0=\left[\frac{1}{(n^\alpha)^2}-\frac{1}{(n^\beta)^2}\right]\mathfrak{E}^\alpha\cdot\boldsymbol{\kappa}\cdot\mathfrak{E}^\beta. \tag{9.3.4}$$

If $n^\alpha \neq n^\beta$, we have

$$\mathcal{E}^\alpha \cdot \boldsymbol{\kappa} \cdot \mathcal{E}^\beta = 0, \qquad (\alpha \neq \beta). \tag{9.3.5}$$

If $n^\alpha = n^\beta$, it is still possible to choose the eigenvectors to satisfy this equation. We may choose the normalization of the eigenvectors to be

$$\mathcal{E}^\alpha \cdot \boldsymbol{\kappa} \cdot \mathcal{E}^\alpha = 1, \qquad (\alpha = 1,2,3). \tag{9.3.6}$$

Combining the last two equations yields a *weighted orthonormality condition*

$$\mathcal{E}^\alpha \cdot \boldsymbol{\kappa} \cdot \mathcal{E}^\beta = \delta^{\alpha\beta}. \tag{9.3.7}$$

If an *electric displacement eigenvector* $\mathcal{D}^\alpha$ is defined as

$$\mathcal{D}^\alpha \equiv \boldsymbol{\kappa} \cdot \mathcal{E}^\alpha, \tag{9.3.8}$$

then the weighted orthonormality condition becomes

$$\mathcal{E}^\alpha \cdot \mathcal{D}^\beta = \delta^{\alpha\beta}, \tag{9.3.9}$$

which is called a *biorthonormality condition*. We thus have a biorthogonal set of six vectors, $\mathcal{E}^\alpha$ $(\alpha = 1,2,3)$ and $\mathcal{D}^\alpha$ $(\alpha = 1,2,3)$, in which each vector is orthogonal to the vectors of the other type whose index differs from its own.

An arbitrary vector $\mathbf{V}$ can be expressed as a linear combination of the electric field eigenvectors as

$$\mathbf{V} = \sum_{\alpha=1}^{3} C_\alpha \mathcal{E}^\alpha. \tag{9.3.10}$$

With the use of the biorthonormality condition the expansion constants C_α are found to be

$$C_\alpha = \mathbf{V} \cdot \mathcal{D}^\alpha. \tag{9.3.11}$$

This allows $\mathbf{V}$ to be expressed as

$$\mathbf{V} = \mathbf{V} \cdot \sum_\alpha \mathcal{D}^\alpha \mathcal{E}^\alpha. \tag{9.3.12}$$

Since $\mathbf{V}$ is an arbitrary vector, we must have

$$\sum_\alpha \mathcal{D}_i^\alpha \mathcal{E}_j^\alpha = \delta_{ij}. \tag{9.3.13}$$

Clearly, the roles of $\mathcal{E}^\alpha$ and $\mathcal{D}^\alpha$ can be reversed in the argument above. Equation (9.3.13) is the *completeness relation* of the biorthogonal set of eigenvectors.

Note that one eigenvector and its eigenvalue can be found by examining (9.3.1) without knowing the specific form of $\boldsymbol{\kappa}(\omega)$. If $\mathcal{E}^{(3)}$ is parallel to $\mathbf{s}$, the left side vanishes. Unless $\boldsymbol{\kappa}\cdot\mathbf{s}=0$ (which can happen only at a few discrete frequencies as we see in Section 9.8), we have $(1/n^{(3)})^2=0$ or $n^{(3)}=\infty$. This is the *longitudinal eigenmode* which cannot propagate as a *free wave*. We see later that it can propagate as a *driven wave*. From (9.3.9) we then find that

$$\mathcal{D}^{(1)}\perp\mathbf{s}, \qquad \mathcal{D}^{(2)}\perp\mathbf{s}, \tag{9.3.14}$$

that is, the *electric displacement vectors of the two propagating modes propagate as transverse waves.*

Consider next the scalar product of $\mathcal{D}^\beta$ with (9.3.1),

$$\mathcal{D}^\beta\cdot(\mathbf{1}-\mathbf{ss})\cdot\mathcal{E}^\alpha=\frac{1}{(n^\alpha)^2}\mathcal{D}^\beta\cdot\mathcal{D}^\alpha, \tag{9.3.15}$$

where $\alpha\neq\beta\neq3$, 3 denoting the longitudinal mode. The left side vanishes because of conditions (9.3.9) and (9.3.14). If $\alpha\neq3$, the right side can vanish only if

$$\mathcal{D}^\beta\cdot\mathcal{D}^\alpha=0, \qquad (3\neq\alpha\neq\beta\neq3). \tag{9.3.16}$$

Thus

$$\mathcal{D}^{(1)}\perp\mathcal{D}^{(2)} \tag{9.3.17}$$

and we find that the *electric displacement vectors of the two freely propagating modes are orthogonal.* This condition is separate from the orthogonalities contained in (9.3.9). In conjunction with (9.3.14) this shows that $\mathcal{D}^{(1)}$, $\mathcal{D}^{(2)}$, and $\mathbf{s}$ form an *orthogonal triplet.*

The various relationships of the vectors in the biorthogonal set are illustrated in Fig. 9.1. In examining the figure note that $\mathcal{E}^{(3)}$, $\mathcal{E}^{(1)}$, and $\mathcal{D}^{(1)}$ are coplanar, that $\mathcal{E}^{(3)}$, $\mathcal{E}^{(2)}$, and $\mathcal{D}^{(2)}$ are coplanar, and that these two planes are perpendicular. The six orthogonalities contained in (9.3.9) are shown as solid lines forming right angles and the additional orthogonality (9.3.17) is shown as dashed lines forming a right angle. Note that neither $\mathcal{D}^{(1)}$, $\mathcal{D}^{(2)}$, $\mathcal{D}^{(3)}$ nor $\mathcal{E}^{(1)}$, $\mathcal{E}^{(2)}$, $\mathcal{E}^{(3)}$ are orthogonal triplets.

Though we always use the eigenvectors $\mathcal{E}^\alpha$ and $\mathcal{D}^\alpha$ with the normalization of (9.3.9) whenever it is necessary to consider all three eigenmodes

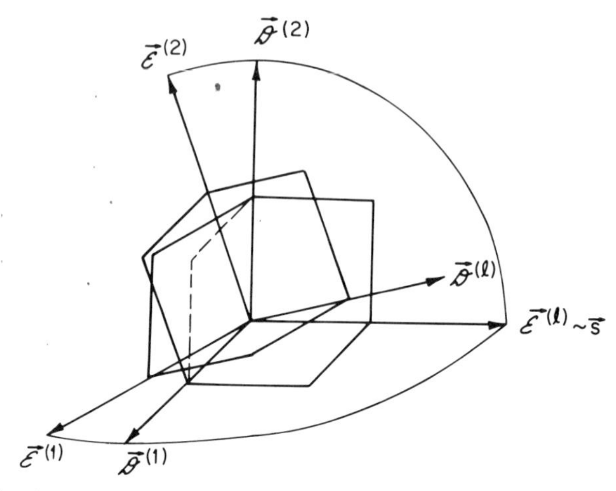

Fig. 9.1. Electric field and electric displacement field eigenvectors. Orthogonalities of eigenvectors are indicated by the lines that meet at a right angle. The six of these that represent (9.3.9) are shown with solid lines. The dotted one represents condition (9.3.17). The arcs join coplanar eigenvectors. The unit propagation vector is **s**; l denotes the longitudinal nonpropagating eigenmode.

$(\alpha=1,2,3)$, there are occasions when we consider only the two propagating modes where use of unit vectors along the directions of $\mathcal{E}^\alpha$ and $\mathcal{D}^\alpha$ has advantages. Thus we need to obtain expressions for the magnitudes of $\mathcal{E}^\alpha$ and $\mathcal{D}^\alpha$ for $\alpha=1,2$. To do this we form the scalar product of $\mathcal{E}^\alpha$ with the eigenvector equation (9.3.1),

$$(n^\alpha)^2\left[\mathcal{E}^\alpha\cdot\mathcal{E}^\alpha-(\mathbf{s}\cdot\mathcal{E}^\alpha)^2\right]=\mathcal{E}^\alpha\cdot\boldsymbol{\kappa}\cdot\mathcal{E}^\alpha=1. \qquad (9.3.18)$$

We now define the angle between $\mathcal{D}^\alpha$ and $\mathcal{E}^\alpha$ to be δ^α. Thus we have

$$|\mathcal{E}^\alpha|\cdot|\mathcal{D}^\alpha|\cos\delta^\alpha=1 \qquad (9.3.19)$$

from the normalization (9.3.9). Because of the orthogonality conditions (9.3.14), the angle between **s** and $\mathcal{E}^\alpha$ is $(\pi/2-\delta^\alpha)$ (see Fig. 9.2) and so we have for the two propagating modes

$$\mathbf{s}\cdot\mathcal{E}^\alpha=|\mathcal{E}^\alpha|\sin\delta^\alpha \qquad (\alpha=1,2). \qquad (9.3.20)$$

Combining this with (9.3.18) yields

$$|\mathcal{E}^\alpha|=\frac{1}{n^\alpha\cos\delta^\alpha} \qquad (\alpha=1,2) \qquad (9.3.21)$$

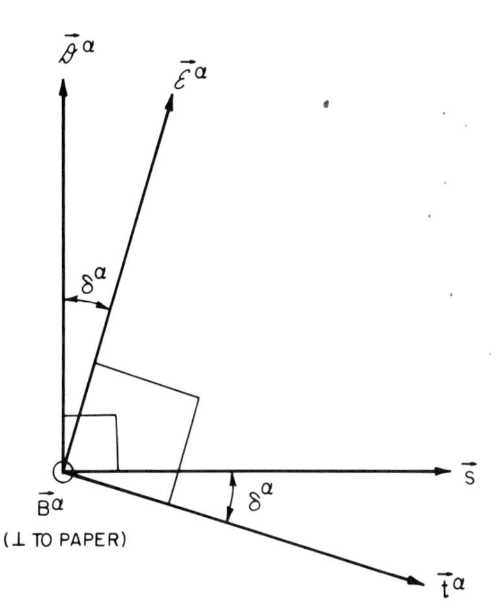

Fig. 9.2. Orientation of eigenvectors of propagating modes. The electric displacement vector $\mathfrak{D}^{\alpha}$, the electric field vector $\mathfrak{E}^{\alpha}$, the unit propagation vector **s**, and the unit group velocity vector $\mathbf{t}^{\alpha}$ (see Section 9.10) are all in a plane perpendicular to $\mathbf{B}^{\alpha}$, the magnetic induction. Note that $\mathfrak{E}^{\alpha} \perp \mathbf{t}^{\alpha}$ and that $\mathfrak{D}^{\alpha} \perp \mathbf{s}$.

and thus

$$\mathfrak{E}^{\alpha} = \frac{\mathbf{e}^{\alpha}}{n^{\alpha} \cos \delta^{\alpha}} \qquad (\alpha = 1, 2) \tag{9.3.22}$$

where $\mathbf{e}^{\alpha}$ is a unit vector. Equation (9.3.19) then gives

$$|\mathfrak{D}^{\alpha}| = n^{\alpha} \qquad (\alpha = 1, 2) \tag{9.3.23}$$

and thus

$$\mathfrak{D}^{\alpha} = n^{\alpha} \mathbf{d}^{\alpha} \qquad (\alpha = 1, 2) \tag{9.3.24}$$

where $\mathbf{d}^{\alpha}$ is a unit vector.

9.4 Dispersion Relation

The equation determining the eigenvalues in terms of **s** and $\boldsymbol{\kappa}(\omega)$ is equivalent to the dispersion relation because of the definition of n (9.2.8). It can be found by setting the determinant of the coefficients of $\mathfrak{E}$ in

(9.3.1) to zero. However, we use a different but equivalent procedure.

Rearrange (9.3.1) into

$$(n^2\mathbf{1}-\boldsymbol{\kappa})\cdot\mathcal{E}=n^2\mathbf{ss}\cdot\mathcal{E} \tag{9.4.1}$$

and refer the vector components to the principal axes of the dielectric tensor $\boldsymbol{\kappa}$. By definition the dielectric tensor is diagonal when referred to these axes. Thus

$$\mathcal{E}_i=\frac{n^2 s_i(\mathbf{s}\cdot\mathcal{E})}{n^2-\kappa_{ii}} \tag{9.4.2}$$

where no summation over repeated indices is implied. Form the scalar product with $\mathbf{s}$,

$$\mathbf{s}\cdot\mathcal{E}=n^2\sum_{i=1}^{3}\frac{(\mathbf{s}\cdot\mathcal{E})s_i s_i}{n^2-\kappa_{ii}}, \tag{9.4.3}$$

and divide by $n^2(\mathbf{s}\cdot\mathcal{E})$,

$$\frac{1}{n^2}=\sum_{i=1}^{3}\frac{s_i^2}{n^2-\kappa_{ii}}. \tag{9.4.4}$$

This is the *dispersion relation* (because $n=ck/\omega$) or one form of *Fresnel's equation of wave normals*. An alternate form is obtained by subtracting

$$\frac{1}{n^2}=\frac{1}{n^2}\sum_i s_i^2 \tag{9.4.5}$$

from (9.4.4). This yields

$$0=\sum_i s_i^2\left(\frac{1}{n^2-\kappa_{ii}}-\frac{1}{n^2}\right) \tag{9.4.6}$$

or

$$0=\sum_i\frac{s_i^2\kappa_{ii}}{n^2-\kappa_{ii}}. \tag{9.4.7}$$

Note that both this form and that of (9.4.4) contain the $n=\infty$ solution for the longitudinal eigenmode. A useful form applying only to the free wave solutions is obtained by multiplying the equation by $-n^2/c^2$ and dividing the numerator and denominator by $n^2\kappa_{ii}$ to obtain

$$0=\sum_i \frac{s_i^2}{v^2-v_i^2} \tag{9.4.8}$$

where

$$v_i \equiv \frac{c}{(\kappa_{ii})^{1/2}} \qquad (i=1,2,3) \tag{9.4.9}$$

are called the *principal velocities* (they do not form a vector) and v is the phase velocity. Written out explicitly with the denominators eliminated, (9.4.8) is

$$s_1^2(v^2-v_2^2)(v^2-v_3^2)+s_2^2(v^2-v_1^2)(v^2-v_3^2)+s_3^2(v^2-v_1^2)(v^2-v_2^2)=0 \tag{9.4.10}$$

or

$$v^4-c_1v^2+c_2=0 \tag{9.4.11}$$

where

$$c_1 \equiv s_1^2(v_2^2+v_3^2)+s_2^2(v_3^2+v_1^2)+s_3^2(v_1^2+v_2^2), \tag{9.4.12a}$$

$$c_2 \equiv s_1^2v_2^2v_3^2+s_2^2v_3^2v_1^2+s_3^2v_1^2v_2^2. \tag{9.4.12b}$$

The phase velocities for the two freely propagating modes are then

$$v=\left\{\tfrac{1}{2}\left[c_1 \pm (c_1^2-4c_2)^{1/2}\right]\right\}^{1/2}. \tag{9.4.13}$$

We discuss in Section 9.8 circumstances when real solutions of this equation do not exist.

9.5 Optically Isotropic Crystals

Cubic crystals are optically isotropic (anaxial) for *linear interactions*. This means that crystal symmetry requires all three principal values of the dielectric tensor to be equal,

$$\kappa_{11}=\kappa_{22}=\kappa_{33}\equiv\kappa. \tag{9.5.1}$$

Since this makes $c_1^2 - 4c_2 = 0$ in (9.4.13), the phase velocities of the two propagating eigenmodes are identical in all directions, that is, isotropic, and given by

$$v = \frac{c}{(\kappa)^{1/2}}, \tag{9.5.2}$$

which gives the index of refraction as

$$n = (\kappa)^{1/2}. \tag{9.5.3}$$

The phase velocity of the nonpropagating mode is zero, as always, because its refractive index is infinite.

The eigenvectors for the three modes are found as follows. The longitudinal mode, we found previously, has eigenvectors of the form

$$\mathcal{E}^{(3)} = \frac{\mathbf{s}}{N^{(3)}}, \tag{9.5.4}$$

$$\mathfrak{D}^{(3)} = \boldsymbol{\kappa} \cdot \mathcal{E}^{(3)} = \frac{\boldsymbol{\kappa} \cdot \mathbf{s}}{N^{(3)}} = \frac{\kappa \mathbf{s}}{N^{(3)}} \tag{9.5.5}$$

where $N^{(3)}$ is the normalization constant. From the normalization condition (9.3.6) we find

$$N^{(3)} = (\kappa)^{1/2}. \tag{9.5.6}$$

Let $\mathbf{e}$ be a unit vector in the direction of the electric field for one propagating mode. Then we have

$$\mathcal{E}^{(1)} = \frac{\mathbf{e}}{N^{(1)}}, \tag{9.5.7}$$

$$\mathfrak{D}^{(1)} = \frac{\boldsymbol{\kappa} \cdot \mathbf{e}}{N^{(1)}} = \frac{\kappa \mathbf{e}}{N^{(1)}}. \tag{9.5.8}$$

Since $\mathfrak{D}^{(1)} \perp \mathcal{E}^{(3)}$, we must have

$$\mathbf{e} \cdot \mathbf{s} = 0. \tag{9.5.9}$$

The normalization condition (9.3.6) gives

$$N^{(1)} = (\kappa)^{1/2} = n. \tag{9.5.10}$$

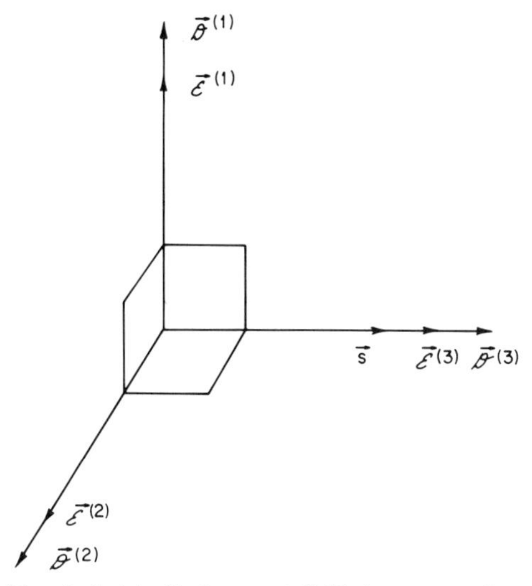

Fig. 9.3. Electric field and electric displacement field eigenvectors in an optically isotropic (cubic) crystal. The unit propagation vector **s** is also shown. The orthogonalities are indicated by lines at right angles joining the eigenvectors. 3 denotes the longitudinal nonpropagating eigenmode.

For the other propagating mode $\mathfrak{D}^{(2)} \perp \mathfrak{E}^{(1)}$ and $\mathfrak{D}^{(2)} \perp \mathfrak{E}^{(3)}$ and so

$$\mathfrak{D}^{(2)} = \frac{\mathbf{s}\times\mathbf{e}}{N^{(2)}}, \tag{9.5.11}$$

$$\mathfrak{E}^{(2)} = \frac{\boldsymbol{\kappa}^{-1}\cdot(\mathbf{s}\times\mathbf{e})}{N^{(2)}} = \frac{\mathbf{s}\times\mathbf{e}}{\kappa N^{(2)}}. \tag{9.5.12}$$

This time the normalization condition yields

$$N^{(2)} = (\kappa)^{-1/2} = n^{-1}. \tag{9.5.13}$$

We see from these equations that $\mathfrak{E}$ as well as $\mathfrak{D}$ is perpendicular to **s** for the two propagating modes in an optically isotropic medium (see Fig. 9.3).

9.6 Uniaxial Crystals

Crystal symmetry of trigonal, tetragonal, and hexagonal crystals requires an axis of symmetry (taken as the 3-direction) to exist and the principal values of the dielectric tensor for any two orthogonal axes lying in the plane perpendicular to the 3-axis to be equal,

$$\kappa_{11} = \kappa_{22}. \tag{9.6.1}$$

Such crystals are called *uniaxial crystals*. For these reasons we denote the velocities (9.4.9) by

$$v_1 = v_2 \equiv v_o, \qquad v_3 \equiv v_e \tag{9.6.2}$$

and refer the components of the unit propagation vector $\mathbf{s}$ to spherical angles θ, φ such that

$$\mathbf{s} = [\sin\theta\cos\varphi, \sin\theta\sin\varphi, \cos\theta]. \tag{9.6.3}$$

With these definitions the phase velocities of the two freely propagating modes are

$$v = \left\{ \tfrac{1}{2}\left[2v_o^2 - (v_o^2 - v_e^2)\sin^2\theta \pm (v_o^2 - v_e^2)\sin^2\theta \right] \right\}^{1/2} \tag{9.6.4}$$

which for the + sign reduces to

$$v = v_o \tag{9.6.5}$$

and for the − sign reduces to

$$v = \left\{ v_o^2\cos^2\theta + v_e^2\sin^2\theta \right\}^{1/2}. \tag{9.6.6}$$

These two solutions correspond to the *ordinary* and *extraordinary waves* respectively. The phase velocity of the ordinary wave is independent of orientation while that of the extraordinary wave depends on the polar angle θ but not on the azimuthal angle φ. This is consistent with the 3-axis being an axis of symmetry. For $\mathbf{s}$ along this axis ($\theta = 0$) the extraordinary wave velocity becomes equal to v_0, the ordinary wave velocity as seen by the vanishing of the double valued term in (9.6.4). For this reason the 3-axis is called the *optic axis*. From (9.6.5) and (9.6.6) the refractive indices are given by

$$n^o = (\kappa_{11})^{1/2}, \tag{9.6.7}$$

$$n^e(\theta) = \left(\frac{\kappa_{11}\kappa_{33}}{\kappa_{11}\sin^2\theta + \kappa_{33}\cos^2\theta} \right)^{1/2} \tag{9.6.8}$$

for the ordinary and extraordinary waves respectively. The nonpropagating longitudinal solution once again has $n^{(3)} = \infty$.

The eigenvectors for the nonpropagating solution can be found by the procedure used for optically isotropic crystals. The properly normalized vectors are

$$\mathfrak{E}^{(3)} = \frac{\mathbf{s}}{(\mathbf{s}\cdot\boldsymbol{\kappa}\cdot\mathbf{s})^{1/2}} \tag{9.6.9}$$

$$\mathfrak{D}^{(3)} = \frac{\boldsymbol{\kappa}\cdot\mathbf{s}}{(\mathbf{s}\cdot\boldsymbol{\kappa}\cdot\mathbf{s})^{1/2}}. \tag{9.6.10}$$

The eigenvectors for the ordinary wave can be found from (9.4.1). Substituting $n^2 = \kappa_{11}$ into the equation makes the left side vanish for either the $i=1$ or 2 components of the vector equation. This makes the right side zero or

$$\mathbf{s}\cdot\mathfrak{E}^{(o)} = 0, \tag{9.6.11}$$

where the index o denotes ordinary. The $i=3$ component of the vector equation gives

$$(\kappa_{11} - \kappa_{33})\mathfrak{E}_3^{(o)} = \kappa_{11} s_3 \mathbf{s}\cdot\mathfrak{E}^{(o)} = 0. \tag{9.6.12}$$

Since $\kappa_{11} - \kappa_{33} \neq 0$, we conclude that

$$\mathfrak{E}_3^{(o)} = 0 \tag{9.6.13}$$

or, if $\mathbf{c}$ is a unit vector in the 3-direction,

$$\mathbf{c}\cdot\mathfrak{E}^{(o)} = 0. \tag{9.6.14}$$

Equations (9.6.11) and (9.6.14) require $\mathfrak{E}^{(o)}$ to be of the form

$$\mathfrak{E}^{(o)} = \frac{\mathbf{s}\times\mathbf{c}}{N^{(o)}}. \tag{9.6.15}$$

Since $\mathfrak{E}^{(o)}$ has no 3-component, $\mathfrak{D}^{(o)}$ is given by

$$\mathfrak{D}^{(o)} = \boldsymbol{\kappa}\cdot\mathfrak{E}^{(o)} = (n^o)^2\mathfrak{E}^{(o)} = (n^o)^2\frac{\mathbf{s}\times\mathbf{c}}{N^{(o)}}. \tag{9.6.16}$$

Use of the normalization condition (9.3.6) gives

$$N^{(o)} = n^o\left[1 - (\mathbf{s}\cdot\mathbf{c})^2\right]^{1/2}. \tag{9.6.17}$$

To find the eigenvectors for the extraordinary wave we see from the biorthogonality condition (9.3.9) that $\mathfrak{D}^{(e)} \perp \mathfrak{E}^{(3)}$ and $\mathfrak{D}^{(e)} \perp \mathfrak{E}^{(o)}$. From the results (9.6.9) and (9.6.15) these give $\mathfrak{D}^{(e)} \perp \mathbf{s}$ and $\mathfrak{D}^{(e)} \perp (\mathbf{s} \times \mathbf{c})$. Thus we set

$$\mathfrak{D}^{(e)} = \frac{\mathbf{s} \times (\mathbf{s} \times \mathbf{c})}{N^{(e)}} \tag{9.6.18}$$

and

$$\mathcal{E}^{(e)} = \frac{\boldsymbol{\kappa}^{-1} \cdot [\mathbf{s} \times (\mathbf{s} \times \mathbf{c})]}{N^{(e)}}. \tag{9.6.19}$$

Substitution of these into the normalization condition gives

$$\begin{aligned}(N^{(e)})^2 &= [\mathbf{s} \times (\mathbf{s} \times \mathbf{c})] \cdot \boldsymbol{\kappa}^{-1} \cdot [\mathbf{s} \times (\mathbf{s} \times \mathbf{c})] \\ &= [\mathbf{s}(\mathbf{s} \cdot \mathbf{c}) - \mathbf{c}] \cdot \boldsymbol{\kappa}^{-1} \cdot [\mathbf{s}(\mathbf{s} \cdot \mathbf{c}) - \mathbf{c}] \\ &= [(s_1^2 + s_2^2)\kappa_{11}^{-1} + s_3^2 \kappa_{33}^{-1}] s_3^2 - 2 s_3^2 \kappa_{33}^{-1} + \kappa_{33}^{-1} \\ &= (1 - s_3^2)[\kappa_{11}^{-1} s_3^2 + \kappa_{33}^{-1}(1 - s_3^2)]\end{aligned} \tag{9.6.20}$$

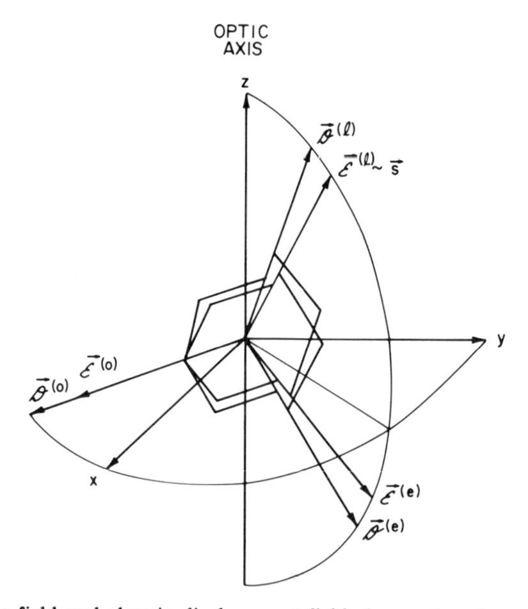

Fig. 9.4. Electric field and electric displacement field eigenvectors for an uniaxial crystal. The superscripts o, e, l denote ordinary, extraordinary, and longitudinal. Orthogonalities are indicated by lines at right angles joining the eigenvectors. The tips of coplanar vectors are joined by arcs.

and finally

$$N^{(e)} = \frac{1}{n^e(\mathbf{s})}\left[1-(\mathbf{s}\cdot\mathbf{c})^2\right]^{1/2}. \tag{9.6.21}$$

Note that, though $\mathcal{E}^{(o)} \| \mathcal{D}^{(o)}$, $\mathcal{E}^{(e)}$ and $\mathcal{D}^{(e)}$ are not parallel in an uniaxial crystal (see Fig. 9.4).

9.7 Biaxial Crystals

Crystals whose dielectric tensor has three distinct principal values are called *biaxial crystals* because, as we see presently, they have two optic axes. Crystals belonging to the orthorhombic, monoclinic, and triclinic systems are biaxial. The symmetry of orthorhombic crystals is high enough that all three principal axes of the dielectric tensor are fixed along the crystallographic axes. However, in monoclinic crystals only one axis is so fixed and in triclinic crystals none are. Principal axes that are not fixed with respect to crystallographic axes may vary in orientation with frequency because of the differing dispersion in the principal values of the dielectric tensor. By convention the principal axes are chosen so that

$$\kappa_{11} < \kappa_{22} < \kappa_{33}. \tag{9.7.1}$$

An optic axis is a direction along which the two propagating modes have the same phase velocity. From (9.4.13) we thus see that along an optic axis $c_1^2 - 4c_2$ must vanish. This quantity may be reexpressed as

$$\begin{aligned}
c_1^2 - 4c_2 &= \left[s_1^2(v_2^2+v_3^2) + s_2^2(v_3^2+v_1^2) + s_3^2(v_1^2+v_2^2)\right]^2 \\
&\quad - 4s_1^2v_2^2v_3^2 - 4s_2^2v_3^2v_1^2 - 4s_3^2v_1^2v_2^2 \\
&= \left[s_1^2(v_2^2-v_3^2) + s_2^2(v_1^2-v_3^2) + s_3^2(v_1^2-v_2^2)\right]^2 \\
&\quad - 4s_1^2s_3^2(v_2^2-v_3^2)(v_1^2-v_2^2) \\
&= \left\{s_2^2(v_1^2-v_3^2) + \left[s_3(v_1^2-v_2^2)^{1/2} + s_1(v_2^2-v_3^2)^{1/2}\right]^2\right\} \\
&\quad \cdot \left\{s_2^2(v_1^2-v_3^2) + \left[s_3(v_1^2-v_2^2)^{1/2} - s_1(v_2^2-v_3^2)^{1/2}\right]^2\right\}.
\end{aligned} \tag{9.7.2}$$

Each factor is seen to be positive definite. For later reference we note that this means that the refractive index in a general direction is either real or imaginary but not complex (in the absence of absorption) as can be seen from (9.4.13). The expression (9.7.2) can vanish for two orientations of **s**, which are the optic axes of the crystal, given by

$$s_2=0, \qquad \frac{s_1}{s_3}=\pm\left(\frac{v_1^2-v_2^2}{v_2^2-v_3^2}\right)^{1/2}. \tag{9.7.3}$$

From this we see that the two optic axes lie in the $z_2=0$ plane at angles of β and $-\beta$ to 3-axis (see Fig. 9.5) where β is given by

$$\tan\beta\equiv\left(\frac{v_1^2-v_2^2}{v_2^2-v_3^2}\right)^{1/2}. \tag{9.7.4}$$

In terms of β we can write

$$c_1^2-4c_2=(v_1^2-v_3^2)^2\left[s_2^2+(s_3\sin\beta+s_1\cos\beta)^2\right] \cdot\left[s_2^2+(s_3\sin\beta-s_1\cos\beta)^2\right]. \tag{9.7.5}$$

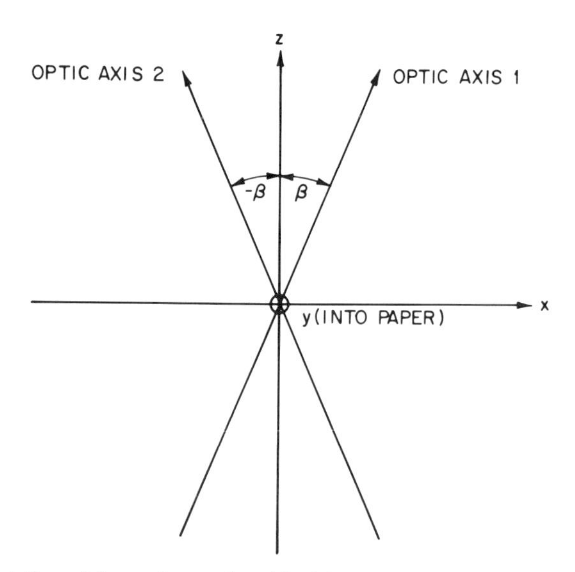

Fig. 9.5. Orientation of the optic axes in a biaxial crystal. The x,y,z axes are the principal axes of the dielectric tensor with the convention that the principal components of the dielectric tensor obey $\kappa_{11}<\kappa_{22}<\kappa_{33}$.

Since the optic axes are important geometric quantities of the problem, it is natural to attempt to relate the phase velocities to them. Define unit vectors $\mathbf{a}_1$ and $\mathbf{a}_2$ along the optic axes by

$$\mathbf{a}_1 \equiv [\sin\beta, 0, \cos\beta], \tag{9.7.6a}$$

$$\mathbf{a}_2 \equiv [-\sin\beta, 0, \cos\beta]. \tag{9.7.6b}$$

Note that the magnitudes of their vector products with $\mathbf{s}$ are

$$|\mathbf{s}\times\mathbf{a}_1| = \left[s_2^2 + (s_3\sin\beta - s_1\cos\beta)^2\right]^{1/2} \equiv \sin\theta_1, \tag{9.7.7a}$$

$$|\mathbf{s}\times\mathbf{a}_2| = \left[s_2^2 + (s_3\sin\beta + s_1\cos\beta)^2\right]^{1/2} \equiv \sin\theta_2 \tag{9.7.7b}$$

where θ_1 and θ_2 are the angles between the unit propagation vector $\mathbf{s}$ and $\mathbf{a}_1$ and $\mathbf{a}_2$ respectively (see Fig. 9.6). These now give us a convenient

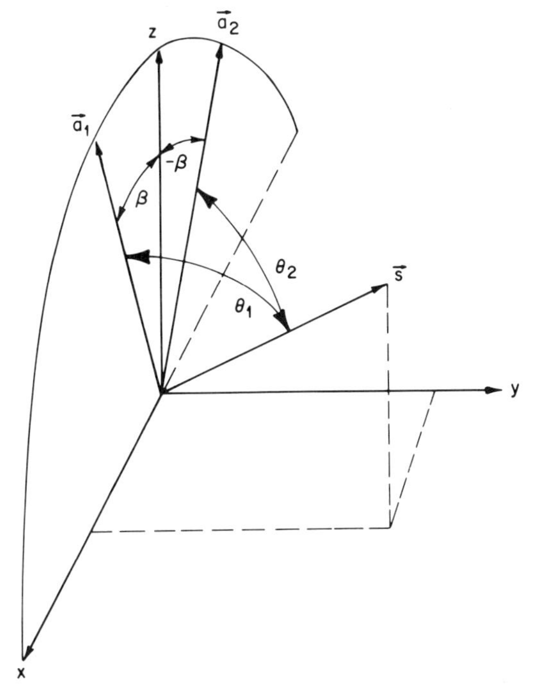

Fig. 9.6. Propagation in a biaxial crystal. The angles between the two optic axes, represented by the unit vectors $\mathbf{a}_1$ and $\mathbf{a}_2$, and the unit propagation vector $\mathbf{s}$ are labeled by θ_1 and θ_2. The x,y,z coordinate system is the principal coordinate system of the dielectric tensor.

expression,

$$c_1^2-4c_2=(v_1^2-v_3^2)^2\sin^2\theta_1\sin^2\theta_2, \tag{9.7.8}$$

for the quantity $c_1^2-4c_2$.

The scalar products of $\mathbf{a}_1$ and $\mathbf{a}_2$ with $\mathbf{s}$,

$$\mathbf{s}\cdot\mathbf{a}_1=s_1\sin\beta+s_3\cos\beta\equiv\cos\theta_1, \tag{9.7.9a}$$

$$\mathbf{s}\cdot\mathbf{a}_2=-s_1\sin\beta+s_3\cos\beta\equiv\cos\theta_2, \tag{9.7.9b}$$

give the relation

$$\cos\theta_1\cos\theta_2=s_3^2\cos^2\beta-s_1^2\sin^2\beta. \tag{9.7.10}$$

The remaining quantity in the phase velocity equation can now be expressed with the use of (9.7.3) and (9.7.4) as

$$\begin{aligned} c_1&=v_1^2(1-s_1^2)+v_2^2(s_1^2+s_3^2)+v_3^2(1-s_3^2)\\ &=v_1^2+v_3^2+(v_1^2-v_3^2)(s_3^2\cos^2\beta-s_1^2\sin^2\beta)\\ &=v_1^2+v_3^2+(v_1^2-v_3^2)\cos\theta_1\cos\theta_2. \end{aligned} \tag{9.7.11}$$

The phase velocities (9.4.13) for the two propagating modes can be written in final form as

$$v^{\pm}=\left\{\tfrac{1}{2}\left[v_1^2+v_3^2+(v_1^2-v_3^2)\cos(\theta_1\pm\theta_2)\right]\right\}^{1/2}. \tag{9.7.12}$$

The dependence on the principal velocity v_2 is implicit in the θ_1 and θ_2 functions. Since $v=c/n$, the last equation gives us the refractive indices for the two propagating modes,

$$n^{\pm}=\left[\frac{\kappa_{11}\kappa_{33}}{\kappa_{11}\sin^2\frac{1}{2}(\theta_1\pm\theta_2)+\kappa_{33}\cos^2\frac{1}{2}(\theta_1\pm\theta_2)}\right]^{1/2}. \tag{9.7.13}$$

This form has a similarity to the uniaxial case. When $\theta_1=\theta_2$, which corresponds to the two optic axes coinciding, $n^+=n^e$ and $n^-=n^o$.

The three electric field eigenvectors in a biaxial crystal can be found from (9.3.1) and (9.3.9). In components referred to the principal coordinate system (9.3.1) is

$$\left[(n^\alpha)^2(1-s_1^2)-\kappa_{11}\right]\mathcal{E}_1^\alpha+\left[-(n^\alpha)^2 s_1 s_2\right]\mathcal{E}_2^\alpha+\left[-(n^\alpha)^2 s_1 s_3\right]\mathcal{E}_3^\alpha=0, \tag{9.7.14a}$$

$$\left[-(n^\alpha)^2 s_1 s_2\right]\mathcal{E}_1^\alpha+\left[(n^\alpha)^2(1-s_2^2)-\kappa_{22}\right]\mathcal{E}_2^\alpha+\left[-(n^\alpha)^2 s_2 s_3\right]\mathcal{E}_3^\alpha=0, \tag{9.7.14b}$$

$$\left[-(n^\alpha)^2 s_1 s_3\right]\mathcal{E}_1^\alpha+\left[-(n^\alpha)^2 s_2 s_3\right]\mathcal{E}_2^\alpha+\left[(n^\alpha)^2(1-s_3^2)-\kappa_{33}\right]\mathcal{E}_3^\alpha=0. \tag{9.7.14c}$$

It is straightforward to solve these equations in conjunction with the normalization condition (9.3.9) to obtain each component ($i=1,2,3$) of each eigenvector ($\alpha=1,2,3$) to be

$$\mathcal{E}_i^\alpha=\frac{s_i}{\left[(n^\alpha)^2-\kappa_{ii}\right]N^\alpha} \tag{9.7.15}$$

where summation over repeated indices is not implied and where N^α denotes

$$N^\alpha \equiv \left[\sum_{i=1}^{3}\frac{s_i^2\kappa_{ii}}{\left[(n^\alpha)^2-\kappa_{ii}\right]^2}\right]^{1/2}. \tag{9.7.16}$$

For the longitudinal mode, for which $n^{(3)}=\infty$, the denominator of (9.7.15) becomes

$$\left[(n^{(3)})^2-\kappa_{ii}\right]N^{(3)}=\left[\sum_{i=1}^{3}s_i^2\kappa_{ii}\right]^{1/2}=\left[\mathbf{s}\cdot\boldsymbol{\kappa}\cdot\mathbf{s}\right]^{1/2}. \tag{9.7.17}$$

The electric displacement eigenvectors are

$$\mathcal{D}_i^\alpha=\frac{\kappa_{ii}s_i}{\left[(n^\alpha)^2-\kappa_{ii}\right]N^\alpha}. \tag{9.7.18}$$

This completes the evaluation of the eigenvalues and eigenvectors for the three categories of crystals—isotropic, uniaxial, and biaxial.

9.8 Polariton Dispersion

We derive in Section 9.1 the frequency dispersion of the linear electric susceptibility and consequently of the dielectric tensor from the dynamical behavior of the internal motions of the crystal. In particular we show that the internal motions possess resonant frequencies and that strong frequency dispersion is present near such frequencies. We now wish to consider what effects this strong dispersion has on the eigenmodes of linear optical propagation.

When the frequency of an electromagnetic wave is close to a resonant frequency of an internal vibration of the medium, that vibration is strongly excited, a sizable portion of the energy is transferred to the mechanical vibration, and the propagation of electromagnetic energy is slowed down. In fact, since the wave is partly a mechanical vibration and partly an electromagnetic vibration, it is usually not called an electromagnetic wave but rather a polariton wave. *Polariton* is the name given to a hybrid quantum, partly phonon and partly photon in the region near a resonance where there is substantial wavevector dispersion.

In order to keep the algebra as simple as possible while illustrating polariton dispersion effects we consider a cubic crystal with two particles per unit cell. The position coordinates $\mathbf{x}^{(1)}$ and $\mathbf{x}^{(2)}$ of particles of type 1 and 2 can be replaced with the center-of-mass position $\mathbf{x}$

$$\mathbf{x}=\frac{\rho^{(1)}\mathbf{x}^{(1)}+\rho^{(2)}\mathbf{x}^{(2)}}{\rho^{0}}, \tag{9.8.1}$$

$$\rho^{0}\equiv\rho^{(1)}+\rho^{(2)}, \tag{9.8.2}$$

and one internal coordinate $\mathbf{y}$

$$\mathbf{y}=\mathbf{x}^{(2)}-\mathbf{x}^{(1)}. \tag{9.8.3}$$

These coordinates are the vector analogs of the coordinates introduced in Section 2.3 for studying the one-dimensional diatomic lattice. The inverse transformation of (9.8.1) and (9.8.3) is

$$\mathbf{x}^{(1)}=\mathbf{x}-\frac{\rho^{(2)}}{\rho^{0}}\mathbf{y}, \tag{9.8.4}$$

$$\mathbf{x}^{(2)}=\mathbf{x}+\frac{\rho^{(1)}}{\rho^{0}}\mathbf{y}. \tag{9.8.5}$$

The transformation matrices defined in (4.6.1) and (4.6.4) are given in this case by

$$U=\begin{bmatrix} \rho^{(1)}/\rho^0 & \rho^{(2)}/\rho^0 \\ -1 & 1 \end{bmatrix}, \tag{9.8.6}$$

$$V=\begin{bmatrix} 1 & -\rho^{(2)}/\rho^0 \\ 1 & \rho^{(1)}/\rho^0 \end{bmatrix}. \tag{9.8.7}$$

From (4.6.17) we find that ρ^0 is the mass density associated with the center of mass and

$$m \equiv \frac{\rho^{(1)}\rho^{(2)}}{\rho^0} \tag{9.8.8}$$

is the mass associated with $\mathbf{y}$. It is simple to show that the matrices U and V satisfy (4.6.3), (4.6.10), (4.6.12), and (4.6.13); $\mathbf{y}$ is thus an internal coordinate. From (4.6.22) we find that the charge density associated with the center of mass is zero and that

$$q \equiv \frac{\rho^{(1)}q^{(2)} - \rho^{(2)}q^{(1)}}{\rho^0} \tag{9.8.9}$$

is the charge density associated with the internal coordinate vector. Here $q^{(1)}$ and $q^{(2)}$ are the charge densities of the sublattices of particle types 1 and 2.

In a cubic crystal the dielectric tensor $\boldsymbol{\kappa}$ has equal diagonal components and this in turn requires the linear electric susceptibility $\boldsymbol{\chi}$, the mechanical admittance $\boldsymbol{\Upsilon}$, and the material descriptor ${}^{20}\mathbf{M}$ to have equal diagonal components. Thus in this simple crystal the three components of the internal coordinate vector are normal coordinates (see Prob. 9.9) of the three optic modes. Since only one internal coordinate is involved, no superscripts are needed on $\boldsymbol{\Upsilon}$ and ${}^{20}\mathbf{M}$. The solution for the internal coordinate becomes

$$\mathbf{y} = \Upsilon(\omega) q \mathbf{E} \tag{9.8.10}$$

where

$$\Upsilon(\omega) = \frac{1}{[2^{20}M - m\omega^2]}. \tag{9.8.11}$$

The susceptibility becomes

$$\epsilon_0 \chi(\omega) = q^2 \Upsilon(\omega) \tag{9.8.12}$$

for the simple crystal considered.

The charge q can be seen to play an essential role in the coupling of an electric field to the optic mode. If crystal symmetry requires $q=0$, the associated optic mode(s) (in the case above, three) are said to be *infrared inactive*; if crystal symmetry allows $q\neq 0$, the associated optic mode(s) are said to be *infrared active*.

We consider the *transverse optic eigenmodes* first. In a cubic crystal their refractive indices are the same and given by

$$n^2=\kappa(\omega)=1+\chi(\omega). \tag{9.8.13}$$

It is convenient to reexpress the parameters in the susceptibility in terms of the dielectric constant κ_0 at zero frequency,

$$\kappa_0\equiv 1+\frac{q^2}{2\epsilon_0{}^{20}M}, \tag{9.8.14}$$

and the resonant frequency ω_T of the internal vibration (the *transverse optic mode frequency*),

$$\omega_T^2\equiv\frac{2^{20}M}{m}. \tag{9.8.15}$$

These definitions allow the refractive index to be expressed as

$$n^2=\left(\frac{kc}{\omega}\right)^2=\kappa(\omega)=1+\frac{(\kappa_0-1)\omega_T^2}{\omega_T^2-\omega^2}. \tag{9.8.16}$$

Note that the dielectric constant in this equation has the value

$$\kappa_\infty=1 \tag{9.8.17}$$

at $\omega=\infty$.

In order to understand the frequency dispersion of (9.8.16), form from it the quartic equation

$$\omega^4-\left[k^2c^2+\kappa_0\omega_T^2\right]\omega^2+k^2c^2\omega_T^2=0, \tag{9.8.18}$$

whose solution is

$$\omega=\left\{\tfrac{1}{2}\left\{k^2c^2+\kappa_0\omega_T^2\pm\left[\left(k^2c^2+\kappa_0\omega_T^2\right)^2-4k^2c^2\omega_T^2\right]^{1/2}\right\}\right\}^{1/2}. \tag{9.8.19}$$

For small k the + and − solutions behave as

$$\omega_+ = \kappa_0^{1/2}\omega_T\left[1 + \frac{k^2c^2(\kappa_0 - 1)}{2\kappa_0^2\omega_T^2}\right], \tag{9.8.20}$$

$$\omega_- = \frac{kc}{\kappa_0^{1/2}}, \tag{9.8.21}$$

the first having a constant frequency $\kappa_0^{1/2}\omega_T$ at $k=0$ and the second having a constant slope near $k=0$. The latter is the usual result for a propagating long wavelength light wave. For large k (but still small compared to Brillouin zone boundary wavevectors, since this is a long wavelength theory) we find

$$\omega_+ = kc, \tag{9.8.22}$$

$$\omega_- = \omega_T, \tag{9.8.23}$$

the first having a constant slope as expected from (9.8.17) and the second having a constant frequency ω_T. The two branches, just discussed, of the transverse optic mode are shown in Fig. 9.7. The region on each branch between the two limits found above, that is, the region where the $\omega(k)$ curves show the most curvature, is called the *polariton regime*. It is the region where the energies contained in the mechanical vibrations and electromagnetic vibrations have the same order of magnitude.

Between the two transverse optic branches there is a band of frequencies for which no real solution exists and hence no propagation can occur, a situation alluded to in Section 9.7. The existence of this *stop-band* is most easily seen by solving Eq. (9.8.18) for k^2c^2,

$$k^2c^2 = \frac{(\kappa_0\omega_T^2 - \omega^2)\omega^2}{\omega_T^2 - \omega^2}. \tag{9.8.24}$$

The right side is seen to be negative, a condition preventing a real solution for k, only when

$$\omega_T < \omega < \kappa_0^{1/2}\omega_T. \tag{9.8.25}$$

A wave whose frequency falls within this stop-band must be totally reflected from a crystal surface since there is no vibration in the medium to which it can couple.

Before considering the *longitudinal optic eigenmode* we must develop the relation between polarization and electric field implied by the Maxwell

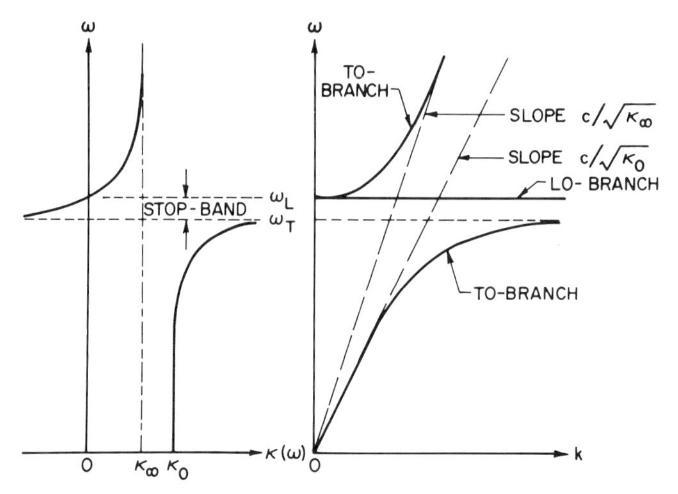

Fig. 9.7. Polariton dispersion curves in relation to dielectric constant dispersion. A resonance of the one internal vibration of a diatomic cubic crystal gives rise to a pole (damping has been neglected) in the dielectric constant κ and a nonpropagating transverse optic mode at frequency ω_T. Below this frequency a propagating transverse optic wave exists and is represented by the solid line labeled *TO*-branch in the frequency ω versus wavevector k plot. In the strongly curved part of this plot the *TO* mode is called a *polariton wave*. A nonpropagating nondispersive longitudinal optic mode, labeled *LO*-branch, is produced at the frequency ω_L where the dielectric constant vanishes. Above this frequency a second transverse optic branch exists. In its region of strong curvature this mode is also called a polariton wave.

equations. The wave equation (9.2.10a) can be rewritten as

$$\mathbf{E}-\mathbf{s}(\mathbf{s}\cdot\mathbf{E})=\frac{1}{n^2}\left(\mathbf{E}+\frac{\mathbf{P}}{\epsilon_0}\right). \tag{9.8.26}$$

From the Maxwell equation,

$$\nabla\cdot(\epsilon_0\mathbf{E}+\mathbf{P})=0, \tag{9.8.27}$$

we find for plane wave propagation

$$\mathbf{s}\cdot\mathbf{E}=-\frac{1}{\epsilon_0}\mathbf{s}\cdot\mathbf{P}. \tag{9.8.28}$$

Substituting this into (9.8.26) and solving for $\mathbf{E}$ yields

$$\mathbf{E}=\frac{1}{\epsilon_0(n^2-1)}\left[\mathbf{P}-n^2\mathbf{s}(\mathbf{s}\cdot\mathbf{P})\right], \tag{9.8.29}$$

which holds for transverse as well as longitudinal modes. For the longitudinal mode ($n=\infty$) this equation reduces to

$$\mathbf{E}=-\frac{1}{\epsilon_0}\mathbf{s}(\mathbf{s}\cdot\mathbf{P}). \tag{9.8.30}$$

From the response of the crystal to an optical electric field (9.1.12) we have for a cubic crystal

$$\begin{aligned}\mathbf{P}&=\epsilon_0\chi(\omega)\mathbf{E}\\ &=-\chi(\omega)\mathbf{s}(\mathbf{s}\cdot\mathbf{P})\end{aligned} \tag{9.8.31}$$

where the last expression applies only to the longitudinal mode. Forming the scalar product with **s** and dividing **s·P** from the equation gives

$$0=1+\chi(\omega)=\kappa(\omega)=1+\frac{(\kappa_0-1)\omega_T^2}{\omega_T^2-\omega^2} \tag{9.8.32}$$

for the longitudinal optic mode. If we denote the *longitudinal optic mode frequency*, which is the solution of this equation, by ω_L, we find

$$\omega_L=\omega_T\kappa_0^{1/2}. \tag{9.8.33}$$

There is no propagation and no dispersion in this mode. It also is shown in Fig. 9.7.

Remembering that $\kappa_\infty=1$ we can write the last equation as

$$\left(\frac{\omega_L}{\omega_T}\right)^2=\frac{\kappa_0}{\kappa_\infty}. \tag{9.8.34}$$

This is the *Lyddane–Sachs–Teller relation* first found by them for cubic crystals. Because it relates four measurable numbers, it is a useful check on them and a means for deriving one when the other three are known. We consider it further in the next section.

The first direct measurement of polariton dispersion was made by Henry and Hopfield [1965] using Raman scattering in GaP crystals. Their results are shown in Fig. 9.8.

It is instructive to fix the time-averaged energy density $\langle H\rangle$ in a propagating polariton wave (transverse optic branches) in order to see the changing partition of energy between mechanical and electromagnetic parts versus frequency. Equation (8.7.11) applied to linear optical propagation in a diatomic cubic crystal becomes

$$\langle H\rangle=\frac{m}{2}\left\langle\left(\frac{\partial\mathbf{y}}{\partial t}\right)^2\right\rangle+{}^{20}M\langle\mathbf{y}^2\rangle+\frac{\epsilon_0}{2}\langle\mathbf{E}^2\rangle+\frac{1}{2\mu_0}\langle\mathbf{B}^2\rangle \tag{9.8.35}$$

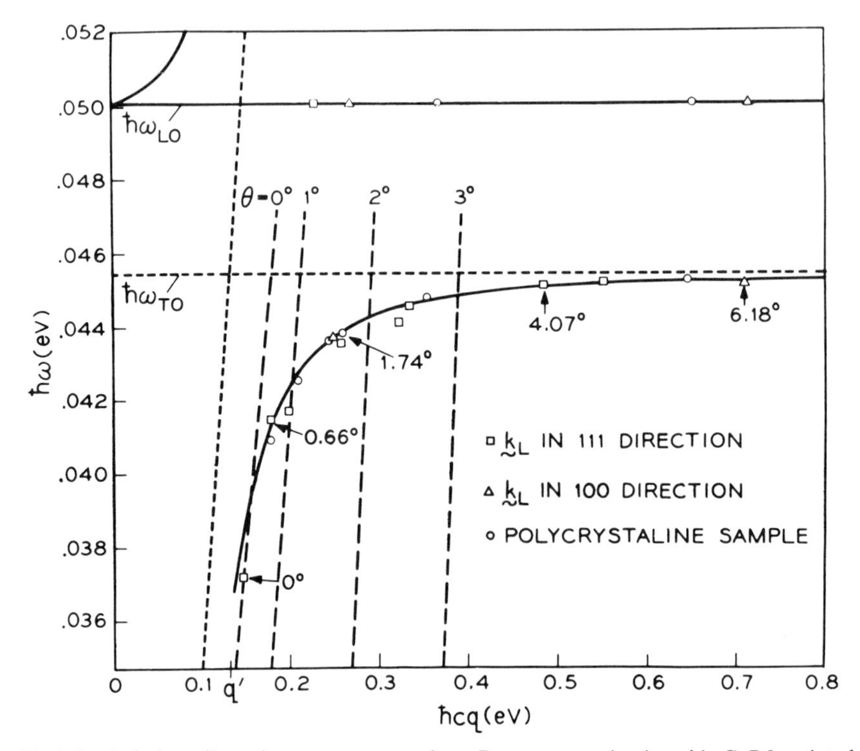

Fig. 9.8. Polariton dispersion measurements from Raman scattering in cubic GaP [reprinted by permission from Henry and Hopfield, 1965]. The dotted lines represent the uncoupled transverse optic and electromagnetic modes; the dashed curves represent values of frequencies and wavevectors kinematically possible at various scattering angles.

where the first two terms are mechanical, the third electric, and the fourth magnetic. Consider a plane wave electric field of the form

$$\mathbf{E} = A\mathcal{E}^{(t)}\cos(k\mathbf{s}\cdot\mathbf{z} - \omega t) \tag{9.8.36}$$

where A is a normalization constant and $\mathcal{E}^{(t)}$ represents either of the transverse eigenmodes of a cubic crystal. From the Maxwell equation (7.1.11) the magnetic induction

$$\mathbf{B} = \frac{k}{\omega}A\mathbf{s}\times\mathcal{E}^{(t)}\cos(k\mathbf{s}\cdot\mathbf{z} - \omega t) \tag{9.8.37}$$

is easily found. From (9.8.10) for $\mathbf{y}$ we find

$$\mathbf{y} = \epsilon_0 q\Upsilon(\omega)A\mathcal{E}^{(t)}\cos(k\mathbf{s}\cdot\mathbf{z} - \omega t). \tag{9.8.38}$$

Substitution of $\mathbf{E}$, $\mathbf{B}$, $\mathbf{y}$, and $\partial\mathbf{y}/\partial t$ into $\langle H\rangle$ now gives

$$\langle H\rangle = \left[(2^{20}M + m\omega^2)\left(\epsilon_0 q\Upsilon(\omega)\mathcal{E}^{(t)}\right)^2 + \epsilon_0(\mathcal{E}^{(t)})^2 + \frac{k^2}{\mu_0\omega^2}(\mathbf{s}\times\mathcal{E}^{(t)})^2 \right]\frac{A^2}{4}. \tag{9.8.39}$$

From Section 9.5 we have for a cubic crystal

$$(\mathcal{E}^{(t)})^2 = \frac{1}{\kappa(\omega)}, \qquad \mathbf{s}\cdot\mathcal{E}^{(t)} = 0 \tag{9.8.40}$$

and so

$$(\mathbf{s}\times\mathcal{E}^{(t)})^2 = (\mathcal{E}^{(t)})^2 - (\mathbf{s}\cdot\mathcal{E}^{(t)})^2 = (\mathcal{E}^{(t)})^2 = \frac{1}{\kappa(\omega)}. \tag{9.8.41}$$

Thus we find

$$A = \left[\frac{4\kappa(\omega)\langle H\rangle}{\epsilon_0\left(1 + \kappa(\omega) + \dfrac{(\omega_T^2 + \omega^2)\chi^2(\omega)}{\omega_T^2(\kappa_0 - 1)}\right)} \right]^{1/2}. \tag{9.8.42}$$

With the use of this expression the mechanical, electrical, and magnetic energy terms of (9.8.35) are plotted versus frequency for a fixed $\langle H\rangle$ in Fig. 9.9.

Several things should be noted in this figure. At very low frequencies the energy is partitioned among the three forms. As ω_T is approached, more energy resides in the mechanical vibrations. At ω_T all of the energy is mechanical and this point represents a purely mechanical vibration or optic phonon. Above ω_T there is a band of frequencies for which no propagation can occur. At the first frequency ω_L above the resonant frequency ω_T where a solution exists the transverse mode is nonpropagating leading to a lack of a magnetic energy even though an electrical energy is present. As the frequency is raised further the mechanical energy decreases to zero because of the increasing difficulty of exciting an oscillator above its resonant frequency. At the same time the magnetic energy increases until it is equal to the electrical energy, a condition characteristic of vacuum propagation. Since the velocity of the wave is approaching c in this simple crystal, this is expected.

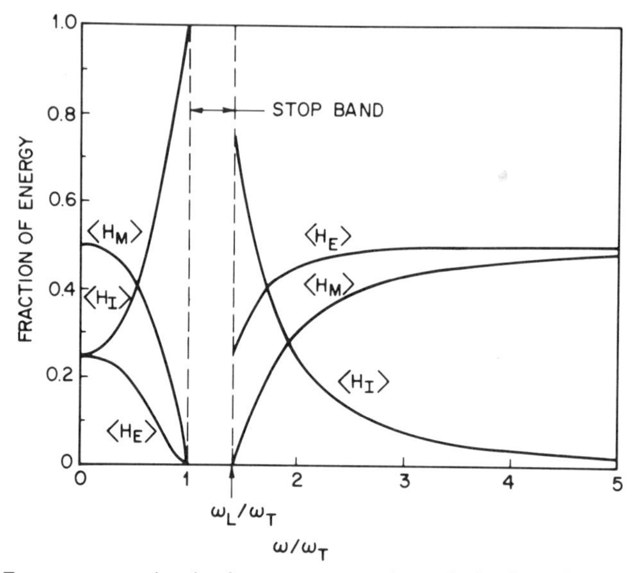

Fig. 9.9. Energy propagation in the transverse optic mode in the polariton regime in a diatomic cubic crystal. The fractions of the time-averaged energy density in the internal (mechanical) vibrations $\langle H_I \rangle$, in the electric field $\langle H_E \rangle$, and in the magnetic field $\langle H_M \rangle$ are plotted versus the frequency normalized to the transverse optic resonant frequency ω_T. At $\omega = 0$, $\langle H_I \rangle = (\kappa_0 - 1)/2\kappa_0, \langle H_E \rangle = \frac{1}{2}\kappa_0, \langle H_M \rangle = \frac{1}{2}$. At the longitudinal optic resonant frequency $\omega_L, \langle H_I \rangle = (\kappa_0 + 1)/2\kappa_0, \langle H_E \rangle = (\kappa_0 - 1)/2\kappa_0, \langle H_M \rangle = 0$. The value $\kappa_0 = 2$ is used for the plot.

9.9 Generalized Lyddane–Sachs–Teller Relation

In the preceding section the original Lyddane–Sachs–Teller relation for cubic crystals is derived. It states that the ratio of the dielectric constant well below to that well above an internal motion resonance is equal to the square of the ratio of the longitudinal optic frequency to the transverse optic frequency. In this section we wish to consider generalizations of this relation to many resonances and to anisotropic media.

Consider first a cubic crystal with many internal motion resonances. We know from the preceding section that the transverse optic frequency produces a pole in the dielectric constant (9.8.16) (damping being neglected) while the longitudinal optic frequency produces a zero in the dielectric constant (9.8.32). This connection may be exploited as follows. If $\kappa(\omega)$ is assumed to be an even function of ω having m simple poles at the frequencies ω_p and to have a finite high frequency limit, it may be written as

$$\frac{\kappa(\omega)}{\kappa(\infty)} = 1 + \sum_{p=1}^{m} \frac{A_p}{\left(\omega_p^2 - \omega^2\right)}. \tag{9.9.1}$$

Rationalizing leads to

$$\frac{\kappa(\omega)}{\kappa(\infty)}=\frac{\prod_{p=1}^{m}\left(\omega_p^2-\omega^2\right)+\sum_{p=1}^{m}A_p\prod_{j=1(\neq p)}^{m}\left(\omega_j^2-\omega^2\right)}{\prod_{p=1}^{m}\left(\omega_p^2-\omega^2\right)}. \tag{9.9.2}$$

The polynomial in ω^2 in the numerator can be rewritten in terms of its roots as

$$\frac{\kappa(\omega)}{\kappa(\infty)}=\frac{\prod_{r=1}^{m}\left(\omega_r^2-\omega^2\right)}{\prod_{p=1}^{m}\left(\omega_p^2-\omega^2\right)} \tag{9.9.3}$$

where ω_r is one of the m roots of $\kappa(\omega)$. Setting $\omega=0$ now leads to the desired generalization [Kurosawa, 1961]

$$\frac{\kappa(0)}{\kappa(\infty)}=\frac{\prod_{r=1}^{m}\omega_r^2}{\prod_{p=1}^{m}\omega_p^2}. \tag{9.9.4}$$

From the development of the preceding section we know that each ω_r corresponds to a longitudinal optic frequency and each ω_p corresponds to a transverse optic frequency.

Next we consider propagation in a general direction in a crystal of general optical anisotropy. Under such circumstances the optic modes are not purely longitudinal or transverse but are hybrids of the two types. A simple association between roots and poles of $\boldsymbol{\kappa}(\omega)$ with these two types of waves thus cannot be expected under general circumstances. A unifying feature of the longitudinal and transverse optic frequencies found in the preceding section, however, was an infinite refractive index occurring at both types of frequencies. This leads us to multiply (9.4.7) by n^2,

$$0=\sum_i \frac{n^2 s_i^2 \kappa_{ii}}{n^2-\kappa_{ii}}, \tag{9.9.5}$$

and let $n\to\infty$. The result is simply

$$0=\mathbf{s}\cdot\boldsymbol{\kappa}(\omega)\cdot\mathbf{s}, \tag{9.9.6}$$

which we expect to hold for all resonant optic mode frequencies regardless of their type (longitudinal, transverse, or hybrid). Thus we find that *all* resonant optic mode frequencies are *roots of the longitudinal dielectric tensor* given in the last equation.

Let us consider the meaning of this equation for propagation in a principal direction. In the principal coordinate system (9.9.6) becomes

$$0=s_1^2\kappa_{11}(\omega)+s_2^2\kappa_{22}(\omega)+s_3^2\kappa_{33}(\omega) \tag{9.9.7}$$

or

$$0=\frac{s_1^2}{\kappa_{33}(\omega)}+\frac{s_2^2\kappa_{22}(\omega)}{\kappa_{11}(\omega)\kappa_{33}(\omega)}+\frac{s_3^2}{\kappa_{11}(\omega)}. \tag{9.9.8}$$

Consider propagation in the 2-direction, that is, $s_2=1$, $s_1=s_3=0$. We conclude that for this principal direction of propagation the resonant optic mode frequencies are solutions of

$$0=\frac{\kappa_{22}(\omega)}{\kappa_{11}(\omega)\kappa_{33}(\omega)}. \tag{9.9.9}$$

This states that the zeroes of the principal longitudinal dielectric constant κ_{22} produce longitudinal optic resonances and that the poles of the principal transverse dielectric constants, κ_{11} and κ_{33}, produce transverse optic resonances.

With this understanding let us obtain a generalized Lyddane–Sachs–Teller relation for more general anisotropy and for an arbitrary number of internal resonances. Consider crystals whose symmetry is high enough so that the principal coordinate system of the dielectric tensor coincides with the crystallographic coordinate system and so does not vary in orientation with frequency (which excludes monoclinic and triclinic crystals). In the principal coordinate system (9.9.6) consists of three terms,

$$\mathbf{s}\cdot\boldsymbol{\kappa}(\omega)\cdot\mathbf{s}=\sum_{i=1}^{3}s_i^2\frac{\prod_{r=1}^{N-1}(\omega_{ri}^2-\omega^2)}{\prod_{p=1}^{N-1}(\omega_{pi}^2-\omega^2)}, \tag{9.9.10}$$

where an expression of the type in (9.9.3) is written for each of the principal dielectric components. Here $N-1$ is the number of internal coordinates. We see that at high frequencies this reduces to

$$\mathbf{s}\cdot\boldsymbol{\kappa}(\infty)\cdot\mathbf{s}=\sum_{i=1}^{3}s_i^2\frac{(-1)^{N-1}}{(-1)^{N-1}}=1. \tag{9.9.11}$$

Forming a common denominator for (9.9.10) and introducing (9.9.11) we get

$$\frac{\mathbf{s}\cdot\boldsymbol{\kappa}(\omega)\cdot\mathbf{s}}{\mathbf{s}\cdot\boldsymbol{\kappa}(\infty)\cdot\mathbf{s}}=\frac{s_1^2\prod_{r=1}^{N-1}(\omega_{r1}^2-\omega^2)\prod_{p=1}^{N-1}(\omega_{p2}^2-\omega^2)\prod_{p'=1}^{N-1}(\omega_{p'3}^2-\omega^2)+\text{cyclic interch.}}{\prod_{i=1}^{3}\prod_{p=1}^{N-1}(\omega_{pi}^2-\omega^2)}$$

$$=\frac{\prod_{r=1}^{3N-3}\left[\omega_r^2(\mathbf{s})-\omega^2\right]}{\prod_{i=1}^{3}\prod_{p=1}^{N-1}(\omega_{pi}^2-\omega^2)} \tag{9.9.12}$$

where $\omega_r(\mathbf{s})$ are the roots of the numerator in the previous expression. They represent all the resonant frequencies (longitudinal, transverse, and hybrid) of a wave traveling in the direction $\mathbf{s}$. From the discussion concerning (9.9.9) we can see that the poles ω_{pi} correspond to the observable transverse resonant frequencies occurring for a wave polarized in the principal direction i and propagating in either of the other principal directions. Setting $\omega=0$ in (9.9.12) now gives us

$$\frac{\mathbf{s}\cdot\boldsymbol{\kappa}(0)\cdot\mathbf{s}}{\mathbf{s}\cdot\boldsymbol{\kappa}(\infty)\cdot\mathbf{s}}=\frac{\prod_{r=1}^{3N-3}\omega_r^2(\mathbf{s})}{\prod_{i=1}^{3}\prod_{p=1}^{N-1}\omega_{pi}^2}, \tag{9.9.13}$$

the *generalized Lyddane–Sachs–Teller relation*. It relates the values of the high and low frequency longitudinal dielectric constants to certain measurable resonant frequencies. This relation applies to a crystal having orthorhombic or higher symmetry and an arbitrary number of resonances and to propagation in an arbitrary direction $\mathbf{s}$.

9.10 Group Velocity in Crystal Optics

In this section we develop general expressions for the group velocity in a linear medium possessing arbitrary anisotropy and frequency dispersion. We then illustrate the effects of frequency dispersion and anisotropy on the group velocity with special cases.

The definition of the refractive index allows us to write

$$\omega(\mathbf{k})=\frac{ck(\mathbf{k})}{n(\omega,\mathbf{s})} \tag{9.10.1}$$

where the functional dependence of the refractive index on the frequency ω and the unit propagation vector $\mathbf{s}$ is given in (9.2.11). The group velocity definition (8.8.1) then yields

$$\begin{aligned}\mathbf{v}_g &= \nabla_k \omega(\mathbf{k}) \\ &= \frac{c}{n}\nabla_k k - \frac{ck}{n^2}\nabla_k n \\ &= \frac{c}{n}\nabla_k k - \frac{\omega}{n}\left(\frac{\partial n}{\partial \omega}\nabla_k \omega + \nabla_k \mathbf{s}\cdot\nabla_s n\right)\end{aligned} \tag{9.10.2}$$

where the subscripts k and s on the gradient operator denote differentiation with respect to the components of $\mathbf{k}$ and $\mathbf{s}$ respectively. Solving the last expression for $\nabla_k \omega$ yields

$$\mathbf{v}_g = \nabla_k \omega = \frac{1}{1+(\omega/n)(\partial n/\partial \omega)}\left(\frac{c}{n}\nabla_k k - \frac{\omega}{n}\nabla_k \mathbf{s}\cdot\nabla_s n\right). \tag{9.10.3}$$

Since

$$k \equiv \sqrt{\mathbf{k}\cdot\mathbf{k}}\ , \tag{9.10.4}$$

we obtain

$$\nabla_k k = \frac{\mathbf{k}}{k} \equiv \mathbf{s}. \tag{9.10.5}$$

In order to obtain $\nabla_k \mathbf{s}$ we differentiate

$$\mathbf{k} = k\mathbf{s} \tag{9.10.6}$$

with respect to a component of $\mathbf{k}$. This gives

$$\nabla_k \mathbf{k} = \mathbf{s}\nabla_k k + k\nabla_k \mathbf{s}, \tag{9.10.7}$$

which becomes

$$\mathbf{1} = \mathbf{ss} + k\nabla_k \mathbf{s}. \tag{9.10.8}$$

The desired derivative is then

$$\nabla_k \mathbf{s} = \frac{1}{k}(\mathbf{1} - \mathbf{ss}). \tag{9.10.9}$$

We now substitute (9.10.9) and (9.10.5) into (9.10.3). Since the gradient with respect to a unit vector is normal to the unit vector, $\mathbf{s}\cdot\nabla_s = 0$ and so the final expression for the group velocity in crystal optics becomes

$$\mathbf{v}_g = \frac{(c/n)}{1+(\omega/n)(\partial n/\partial\omega)}\left[\mathbf{s} - \frac{1}{n}\nabla_s n\right]. \tag{9.10.10}$$

The gradient ∇_s is a transverse gradient calculated by

$$\nabla_s = \boldsymbol{\theta}\frac{\partial}{\partial\theta} + \frac{\boldsymbol{\phi}}{\sin\theta}\frac{\partial}{\partial\phi} \tag{9.10.11}$$

where $\boldsymbol{\theta}$ and $\boldsymbol{\phi}$ are unit vectors in the directions of the spherical angles θ and ϕ that specify the rectangular components of $\mathbf{s}$ as

$$\mathbf{s} = [\cos\phi\sin\theta, \sin\phi\sin\theta, \cos\theta]. \tag{9.10.12}$$

Several things should be said about this relation. The magnitude of the group velocity differs from the phase velocity v, defined by

$$v \equiv \frac{\omega}{k} = \frac{c}{n}, \tag{9.10.13}$$

as a result of two phenomena: frequency dispersion and anisotropy. The factor $[1+(\omega/n)\partial n/\partial\omega]^{-1}$ results from the frequency dispersion of the refractive index. Note that this factor affects the group velocity, which is the energy propagation velocity (see Section 8.8), even for a wave having only a single frequency. The term proportional to $\nabla_s n$ represents the dependence of group velocity on the anisotropy of the medium. For amorphous media such as glass, cubic crystals, and the ordinary ray in uniaxial crystals

$$\nabla_s n = 0 \tag{9.10.14}$$

and the group velocity is then

$$\mathbf{v}_g = \frac{v\mathbf{s}}{1+(\omega/n)(\partial n/\partial\omega)}. \tag{9.10.15}$$

For any of these particular media in a dispersionless region, where

$$\frac{\partial n}{\partial\omega} = 0, \tag{9.10.16}$$

the group velocity is simply

$$\mathbf{v}_g = v\mathbf{s}. \tag{9.10.17}$$

The general expression for the group velocity (9.10.10) can be written as the magnitude v_g multiplied by a unit vector $\mathbf{t}$,

$$\mathbf{v}_g \equiv v_g \mathbf{t}. \tag{9.10.18}$$

The unit energy propagation vector is then

$$\mathbf{t} = \cos\delta\left[\mathbf{s} - \frac{1}{n}\nabla_s n\right] \tag{9.10.19}$$

where $\cos\delta$ can be variously expressed as

$$\cos\delta \equiv \mathbf{s}\cdot\mathbf{t} = \frac{\mathbf{s}\cdot\mathbf{v}_g}{v_g} = \frac{v}{v_g[1+(\omega/n)(\partial n/\partial\omega)]} = \left[1 + \frac{1}{n^2}(\nabla_s n\cdot\nabla_s n)\right]^{-1/2}. \tag{9.10.20}$$

We now wish to develop an alternate expression for the group velocity in linear crystal optics. This expression follows from the equality, proved in Section 8.8, of the group velocity and the energy propagation velocity. From (8.8.21) we have

$$\mathbf{v}_g = v_g \mathbf{t} = \frac{\langle \mathbf{S}\rangle}{\langle H\rangle} \tag{9.10.21}$$

where the unit vector $\mathbf{t}$ is given by

$$\mathbf{t} = \frac{\langle \mathbf{S}\rangle}{[\langle \mathbf{S}\rangle\cdot\langle \mathbf{S}\rangle]^{1/2}}. \tag{9.10.22}$$

From the general expression for $\mathbf{S}$ in (8.7.12) we see that in the optical frequency region, where $\mathbf{u}\cong 0$, $\mathbf{S}$ is given by

$$\mathbf{S} = \frac{1}{\mu_0}\mathbf{E}\times\mathbf{B}, \tag{9.10.23}$$

which is just the familiar Poynting vector for a dielectric. We see that $\mathbf{S}$ and therefore $\mathbf{t}$ are perpendicular to both $\mathbf{E}$ and $\mathbf{B}$. Thus the angle δ introduced in (9.10.20) is the same angle as δ^α, the angle between $\mathfrak{E}^\alpha$ and $\mathfrak{D}^\alpha$ in Fig. 9.2.

For a plane wave we have

$$\mathbf{E} = A\mathbf{e}^\alpha \cos(\mathbf{k}^\alpha\cdot\mathbf{z} - \omega t) \tag{9.10.24}$$

where A is an amplitude constant and $\mathbf{e}^{\alpha}$ an electric field unit vector. The corresponding magnetic induction is

$$\mathbf{B}=\frac{Ak^{\alpha}}{\omega}\mathbf{s}\times\mathbf{e}^{\alpha}\cos(k^{\alpha}\mathbf{s}\cdot\mathbf{z}-\omega t). \tag{9.10.25}$$

The Poynting vector is then

$$\mathbf{S}=\frac{A^2k^{\alpha}}{\mu_0\omega}\mathbf{e}^{\alpha}\times(\mathbf{s}\times\mathbf{e}^{\alpha})\cos^2(k^{\alpha}\mathbf{s}\cdot\mathbf{z}-\omega t). \tag{9.10.26}$$

Averaging over one period and expanding the vector triple product gives

$$\langle\mathbf{S}\rangle=\frac{A^2k^{\alpha}}{2\mu_0\omega}\left[\mathbf{s}-\mathbf{e}^{\alpha}(\mathbf{s}\cdot\mathbf{e}^{\alpha})\right]=\tfrac{1}{2}\epsilon_0 cA^2n^{\alpha}\cos\delta^{\alpha}\mathbf{t}^{\alpha} \tag{9.10.27}$$

where the unit vector $\mathbf{t}^{\alpha}$ is given by

$$\mathbf{t}^{\alpha}=\sec\delta^{\alpha}(\mathbf{s}-\mathbf{e}^{\alpha}\sin\delta^{\alpha}) \qquad (\alpha=1,2). \tag{9.10.28}$$

Figure 9.2 illustrates the orientation of $\mathbf{t}^{\alpha}$.

We now need an expression for $\langle H\rangle$ for substitution into (9.10.21). Since center-of-mass variables play no role at optical frequencies, as discussed in Section 9.1, the general expression (8.7.11) for the energy density yields

$$\langle H\rangle=\left\langle\sum_{\nu}\frac{m^{\nu}}{2}(\dot{\mathbf{y}}^{\nu})^2+\sum_{\nu\mu}{}^{20}M_{jk}^{\nu\mu}y_j^{\nu}y_k^{\mu}+\frac{\epsilon_0}{2}(\mathbf{E})^2+\frac{1}{2\mu_0}(\mathbf{B})^2\right\rangle. \tag{9.10.29}$$

The internal coordinates are given by (9.1.9),

$$y_j^{\nu}=\sum_{\lambda}\Upsilon_{jk}^{\nu\lambda}(\omega)q^{\lambda}E_k. \tag{9.10.30}$$

Substitution of this equation and the equations for $\mathbf{E}$ and $\mathbf{B}$ into the expression for $\langle H\rangle$ leads to

$$\langle H\rangle=\frac{\epsilon_0A^2}{4}\left\{\mathbf{e}^{\alpha}\cdot\mathbf{G}\cdot\mathbf{e}^{\alpha}+\mathbf{e}^{\alpha}\cdot\boldsymbol{\kappa}\cdot\mathbf{e}^{\alpha}+\frac{c^2(k^{\alpha})^2}{\omega^2}\left[1-(\mathbf{s}\cdot\mathbf{e}^{\alpha})^2\right]\right\} \tag{9.10.31}$$

where

$$G_{mn} \equiv \frac{2\omega^2}{\epsilon_0} \sum_{\sigma\nu\rho} q^\sigma \Upsilon_{mk}^{\sigma\nu}(\omega) m^\nu \Upsilon_{kn}^{\nu\rho}(\omega) q^\rho. \tag{9.10.32}$$

In order to interpret **G**, we differentiate the definition of $\Upsilon^{\lambda\nu}$ (9.1.8)

$$\sum_\nu \frac{\partial \Upsilon_{ji}^{\lambda\nu}(\omega)}{\partial\omega}\left[2^{20}M_{ib}^{\nu\mu} - m^\nu\omega^2\delta^{\nu\mu}\delta_{ib}\right] + \sum_\nu \Upsilon_{ji}^{\lambda\nu}(\omega)\left[-2m^\nu\omega\delta^{\nu\mu}\delta_{ib}\right] = 0. \tag{9.10.33}$$

We now form a scalar product of this equation with $\omega q^\lambda \Upsilon_{bk}^{\mu\rho}(\omega) q^\rho/\epsilon_0$ and sum over λ, μ, and ρ. Use of the definition of $\Upsilon^{\lambda\nu}$ to contract the first term leads to

$$\frac{2\omega^2}{\epsilon_0} \sum_{\lambda\nu\rho} q^\lambda \Upsilon_{ji}^{\lambda\nu}(\omega) m^\nu \Upsilon_{ik}^{\nu\rho}(\omega) q^\rho = \frac{\omega}{\epsilon_0} \sum_{\lambda\nu} q^\lambda \frac{\partial \Upsilon_{jk}^{\lambda\nu}(\omega)}{\partial\omega} q^\nu. \tag{9.10.34}$$

We now recognize the left side as **G** and the right side as a derivative of the susceptibility $\boldsymbol{\chi}$ defined in (9.1.13), or equivalently, a derivative of the dielectric tensor,

$$\mathbf{G} = \omega \frac{\partial \boldsymbol{\chi}(\omega)}{\partial\omega} = \omega \frac{\partial \boldsymbol{\kappa}(\omega)}{\partial\omega}. \tag{9.10.35}$$

This expression for **G** can now be substituted into the equation (9.10.31). We can also identify the last pair of terms in that equation as $\mathbf{e}^\alpha\cdot\boldsymbol{\kappa}\cdot\mathbf{e}^\alpha$ by taking a scalar product of the eigenvector equation (9.3.1) with $\mathbf{e}^\alpha$. The time-averaged energy density may now be written as

$$\langle H \rangle = \frac{\epsilon_0 A^2}{2} \mathbf{e}^\alpha\cdot\left(\boldsymbol{\kappa} + \frac{\omega}{2}\frac{\partial\boldsymbol{\kappa}}{\partial\omega}\right)\cdot\mathbf{e}^\alpha. \tag{9.10.36}$$

We may now obtain the alternate expression of the group velocity by substituting (9.10.27) and (9.10.36) into (9.10.21) with the result

$$\mathbf{v}_g = \frac{c^2 k^\alpha[\mathbf{s} - \mathbf{e}^\alpha(\mathbf{s}\cdot\mathbf{e}^\alpha)]}{\omega\mathbf{e}^\alpha\cdot[\boldsymbol{\kappa} + (\omega/2)(\partial\boldsymbol{\kappa}/\partial\omega)]\cdot\mathbf{e}^\alpha}. \tag{9.10.37}$$

This expression for the group velocity has been obtained by Kleinman [1968] by a different derivation.

It is interesting to note that the group velocity (and the energy density) contains a dispersive term $\partial \boldsymbol{\kappa}/\partial\omega$ even though the derivation specifically employs a *single frequency* wave. Thus the velocity of energy transport for a monochromatic wave depends not only on the dielectric tensor but also on the frequency derivative of the dielectric tensor, both evaluated at the frequency of the wave. The physical origin of the dispersive term in the energy density and the group velocity may be described as the need to supply that extra energy in order to establish the light wave in the medium. If in the distant past we applied a single frequency wave of infinitesimal amplitude to the crystal and then caused its amplitude to grow very slowly to some finite final value, the wave during the growth period, as seen from Fourier analysis, would not be a single frequency wave but would have a small spread of frequencies. Thus the establishment of a final constant amplitude, truly single frequency wave would have had to sense the variation of the optical propagation properties (the optical frequency dielectric tensor or, alternatively, the refractive index) with frequency in a small neighborhood about the chosen frequency of the light wave.

The angle δ between the group velocity and the propagation vector is a measure of the effect of anisotropy on the group velocity. It becomes larger as the optical anisotropy becomes larger, that is, as the differences between the principal components of the dielectric tensor grow larger. However, the most birefringent crystals in the optical frequency region have a δ of only a few degrees. An example is shown in Fig. 9.10.

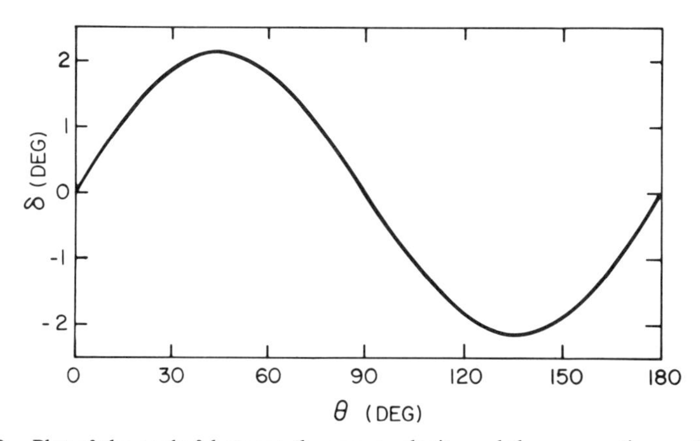

Fig. 9.10. Plot of the angle δ between the group velocity and the propagation vector versus the angle θ that the propagation vector makes with the optic axis. The plot is for $LiNb0_3$ at 632.8 nm wavelength.

Next we consider a specific case of the second phenomenon that causes the group velocity to differ from the phase velocity, namely frequency dispersion. The cubic crystal with two particles per unit cell that is discussed in Section 9.8 in the region of polariton dispersion is used as an example. Because the crystal is cubic, it has no optical anisotropy. Thus in this case we have

$$\nabla_s n = 0 \tag{9.10.38}$$

and (9.10.10) becomes simply

$$\mathbf{v}_g = \frac{c\mathbf{s}}{n[1+(\omega/n)(\partial n/\partial\omega)]}. \tag{9.10.39}$$

We found the refractive index for the transverse waves in (9.8.16) to be

$$n = \left[1 + \frac{(\kappa_0 - 1)\omega_T^2}{\omega_T^2 - \omega^2}\right]^{1/2}. \tag{9.10.40}$$

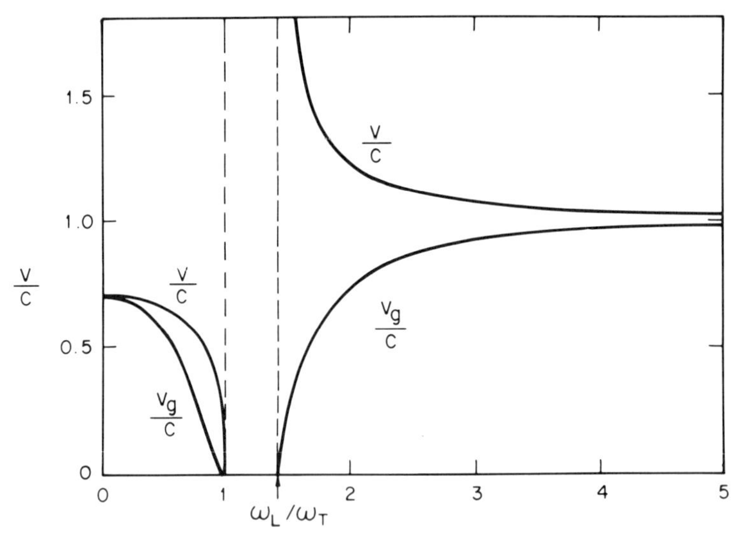

Fig. 9.11. Plot of group and phase velocities in the region of the transverse and longitudinal optic mode frequencies ω_T and ω_L. Note that $v_g \leq v$ and $v_g \leq c$ throughout the region. On the high frequency side of the resonance $v \geq c$. The plot applies to a cubic crystal having two particles per unit cell (see Section 9.8) and ignores the effects of absorption. $\kappa_0 = 2$ is chosen for the plot.

Substitution of this into the last equation yields

$$v_g = c\left[1+\frac{(\kappa_0-1)\omega_T^2}{\omega_T^2-\omega^2}\right]^{1/2}\cdot\left[1+\frac{(\kappa_0-1)\omega_T^4}{\left(\omega_T^2-\omega^2\right)^2}\right]^{-1} \tag{9.10.41}$$

for the magnitude of the group velocity. In contrast the phase velocity is

$$v=\frac{c}{n}=c\left[1+\frac{(\kappa_0-1)\omega_T^2}{\omega_T^2-\omega^2}\right]^{-1/2}. \tag{9.10.42}$$

These two equations are plotted in Fig. 9.11. Note that the group velocity of an optically isotropic medium is never greater than the phase velocity and never greater than the speed of light in a vacuum. However, the phase velocity is greater than the speed of light in vacuum on the high frequency side of the resonance. In this region the refractive index is less than one.

PROBLEMS

9.1. Show directly that the eigenvectors for the uniaxial crystal satisfy the completeness relation (9.3.13).

9.2. Show that the direction of energy propagation of an extraordinary wave having a unit propagation vector $\mathbf{s}$ in a uniaxial crystal is

$$\mathbf{t}=\frac{\boldsymbol{\kappa}\cdot\mathbf{s}}{\left[\mathbf{s}\cdot\boldsymbol{\kappa}\cdot\boldsymbol{\kappa}\cdot\mathbf{s}\right]^{1/2}}. \tag{9.P.1}$$

Also show for this case that

$$\tan\delta=\tfrac{1}{2}\left[n^e(\theta)\right]^2\left[\frac{1}{\kappa_{33}}-\frac{1}{\kappa_{11}}\right]\sin 2\theta. \tag{9.P.2}$$

9.3. Show that

$$\langle\mathfrak{L}\rangle=0 \tag{9.P.3}$$

for the specific fields found in Section 9.8 for the transverse optic branches in a cubic crystal having two particles per unit cell.

9.4. Derive the expression (9.10.41) for the group velocity of a plane electromagnetic wave in a cubic crystal having two particles per unit cell beginning from (9.10.37).

9.5. Consider a uniaxial crystal having two particles per primitive unit cell and thus one internal coordinate. Its resonant behavior has uniaxial anisotropy. Find expressions for the ordinary and extraordinary transverse optic mode frequencies and the longitudinal optic mode frequency. Derive a Lyddane–Sachs–Teller relation for this crystal. [Do not begin from (9.9.13).]

9.6. Show that the time-averaged Poynting vector $\langle \mathbf{S} \rangle$ is given by

$$\langle \mathbf{S} \rangle = \tfrac{1}{2} \mathcal{R}\{\mathbf{E} \times \mathbf{H}^*\} \tag{9.P.4}$$

when the $\mathbf{E}$ and $\mathbf{H}$ are expressed as complex fields with the time dependence $\exp - i\omega t$. $\mathcal{R}$ means "real part of."

9.7. A light wave in a dielectric crystal can create an *exciton* which is an electron, removed from the valence structure of the crystal, bound to the hole left in that structure. It is a solid state analogue of positronium. An exciton in the long wavelength limit can be described by an internal coordinate. Because of its sizable mobility a propagation term must be included in the internal motion equation even in the long wavelength limit (such a term is a wavevector dispersion correction term). Find the transverse optic mode dispersion relation for an exciton modeled by the equations

$$\frac{\partial^2 E}{\partial z^2} - \frac{\kappa_h}{c^2} \frac{\partial^2 E}{\partial t^2} = \frac{q}{c^2} \frac{\partial^2 y}{\partial t^2}, \tag{9.P.5}$$

$$m \frac{\partial^2 y}{\partial t^2} = qE - 2^{20} My + m v_o^2 \frac{\partial^2 y}{\partial z^2} \tag{9.P.6}$$

where y and E are scalar quantities, κ_h is the high frequency limit of the dielectric constant and v_o a material constant. Sketch the $\omega(k)$ dispersion relation. How does it differ from that of a transverse optic mode? Establish whether it has a stop-band or not.

9.8. Show the equality of the two expressions for the group velocity (9.10.10) and (9.10.37) directly (do not use the theorem in Section 8.8).

9.9. *Normal coordinates* are those that decouple all the independent excitations of a system. In lattice dynamics coordinates that decouple all the internal motions of a crystal from one another while leaving coupling of these motions to the electric field are called normal coordinates. Such coordinates can be defined by

$$\eta^k \equiv \sum_{\mu} (m^\mu)^{1/2} \mathbf{i}^{k\mu} \cdot \mathbf{y}^\mu \tag{9.P.7}$$

where η^k $(k=1,2,\ldots 3N-3)$ are scalars. Show that the normal mode eigenvectors $\mathbf{i}^{k\mu}$ satisfy

$$2\sum_{\mu} i_a^{k\mu} \frac{{}^{20}M_{ab}^{\mu\nu}}{(m^{\mu}m^{\nu})^{1/2}} = \Omega_k^2 i_b^{k\nu} \tag{9.P.8}$$

in conjunction with

$$\sum_{\nu} \mathbf{i}^{k\nu}\cdot\mathbf{i}^{l\nu} = \delta^{kl}, \tag{9.P.9}$$

$$\sum_{k} i_i^{k\mu} i_j^{k\nu} = \delta_{ij}\delta^{\mu\nu} \tag{9.P.10}$$

where Ω_k^2 are eigenvalues to be found from (9.P.8). Show that the normal coordinate equations are

$$\ddot{\eta}^k = \mathbf{c}^k\cdot\mathbf{E} - \Omega_k^2\eta^k \tag{9.P.11}$$

where

$$\mathbf{c}^k \equiv \sum_{\mu} \frac{q^{\mu}}{(m^{\mu})^{1/2}}\mathbf{i}^{k\mu}. \tag{9.P.12}$$

Note that the k mode is infrared inactive if and only if $\mathbf{c}^k = 0$.

9.10. Using the normal coordinate equations of the last problem find new expressions for the mechanical admittance and the linear electric susceptibility,

$$\Upsilon_{ij}^{\mu\nu}(\omega) = \sum_{k} \frac{i_i^{k\mu} i_j^{k\nu}}{(m^{\mu}m^{\nu})^{1/2}(\Omega_k^2 - \omega^2)}, \tag{9.P.13}$$

$$\epsilon_0\chi_{ij}(\omega) = \sum_{k} \frac{c_i^k c_j^k}{\Omega_k^2 - \omega^2}. \tag{9.P.14}$$

Note that Ω_k $(k=1,2,\ldots 3N-3)$ are the set of transverse optic mode frequencies.

9.11. From the result of the last problem and the methods of Section 9.8 find a polynominal whose solutions are the longitudinal optic mode frequencies for any direction of propagation $\mathbf{s}$.

CHAPTER 10

Linear Crystal Acoustics

In this chapter we derive the basic equations of linear elasticity and linear piezoelectricity. We treat these two subjects together because in crystal physics they are so often intimately connected, a fact that results from piezoelectricity occurring in 20 of the 32 crystal classes. *Linear elasticity* is the study of small deformations as described by the center-of-mass equation and, of course, exists in all solids. Indeed, the elasticity of isotropic solids, so important to engineering sciences, fills many books. The emphasis here is on the elasticity of anisotropic media. *Piezoelectricity* is the study of the creation of an electric field in a crystal of appropriate symmetry by straining it (the *direct effect*) and of the deformation of such a crystal by the application of an electric field (the *converse effect*).

The treatment in this chapter applies to simple dielectrics, piezoelectrics, and pyroelectrics. To apply any final result to a simple dielectric one need only set the piezoelectric tensor and the spontaneous polarization and spontaneous electric field equal to zero. For a nonpyroelectric piezoelectric the spontaneous polarization and spontaneous electric field must be set equal to zero while for a pyroelectric whose spontaneous electric field is canceled by extrinsic surface charge that electric field may be set equal to zero. We hope that the economy in space that results from treating all three types of crystals at once outweighs the somewhat more complex appearance of the equations.

Because this theory gives a new basis for the theory of piezoelectricity in pyroelectrics, we devote considerable space to this subject. In this chapter we obtain the linearized equations from the general nonlinear equations of Chapter 7. These include the various linear constitutive relations for polarization, stress, and so on, as well as the dynamical equations. The treatment is based on Nelson and Lax [1976b] though the use of a different rotational invariant in the expansion of the stored energy makes the expressions here appear somewhat different and simpler. We then consider certain plane wave phenomena from an eigenvector point of view. Chapter 11 continues the treatment of piezoelectricity particularly with respect to boundary value problems.

As pointed out in Chapter 3 there is a duality in the expression of elastic deformation theory, the spatial description and the material description. Of course both, when correctly applied, make the same prediction to be compared to an experiment. The choice is thus made on the basis of convenience of calculation. For ordinary dielectrics and for piezoelectrics there is no difference in convenience at the linear level; for pyroelectrics there is some preference for a material description. To understand this preference we must consider both descriptions. In this chapter we use the spatial description and in Section 17.2 we use the material description for comparison. For nonlinear elastic phenomena the material description is much more preferable and so the treatment of such phenomena in Chapter 17 uses that description.

10.1 Adiabatic Elimination of Internal Coordinates

Elasticity and piezoelectricity can be excited significantly only at frequencies well below the resonant frequencies of the internal motions of the crystal which typically lie in the infrared. At these low frequencies the internal motions follow the elastic deformation adiabatically. The inertial force term in the internal motion equations (7.3.3) can thus be dropped prior to using the equations for eliminating the internal coordinates from the center-of-mass and electromagnetic equations. The procedure used here is a generalization of the adiabatic elimination introduced in Section 2.3.

Since linear dynamical equations are sought, we need carry the series expansion of the stored energy (6.5.8) only to bilinear terms,

$$\rho^0\Sigma = {}^{01}M_{AB}E_{AB} + {}^{02}M_{ABCD}E_{AB}E_{CD} + \sum_{\nu} {}^{10}M_A^{\nu}\Pi_A^{\nu}$$

$$+ \sum_{\nu\mu} {}^{20}M_{AB}^{\nu\mu}\Pi_A^{\nu}\Pi_B^{\mu} + \sum_{\nu} {}^{11}M_{ABC}^{\nu}\Pi_A^{\nu}E_{BC}. \tag{10.1.1}$$

In contrast to the procedure used in the chapter on crystal optics, terms linear in the center-of-mass coordinate are kept in the internal motion equations (7.3.3). These equations become

$$0 = q^{\nu}\mathscr{E}_i - {}^{10}M_A^{\nu}X_{A,i} - 2\sum_{\mu} {}^{20}M_{AB}^{\nu\mu}\Pi_B^{\mu}X_{A,i} - {}^{11}M_{ABC}^{\nu}E_{BC}X_{A,i}. \tag{10.1.2}$$

Next we form a scalar product of this equation with $x_{i,D}$ and subtract the intrinsic natural-state condition (7.3.7). The result is

$$2\sum_{\mu} {}^{20}M_{DB}^{\nu\mu}\Pi_B^{\mu} = q^{\nu}F_D - {}^{11}M_{DBC}^{\nu}E_{BC} \tag{10.1.3}$$

where

$$F_D \equiv \mathscr{E}_i x_{i,D} - E_i^S \delta_{iD}. \tag{10.1.4}$$

Equation (10.1.3) can be solved for Π_B^{μ} with the use of the same mechanical admittance $\boldsymbol{\Upsilon}$ defined in (9.1.8). Since the frequency term in that equation is now set equal to zero, the $\boldsymbol{\Upsilon}$ used here is just the low frequency limit of the $\boldsymbol{\Upsilon}(\omega)$ used before. Thus we obtain

$$\Pi_E^{\lambda} = \sum_{\nu} \Upsilon_{ED}^{\lambda\nu}\left[q^{\nu}F_D - {}^{11}M_{DBC}^{\nu}E_{BC}\right], \tag{10.1.5}$$

which is useful for eliminating Π_E^{λ} from stored energy terms. The nonlinear parts of F_D and E_{BC} are dropped later in the development. Another useful expression in the adiabatic elimination is obtained by combining (6.5.5) and (6.5.6) into

$$y_i^{T\nu} = x_{i,A}\left(Y_A^{\nu} + \Pi_A^{\nu}\right). \tag{10.1.6}$$

10.2 Linearized Electromagnetic Equations

Two of the electromagnetic equations contain properties of the material medium, the dielectric charge density and the dielectric current density. Since these quantities are functions only of the polarization, we need to linearize only it.

The polarization may be reexpressed as

$$\begin{aligned}
P_i &= \frac{1}{J}\sum_{\nu} q^{\nu} y_i^{T\nu} \\
&= \frac{x_{i,A}}{J}\sum_{\nu} q^{\nu}\left(Y_A^{\nu} + \Pi_A^{\nu}\right) \\
&= \frac{x_{i,A}}{J}\sum_{\nu} q^{\nu}\left[Y_A^{\nu} + \sum_{\mu}\Upsilon_{AD}^{\nu\mu}\left(q^{\mu}F_D - {}^{11}M_{DBC}^{\mu}E_{BC}\right)\right] \\
&= \frac{x_{i,A}}{J}\left[P_A^S + e_{ABC}E_{BC} + \epsilon_0\chi_{AD}F_D\right]
\end{aligned} \tag{10.2.1}$$

by using (10.1.5) and (10.1.6), the definitions of $\mathbf{P}^S$ (7.3.6) and $\boldsymbol{\chi}$ (9.1.13), and by defining e_{ABC} as

$$e_{ABC} \equiv -\sum_{\nu\mu} q^{\nu} \Upsilon^{\nu\mu}_{AD}\, {}^{11}M^{\mu}_{DBC} = e_{ACB}. \tag{10.2.2}$$

This is the *piezoelectric stress tensor* that is commonly measured. Because Υ is the low frequency limit of $\Upsilon(\omega)$, e_{ABC} has no frequency dispersion.

We now introduce expressions for the quantities in (10.2.1) linearized with respect to the displacement and the electromagnetic fields,

$$J^{-1} = 1 - u_{l,l}, \tag{10.2.3}$$

$$x_{i,A} = \delta_{iA} + u_{i,A}, \tag{10.2.4}$$

$$E_{BC} = u_{(B,C)}, \tag{10.2.5}$$

$$F_D = E_D + E_i^S u_{i,D}. \tag{10.2.6}$$

Now that the displacement is introduced and space derivatives are expanded to first order, the distinction between uppercase and lowercase subscripts can be dropped. All space derivatives are now taken with respect to $\mathbf{z}$. The polarization becomes

$$\begin{aligned} P_i &= P_i^S + \epsilon_0 \chi_{ij} E_j + \left(e_{ijk} + \delta_{ij} P_k^S - P_i^S \delta_{jk} + \epsilon_0 \chi_{ik} E_j^S\right) u_{j,k} \\ &= P_i^S + \epsilon_0 \chi_{ij} E_j + \left(e^F_{ijk} + 2\delta_{j[i} P^S_{k]}\right) u_{j,k} \end{aligned} \tag{10.2.7}$$

where we use the definition,

$$e^F_{ijk} \equiv e_{ijk} + \epsilon_0 \chi_{ik} E_j^S. \tag{10.2.8}$$

The quantity e^F_{ijk} is the piezoelectric stress tensor that would be measured in a pyroelectric in possession of its spontaneous electric field. Note that another constitutive relation, this time the polarization in the low frequency regime, is derived from the theory.

The expression given above for the polarization contains two new terms. The first is the term $\epsilon_0 \chi_{ik} E_j^S u_{j,k}$ which depends on the presence of the spontaneous electric field. Note that it has no interchange symmetry on the pair of indices where e_{ijk} has interchange symmetry. The second is the term $2\delta_{j[i} P^S_{k]} u_{j,k}$ which depends on the spontaneous polarization, present even when $\mathbf{E}^S = 0$, and which has different interchange symmetry than e_{ijk}.

In terms of the polarization the linearized dielectric charge density q^D is

$$\begin{aligned} q^D &= -\nabla \cdot \mathbf{P} \\ &= -\epsilon_0 \chi_{ij} E_{j,i} - e^F_{ijk} u_{j,ki}. \end{aligned} \tag{10.2.9}$$

Note that the term $2\delta_{j[i}P^S_{k]}u_{j,ki}$ has disappeared from q^D and therefore from the Maxwell charge equation because a symmetric quantity is being summed over an antisymmetric quantity. The term depending on $\mathbf{E}^S$ has, however, remained. The dielectric current density is

$$\mathbf{j}^D = \frac{\partial \mathbf{P}}{\partial t} + \nabla \times (\mathbf{P}^S \times \dot{\mathbf{u}}) \tag{10.2.10}$$

to linear order, that is, only $\mathbf{P}^S$ remains for $\mathbf{P}$ in the second term. This equation becomes

$$\begin{aligned} j_i^D &= \epsilon_0 \chi_{ij} \frac{\partial E_j}{\partial t} + \left(e^F_{ijk} + 2\delta_{j[i]}P^S_{[k]}\right)\frac{\partial u_{j,k}}{\partial t} + 2\delta_{j[k]}P^S_{[i]}\frac{\partial u_{j,k}}{\partial t} \\ &= \epsilon_0 \chi_{ij} \frac{\partial E_j}{\partial t} + e^F_{ijk}\frac{\partial u_{j,k}}{\partial t}. \end{aligned} \tag{10.2.11}$$

Note the disappearance here also of the term proportional to $2\delta_{j[i}P^S_{k]}$. This term from the constitutive relation for $\mathbf{P}$, having disappeared from both q^D and $\mathbf{j}^D$, cannot enter the Maxwell equations. Its measurability is thus in question; we consider it again in later sections.

Combining (10.2.9) and (7.1.3) gives the linearized Maxwell charge equation as

$$\epsilon_0 \kappa_{ij} E_{j,i} = -e^F_{ijk} u_{j,ki}. \tag{10.2.12}$$

Combining (10.2.11) and (7.1.6) yields the linearized Maxwell current equation

$$\frac{(\nabla \times \mathbf{B})_i}{\mu_0} - \epsilon_0 \kappa_{ij} \frac{\partial E_j}{\partial t} = e^F_{ijk}\frac{\partial u_{j,k}}{\partial t}. \tag{10.2.13}$$

We may now introduce the electric displacement vector

$$\mathbf{D} \equiv \epsilon_0 \mathbf{E} + \mathbf{P} \tag{10.2.14}$$

and the magnetic intensity vector

$$\mathbf{H} \equiv \frac{1}{\mu_0}\mathbf{B} - \mathbf{P} \times \dot{\mathbf{x}}. \tag{10.2.15}$$

With the use of (10.2.7) these become to linear order

$$D_i = \epsilon_0 E_i^S + P_i^S + \epsilon_0 \kappa_{ij} E_j + \left(e^F_{ijk} + 2\delta_{j[i}P^S_{k]}\right)u_{j,k}, \tag{10.2.16}$$

$$\mathbf{H} = \frac{1}{\mu_0}\mathbf{B} - \mathbf{P}^S \times \dot{\mathbf{u}}. \tag{10.2.17}$$

These fields may be used with the boundary conditions (7.2.9) and (7.2.22) found previously. In terms of these fields (10.2.12) and (10.2.13) are given in the Maxwell form by

$$\nabla \cdot \mathbf{D} = 0, \tag{10.2.18}$$

$$\nabla \times \mathbf{H} - \frac{\partial \mathbf{D}}{\partial t} = 0. \tag{10.2.19}$$

The constitutive relations for **D** and **H** above and that for **P** earlier differ from the traditionally accepted ones for studies of piezoelectricity in pyroelectrics. They are in agreement, however, with the traditional constitutive relations for nonpyroelectric piezoelectrics. In **P** and **D** there are two new linear piezoelectric-like terms, one involving $\mathbf{E}^S$ and the other involving $\mathbf{P}^S$, and in **H** there is a new linear term involving $\mathbf{P}^S$. That involving $\mathbf{E}^S$ enters the differential equations and so can be expected to be observable. The linear terms involving $\mathbf{P}^S$, however, disappear from the differential equations and so would not seem to be observable. To be sure of this conclusion we must first examine the linearized boundary conditions. This is done in Section 11.4.

10.3 Linearized Dynamic Elasticity Equation

We next proceed to linearize the center-of-mass equation,

$$\rho \ddot{x}_i = t^E_{ij,j} + q^D E_i + (\mathbf{j}^D \times \mathbf{B})_i, \tag{10.3.1}$$

in terms of the displacement and the electric field, the internal coordinates being adiabatically eliminated in the process. An examination of (10.2.10) for the dielectric current shows that the body force term $\mathbf{j}^D \times \mathbf{B}$ in the center-of-mass equation contains no linear terms. The body force $q^D\mathbf{E}$ contributes

$$q^D E_i^S = -\epsilon_0 \chi_{kl} E_{l,k} E_i^S - e^F_{klm} u_{l,mk} E_i^S. \tag{10.3.2}$$

The elastic stress tensor can be seen from (7.6.7) to be given simply by

$$t^E_{ij} = t^{\Pi}_{ij} \equiv \rho x_{i,A} x_{j,B} \left.\frac{\partial \Sigma}{\partial E_{AB}}\right|_{\Pi^r \text{ constant}} \tag{10.3.3}$$

when the inertial term of the internal coordinates is dropped by the adiabatic approximation.

We now need to find the linear form of t_{ij}^{Π}. Carrying out the derivative on $\rho^0\Sigma$ given by (10.1.1) yields

$$t_{ij}^{\Pi}=\frac{x_{i,A}x_{j,B}}{J}\left[{}^{01}M_{AB}+2{}^{02}M_{ABCD}E_{CD}+\sum_{\nu}{}^{11}M_{CAB}^{\nu}\Pi_C^{\nu}\right]. \qquad (10.3.4)$$

Equation (10.1.5) can now be used to eliminate Π_C^{ν} and so give

$$t_{ij}^{\Pi}=\frac{x_{i,A}x_{j,B}}{J}\left[{}^{01}M_{AB}+\left(2{}^{02}M_{ABCD}-\sum_{\nu\mu}{}^{11}M_{EAB}^{\nu}\Upsilon_{EF}^{\nu\mu}\,{}^{11}M_{FCD}^{\mu}\right)E_{CD}\right.$$
$$\left.+\sum_{\mu\nu}q^{\nu}\Upsilon_{DC}^{\nu\mu}\,{}^{11}M_{CAB}^{\mu}F_D\right]. \qquad (10.3.5)$$

We define the coefficient of the strain in this linear stress expression as the elastic *stiffness tensor* c_{ABCD},

$$c_{ABCD}\equiv 2{}^{02}M_{ABCD}-\sum_{\mu\nu}{}^{11}M_{EAB}^{\nu}\Upsilon_{EF}^{\nu\mu}\,{}^{11}M_{FCD}^{\mu}, \qquad (10.3.6)$$

and the constant appearing in the stress as the *spontaneous stress tensor* t_{AB}^S,

$$t_{AB}^S\equiv{}^{01}M_{AB}=t_{BA}^S. \qquad (10.3.7)$$

Using the definition of e_{ABC} (10.2.2) and linearizing the remaining factors in (10.3.5), we find

$$t_{ij}^{\Pi}=\frac{x_{i,A}x_{j,B}}{J}\left[t_{AB}^S+c_{ABCD}E_{CD}-e_{DAB}F_D\right]$$
$$=(\delta_{iA}+u_{i,A})(\delta_{jB}+u_{j,B})(1-u_{l,l})\left[t_{AB}^S+c_{ABCD}u_{C,D}\right.$$
$$\left.-e_{DAB}\left(E_D+E_i^S u_{i,D}\right)\right]$$
$$=t_{ij}^S+2t_{k(i}^S u_{j),k}-t_{ij}^S u_{k,k}-e_{lij}E_k^S u_{k,l}$$
$$+c_{ijkl}u_{k,l}-e_{kij}E_k. \qquad (10.3.8)$$

The stiffness tensor defined above in (10.3.6) is the stiffness tensor operative in dielectrics, piezoelectrics, and pyroelectrics (provided in the latter materials that the spontaneous electric field is canceled). We see this more clearly later in this section. The first term in its definition is the stiffness resulting from the direct response of the crystal lattice to the strain while the second term accounts for the lowering of the stiffness caused by the repositioning of the constituent particles of the unit cell by way of the internal motion degrees of freedom. It is the same effect as found in the simple diatomic one dimensional lattice in Section 2.3 and illustrated for that case in Fig. 2.1. From the interchange symmetry of $^{02}\mathbf{M}$ and $^{11}\mathbf{M}$ coefficients implied by the stored energy expansion (10.1.1) and the interchange symmetry of $\boldsymbol{\Upsilon}$ we easily deduce

$$c_{ABCD}=c_{(AB)(CD)}=c_{(CD)(AB)} \tag{10.3.9}$$

for the interchange symmetry of the elastic stiffness tensor.

We now substitute (10.3.2) and (10.3.8) into the center-of-mass equation and obtain

$$\rho^0\frac{\partial^2 u_i}{\partial t^2}=c^F_{ijkl}u_{k,lj}-e^F_{kij}E_{k,j} \tag{10.3.10}$$

for the *linearized dynamic elasticity equation.* The elastic stiffness tensor c^F_{ijkl} is the stiffness tensor appropriate to a pyroelectric crystal in possession of its spontaneous electric field and is defined by

$$c^F_{abcd}=c_{abcd}+t^S_{bd}\delta_{ac}-\epsilon_0\chi_{bd}E^S_aE^S_c-\tfrac{1}{2}E^S_ae_{bcd}-\tfrac{1}{2}E^S_ae_{dcb}-\tfrac{1}{2}E^S_ce_{dab}-\tfrac{1}{2}E^S_ce_{bad}. \tag{10.3.11}$$

Note the many terms in c^F_{abcd} depending on the presence of $\mathbf{E}^S$ and the much lowered interchange symmetry compared to c_{abcd},

$$c^F_{abcd}=c^F_{cdab}. \tag{10.3.12}$$

In particular there are terms in c^F_{abcd} that possess no interchange symmetry within the a,b or c,d pairs of indices. None of the terms depending on $\mathbf{E}^S$ have been experimentally verified as yet. Because of the derivatives present on the displacement in (10.3.10), we have symmetrized the spontaneous stress and spontaneous electric field terms with respect to j,l. Note also that a spontaneous electric field term enters the piezoelectric term in the elasticity equation through the definition (10.2.8) of e^F_{kij}.

In the absence of a spontaneous electric field and a spontaneous stress we have

$$c^F_{abcd}=c_{abcd} \qquad (\mathbf{E}^S=0,\mathbf{t}^S=0), \tag{10.3.13}$$

$$e^F_{kij}=e_{kij} \qquad (\mathbf{E}^S=0,\mathbf{t}^S=0), \tag{10.3.14}$$

and so

$$\rho^0\frac{\partial^2 u_i}{\partial t^2}=c_{ijkl}u_{k,lj}-e_{kij}E_{k,j}, \tag{10.3.15}$$

which is the conventional result. This confirms that the c_{ijkl} defined in (10.3.6) is the ordinary elastic stiffness tensor operative in dielectrics, piezoelectrics, and pyroelectrics for which $\mathbf{E}^S = 0$. For piezoelectrics c_{ijkl} is the stiffness tensor operative when the piezoelectrically generated electric field is shorted out, that is, when the last term in (10.3.15) can be dropped. We consider in later sections the modification of the effective stiffness tensor in piezoelectrics when a piezoelectrically generated electric field is present.

10.4 Linearized Total Stress Tensor

In Section 8.3 the criteria for the definition of the total stress tensor are developed. They lead us to define it as that stress tensor whose scalar product with the outward unit normal is continuous at a stationary surface where material properties are discontinuous. This makes it the most important stress tensor of the many that we encounter.

The total stress tensor is given in the adiabatic approximation by

$$t^T_{ij} = t^{\Pi}_{ij} + m_{ij} - \rho \dot{x}_i \dot{x}_j. \tag{10.4.1}$$

The last term, the flow of center-of-mass momentum, has no linear part. The Maxwell stress tensor m_{ij} has linear and constant parts given by

$$m_{ij} = \epsilon_0\left(2E^S_{(i}E_{j)} - \tfrac{1}{2}E^S_k E^S_k \delta_{ij} - E^S_k E_k \delta_{ij} + E^S_i E^S_j\right) \tag{10.4.2}$$

as readily apparent from examining its definition (8.1.8). The linear form of $\mathbf{t}^{\Pi}$ is obtained in the preceding section. Combining these equations yields

$$\begin{aligned} t^T_{ij} = &\left(t^S_{ij} + \epsilon_0 E^S_i E^S_j - \tfrac{1}{2}\epsilon_0 E^S_k E^S_k \delta_{ij}\right) \\ &+ \left(c_{ijkl} + 2t^S_{l(i}\delta_{j)k} - t^S_{ij}\delta_{kl} - E^S_k e_{lij}\right) u_{k,l} \\ &- \left(e_{kij} - 2\epsilon_0 E^S_{(i}\delta_{j)k} + \epsilon_0 \delta_{ij} E^S_k\right) E_k = t^T_{ji} \end{aligned} \tag{10.4.3}$$

for the linearized total stress tensor. We have already discussed its constant part in Section 8.5 in regard to the stress present in the natural state of the crystal. Note that most of the terms in (10.4.3) depend on the existence of a spontaneous electric field in a pyroelectric and that a few couple to rotations $u_{[k,l]}$ as well as to strains $u_{(k,l)}$. For a pyroelectric for which $\mathbf{E}^S = 0$ or for a piezoelectric the linearized total stress becomes

$$t^T_{ij} = c_{ijkl} u_{k,l} - e_{kij} E_k = t^T_{ji}. \tag{10.4.4}$$

and for a nonpiezoelectric dielectric it becomes

$$t_{ij}^T = c_{ijkl} u_{k,l} = t_{ji}^T. \tag{10.4.5}$$

By comparison with (10.3.15) we see that the divergence of the stress tensors of (10.4.4) and (10.4.5) appear in the linearized dynamic elasticity equation for these respective types of crystals. Note that a similar comparison carried out for the total stress tensor of a pyroelectric in possession of $\mathbf{E}^S$ as given in (10.4.3) shows that it is not the stress tensor whose divergence enters the general linearized dynamic elasticity equation (10.3.10).

10.5 Matrix Notation in Linear Elasticity and Piezoelectricity

Because the piezoelectric stress tensor is a third rank tensor and the elastic stiffness tensor is a fourth rank tensor, it is complicated to record and to manipulate these quantities for crystals of various symmetries. The interchange symmetries that these tensors possess, however, allow the introduction of a *contracted* or *matrix notation* by which a convenient two dimensional array or matrix representation of these tensors can be made. This notation is commonly used when doing calculations for specific crystals.

Before introducing the matrix notation let us consider rearrangements of the linear stress-strain-electric field relation in piezoelectrics. Here we rewrite (10.4.4) as

$$t_{ij} = c_{ijkl} S_{kl} - e_{kij} E_k \tag{10.5.1}$$

where the superscript T has been dropped from the stress and the infinitesimal strain tensor S_{kl}, introduced in (3.12.3), is inserted for $u_{(k,l)}$. This equation may be solved for the strain,

$$S_{mn} = s_{mnij} t_{ij} + d_{kmn} E_k, \tag{10.5.2}$$

provided that we define the elastic *compliance tensor* s_{mnij} by

$$s_{mnij} c_{ijkl} \equiv \tfrac{1}{2}(\delta_{mk}\delta_{nl} + \delta_{ml}\delta_{nk}) \tag{10.5.3}$$

and the *piezoelectric strain tensor* d_{kmn} by

$$d_{kmn} \equiv s_{mnij} e_{kij}. \tag{10.5.4}$$

From (10.5.3) and the interchange symmetry of c_{ijkl} we conclude that

$$s_{mnij} = s_{(mn)(ij)} = s_{(ij)(mn)}. \tag{10.5.5}$$

This leads to

$$d_{kmn} = d_{k(mn)} \tag{10.5.6}$$

for the interchange symmetry for d_{kmn}.

We now contract the tensor index notation to a matrix index notation for each pair of tensor indices for which interchange symmetry exists. Any such pair (ij) is denoted by m where the association between the notations is

$$\begin{array}{lcccccc} (ij)\to & 11 & 22 & 33 & 23 \text{ or } 32 & 13 \text{ or } 31 & 12 \text{ or } 21 \\ m\to & 1 & 2 & 3 & 4 & 5 & 6. \end{array} \tag{10.5.7}$$

This makes the stress and strain tensors 1×6 matrices, the stiffness and compliance tensors 6×6 matrices, and the piezoelectric stress and strain tensors 3×6 matrices.

In order to preserve the normal rules of matrix multiplication some factors of 2 and 4 must be introduced into the definitions of the quantities in contracted notation. For the strain tensor

$$\begin{aligned} S_{(ij)} &= S_m \qquad (m=1,2,3) \\ &= \tfrac{1}{2}S_m \qquad (m=4,5,6). \end{aligned} \tag{10.5.8}$$

For the stress tensor

$$t_{(ij)} = t_m \qquad (m=1,2,3,4,5,6). \tag{10.5.9}$$

For the piezoelectric stress tensor

$$e_{k(ij)} = e_{km} \qquad (m=1,2,3,4,5,6). \tag{10.5.10}$$

For the piezoelectric strain tensor

$$\begin{aligned} d_{k(ij)} &= d_{km} \qquad (m=1,2,3) \\ &= \tfrac{1}{2}d_{km} \qquad (m=4,5,6). \end{aligned} \tag{10.5.11}$$

For the elastic stiffness tensor

$$c_{(ij)(kl)} = c_{mn} \qquad (m \text{ and } n=1,2,3,4,5,6). \tag{10.5.12}$$

For the elastic compliance tensor

$$
\begin{aligned}
s_{(ij)(kl)} &= s_{mn} && (m \text{ and } n = 1, 2, 3) \\
&= \tfrac{1}{2} s_{mn} && (m \text{ or } n = 4, 5, 6) \\
&= \tfrac{1}{4} s_{mn} && (m \text{ and } n = 4, 5, 6).
\end{aligned} \tag{10.5.13}
$$

In matrix notation (10.5.1) to (10.5.4) are written as

$$t_m = c_{mn} S_n - e_{km} E_k, \tag{10.5.14}$$

$$S_n = s_{nm} t_m + d_{kn} E_k, \tag{10.5.15}$$

$$s_{lm} c_{mn} = \delta_{ln}, \tag{10.5.16}$$

$$d_{km} = e_{kn} s_{nm} \tag{10.5.17}$$

where summation over repeated subscripts is implied. The interchange symmetries (10.3.9) and (10.5.5) require

$$c_{mn} = c_{nm}, \tag{10.5.18}$$

$$s_{mn} = s_{nm}. \tag{10.5.19}$$

If (10.5.14) is written out, for example, in terms of matrix arrays, we have

$$
\begin{bmatrix} t_1 \\ t_2 \\ t_3 \\ t_4 \\ t_5 \\ t_6 \end{bmatrix}
=
\begin{bmatrix}
c_{11} & c_{12} & c_{13} & c_{14} & c_{15} & c_{16} \\
c_{21} & c_{22} & c_{23} & c_{24} & c_{25} & c_{26} \\
c_{31} & c_{32} & c_{33} & c_{34} & c_{35} & c_{36} \\
c_{41} & c_{42} & c_{43} & c_{44} & c_{45} & c_{46} \\
c_{51} & c_{52} & c_{53} & c_{54} & c_{55} & c_{56} \\
c_{61} & c_{62} & c_{63} & c_{64} & c_{65} & c_{66}
\end{bmatrix}
\cdot
\begin{bmatrix} S_1 \\ S_2 \\ S_3 \\ S_4 \\ S_5 \\ S_6 \end{bmatrix}
-
\begin{bmatrix}
e_{11} & e_{21} & e_{31} \\
e_{12} & e_{22} & e_{32} \\
e_{13} & e_{23} & e_{33} \\
e_{14} & e_{24} & e_{34} \\
e_{15} & e_{25} & e_{35} \\
e_{16} & e_{26} & e_{36}
\end{bmatrix}
\cdot
\begin{bmatrix} E_1 \\ E_2 \\ E_3 \end{bmatrix}.
\tag{10.5.20}
$$

Note that the matrix notation convention of having the first subscript denote the row is not obeyed for e_{km} (or d_{km} either).

Crystal symmetry places restrictions among the elements of any tensor that characterizes the material properties of a crystal. In high symmetry crystals many elements are required to vanish. These restrictions for the elastic and piezoelectric properties, c_{mn}, s_{mn}, e_{km}, and d_{km}, of the 32 crystal classes are listed in the Appendix.

10.6 Elasticity of Isotropic Solids

In the formulation of this theory the periodicity of a crystal lattice produces a discrete number of internal modes of motion of the lattice. This theory would give a poor representation of a structurally amorphous solid such as glass or plastic in the region near the resonances of these internal motions, since such a material would not have discrete resonances. However, for frequencies far below the resonance region the lack of lattice regularity makes little difference, since the material motion need be characterized only by the center of mass. Thus the elastic motion of a structurally amorphous solid is adequately described by this theory provided only that any lack of homogeneity or isotropy occurs within dimensions small compared to the wavelength of the shortest wavelength acoustic disturbance considered.

The symmetry requirement of elastic isotropy requires the stiffness tensor to have the simple form

$$c_{ijkl}=\lambda\delta_{ij}\delta_{kl}+\mu(\delta_{ik}\delta_{jl}+\delta_{il}\delta_{jk}) \tag{10.6.1}$$

where λ and μ are called the *Lamé coefficients.* The need for only two coefficients in an isotropic medium is reasonable since the elastic response can be different only for longitudinal (compressional) and transverse (shear) disturbances. This becomes clearer when we derive the acoustic eigenmodes of an elastically isotropic solid in Section 10.9. From (10.6.1) we find the only nonzero stiffness coefficients (in matrix notation) to be

$$c_{11}=c_{22}=c_{33}=2\mu+\lambda, \tag{10.6.2a}$$

$$c_{(12)}=c_{(23)}=c_{(31)}=\lambda, \tag{10.6.2b}$$

$$c_{44}=c_{55}=c_{66}=\mu. \tag{10.6.2c}$$

From these values and the relation (10.5.16) we obtain the only nonzero compliance coefficients (in matrix notation) to be

$$s_{11}=s_{22}=s_{33}=\frac{\mu+\lambda}{\mu(2\mu+3\lambda)}, \tag{10.6.3a}$$

$$s_{(12)}=s_{(23)}=s_{(31)}=\frac{-\lambda}{2\mu(2\mu+3\lambda)}, \tag{10.6.3b}$$

$$s_{44}=s_{55}=s_{66}=\frac{1}{\mu}. \tag{10.6.3c}$$

Other measures of the elasticity of isotropic solids are often used. Among these are Young's modulus and Poisson's ratio. To understand their usefulness consider the compression of a rod along its length. If the 3-direction is along the rod length and the pressure on its ends denoted by $-p$, the stress may be taken as homogeneous and given by

$$t_j = -p\delta_{j3}. \tag{10.6.4}$$

The strain is given by

$$S_i = s_{ij} t_j \tag{10.6.5}$$

from (10.5.15) (the piezoelectric term is set equal to zero because isotropic solids are nonpiezoelectric). The strain S_3 is thus

$$S_3 = -s_{33} p = -\frac{p}{E} \tag{10.6.6}$$

where

$$E \equiv \frac{1}{s_{33}} \tag{10.6.7}$$

is called *Young's modulus.* It is seen to be a stiffness coefficient that measures the resistance of the rod to longitudinal compression. The larger Young's modulus is, the less the compression is for a fixed pressure. Since $s_{13} = s_{23}$ for an isotropic solid, the strain components S_1 and S_2 are equal and given by

$$S_1 = S_2 = -s_{13} p. \tag{10.6.8}$$

Poisson's ratio ν is defined as the negative of the ratio of the strain for a direction perpendicular to the rod axis to that along the direction of the rod axis and so measures lateral expansion during length contraction. Thus

$$\nu \equiv -\frac{S_1}{S_3} = -\frac{S_2}{S_3} = -\frac{s_{13}}{s_{33}} = -E s_{13}. \tag{10.6.9}$$

The remaining compliance coefficient s_{44} can be expressed in terms of E and ν by noting from (10.6.3) that

$$s_{44} = 2(s_{11} - s_{12}). \tag{10.6.10}$$

In summary, the compliance coefficients can be expressed in terms of Young's modulus E and Poisson's ratio ν by

$$s_{11}=s_{22}=s_{33}=\frac{1}{E}, \tag{10.6.11a}$$

$$s_{(12)}=s_{(23)}=s_{(31)}=-\frac{\nu}{E}, \tag{10.6.11b}$$

$$s_{44}=s_{55}=s_{66}=\frac{2(1+\nu)}{E}. \tag{10.6.11c}$$

By (10.5.16) these imply that

$$c_{11}=c_{22}=c_{33}=\frac{E(1-\nu)}{(1+\nu)(1-2\nu)}, \tag{10.6.12a}$$

$$c_{(12)}=c_{(23)}=c_{(31)}=\frac{E\nu}{(1+\nu)(1-2\nu)}, \tag{10.6.12b}$$

$$c_{44}=c_{55}=c_{66}=\frac{E}{2(1+\nu)}. \tag{10.6.12c}$$

10.7 Piezoelectrically Generated Electric Field

In the linearized dynamic elasticity equation (10.3.10) a piezoelectric term depending on the electric field appears. On the other hand, the dielectric current (10.2.11) gives the electric field wave equation (9.2.3) the form

$$[\nabla\times(\nabla\times\mathbf{E})]_i+\frac{\kappa_{ij}}{c^2}\frac{\partial^2 E_j}{\partial t^2}=-\mu_0 e^F_{ijk}\frac{\partial^2 u_{j,k}}{\partial t^2}, \tag{10.7.1}$$

which contains a piezoelectric term depending on the displacement gradient. We thus have a coupled pair of differential equations for the displacement and the electric field. Consider the propagation of an acoustic plane wave in a piezoelectric crystal. The acoustic wave drives the electric field wave equation through the term on the right side. This piezoelectrically generated electric field, which we deduce from this equation in this section, then affects the propagation of the acoustic wave that generates it.

Consider a plane wave whose displacement and electric field each contain the factor

$$e^{i(k\mathbf{s}\cdot\mathbf{z}-\omega t)}, \tag{10.7.2}$$

k being the propagation constant, $\mathbf{s}$ the unit vector in the direction of propagation, and ω the angular frequency. Substitution into the wave equation gives

$$-k^2[\mathbf{s}\times(\mathbf{s}\times\mathbf{E})]_i - \frac{\omega^2}{c^2}\kappa_{ij}E_j = \mu_0\omega^2 e^F_{ijk}u_{j,k} \tag{10.7.3}$$

or

$$\left[n^2(\delta_{ij} - s_i s_j) - \kappa_{ij}\right]E_j = \frac{1}{\epsilon_0} e^F_{ijk}u_{j,k} \tag{10.7.4}$$

where

$$n \equiv \frac{kc}{\omega} \tag{10.7.5}$$

is the index of refraction of the driven wave. Formally the solution for the piezoelectrically generated electric field is

$$E_l = \frac{1}{\epsilon_0}[\alpha^{-1}]_{li} e^F_{ijk}u_{j,k} \tag{10.7.6}$$

where the $\boldsymbol{\alpha}$ tensor is defined by

$$\boldsymbol{\alpha} \equiv n^2(\mathbf{1} - \mathbf{ss}) - \boldsymbol{\kappa}. \tag{10.7.7}$$

The problem is now reduced to finding an expression for $\boldsymbol{\alpha}^{-1}$. A natural choice is to express $\boldsymbol{\alpha}^{-1}$ in terms of the electric field eigenvectors of the undriven wave equation as done for a vector in (9.3.10) and (9.3.11). This gives

$$\boldsymbol{\alpha}^{-1} = \sum_{\beta,\delta=1}^{3} C_{\beta\delta}\mathcal{E}^{\beta}\mathcal{E}^{\delta} \tag{10.7.8}$$

with

$$C_{\beta\delta} = \mathfrak{D}^{\beta}\cdot\boldsymbol{\alpha}^{-1}\cdot\mathfrak{D}^{\delta}. \tag{10.7.9}$$

The problem is now to find $C_{\beta\delta}$.

Consider the definition of $\boldsymbol{\alpha}^{-1}$,

$$[\alpha^{-1}]_{ij}\left[n^2(\delta_{jk} - s_j s_k) - \kappa_{jk}\right] = \delta_{ik}. \tag{10.7.10}$$

Form the scalar product over i with $\mathfrak{D}_i^\beta$ and over k with $\mathcal{E}_k^\delta$ to obtain

$$\mathfrak{D}_i^\beta[\alpha^{-1}]_{il}\delta_{lj}[n^2(\delta_{jk}-s_js_k)-\kappa_{jk}]\mathcal{E}_k^\delta=\delta^{\beta\delta} \tag{10.7.11}$$

in which the biorthonormality condition (9.3.9) is used on the right side. Next substitute the completeness relation (9.3.13) for δ_{lj},

$$\sum_\gamma \mathfrak{D}_i^\beta[\alpha^{-1}]_{il}\mathfrak{D}_l^\gamma\mathcal{E}_j^\gamma\,[n^2(\delta_{jk}-s_js_k)-\kappa_{jk}]\mathcal{E}_k^\delta=\delta^{\beta\delta}. \tag{10.7.12}$$

The eigenvector equation (9.3.1) and the electric displacement eigenvector definition (9.3.8) are now substituted, leading to

$$\sum_\gamma \mathfrak{D}_i^\beta[\alpha^{-1}]_{il}\mathfrak{D}_l^\gamma\mathcal{E}_j^\gamma\left[\left(\frac{n}{n^\delta}\right)^2-1\right]\mathfrak{D}_j^\delta=\delta^{\beta\delta} \tag{10.7.13}$$

where n^δ is the refractive index of a free wave in the same direction $\mathbf{s}$ and at the same frequency as the forced wave of refractive index n. Use of the biorthonormality condition gives

$$\sum_\gamma \mathfrak{D}_i^\beta[\alpha^{-1}]_{il}\mathfrak{D}_l^\gamma\delta^{\gamma\delta}\left[\left(\frac{n}{n^\delta}\right)^2-1\right]=\delta^{\beta\delta} \tag{10.7.14}$$

and finally

$$C_{\beta\delta}=\boldsymbol{\mathfrak{D}}^\beta\cdot\boldsymbol{\alpha}^{-1}\cdot\boldsymbol{\mathfrak{D}}^\delta=\frac{\delta^{\beta\delta}}{(n/n^\delta)^2-1}. \tag{10.7.15}$$

Combining (10.7.8), (10.7.9), and (10.7.15) yields

$$\boldsymbol{\alpha}^{-1}=\sum_{\beta=1}^{3}\frac{\boldsymbol{\mathcal{E}}^\beta\boldsymbol{\mathcal{E}}^\beta}{(n/n^\beta)^2-1} \tag{10.7.16a}$$

$$=\frac{\boldsymbol{\mathcal{E}}^{(1)}\boldsymbol{\mathcal{E}}^{(1)}}{(n/n^{(1)})^2-1}+\frac{\boldsymbol{\mathcal{E}}^{(2)}\boldsymbol{\mathcal{E}}^{(2)}}{(n/n^{(2)})^2-1}-\frac{\mathbf{ss}}{\mathbf{s}\cdot\boldsymbol{\kappa}\cdot\mathbf{s}} \tag{10.7.16b}$$

where in the expanded form the longitudinal eigenvector,

$$\boldsymbol{\mathcal{E}}^{(3)}=\frac{\mathbf{s}}{(\mathbf{s}\cdot\boldsymbol{\kappa}\cdot\mathbf{s})^{1/2}}, \tag{10.7.17}$$

and eigenvalue,

$$n^{(3)}=\infty, \tag{10.7.18}$$

are substituted since their forms are the same for dielectrics of any anisotropy. We emphasize that n is the index of the forced or driven wave while $n^{(1)}$ and $n^{(2)}$ are the refractive indices of the freely propagating waves of frequency ω and propagation direction $\mathbf{s}$. Equation (10.7.16) is a general result useful for studying any *forced wave solution* of the electric field wave equation. We find use for it in studying nonlinear optics in later chapters. Note that the derivation is made possible by the use of a complete set of three eigenvectors that includes the longitudinal mode eigenvector.

For a piezoelectrically generated electric wave $\boldsymbol{\alpha}^{-1}$ can be simplified by an approximation. The acoustic wave that generates and carries this electric wave with it has a velocity that is typically

$$v=\frac{\omega}{k}\sim 3\times 10^5\ \text{cm/sec} \tag{10.7.19}$$

leading to a refractive index of

$$n=\frac{kc}{\omega}=\frac{c}{v}\sim 10^5, \tag{10.7.20}$$

an exceptionally large value. Since $n^{(1)}$ and $n^{(2)}$ are in the range of 1 to 10 typically, the first two terms in (10.7.16b) are about 10^8 to 10^{10} times smaller than the longitudinal mode term and so may be dropped to an excellent level of approximation. The interpretation of this is simply that an electric wave traveling as slowly as an acoustic wave is much more like the longitudinal "nonpropagating" mode than either of the transverse propagating modes. The *piezoelectrically generated electric field* is now given by

$$E_l=-\frac{s_l s_i}{\epsilon_0 \mathbf{s}\cdot\boldsymbol{\kappa}\cdot\mathbf{s}}\,e_{ijk}^F u_{j,k} \tag{10.7.21}$$

and is a *longitudinal, propagating electric wave.* Thus the longitudinal eigenmode that cannot propagate as a free wave can propagate as a forced wave.

10.8 Acoustic Eigenmodes

The expression for the piezoelectrically generated electric field (10.7.21) may be used to eliminate the electric field from the linearized dynamic elasticity equation (10.3.10) which is then referred to as the *acoustic wave equation.* We use it to determine the plane wave acoustic eigenmodes of crystals.

If both the displacement and the electric field are regarded as containing the plane wave factor (10.7.2), then substitution of (10.7.21) into (10.3.10) results in

$$-\rho^0\omega^2 u_i = -k^2 s_l s_j c^F_{ijkl} u_k - k^2 s_l s_j \frac{e^F_{mij} s_m s_n e^F_{nkl}}{\epsilon_0 \mathbf{s}\cdot\boldsymbol{\kappa}\cdot\mathbf{s}} u_k. \tag{10.8.1}$$

This is conveniently reexpressed as

$$\rho^0 v^2 u_i = \overline{c^F_{ijkl}}\, s_j s_l u_k \tag{10.8.2}$$

where the acoustic wave phase velocity v is defined by

$$v \equiv \frac{\omega}{k} \tag{10.8.3}$$

and the *piezoelectrically stiffened stiffness tensor* is defined by

$$\overline{c^F_{ijkl}} \equiv c^F_{ijkl} + \frac{e^F_{mij} s_m s_n e^F_{nkl}}{\epsilon_0 \mathbf{s}\cdot\boldsymbol{\kappa}\cdot\mathbf{s}}. \tag{10.8.4}$$

The second term in this equation is always positive and so increases or stiffens the stiffness tensor. It represents the effect that the piezoelectrically created electric field has on the propagation of the wave that creates it. It is an *indirect* contribution to the stiffness tensor and is experimentally distinguishable from the *direct* contribution c^F_{ijkl}. It is distinguishable because it is a function of the unit propagation vector $\mathbf{s}$, a property of the wave, not of the crystal.

If the eigenvectors of the displacement are denoted by $\mathbf{b}^\alpha$ ($\alpha = 1,2,3$), (10.8.2) becomes *Christoffel's equation* or the *eigenvector equation*

$$\frac{1}{\rho^0}\overline{c^F_{ijkl}}\, s_j s_l b^\alpha_k = v^2_\alpha b^\alpha_i \tag{10.8.5}$$

with the square of the phase velocity being the eigenvalue. A solution of these equations exists provided that

$$\det\left(\frac{1}{\rho^0}\overline{c^F_{ijkl}} s_j s_l - v^2\delta_{ik}\right) = 0. \tag{10.8.6}$$

This in conjunction with (10.8.3) is the *dispersion relation* for acoustic waves in crystals which may be simple dielectrics, piezoelectrics, or pyroelectrics (the latter with or without their spontaneous electric field). It is a cubic equation for v^2 and so gives three values of $|v|$ for a given frequency ω and propagation direction **s**. Thus, in contrast to optical propagation, there are *three freely propagating eigenmodes of acoustic waves in crystals.* It is apparent from (10.8.6) that the solutions for v depend only on the components of $\overline{c_{ijkl}^{F}}$ and the mass density ρ^0 and so are *constants*, since neither of these quantities has any appreciable frequency dependence. Thus ω is a linear function of k for long wavelength acoustic waves just as is found in Chapters 1 and 2.

It is simple to show (see Section 3.5) that the displacement eigenvectors $\mathbf{b}^{\alpha}$ satisfy an *orthonormality condition*

$$\mathbf{b}^{\alpha}\cdot\mathbf{b}^{\beta}=\delta^{\alpha\beta}. \tag{10.8.7}$$

This states that, regardless of the symmetry of the crystal and regardless of the direction of propagation, the *displacement vectors of the three eigenmodes are perpendicular* (see Fig. 10.1). In special cases one mode may be purely *longitudinal* and the other two purely *transverse*. In more general circumstances all three modes contain longitudinal and transverse components and are termed *hybrid* or *mixed* modes. By the methods of Section

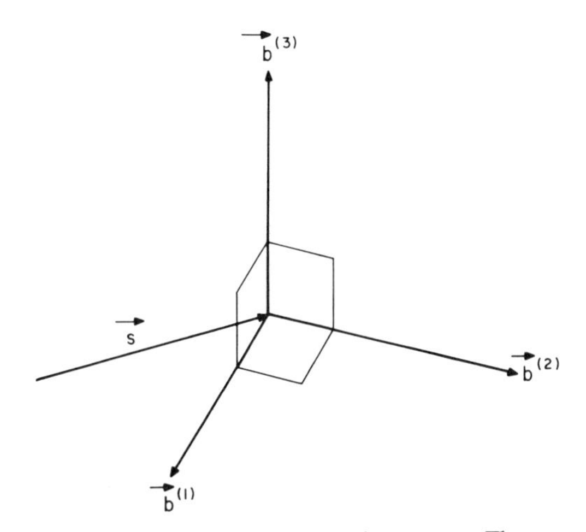

Fig. 10.1. Orthogonality of acoustic displacement eigenvectors. The general case where no eigenvector is parallel (longitudinal) or perpendicular (transverse) to the unit propagation vector **s** is shown.

3.5 the displacement eigenvectors may also be shown to satisfy

$$\sum_{\alpha} b_i^{\alpha} b_j^{\alpha} = \delta_{ij}, \tag{10.8.8}$$

which is the *completeness relation.*

As an example of the usefulness of the eigenvectors consider the problem of the driven acoustic wave equation, analogous to the problem of the driven optical wave equation of Section 10.7. Let G_i be the amplitude of some driving plane wave in the acoustic wave equation,

$$\beta_{ik} u_k = G_i, \tag{10.8.9}$$

where

$$\beta_{ik} \equiv \frac{1}{\rho^0} \overline{c_{ijkl}^F} \, s_j s_l - v^2 \delta_{ik}. \tag{10.8.10}$$

The driving wave propagates in the direction $\mathbf{s}$ (a unit vector) with frequency ω and velocity v. Formally the amplitude of the displacement vector is given by

$$u_j = (\beta^{-1})_{ji} G_i. \tag{10.8.11}$$

The tensor $\boldsymbol{\beta}^{-1}$ can be expressed in terms of the eigenvectors of the undriven wave equation as

$$\boldsymbol{\beta}^{-1} = \sum_{\alpha} \frac{\mathbf{b}^{\alpha} \mathbf{b}^{\alpha}}{v_{\alpha}^2 - v^2}. \tag{10.8.12}$$

We leave the demonstration of this for an exercise. The solution for the displacement vector amplitude is then

$$\mathbf{u} = \sum_{\alpha} \frac{\mathbf{b}^{\alpha} (\mathbf{b}^{\alpha} \cdot \mathbf{G})}{v_{\alpha}^2 - v^2}. \tag{10.8.13}$$

We make use of this expression when studying nonlinear acoustics in Chapter 17.

10.9 Acoustic Eigenmodes of Isotropic Solids

Isotropic solids such as glasses offer a simple but important example of acoustic eigenmodes. If the stiffness tensor, expressed in terms of the

Lamé coefficients in (10.6.1), is substituted into the eigenvector equation (10.8.5), we have

$$[\lambda s_i s_k + \mu(\delta_{ik} + s_i s_k)]b_k = \rho^0 v^2 b_i. \tag{10.9.1}$$

If we choose a coordinate system with the 3-direction along $\mathbf{s}$,

$$\mathbf{s} = [0,0,1], \tag{10.9.2}$$

then the eigenvector equation gives three decoupled equations,

$$\mu b_1 = \rho^0 v^2 b_1, \tag{10.9.3a}$$

$$\mu b_2 = \rho^0 v^2 b_2, \tag{10.9.3b}$$

$$(\lambda + 2\mu) b_3 = \rho^0 v^2 b_3. \tag{10.9.3c}$$

The eigenvectors thus are

$$\mathbf{b}^\alpha = [1,0,0] \qquad (\alpha=1), \tag{10.9.4a}$$

$$= [0,1,0] \qquad (\alpha=2), \tag{10.9.4b}$$

$$= [0,0,1] \qquad (\alpha=3), \tag{10.9.4c}$$

and the eigenvalues are

$$v_\alpha = \left[\frac{\mu}{\rho^0}\right]^{1/2} \qquad (\alpha=1), \tag{10.9.5a}$$

$$= \left[\frac{\mu}{\rho^0}\right]^{1/2} \qquad (\alpha=2), \tag{10.9.5b}$$

$$= \left[\frac{\lambda+2\mu}{\rho^0}\right]^{1/2} \qquad (\alpha=3). \tag{10.9.5c}$$

The orthogonality of the eigenvectors is readily apparent from (10.9.4). It is seen that the $\alpha=1$ and $\alpha=2$ eigenvectors are purely transverse ($\mathbf{b}^\alpha \perp \mathbf{s}$) and that the $\alpha=3$ eigenvector is purely longitudinal ($\mathbf{b}^\alpha \| \mathbf{s}$) as illustrated in Fig. 10.2. The eigenvalues (the velocities) of the two transverse modes are the same and determined by the Lamé coefficient μ while the eigenvalue of the longitudinal mode is determined by a combination of the Lamé coefficients.

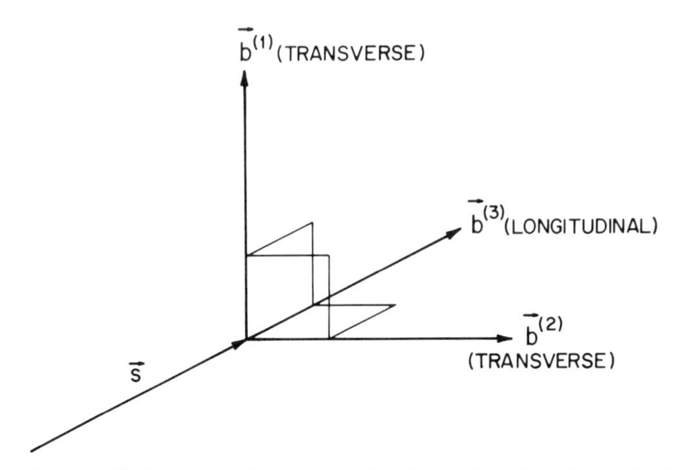

Fig. 10.2. Acoustic displacement eigenvectors of an isotropic solid. The longitudinal eigenvector is parallel to the unit propagation vector **s** while the two transverse eigenvectors are perpendicular to **s**.

By including the space and time dependence the three eigenmodes are given by

$$\mathbf{u}^{(1)} = \mathbf{b}^{(1)} e^{i\omega(z/v_1 - t)}, \tag{10.9.6a}$$

$$\mathbf{u}^{(2)} = \mathbf{b}^{(2)} e^{i\omega(z/v_2 - t)}, \tag{10.9.6b}$$

$$\mathbf{u}^{(3)} = \mathbf{b}^{(3)} e^{i\omega(z/v_3 - t)}. \tag{10.9.6c}$$

Calculating the nonzero displacement gradients yields only

$$u_{1,3} = \frac{i\omega}{v_1} b_1 e^{i\omega(z/v_1 - t)} = u_{(1,3)} + u_{[1,3]}, \tag{10.9.7a}$$

$$u_{2,3} = \frac{i\omega}{v_2} b_2 e^{i\omega(z/v_2 - t)} = u_{(2,3)} + u_{[2,3]}, \tag{10.9.7b}$$

$$u_{3,3} = \frac{i\omega}{v_3} b_3 e^{i\omega(z/v_3 - t)} = u_{(3,3)}. \tag{10.9.7c}$$

The point of this is the demonstration that while both longitudinal and transverse acoustic waves possess strain, $u_{(i,j)}$, *only the transverse waves possess rotation*, $u_{[i,j]}$. Figure 10.3 illustrates this point. It is further apparent from (10.9.7) that for purely transverse waves the strain $u_{(i,j)}$ and rotation $u_{[i,j]}$ are numerically equal.

(b)

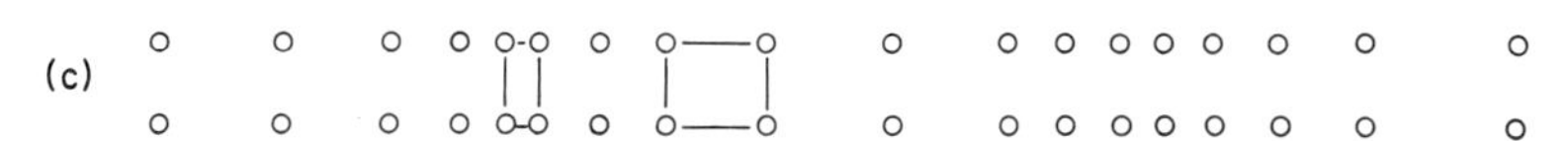

Fig. 10.3. Rotation in shear waves. (*a*) No excitation. (*b*) Transverse wave having strain plus rotation. (*c*) Longitudinal wave having strain only. [Reprinted with permission from Lax and Nelson, 1972.]

10.10 Reflection and Refraction of Acoustic Waves

The laws governing the directions of reflection and refraction of acoustic waves are simple and worth deriving. The laws governing the amplitudes of the reflected and refracted waves are very complicated under general circumstances and so are not derived here. It is better to calculate the amplitudes for each particular geometry studied.

Consider a nonpyroelectric, nonpiezoelectric medium in which a plane acoustic wave is incident on an interface plane between two different crystals that are in adhesive contact. If the origin of coordinates is placed in the interface plane, the equation of the plane is simply

$$\mathbf{n}\cdot\mathbf{z}=0 \tag{10.10.1}$$

where $\mathbf{n}$ is a unit normal and $\mathbf{z}$ the coordinate vector. Across this plane the scalar product of the linearized total stress tensor, which for the medium considered is simply

$$t_{ij}=c_{ijkl}u_{k,l}, \tag{10.10.2}$$

must be continuous,

$$\left(t_{ij}^{(2)}-t_{ij}^{(1)}\right)n_j=0. \tag{10.10.3}$$

Because the two crystals are in adhesive contact, the displacement vector must also be continuous across the interface,

$$u_i^{(2)} - u_i^{(1)} = 0. \tag{10.10.4}$$

If the second medium were a vacuum, this condition would not be needed.

We take the incident plane wave to be a single eigenmode designated α,

$$\mathbf{u}^I = A_I^\alpha \mathbf{b}^\alpha e^{i\varphi_I^\alpha}, \qquad \varphi_I^\alpha \equiv \mathbf{k}_I^\alpha \cdot \mathbf{z} - \omega t, \tag{10.10.5}$$

with A_I^α its amplitude and $\mathbf{b}^\alpha$ its eigenvector. In general this wave excites three reflected wave eigenmodes,

$$\mathbf{u}^R = \sum_{\beta=1}^{3} A_R^\beta \mathbf{b}^\beta e^{i\varphi_R^\beta}, \qquad \varphi_R^\beta \equiv \mathbf{k}_R^\beta \cdot \mathbf{z} - \omega t \tag{10.10.6}$$

and three transmitted wave eigenmodes,

$$\mathbf{u}^T = \sum_{\gamma=1}^{3} A_T^\gamma \mathbf{c}^\gamma e^{i\varphi_T^\gamma}, \qquad \varphi_T^\gamma \equiv \mathbf{k}_T^\gamma \cdot \mathbf{z} - \omega t. \tag{10.10.7}$$

Here $\mathbf{c}^\gamma$ denotes the eigenvectors of the second medium.

These displacement vectors are now substituted into the boundary conditions. Continuity of the displacement vector on the interface gives

$$A_I^\alpha \mathbf{b}^\alpha e^{i\varphi_I^\alpha} + \sum_\beta A_R^\beta \mathbf{b}^\beta e^{i\varphi_R^\beta} = \sum_\gamma A_T^\gamma \mathbf{c}^\gamma e^{i\varphi_T^\gamma} \tag{10.10.8}$$

and continuity of the normal stress on the interface gives

$$c_{ijkl}^{(1)} n_j \left(k_{Il}^\alpha A_I^\alpha b_k^\alpha e^{i\varphi_I^\alpha} + \sum_\beta k_{Rl}^\beta A_R^\beta b_k^\beta e^{i\varphi_R^\beta} \right) = c_{ijkl}^{(2)} n_j \sum_\gamma k_{Tl}^\gamma A_T^\gamma c_k^\gamma e^{i\varphi_T^\gamma}. \tag{10.10.9}$$

Since these two conditions must hold for all $\mathbf{z}$ on the interface, the $\mathbf{z}$ dependent terms in the three phase factors must be equal,

$$\mathbf{k}_I^\alpha \cdot \mathbf{z} = \mathbf{k}_R^\beta \cdot \mathbf{z} = \mathbf{k}_T^\gamma \cdot \mathbf{z} \qquad (\beta, \gamma = 1, 2, 3). \tag{10.10.10}$$

This leads immediately to

$$(\mathbf{k}_I^\alpha - \mathbf{k}_R^\beta) \cdot \mathbf{z} = (\mathbf{k}_I^\alpha - \mathbf{k}_T^\gamma) \cdot \mathbf{z} = (\mathbf{k}_R^\beta - \mathbf{k}_T^\gamma) \cdot \mathbf{z} = 0, \tag{10.10.11}$$

which states that any **z** lying in the interface plane is perpendicular to the difference of any two wavevectors. Hence all these differences are perpendicular to the plane and so parallel to the unit normal **n**. This causes the vector product of these differences with **n** to vanish,

$$(\mathbf{k}_I^\alpha - \mathbf{k}_R^\beta) \times \mathbf{n} = (\mathbf{k}_I^\alpha - \mathbf{k}_T^\gamma) \times \mathbf{n} = (\mathbf{k}_R^\beta - \mathbf{k}_T^\gamma) \times \mathbf{n} = 0 \qquad (10.10.12)$$

or

$$\mathbf{k}_I^\alpha \times \mathbf{n} = \mathbf{k}_R^\beta \times \mathbf{n} = \mathbf{k}_T^\gamma \times \mathbf{n}, \qquad (10.10.13)$$

valid for all three values of β and all three values of γ. The last series of

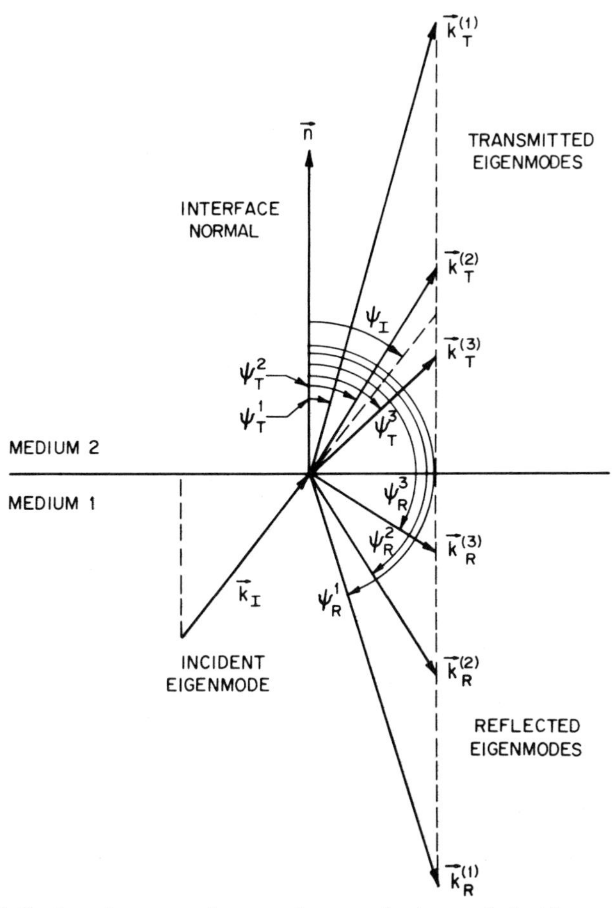

Fig. 10.4. Reflection of an acoustic wave. An acoustic eigenmode incident on an interface between two crystalline media creates in general three reflected waves and three transmitted waves. The propagation vectors of all the waves are in the same plane and have the same projection in that plane on the interface plane as shown by the dotted lines.

equalities constitute the general *laws of reflection and refraction for acoustic waves*.

These general laws, being vector laws, contain two parts, one concerning directions and one concerning magnitudes. The equality of the seven vector products of (10.10.13) means that $\mathbf{n}$, $\mathbf{k}_I^\alpha$, $\mathbf{k}_R^\beta$ ($\beta=1,2,3$), and $\mathbf{k}_T^\gamma$ ($\gamma=1,2,3$) are *all coplanar*, the equation of the plane being

$$(\mathbf{k}_I^\alpha \times \mathbf{n})\cdot\mathbf{z}=0. \tag{10.10.14}$$

The equality of the magnitudes of the seven vector products states

$$\frac{\sin\psi_I^\alpha}{v_I^\alpha}=\frac{\sin\psi_R^\beta}{v_R^\beta}=\frac{\sin\psi_T^\gamma}{v_T^\gamma} \tag{10.10.15}$$

for $\beta=1,2,3$ and $\gamma=1,2,3$. The angles denoted by ψ are those between the respective propagation vector $\mathbf{k}$ and the surface normal $\mathbf{n}$. Figure 10.4 depicts the reflection and refraction of acoustic waves. Reflected and refracted amplitudes can be found from (10.10.8) and (10.10.9).

10.11 Group Velocity in Crystal Acoustics

We determine in this section the form of the group velocity for linear elastic waves in a dielectric or a piezoelectric medium. In Section 8.8 the equality of the group velocity $\mathbf{v}_g$ and energy propagation (ray) velocity $\mathbf{v}_r$ is shown,

$$\mathbf{v}_g \equiv \nabla_k \omega \Big|_{\frac{\langle \mathfrak{L}\rangle}{\omega}} = \frac{\langle \mathbf{S}\rangle}{\langle H\rangle} \equiv v_r, \tag{10.11.1}$$

where $\mathbf{S}$ is the energy flux vector and H is the energy density. It is also shown there that for linear waves $\langle \mathfrak{L}\rangle=0$ and so holding $\langle \mathfrak{L}\rangle/\omega$ constant during differentiation in the present case is trivial. We calculate the group velocity first from the expression $\langle \mathbf{S}\rangle/\langle H\rangle$ and second from the expression $\nabla_k\omega$. Frequency dispersion plays no role in determining the group velocity for long wavelength acoustic waves because the elastic stiffness possesses no frequency dispersion; anisotropy, however, plays a much larger role than it does for light waves in crystals. The considerations in this section assume that $\mathbf{E}^S=0$ and $\mathbf{t}^S=0$ for simplicity.

The general expression for the energy flux (8.7.12) becomes

$$S_j = -\frac{\partial u_i}{\partial t} t_{ij}^E + \frac{1}{\mu_0}(\mathbf{E}\times\mathbf{B})_j \tag{10.11.2}$$

for the present case, since the other terms in that equation are at least trilinear in the fields. Since the electric field created piezoelectrically by the elastic wave is almost exactly longitudinal, as shown in Section 10.7, the magnetic induction **B** is minute and so the Poynting vector term $(\mathbf{E}\times\mathbf{B})/\mu_0$, representing the flux of electromagnetic energy, may be dropped. The elastic stress tensor $\mathbf{t}^E$ may be found from (10.3.3) and (10.3.8) to the linear level to be

$$t_{ij}^E = c_{ijkl}u_{k,l} - e_{kij}E_k. \tag{10.11.3}$$

Thus the energy flux **S** for a linear acoustic wave in a piezoelectric medium is given by

$$S_j = -\frac{\partial u_i}{\partial t}\left(c_{ijkl}u_{k,l} - e_{kij}E_k\right). \tag{10.11.4}$$

The first term is the elastic energy flux and the second term is the piezoelectric energy flux.

The energy density H of (8.7.11) can also be simplified. The procedure of adiabatic elimination of the internal coordinates given in Section 10.1 implies that the kinetic energies of these degrees of freedom are negligible. Since the magnetic field is negligible, as pointed out above, the magnetic field energy can also be dropped. In the absence of a spontaneous electric field and a spontaneous stress the stored energy (10.1.1) to bilinear order becomes

$$\rho^0\Sigma = \tfrac{1}{2}u_{k,l}c_{klmn}u_{m,n} + \frac{\epsilon_0}{2}E_k\chi_{kl}E_l. \tag{10.11.5}$$

Thus the energy density for linear elasticity and piezoelectricity is

$$H = \frac{\rho^0}{2}\left(\frac{\partial u_k}{\partial t}\right)^2 + \tfrac{1}{2}u_{k,l}c_{klmn}u_{m,n} + \frac{\epsilon_0}{2}E_k\kappa_{kl}E_l \tag{10.11.6}$$

where the dielectric tensor $\boldsymbol{\kappa}$ is introduced. Note that no piezoelectric term enters either the stored energy density (10.11.5) or the total energy density (10.11.6).

For a plane acoustic wave in a piezoelectric medium we have

$$\mathbf{u} = A\mathbf{b}\cos(k\mathbf{s}\cdot\mathbf{z} - \omega t) \tag{10.11.7}$$

and

$$\begin{aligned}\mathbf{E} &= -\frac{\mathbf{s}}{\epsilon_0\mathbf{s}\cdot\boldsymbol{\kappa}\cdot\mathbf{s}}s_ie_{ijk}u_{j,k}\\ &= \frac{\mathbf{s}}{\mathbf{s}\cdot\boldsymbol{\kappa}\cdot\mathbf{s}}\frac{Ak}{\epsilon_0}s_ie_{ijk}b_js_k\sin(k\mathbf{s}\cdot\mathbf{z} - \omega t),\end{aligned} \tag{10.11.8}$$

where A is an amplitude constant, $\mathbf{b}$ is an acoustic eigenvector, and the piezoelectrically generated electric field was obtained from (10.7.21). The energy flux averaged over one period is

$$\langle S_j \rangle = \frac{A^2 k\omega}{2} b_i b_k s_l \bar{c}_{ijkl} \tag{10.11.9}$$

where the piezoelectrically stiffened stiffness tensor is defined in (10.8.4). The energy density averaged over one period becomes

$$\begin{aligned}\langle H \rangle &= \frac{A^2 k^2}{4}\left(\rho^0 v^2 + b_k s_l \bar{c}_{klmn} b_m s_n\right)\\ &= \frac{A^2 k^2 \rho^0 v^2}{2}\end{aligned} \tag{10.11.10}$$

if the phase velocity v is introduced and if the eigenvector equation is used to simplify the expression. The group velocity may now be written in final form as

$$(v_g)_j = \frac{1}{\rho^0 v} \bar{c}_{ijkl} b_i b_k s_l. \tag{10.11.11}$$

The noncollinearity of the unit group velocity vector $\mathbf{t} \equiv \mathbf{v}_g / v_g$ and the unit propagation vector $\mathbf{s}$ is measured by the angle δ between them,

$$\mathbf{s} \cdot \mathbf{t} = \cos\delta. \tag{10.11.12}$$

The last two equations may be combined to yield expressions for the direction $\mathbf{t}$ and magnitude v_g of the group velocity,

$$t_j = \frac{\cos\delta}{\rho^0 v^2} \bar{c}_{ijkl} b_i b_k s_l, \tag{10.11.13}$$

$$v_g = \frac{v}{\cos\delta}, \tag{10.11.14}$$

where

$$\cos\delta = \frac{\rho^0 v^2}{\left(\bar{c}_{ijkl} b_i b_k s_l \bar{c}_{pjmn} b_p b_m s_n\right)^{1/2}}. \tag{10.11.15}$$

The expression for $\mathbf{v}_g$ allows the time-averaged energy flux vector (10.11.9) to be written as

$$\langle \mathbf{S} \rangle = \frac{\rho^0 \omega^2 A^2}{2} \mathbf{v}_g = \frac{\rho^0 \omega^2 v A^2}{2\cos\delta} \mathbf{t}. \tag{10.11.16}$$

Next we obtain an expression for the group velocity in crystal acoustics from

$$\mathbf{v}_g = \nabla_k \omega(\mathbf{k}) \tag{10.11.17}$$

which is useful to us in a later chapter. The result resembles one of the group velocity expressions we obtained for crystal optics except for two things: (1) there is no frequency dispersion for long wavelength acoustic plane waves and (2) the expression involves the phase velocity of sound rather than the index of refraction (an inverse relative phase velocity). From the definition of the acoustic phase velocity v we have

$$\omega(\mathbf{k}) = v(\mathbf{s})\, k(\mathbf{k}). \tag{10.11.18}$$

This yields

$$\begin{aligned}\mathbf{v}_g &= v\nabla_k k + k\nabla_k \mathbf{s}\cdot\nabla_s v \\ &= v\mathbf{s} + (\mathbf{1} - \mathbf{ss})\cdot\nabla_s v \\ &= v\left(\mathbf{s} + \frac{1}{v}\nabla_s v\right)\end{aligned} \tag{10.11.19}$$

with the help of the derivative relations (9.10.5) and (9.10.9) and the fact that a gradient with respect to a unit vector must be perpendicular to the unit vector [see (9.10.11)]. If the group velocity is expressed in terms of the unit vector $\mathbf{t}$,

$$\mathbf{v}_g = v_g \mathbf{t}, \tag{10.11.20}$$

we find

$$\mathbf{t} = \cos\delta\left(\mathbf{s} + \frac{1}{v}\nabla_s v\right), \tag{10.11.21}$$

where now $\cos\delta$ is given by

$$\cos\delta \equiv \mathbf{s}\cdot\mathbf{t} = \frac{\mathbf{s}\cdot\mathbf{v}_g}{v_g} = \frac{v}{v_g} = \frac{1}{\left\{1 + \left[(1/v)\nabla_s v\right]^2\right\}^{1/2}}. \tag{10.11.22}$$

Note that the last equality shows that the group velocity is never less than the phase velocity in plane wave acoustics. This results from the lack of frequency dispersion in acoustic plane waves. We emphasize plane waves because guided waves can possess frequency dispersion resulting from the waveguide structure.

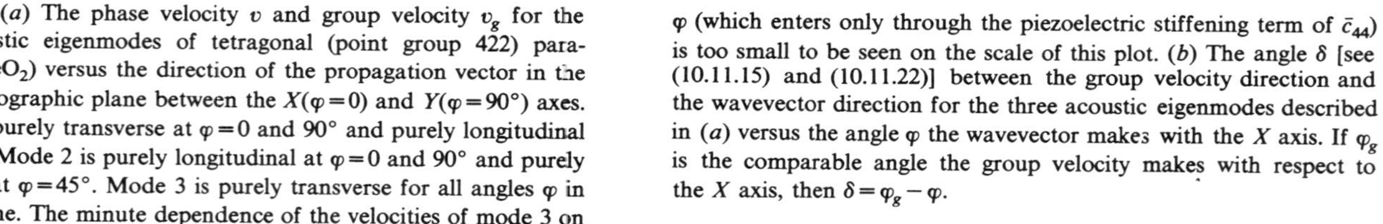

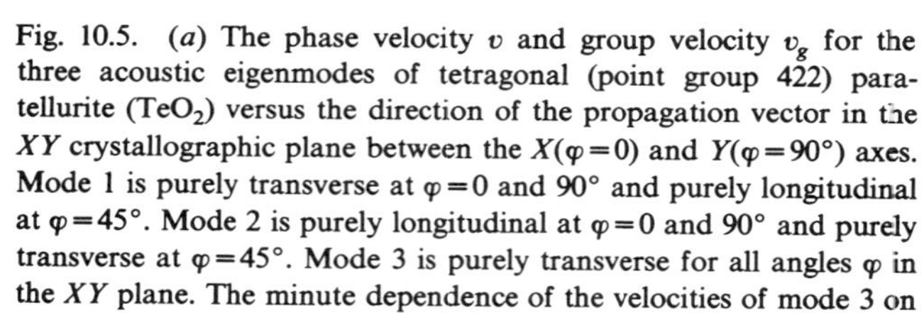

(a)

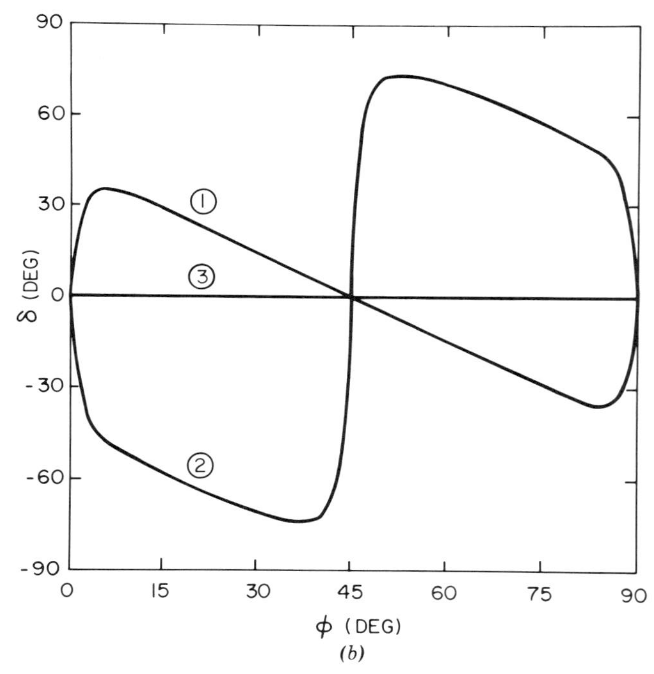

(b)

Fig. 10.5. (*a*) The phase velocity v and group velocity v_g for the three acoustic eigenmodes of tetragonal (point group 422) paratellurite (TeO_2) versus the direction of the propagation vector in the XY crystallographic plane between the $X(\varphi=0)$ and $Y(\varphi=90°)$ axes. Mode 1 is purely transverse at $\varphi=0$ and 90° and purely longitudinal at $\varphi=45°$. Mode 2 is purely longitudinal at $\varphi=0$ and 90° and purely transverse at $\varphi=45°$. Mode 3 is purely transverse for all angles φ in the XY plane. The minute dependence of the velocities of mode 3 on φ (which enters only through the piezoelectric stiffening term of $\bar{c}_{44}$) is too small to be seen on the scale of this plot. (*b*) The angle δ [see (10.11.15) and (10.11.22)] between the group velocity direction and the wavevector direction for the three acoustic eigenmodes described in (*a*) versus the angle φ the wavevector makes with the X axis. If φ_g is the comparable angle the group velocity makes with respect to the X axis, then $\delta=\varphi_g-\varphi$.

The direction of energy flux **t** often differs from the unit propagation vector **s** more markedly for acoustic waves than for optical waves. The reason simply is that the various components of the stiffness tensor differ more strongly from one another than the components of the optical dielectric tensor do. An extreme example of noncollinearity of energy flux and propagation vector is shown in Fig. 10.5 for the case of paratellurite (TeO_2). Note that the group velocity and propagation vector differ in direction by as much as 74°. Near $\phi=45°$ a 1° change in the propagation vector direction causes a 48° change in the direction of the group velocity. When $\delta=0$, $v_g=v$ as expected from (10.11.14) and (10.11.22).

PROBLEMS

10.1. Derive the energy conservation equation from the linear equations of motion.

10.2. Consider a purely longitudinal acoustic wave in an isotropic solid impinging upon a planar solid-vacuum surface. Find the amplitudes and reflection angles of the modes emerging from the surface in terms of the amplitude and angle of incidence of the impinging wave.

10.3. Find the plane wave eigenmodes in the YZ crystallographic plane of $LiNbO_3$ (point group $3m$) as a function of a unit propagation vector $\mathbf{s}=[0, \sin\theta, \cos\theta]$. What is the character of the modes (longitudinal, transverse, or hybrid)? Find expressions for their velocities.

10.4. Consider the pure shear mode that exists throughout the YZ plane in $LiNbO_3$ (Prob. 10.3). What eigenmodes are created in a reflection of the mode on a planar vacuum-solid surface which is perpendicular to the YZ plane? Does the angle of incidence equal the angle of reflection in this case? If not, why not? Is there a special orientation of the surface which allows this equality?

10.5. One wishes to measure the elastic and piezoelectric tensor components in insulating GaAs (point group $\bar{4}3m$) by acoustic wave velocity measurements. By considering simple crystallographic directions first ([100],[110]) show what measurements need be made to obtain all the independent tensor components. Are any subsidiary measurements needed for the analysis?

10.6. Find s_{11}, s_{12}, and s_{44} in terms of c_{11}, c_{12}, and c_{44} for a $\bar{4}3m$ crystal.

10.7. Prove in general that an acoustic eigenmode has unit reflectivity for normal incidence on a vacuum surface and thus that no other eigenmodes are excited at the surface.

10.8. Show that, if the linearized dynamic elasticity equation (10.3.15) is driven through the piezoelectric term by an optical electric field, the stiffness term is $(v_A/v_O)^2 \approx 10^{-10}$ the size of the inertial term. Here v is a velocity and A and O refer to acoustic and optical. Hence show that the acoustic displacement driven by an optical frequency (ω_O) electric field is

$$u_i = \frac{e_{mij} E_{m,j}}{\rho^0 \omega_O^2}. \tag{10.P.1}$$

Finally show that the piezoelectric term in the wave equation (10.7.1) for an optical electric field is of order $(v_A/v_O)^2 \approx 10^{-10}$ of the other terms. This demonstrates the assertion made in Section 9.1 that acoustic displacement terms could be dropped from optical frequency equations.

10.9. Derive (10.8.12).

10.10. Prove the equality of (10.11.11) and (10.11.19) without the use of the theorem (10.11.1).

10.11. A *surface wave* in the strictest sense is a *guided* wave solution made possible by a *single* planar interface of two media. It propagates parallel to the surface and has its intensity localized near the surface. A *Rayleigh wave* is a surface acoustic wave on a non-piezoelectric, elastically isotropic half-space. To find the Rayleigh wave look for a displacement solution in the matter of the form

$$u_1 = Ae^{-bz_2 + ik(z_1 - vt)}, \tag{10.P.2}$$

$$u_2 = Be^{-bz_2 + ik(z_1 - vt)}, \tag{10.P.3}$$

$$u_3 = 0 \tag{10.P.4}$$

where z_2 is measured normal to the surface and positively inward. Determine the allowed positive values of b; using them construct a solution which can match the boundary condition of a stress-free surface. Find the equation for the Rayleigh wave velocity. Is it dispersive or not?

CHAPTER 11

Linear Piezoelectricity

This chapter continues the exploration of linear piezoelectricity based on the dynamical and constitutive equations derived from this theory in Chapter 10. We wish to explore further the terms involving the spontaneous polarization and spontaneous electric field of a pyroelectric that enter the linear polarization, the linear electric displacement, and the magnetic intensity. All such terms differ from the traditional Voigt theory of piezoelectricity. Because the terms that depend only on the spontaneous polarization disappear from the Maxwell equations, they apparently are not measurable. To be sure of this we must show that they disappear from the linearized boundary conditions.

For the consideration of boundary value problems it is convenient to use the *quasi-electrostatic approximation* of the Maxwell equations. This approximation is an accurate representation of the full electromagnetic equations whenever the solid body considered has dimensions small compared to the electromagnetic wavelength. The quasi-electrostatic approximation corresponds to the zeroth order terms in an iterative expansion of the Maxwell equations when only electric (and material motion) phenomena are considered. When magnetic effects are considered, first order iterative terms in the magnetic fields must be considered and are sometimes included in what is termed the quasi-electrostatic approximation. The quasi-electrostatic approximation is then used to study various boundary value problems involving piezoelectrically driven mechanical vibrations.

11.1 Quasi-electrostatic Approximation

In Section 10.7 we found that the piezoelectrically generated electric field of a plane acoustic wave,

$$\mathbf{E} = -\mathbf{s}\frac{s_k e^F_{kij} u_{i,j}}{\epsilon_0 \mathbf{s}\cdot\boldsymbol{\kappa}\cdot\mathbf{s}}, \tag{11.1.1}$$

is longitudinal to a high degree of approximation. Under such circumstances

$$\nabla\times\mathbf{E}=-ik\mathbf{s}\times\mathbf{s}\frac{s_k e^F_{kij}u_{i,j}}{\epsilon_0\mathbf{s}\cdot\boldsymbol{\kappa}\cdot\mathbf{s}}=0 \tag{11.1.2}$$

and so

$$\nabla\times\mathbf{E}+\frac{\partial\mathbf{B}}{\partial t}=\frac{\partial\mathbf{B}}{\partial t}=0. \tag{11.1.3}$$

Equation (11.1.2) states that $\mathbf{E}$ is derivable as a (negative) gradient of a potential Φ,

$$\mathbf{E}=-\nabla\Phi, \tag{11.1.4}$$

and (11.1.3) states that the magnetic induction (more precisely, its time derivative) vanishes. These results are typical of electrostatics. The question naturally arises whether the electric interaction in piezoelectric problems can always be handled as an electrostatic interaction in some low frequency regime. The answer is affirmative and the resulting equations are called the *quasi-electrostatic approximation* to the Maxwell equations.

The key to deriving this approximate set of equations is the realization that the free space wavelength λ of an electromagnetic wave for a typical angular frequency ω is large compared to a typical dimension L of the piezoelectric crystal studied. For instance,

$$\frac{\omega L}{c}=\frac{2\pi L}{\lambda}\ll 1 \tag{11.1.5}$$

holds for $L=1$ cm for all $\omega/2\pi<1$ GHz. We are thus led to scale the independent variables of the problem according to

$$\boldsymbol{\zeta}\equiv\frac{\mathbf{z}}{L}, \tag{11.1.6}$$

$$\tau\equiv\omega t, \tag{11.1.7}$$

which give the derivative relations

$$\frac{\partial}{\partial z_i}=\frac{1}{L}\frac{\partial}{\partial\zeta_i}, \tag{11.1.8}$$

$$\frac{\partial}{\partial t}=\omega\frac{\partial}{\partial\tau}. \tag{11.1.9}$$

We also scale the dependent fields according to

$$\boldsymbol{v}\equiv\frac{\mathbf{u}}{L}, \tag{11.1.10}$$

$$\mathscr{B}\equiv c\mathbf{B}, \tag{11.1.11}$$

the scaling of $\mathbf{B}$ being designed to give $\mathscr{B}$ the same dimensions as $\mathbf{E}$. The elasticity equation and the four electromagnetic equations then become

$$c_{ijkl}^F \frac{\partial^2 v_k}{\partial \zeta_j \partial \zeta_l} - e_{kij}^F \frac{\partial E_k}{\partial \zeta_j} = \rho^0 \omega^2 L^2 \frac{\partial^2 v_i}{\partial \tau^2}, \tag{11.1.12}$$

$$\epsilon_0 \kappa_{ij} \frac{\partial E_j}{\partial \zeta_i} + e_{kij}^F \frac{\partial^2 v_i}{\partial \zeta_j \partial \zeta_k} = 0, \tag{11.1.13}$$

$$\nabla_\zeta \times \mathbf{E} = -\eta \frac{\partial \mathscr{B}}{\partial \tau}, \tag{11.1.14}$$

$$(\nabla_\zeta \times \mathscr{B})_i = \eta \frac{\partial}{\partial \tau}\left(\kappa_{ij} E_j + \frac{e_{ijk}^F}{\epsilon_0} \frac{\partial v_j}{\partial \zeta_k} \right), \tag{11.1.15}$$

$$\nabla_\zeta \cdot \mathscr{B} = 0 \tag{11.1.16}$$

where

$$\eta \equiv \frac{\omega L}{c}. \tag{11.1.17}$$

Since $\eta \ll 1$ by the condition (11.1.5), we are led to expand the three dependent fields, $\mathbf{E}$, $\mathscr{B}$, and $\boldsymbol{v}$, in series of the form

$$\mathbf{f} = \sum_{n=0}^{\infty} \eta^n \mathbf{f}^{(n)}. \tag{11.1.18}$$

Substitution into (11.1.12) to (11.1.16) yields

$$\sum_{n=0}^{\infty} \eta^n \left[c_{ijkl}^F \frac{\partial^2 v_k^{(n)}}{\partial \zeta_j \partial \zeta_l} - e_{kij}^F \frac{\partial E_k^{(n)}}{\partial \zeta_j} - \rho^0 \omega^2 L^2 \frac{\partial^2 v_i^{(n)}}{\partial \tau^2} \right] = 0, \tag{11.1.19}$$

$$\sum_{n=0}^{\infty} \eta^n \left[\epsilon_0 \kappa_{ij} \frac{\partial E_j^{(n)}}{\partial \zeta_i} + e_{kij}^F \frac{\partial^2 v_i^{(n)}}{\partial \zeta_j \partial \zeta_k} \right] = 0, \tag{11.1.20}$$

$$\sum_{n=0}^{\infty} \eta^n \nabla_\zeta \times \mathbf{E}^{(n)} = -\sum_{n=1}^{\infty} \eta^n \frac{\partial \mathscr{B}^{(n-1)}}{\partial \tau}, \tag{11.1.21}$$

$$\sum_{n=0}^{\infty} \eta^n (\nabla_\zeta \times \mathscr{B}^{(n)})_i = \sum_{n=1}^{\infty} \eta^n \frac{\partial}{\partial \tau} \left[\kappa_{ij} E_j^{(n-1)} + \frac{e_{ijk}^F}{\epsilon_0} \frac{\partial v_j^{(n-1)}}{\partial \zeta_k} \right], \tag{11.1.22}$$

$$\sum_{n=0}^{\infty} \eta^n \nabla_\zeta \cdot \mathscr{B}^{(n)} = 0. \tag{11.1.23}$$

The terms in each equation of like power of η are now equated. The zeroth order equations with the scaling removed are then

$$c_{ijkl}^{F}u_{k,lj}^{(0)} - e_{kij}^{F}E_{k,j}^{(0)} = \rho^{0}\frac{\partial^{2}u_{i}^{(0)}}{\partial t^{2}}, \tag{11.1.24}$$

$$\epsilon_0\kappa_{ij}E_{j,i}^{(0)} + e_{ijk}^{F}u_{j,ki}^{(0)} = 0, \tag{11.1.25}$$

$$\nabla\times\mathbf{E}^{(0)} = 0, \tag{11.1.26}$$

$$\nabla\times\mathbf{B}^{(0)} = 0, \tag{11.1.27}$$

$$\nabla\cdot\mathbf{B}^{(0)} = 0. \tag{11.1.28}$$

Note that in the zeroth order equations time derivatives of the electric and magnetic fields disappear while the time derivative of the acoustic displacement remains. As a result $\mathbf{B}^{(0)}$ is not coupled to either of the other fields. Thus it can represent only an applied magnetic field which typically in piezoelectric studies is not used. Equation (11.1.26) shows that the zeroth order electric field is irrotational and so is the (negative) gradient of the electric potential $\Phi^{(0)}$,

$$\mathbf{E}^{(0)} = -\nabla\Phi^{(0)}. \tag{11.1.29}$$

Thus the vector field $\mathbf{E}^{(0)}$ may be eliminated in favor of the scalar field $\Phi^{(0)}$ leading to

$$c_{ijkl}^{F}u_{k,lj}^{(0)} + e_{kij}^{F}\Phi_{,kj}^{(0)} = \rho^{0}\frac{\partial^{2}u_{i}^{(0)}}{\partial t^{2}}, \tag{11.1.30}$$

$$\epsilon_0\kappa_{ij}\Phi_{,ji}^{(0)} - e_{ijk}^{F}u_{j,ki}^{(0)} = 0 \tag{11.1.31}$$

as the four zeroth order equations to be solved for $\Phi^{(0)}$ and $\mathbf{u}^{(0)}$.

The $n=1$ terms of (11.1.19) to (11.1.23) with the scaling removed and with η absorbed into $\mathbf{u}^{(1)}$, $\mathbf{E}^{(1)}$, and $\mathbf{B}^{(1)}$ are

$$c_{ijkl}^{F}u_{k,lj}^{(1)} - e_{kij}^{F}E_{k,j}^{(1)} = \rho^{0}\frac{\partial^{2}u_{i}^{(1)}}{\partial t^{2}}, \tag{11.1.32}$$

$$\epsilon_0\kappa_{ij}E_{j,i}^{(1)} + e_{ijk}^{F}u_{j,k}^{(1)} = 0, \tag{11.1.33}$$

$$\nabla\times\mathbf{E}^{(1)} = -\frac{\partial\mathbf{B}^{(0)}}{\partial t}, \tag{11.1.34}$$

$$\begin{aligned}\frac{1}{\mu_0}(\nabla\times\mathbf{B}^{(1)})_i &= \frac{\partial}{\partial t}\left[\epsilon_0\kappa_{ij}E_j^{(0)} + e_{ijk}^{F}u_{j,k}^{(0)}\right],\\ &= \frac{\partial}{\partial t}\left[-\epsilon_0\kappa_{ij}\Phi_{,j}^{(0)} + e_{ijk}^{F}u_{j,k}^{(0)}\right],\end{aligned} \tag{11.1.35}$$

$$\nabla\cdot\mathbf{B}^{(1)} = 0. \tag{11.1.36}$$

If $\mathbf{B}^{(0)}=0$ as is typical, we see that $\mathbf{E}^{(1)}$ and $\mathbf{u}^{(1)}$ are uncoupled from $\mathbf{B}^{(1)}$ and that $\mathbf{B}^{(1)}$ is driven by $\mathbf{E}^{(0)}$ and $\mathbf{u}^{(0)}$, the zeroth order solutions. Equations (11.1.35) and (11.1.36) in conjunction with (11.1.30) and (11.1.31) constitute the *quasi-electrostatic approximation* as applied to piezoelectricity. It is seldom useful to go beyond this level of iteration; when $\eta \sim 1$ it is usually easier to use the full set of Maxwell equations.

If the other fields, $\mathbf{D}$, $\mathbf{t}^T$, and $\mathbf{H}$, are to be used, $\mathbf{D}$ and $\mathbf{t}^T$ must be accurate to the same order as Φ (or $\mathbf{E}$) and $\mathbf{u}$ while $\mathbf{H}$ must be accurate to the same order as $\mathbf{B}$. Since no time derivatives occur in the constitutive relations (10.2.16) and (10.4.3) for $\mathbf{D}$ and $\mathbf{t}^T$, these relations are good for each iterative order and do not mix orders. The constitutive relation for $\mathbf{H}$ (10.2.17) needs closer examination. If we scale $\mathbf{H}$ to yield a field $\mathscr{H}$ having the dimensions of $\mathbf{E}$,

$$\mathscr{H} = \mu_0 c \mathbf{H}, \tag{11.1.37}$$

and use the previous scalings for t, $\mathbf{u}$, and $\mathbf{B}$, we find by the procedure above that

$$\mathbf{H}^{(0)} = \frac{1}{\mu_0}\mathbf{B}^{(0)}, \tag{11.1.38}$$

$$\mathbf{H}^{(1)} = \frac{1}{\mu_0}\mathbf{B}^{(1)} - \mathbf{P}^S \times \dot{\mathbf{u}}^{(0)}. \tag{11.1.39}$$

The first of these equations is seldom of interest, since usually $\mathbf{B}^{(0)}=0$. The second equation shows that the interesting second term persists in the quasi-electrostatic approximation. We consider its measurability in Section 11.4.

11.2 Piezoelectric Measurements from Acoustic Velocities

As a simple example of the use of the quasi-electrostatic approximation, let us reconsider acoustic wave propagation. For a plane wave with a unit propagation vector $\mathbf{s}$ and a scalar coordinate z measured in the direction of $\mathbf{s}$ the gradient operator may be represented as

$$\frac{\partial}{\partial z_j} = s_j \frac{\partial}{\partial z} \tag{11.2.1}$$

since the only spatial variation in a plane wave is parallel to $\mathbf{s}$. Note that this condition in conjunction with (11.1.29) shows that the piezoelectrically generated electric field is longitudinal as previously shown in Section 10.7.

Substitution of this into (11.1.31) results in

$$\frac{\partial^2 \Phi^{(0)}}{\partial z^2} = \frac{s_n e_{nkl}^F s_l}{\epsilon_0 \mathbf{s} \cdot \boldsymbol{\kappa} \cdot \mathbf{s}} \cdot \frac{\partial^2 u_k^{(0)}}{\partial z^2}. \tag{11.2.2}$$

This may be used to eliminate $\Phi^{(0)}$ from the elasticity equation (11.1.30) with the result

$$\rho^0 \frac{\partial^2 u_i^{(0)}}{\partial t^2} = \left[c_{ijkl}^F + \frac{s_m e_{mij}^F s_n e_{nkl}^F}{\epsilon_0 \mathbf{s} \cdot \boldsymbol{\kappa} \cdot \mathbf{s}} \right] s_j s_l \frac{\partial^2 u_k^{(0)}}{\partial z^2}$$

$$= \overline{c_{ij\,kl}^F} s_j s_l \frac{\partial^2 u_k^{(0)}}{\partial z^2}, \tag{11.2.3}$$

where the stiffened stiffness tensor $\overline{c_{ijkl}^F}$ is defined in (10.8.4).

The derivation above demonstrates the relative ease of obtaining the acoustic propagation equation from the quasi-electrostatic approximation compared with the approach of Sections 10.7 and 10.8 where the electromagnetic wave equation is employed in finding the electric field. As noted in Section 10.8, the velocities of the three eigenmodes of this equation depend on the components of $\overline{c_{ijkl}^F}$ and on ρ^0. A sufficient number of velocity measurements versus propagation direction and eigenmode type can determine $\overline{c_{ijkl}^F}$ if ρ^0 is separately measured. Also, if $\boldsymbol{\kappa}$ is separately determined, such measurements may also be used to determine e_{mij}^F. In turn, measurements of e_{mij}^F and c_{ijkl}^F could determine $\mathbf{E}^S$ on which they both depend.

11.3 Cancellation of $\mathbf{E}^S$ in Pyroelectrics

In this section we discuss the spontaneous electric field $\mathbf{E}^S$ produced by the spontaneous polarization $\mathbf{P}^S$ in pyroelectrics and, in particular, the typical situation in which extrinsic charge has collected on the crystal surface in just sufficient quantity to cancel $\mathbf{E}^S$. Before discussing its cancellation we should recall that $\mathbf{E}^S$ enters this theory as the electric field present in the natural state of a pyroelectric crystal and is left unrelated to the spontaneous polarization in the formulation. This is quite natural because, when $\mathbf{E}^S$ exists, its magnitude and direction depend on the shape of the crystal as well as on $\mathbf{P}^S$. Thus the relation between $\mathbf{E}^S$ and $\mathbf{P}^S$ cannot be specified until the geometry of a particular problem is known. Also, the spontaneous electric field may be canceled by extrinsic charge attracted by

$\mathbf{E}^S$ to the crystal surfaces from the ambiance. Alternatively, a small volume or surface conductivity may allow charge to flow from one side of the crystal to the opposite one and so cancel $\mathbf{E}^S$. As one or both of these mechanisms is bound to occur, the spontaneous electric field exists only for a period of time, often very short, following either a temperature change through a structural phase transition in which $\mathbf{P}^S$ is created or a reversal of the direction of $\mathbf{P}^S$ caused by an electric field applied to a ferroelectric crystal. It should be realized that cancellation of $\mathbf{E}^S$ by extrinsic charge does not affect $\mathbf{P}^S$, it merely creates an equal and opposite extrinsic electric moment.

Now let us determine how the extrinsic charge distributes itself in order to cancel the spontaneous electric field. Consider the expression [Mason and Weaver, 1929] for the potential of a body possessing a volume charge density q and a surface charge density Σ as well as a polarization $\mathbf{P}$,

$$\Phi(\mathbf{z}) = \frac{1}{4\pi\epsilon_0}\int \frac{\left[q(\mathbf{z}') - \nabla'\cdot\mathbf{P}(\mathbf{z}')\right]dv'}{|\mathbf{z}-\mathbf{z}'|} + \frac{1}{4\pi\epsilon_0}\int \frac{\left[\Sigma(\mathbf{z}') + \mathbf{N}\cdot\mathbf{P}(\mathbf{z}')\right]da'}{|\mathbf{z}-\mathbf{z}'|}. \tag{11.3.1}$$

Here $\mathbf{N}$ is the unit outward surface normal, dv' is an element of volume, and da' is an element of surface area. We regard the crystal as being in its natural state except for the effects of the extrinsic charge.

After the passage of a sufficient time the free charge in the volume and on the surface distributes in such a way that the spontaneous electric field is zero everywhere. Since this redistribution of charge is a quasi-electrostatic phenomenon, the electric field is simply the negative gradient of the potential of (11.3.1). If it is to vanish everywhere for a body of arbitrary size and shape, the integrands of the volume and surface integrals must vanish separately. Note that the integrand of the volume integral contains the divergence of the polarization. If we consider structurally homogeneous crystals as we do throughout this treatment, the only inhomogeneity in the polarization that can occur (and so produce a nonzero divergence) arises from the effect of the spontaneous electric field back on the crystal. Thus, as the collection of extrinsic charge proceeds, the spontaneous electric field diminishes, and with it the inhomogeneity of the polarization diminishes, and so the need for an extrinsic volume charge diminishes. Clearly then, when $\mathbf{E}^S$ is completely canceled, the extrinsic surface charge density is

$$\Sigma^S = -\mathbf{N}\cdot\mathbf{P}^S \tag{11.3.2}$$

and *no* extrinsic volume charge,

$$q = \nabla \cdot \mathbf{P}^S = 0 \tag{11.3.3}$$

is needed. Since pyroelectric crystals commonly occur in this condition (the extrinsic natural state), it is important for us to consider the effects that the extrinsic surface charge has on various measurements.

11.4 Linearized Boundary Conditions

In Section 7.2 the general boundary conditions on the electromagnetic fields, **E**, **D**, **B**, and **H**, are found while in Section 8.4 the general stress boundary condition is found. Here we wish to obtain linearized versions of them. We also specialize them for the case of the outside medium being a vacuum.

We begin by linearizing the general expressions for the unit normal **n** and scalar area element *da* of a deformed surface developed in Section 3.3. In terms of the displacement gradient (3.3.16) for the unit normal becomes

$$\begin{aligned} n_i &= (\delta_{Ri} - u_{R,i}) N_R \left[(\delta_{Pj} - u_{P,j}) N_P (\delta_{Qj} - u_{Q,j}) N_Q \right]^{-1/2} \\ &= (\delta_{ri} - u_{r,i}) N_r \left[1 - \delta_{pj} N_p u_{q,j} N_q - \delta_{qj} N_p u_{p,j} N_q \right]^{-1/2} \\ &= (\delta_{ir} \quad u_{r,i} + N_p u_{p,q} N_q \delta_{ir}) N_r \end{aligned} \tag{11.4.1}$$

where **N** is the unit normal of the undeformed surface. Equation (3.3.15) for the area element *da* becomes

$$\begin{aligned} da &= (1 + u_{r,r}) \left[(\delta_{iP} - u_{P,i}) N_P (\delta_{iQ} - u_{Q,i}) N_Q \right]^{1/2} dA \\ &= (1 + u_{r,r}) \left[1 - 2 N_p u_{p,q} N_q \right]^{1/2} dA \\ &= (1 + u_{p,p} - N_p u_{p,q} N_q) \, dA \end{aligned} \tag{11.4.2}$$

where dA is the scalar area element of the undeformed surface.

The tangential electric field boundary condition (7.2.15) for the outside medium being a vacuum is

$$\mathbf{n} \times \left[\mathbf{E}^o - (\mathbf{E} + \dot{\mathbf{x}} \times \mathbf{B})^i \right] = 0 \tag{11.4.3}$$

where o stands for outside and i for inside. The $\dot{\mathbf{x}} \times \mathbf{B}$ term may be dropped because it is nonlinear. If a spontaneous electric field is present, the boundary condition applied to the undeformed or natural state gives

$$\mathbf{N}\times(\mathbf{E}^{So}-\mathbf{E}^{Si})=0. \tag{11.4.4}$$

By subtracting these terms we are left with

$$\mathbf{N}\times(\mathbf{E}^{o}-\mathbf{E}^{i})+(\mathbf{n}-\mathbf{N})\times(\mathbf{E}^{So}-\mathbf{E}^{Si})=0 \tag{11.4.5}$$

with $\mathbf{n}$ given by (11.4.1) for the linearized electric field boundary condition.

The tangential magnetic intensity boundary condition (7.2.22) for the outside medium being a vacuum is

$$\mathbf{n}\times\left[\mathbf{H}^{o}-(\mathbf{H}-\dot{\mathbf{x}}\times\mathbf{D})^{i}\right]=\mathbf{k}^{c} \tag{11.4.6}$$

where $\mathbf{k}^{c}$ is a surface current density. Here the velocity dependent term can produce linear terms in combination with the spontaneous terms $\epsilon_0\mathbf{E}^S+\mathbf{P}^S$ in the dielectric displacement. Thus we obtain

$$\mathbf{N}\times\left[\mathbf{H}^{o}-\left(\mathbf{H}-\dot{\mathbf{u}}\times\mathbf{P}^{S}-\epsilon_0\dot{\mathbf{u}}\times\mathbf{E}^{S}\right)^{i}\right]=\mathbf{k}^{c}. \tag{11.4.7}$$

Though we include both a spontaneous field $\mathbf{E}^S$ and a surface current density $\mathbf{k}^c$, it is unlikely that they are present simultaneously. Since the magnetic induction field $\mathbf{B}$ (not $\mathbf{H}$) is found directly from (11.1.35) and (11.1.36) in the quasi-electrostatic approximation, it is more appropriate to eliminate $\mathbf{H}$ in favor of $\mathbf{B}$ using (11.1.39). This gives

$$\mathbf{N}\times\left[\frac{\mathbf{B}^{o}}{\mu_0}-\left(\frac{\mathbf{B}}{\mu_0}-\epsilon_0\dot{\mathbf{u}}\times\mathbf{E}^{S}\right)^{i}\right]=\mathbf{k}^{c}. \tag{11.4.8}$$

This demonstrates that the interesting term $\mathbf{P}^S\times\dot{\mathbf{u}}$ in the linearized constitutive expression for $\mathbf{H}$ is canceled from the boundary condition by a moving medium correction and so does not produce measurable effects.

The electric displacement boundary condition (7.2.9) for the outside medium being a vacuum is

$$\mathbf{n}\cdot(\mathbf{D}^{o}-\mathbf{D}^{i})=\sigma^{f} \tag{11.4.9}$$

where σ^f is the extrinsic, mobile or immobile, surface charge density of the deformed surface related to Σ^f, the surface charge density of the undeformed surface, by

$$\sigma^{f}=\Sigma^{f}\frac{dA}{da}. \tag{11.4.10}$$

Consider first the case discussed in the preceding section, that is, where an immobile extrinsic surface charge of an amount

$$\Sigma^s = -\mathbf{N}\cdot\mathbf{P}^S \tag{11.4.11}$$

collects and cancels the spontaneous electric field,

$$\mathbf{E}^{So} = \mathbf{E}^{Si} = 0. \tag{11.4.12}$$

Substitution of (11.4.10) and (11.4.11) and the constitutive expression (10.2.16) for **D** into the boundary condition gives

$$\left[\epsilon_0 E_i^o - \left(\epsilon_0 \kappa_{ij} E_j + e_{ijk} u_{j,k}\right)^i\right] N_i = \Sigma^c \tag{11.4.13}$$

where Σ^c is a mobile (conduction) surface charge density. In the case of no surface charge the spontaneous electric field is present and a similar procedure gives

$$\left[\epsilon_0 E_i^o - \left(\epsilon_0 \kappa_{ij} E_j + e_{ijk}^F u_{j,k}\right)^i\right] N_i\, dA$$
$$+\epsilon_0\left(E_i^{So} - E_i^{Si}\right)(n_i\, da - N_i\, dA) = 0 \tag{11.4.14}$$

for the electric displacement boundary condition with **n** and *da* given by the linear expressions (11.4.1) and (11.4.2).

It is particularly important to note that the simple form of the piezoelectric term in (11.4.13) and (11.4.14) results because of the exact cancellation,

$$-2P_{[k}^S \delta_{i]j} u_{j,k} N_i\, dA - P_i^S(n_i\, da - N_i\, dA) = 0, \tag{11.4.15}$$

between the linear piezoelectric-like term involving $\mathbf{P}^S$ in the constitutive relation (10.2.16) for **D** and the constant spontaneous polarization term $\mathbf{P}^S$ in that relation multiplied by the linear changes in the deformed normal and deformed area element. This exact cancellation eliminates $\mathbf{P}^S$ from any explicit appearance in the boundary condition. Thus the term $+2P_{[k}^S \delta_{i]j} u_{j,k}$ that is present in the constitutive relation (10.2.16) for **D** is canceled from the Maxwell differential equations and from the **D** field boundary condition. We are forced to conclude that a term in a classical constitutive relation is unobservable even though necessary! We say necessary because the effects that it cancels in the boundary condition (11.4.15) —the spontaneous polarization multiplied by changes in the unit normal and area element resulting from deformation—are real tangible effects. Thus the inclusion in this theory of both this constitutive term and the boundary deformation terms is necessary, correct, and consistent in spite of the final disappearance of linearly varying terms involving $\mathbf{P}^S$. The Voigt theory included neither of these effects and so presented a correct interpretation only through the good fortune of compensating omissions.

Finally we note that the magnetic induction boundary condition (7.2.4),

$$(\mathbf{B}^o - \mathbf{B}^i)\cdot\mathbf{n} = 0, \qquad (11.4.16)$$

when linearized is simply

$$(\mathbf{B}^o - \mathbf{B}^i)\cdot\mathbf{N} = 0. \qquad (11.4.17)$$

The stress boundary condition at a material surface was found in Section 8.4 to be

$$\left[\left(t_{ij}^T + g_i\dot{x}_j\right)^o - \left(t_{ij}^T + g_i\dot{x}_j\right)^i\right] n_j = 0. \qquad (11.4.18)$$

Since the terms involving $\dot{\mathbf{x}}$ here add only nonlinear contributions, they may be dropped. The constant or spontaneous terms of this equation,

$$\left[\left(t_{ij}^T\right)^o - \left(t_{ij}^T\right)^i\right]^{NS} N_j = 0, \qquad (11.4.19)$$

which hold in the natural state (NS), may be subtracted from (11.4.18). Linear terms in that equation arise in two ways: linear terms in t_{ij}^T multiplied by the constant term in n_j and constant or spontaneous terms in t_{ij}^T multiplied by the linear terms in n_j. The constant and linear terms in t_{ij}^T are given in (10.4.3) while the constant and linear terms of n_j are given in (11.4.1). Because of the length of the resulting expression we do not write it out here. However, the simple and commonly met case where the outside medium is a vacuum and no spontaneous electric field or spontaneous stress is present leads to a linearized stress boundary condition of

$$\left(c_{ijkl}u_{k,l} - e_{kij}E_k\right)N_j = 0. \qquad (11.4.20)$$

11.5 Low Frequency Dielectric Tensor

In Sections 10.7 and 10.8 the piezoelectric coupling between the elasticity equation and the electromagnetic equations is shown to lead to a piezoelectric stiffening of the stiffness tensor and so to an increase in the velocity of propagation of acoustic waves. We now explore the effects of this piezoelectric coupling on the simplest electrical property of the crystal, the dielectric tensor, in a frequency regime low enough that the acoustic displacement can respond fully, that is, without inertia and without constraints, to the piezoelectric driving force.

Since inertia plays an important role near, at, or above a resonant frequency, the frequencies considered here must be well below the frequencies of mechanical resonance of the body. Such resonant frequencies, as we see in later sections, occur whenever some dimension of the crystal is of the order of an acoustic half wavelength or an integral multiple thereof. This means that an infinite dimension has a zero resonant frequency. Thus if we are to have a *region* of frequencies well below the resonant frequencies, all dimensions of the crystal must be finite. Since we prefer the finite crystal to have a homogeneous natural state for simplicity, it cannot possess a spontaneous electric field according to the result of Section 8.5. Thus the crystal considered in this section can be either a simple piezoelectric crystal or a pyroelectric crystal for which $\mathbf{E}^S=0$.

The elasticity equation in the quasi-electrostatic approximation (11.1.30) with the inertial term neglected and $\mathbf{E}^S=0$ is

$$c_{ijkl}u_{k,lj}^{(0)}+e_{kij}\Phi_{,kj}^{(0)}=0. \tag{11.5.1}$$

To obtain the full piezoelectric response of the crystal all of its boundaries must be unconstrained, that is, traction-free. The linearized stress boundary condition (11.4.20) then is simply

$$\left(c_{ijkl}u_{k,l}^{(0)}+e_{kij}\Phi_{,k}^{(0)}\right)N_j=0, \tag{11.5.2}$$

where $\mathbf{N}$ is the unit outward normal of the surface of the undeformed crystal.

In the low frequency region considered the solution that we desire is essentially homogeneous. The only such solution of (11.5.1) and (11.5.2) obeys

$$c_{ijkl}u_{k,l}^{(0)}+e_{kij}\Phi_{,k}^{(0)}=0 \tag{11.5.3}$$

throughout the crystal. This may be solved for the strain,

$$u_{(m,n)}^{(0)}=-d_{kmn}\Phi_{,k}^{(0)}, \tag{11.5.4}$$

by using (10.5.3) and (10.5.4).

The strain may now be eliminated from (11.1.31) with $\mathbf{E}^S=0$ with the result

$$\epsilon_0\left(\kappa_{ij}(0)+\frac{1}{\epsilon_0}e_{ikl}d_{jkl}\right)\Phi_{,ij}^{(0)}=0 \tag{11.5.5}$$

where $\boldsymbol{\kappa}(0)$ refers to the zero frequency value of $\boldsymbol{\kappa}(\omega)$. This leads us to define the *low frequency dielectric tensor* $\boldsymbol{\kappa}^L$ of a piezoelectric crystal as

$$\kappa_{ij}^L \equiv \kappa_{ij}(0) + \frac{1}{\epsilon_0} e_{ikl} d_{jkl}. \tag{11.5.6}$$

The piezoelectric contribution to $\boldsymbol{\kappa}^L$ is always positive. Thus the low frequency dielectric tensor is larger than that above the mechanical resonance frequencies of the crystal. The piezoelectric term in this equation is an example of an *indirect effect*, that is, a contribution to one interaction from a measurably distinguishable combination of other interactions. Here the double use of the piezoelectric interaction is distinguishable from the linear electric interaction by comparing the values of components of the dielectric tensor at frequencies well above the mechanical resonance frequencies to the values of components at frequencies well below them.

11.6 Homogeneous Static Deformation in Piezoelectrics

Let us find the voltage induced between two opposite electroded surfaces of a rectangular parallelepiped, shown in Fig. 11.1, by a uniform traction applied to its top surface characterized by the undeformed unit normal $\mathbf{N}^t$. The bottom surface rests on a reference plane. Two large area side surfaces characterized by the undeformed unit normal $\mathbf{N}^e$ are electroded and at a distance w apart; the other two side surfaces are not electroded; all four are traction-free. We assume an arbitrary crystalline orientation. We also assume that the spontaneous electric field is canceled by surface charge given by (11.3.2).

The boundary condition (11.4.20) on the linearized total stress may be expressed with the convenient artifice of representing the force applied to the body by outside means as simply a compressive surface traction $-\mathbf{n}^t\mathcal{F}$ [which thus must appear on the right side of (11.4.20)] as

$$(c_{ijkl}u_{k,l} + e_{mij}\Phi_{,m})N_j^t = -\frac{n_i^t\mathcal{F}}{a^t}, \tag{11.6.1}$$

$$(c_{ijkl}u_{k,l} + e_{mij}\Phi_{,m})N_j^e = 0, \tag{11.6.2}$$

$$(c_{ijkl}u_{k,l} + e_{mij}\Phi_{,m})N_j^f = 0. \tag{11.6.3}$$

Here a^t is the deformed area, $\mathbf{n}^t$ is the deformed normal of the surface under traction, and $\mathbf{N}^t$, $\mathbf{N}^e$, and $\mathbf{N}^f$ are unit undeformed normals to the surfaces which are under traction, are electroded, and are free, respectively. Since $\mathcal{F}$ is a linear variable, the area a^t and normal $\mathbf{n}^t$ may be

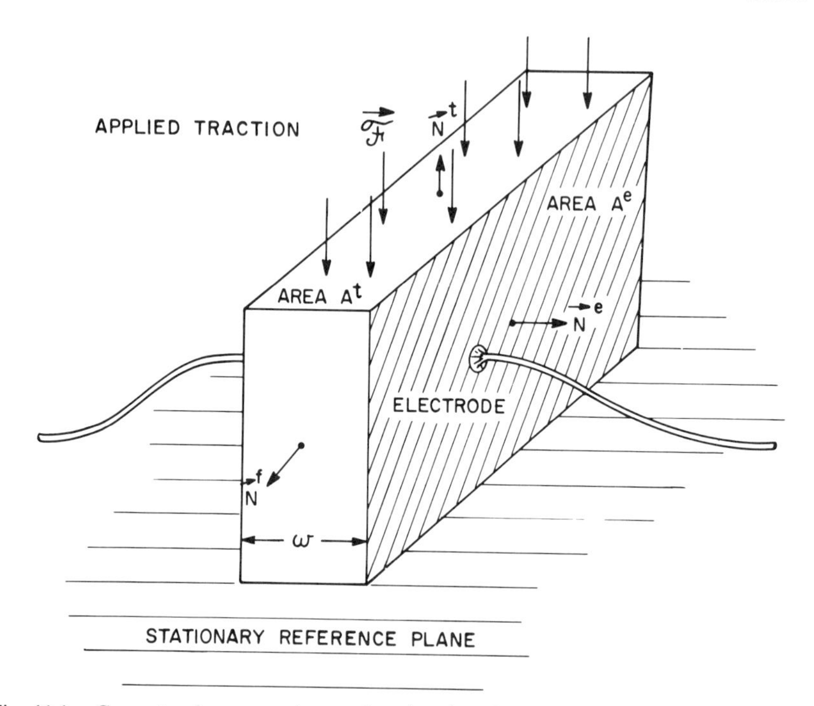

Fig. 11.1. Geometry for measuring a piezoelectric voltage generated by a static homogeneous stress.

replaced to linear accuracy by the corresponding undeformed surface quantities A^t and $\mathbf{N}^t$.

The homogeneous solution satisfying (11.6.1) to (11.6.3) is

$$t_{ij} = c_{ijkl} u_{k,l} + e_{kij} \Phi_{,k} \tag{11.6.4}$$

where t_{ij} represents the applied stress defined by

$$\mathbf{t} = -\frac{\mathbf{N}^t \mathbf{N}^t \mathcal{F}}{A^t}. \tag{11.6.5}$$

With the use of the definitions of s_{mnij} and d_{kmn} in (10.5.3) and (10.5.4) the strain $u_{(m,n)}$ may be found to be

$$u_{(m,n)} = s_{mnij} t_{ij} - d_{kmn} \Phi_{,k}. \tag{11.6.6}$$

Since the electric field tangential to the electrodes vanishes, the electric field must be normal to the electrodes,

$$\mathbf{E} = -\nabla\Phi = -\mathbf{N}^e(\mathbf{N}^e \cdot \nabla\Phi), \tag{11.6.7}$$

where the undeformed normals may be used to linear accuracy.

To obtain the charge-voltage-traction relation we use the boundary condition (11.4.13) on the normal component of **D** with $\mathbf{E}^o$ (in the electrode) set equal to zero. Substitution of (11.6.6) and (11.6.7) into this boundary condition leads to

$$Q = CV + Fd, \tag{11.6.8}$$

where

$$Q \equiv \Sigma^c A^e \tag{11.6.9}$$

$$C \equiv \frac{\epsilon_0 A^e N_i^e N_j^e \kappa_{ij}^L}{w}, \tag{11.6.10}$$

$$V \equiv w N_l^e \Phi_{,l}, \tag{11.6.11}$$

$$F \equiv \frac{A^e \mathcal{F}}{A^t}, \tag{11.6.12}$$

$$d \equiv N_i^e d_{ijk} N_j^t N_k^t \tag{11.6.13}$$

are respectively the charge, the capacitance, the induced voltage, the normalized applied traction, and the measured component of the piezoelectric strain tensor.

To measure d from (11.6.8) the voltage V_1 is measured across a capacitance C_1 connected between the electrodes for a given normalized traction F_1,

$$C_1 V_1 = C V_1 + F_1 d. \tag{11.6.14}$$

A second capacitance C_2 is put in place of C_1 and V_2 and F_2 measured. The measured d is then obtained from

$$d = \frac{C_2 - C_1}{(F_2/V_2) - (F_1/V_1)}. \tag{11.6.15}$$

Use of a sufficient number of crystals with different crystallographic orientations allows the measurement of all the d_{ijk} components.

The experiment just described is called the *direct piezoelectric effect*. When the voltage is applied to the crystal and the resulting strain is measured by way of strain gauges or interferometry of the surfaces, the experiment is said to measure the *converse piezoelectric effect*. Such a strain is given by (11.6.6) with $t_{ij}=0$. Clearly, it measures the same d_{kmn} tensor.

11.7 Piezoelectrically Excited Thickness Vibrations of Plates

Since a frequency of mechanical resonance is determined by a relation between the acoustic wavelength and a dimension of the body, it obviously depends on the acoustic velocity. Since the acoustic velocity in a piezoelectric crystal contains a piezoelectric stiffening term, measurement of the resonant frequencies can be used to determine the piezoelectric tensor components. For example, such measurements were used in the first determination [Warner, Onoe, and Coquin, 1967] of the piezoelectric tensor components of $LiTaO_3$, an important piezoelectric crystal.

In this section we solve the problem of thickness vibrations of an infinite plate excited by an oscillating electric potential applied to the two large area electroded surfaces. The first completely correct treatment of this problem is due to Tiersten [1963]. Thickness vibrations are vibrations consisting of waves traveling normal to the broad area surfaces. They depend only on the coordinate measured perpendicular to these surfaces. However, the displacement vector of thickness vibrations can have any orientation, that is, the thickness vibration can be transverse, longitudinal, or hybrid.

We consider a piezoelectric crystal which may be pyroelectric or nonpyroelectric. If it is pyroelectric, we assume that the spontaneous electric field $\mathbf{E}^S$ is canceled by extrinsic surface charge, since this is the commonly encountered situation for an electroded plate. Since the plate thickness h (see Fig. 11.2) is taken as small compared to the free-space electromagnetic wavelength, the problem may be treated in the quasi-electrostatic approximation. No problem arises from the infinite lateral dimensions, since no coordinate dependence on those directions is allowed. The problem to be solved is thus a model for a thin, broad faced plate all of whose dimensions are small compared to the electromagnetic wavelength.

The elasticity equation,

$$c_{ijkl}u_{k,lj}+e_{kij}\Phi_{,kj}=\rho^0\frac{\partial^2 u_i}{\partial t^2}, \tag{11.7.1}$$

and the $\nabla\cdot\mathbf{D}$ equation,

$$\epsilon_0\kappa_{ij}\Phi_{,ij}-e_{jkl}u_{k,lj}=0, \tag{11.7.2}$$

must be solved simultaneously. Arbitrary crystal symmetry and arbitrary

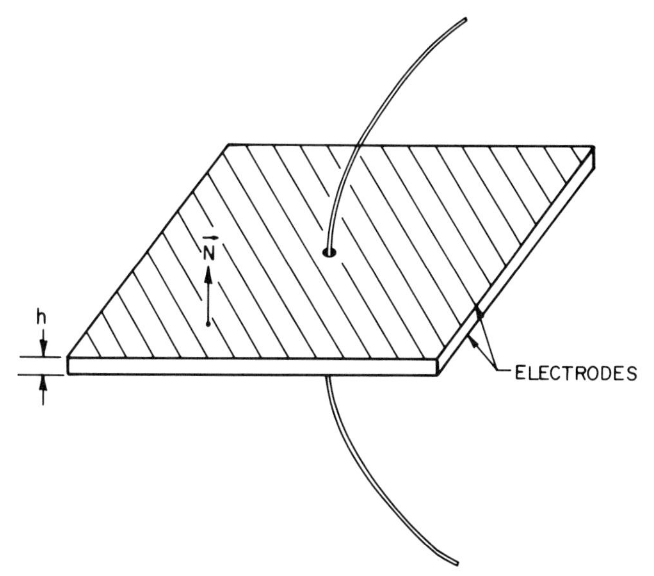

Fig. 11.2. Geometry of an electroded plate used to study piezoelectric thickness vibrations.

crystallographic orientation are allowed. We assume the surfaces to be stress-free,

$$\left(c_{ijkl}u_{k,l}+e_{kij}\Phi_{,k}\right)N_j=0 \qquad \left(z=\pm\frac{h}{2}\right). \tag{11.7.3}$$

Here $\mathbf{N}$ is a unit normal to the plate surfaces in the positive direction and z is a scalar coordinate measured in that direction. We also assume that a potential of amplitude V and angular frequency ω is applied to the plates,

$$\Phi=\pm\frac{V}{2}\cos\omega t \qquad \left(z=\pm\frac{h}{2}\right). \tag{11.7.4}$$

Because of the one-dimensional nature of the problem we set

$$\frac{\partial}{\partial z_i}=N_i\frac{\partial}{\partial z}. \tag{11.7.5}$$

Equations (11.7.1) and (11.7.2) then become

$$c_{ijkl}N_lN_j\frac{\partial^2 u_k}{\partial z^2}+e_{kij}N_kN_j\frac{\partial^2\Phi}{\partial z^2}=\rho^0\frac{\partial^2 u_i}{\partial t^2}, \tag{11.7.6}$$

$$\epsilon_0\kappa_{ij}N_iN_j\frac{\partial^2\Phi}{\partial z^2}-e_{jkl}N_lN_j\frac{\partial^2 u_k}{\partial z^2}=0. \tag{11.7.7}$$

Eliminating $\partial^2\Phi/\partial z^2$ between these equations yields

$$\rho^0 \frac{\partial^2 u_i}{\partial t^2} = \bar{c}_{ijkl} N_j N_l \frac{\partial^2 u_k}{\partial z^2} \tag{11.7.8}$$

where the piezoelectrically stiffened stiffness tensor is

$$\bar{c}_{ijkl} \equiv c_{ijkl} + \frac{e_{mij} N_m e_{nkl} N_n}{\epsilon_0 \mathbf{N} \cdot \boldsymbol{\kappa} \cdot \mathbf{N}}. \tag{11.7.9}$$

A standing wave solution,

$$\mathbf{u} = \mathbf{b}^\alpha (A \cos k^\alpha z + B \sin k^\alpha z) \cos \omega t, \tag{11.7.10}$$

of (11.7.8) is desired where $\mathbf{b}^\alpha$ is one of the three displacement eigenvectors of

$$\rho^0 v_\alpha^2 b_i^\alpha = \bar{c}_{ijkl} N_j N_l b_k^\alpha. \tag{11.7.11}$$

Here the phase velocity,

$$v_\alpha \equiv \frac{\omega}{k^\alpha}, \tag{11.7.12}$$

is one of the three solutions of

$$\det\left(\bar{c}_{ijkl} N_j N_l - \rho^0 v^2 \delta_{ik}\right) = 0 \tag{11.7.13}$$

as discussed in Section 10.8. In order to match boundary conditions the displacement must be taken as a sum of all three eigenvector solutions of the type (11.7.10),

$$\mathbf{u} = \sum_{\alpha=1}^{3} \mathbf{b}^\alpha (A^\alpha \cos k^\alpha z + B^\alpha \sin k^\alpha z) \cos \omega t. \tag{11.7.14}$$

From (11.7.7) we find that Φ is given in terms of this $\mathbf{u}$ by

$$\Phi = \frac{e_{jkl} N_j N_l}{\epsilon_0 \mathbf{N} \cdot \boldsymbol{\kappa} \cdot \mathbf{N}} u_k + (C_1 z + C_2) \cos \omega t \tag{11.7.15}$$

where C_1 and C_2 are integration constants.

The boundary conditions can now be applied to determine the arbitrary constants. The potential conditions (11.7.4) applied to Φ yield

$$C_1 = \frac{V}{h} - \frac{2}{h}\frac{e_{jkl}N_jN_l}{\epsilon_0 \mathbf{N}\cdot\boldsymbol{\kappa}\cdot\mathbf{N}}\sum_\alpha b_k^\alpha B^\alpha \sin\frac{k^\alpha h}{2}, \tag{11.7.16}$$

$$C_2 = -\frac{e_{jkl}N_jN_l}{\epsilon_0 \mathbf{N}\cdot\boldsymbol{\kappa}\cdot\mathbf{N}}\sum_\alpha b_k^\alpha A^\alpha \cos\frac{k^\alpha h}{2}. \tag{11.7.17}$$

Substitution of both the potential and the displacement into the stress boundary conditions (11.7.3) produces two conditions,

$$\sum_\alpha A^\alpha k^\alpha b_k^\alpha \bar{c}_{ijkl}N_jN_l \sin\frac{k^\alpha h}{2} = 0, \tag{11.7.18}$$

$$\sum_\alpha B^\alpha k^\alpha b_k^\alpha N_jN_l\left[\bar{c}_{ijkl}\cos\frac{k^\alpha h}{2} + \left(c_{ijkl} - \bar{c}_{ijkl}\right)\frac{2}{k^\alpha h}\sin\frac{k^\alpha h}{2}\right] = -e_{lij}N_lN_j\frac{V}{h}. \tag{11.7.19}$$

The first of these states that the even vibrations of $\mathbf{u}$ in (11.7.14) are not driven in the plate geometry. They can thus be dropped. The second condition has three component equations to determine the three B^α coefficients. We can solve for these coefficients by substituting the eigenvector equation (11.7.11) into the first term and then forming the scalar product of the equation with $\mathbf{b}^\beta$. The equation can then be written as

$$B^\beta - C^\beta \sum_\alpha D^\alpha B^\alpha = -C^\beta \tag{11.7.20}$$

where

$$C^\beta \equiv \frac{e^\beta V}{\rho^0 v_\beta^2 k^\beta \cos\left(k^\beta h/2\right)}, \tag{11.7.21}$$

$$D^\alpha \equiv \frac{2e^\alpha \sin(k^\alpha h/2)}{\epsilon_0(\mathbf{N}\cdot\boldsymbol{\kappa}\cdot\mathbf{N})hV}, \tag{11.7.22}$$

$$e^\alpha \equiv N_i e_{ijk} b_j^\alpha N_k. \tag{11.7.23}$$

If (11.7.20) is multiplied by D^β and then summed over β, we obtain

$$\sum_\beta D^\beta B^\beta = \frac{-\sum_\beta D^\beta C^\beta}{1 - \sum_\alpha D^\alpha C^\alpha}. \tag{11.7.24}$$

Substitution of this back into (11.7.20) gives

$$B^{\beta}=-\frac{C^{\beta}}{1-\sum_{\alpha}D^{\alpha}C^{\alpha}}$$
$$=-\frac{e^{\beta}V}{\rho^{0}v_{\beta}^{2}k^{\beta}\cos(k^{\beta}h/2)}\left[1-\sum_{\alpha}(K^{\alpha})^{2}\left(\frac{k^{\alpha}h}{2}\right)^{-1}\tan\left(\frac{k^{\alpha}h}{2}\right)\right]^{-1} \tag{11.7.25}$$

where the constant K^{α},

$$K^{\alpha}\equiv\frac{e^{\alpha}}{\left[\epsilon_{0}(\mathbf{N}\cdot\boldsymbol{\kappa}\cdot\mathbf{N})\rho^{0}v_{\alpha}^{2}\right]^{1/2}}, \tag{11.7.26}$$

is called the *electromechanical coupling factor* for the α eigenmode.

The final solutions for the displacement and the electric potential are

$$\mathbf{u}=-\sum_{\alpha}\frac{\mathbf{b}^{\alpha}e^{\alpha}V\sin k^{\alpha}z\cos\omega t}{k^{\alpha}h\rho^{0}v_{\alpha}^{2}\cos(k^{\alpha}h/2)\left[1-\sum_{\epsilon}(K^{\epsilon})^{2}(k^{\epsilon}h/2)^{-1}\tan(k^{\epsilon}h/2)\right]}, \tag{11.7.27}$$

$$\Phi=\frac{V\cos\omega t}{h}\left\{z-\sum_{\alpha}\frac{(K^{\alpha})^{2}\left[\sin k^{\alpha}z-(2z/h)\sin(k^{\alpha}h/2)\right]}{k^{\alpha}\cos(k^{\alpha}h/2)\left[1-\sum_{\epsilon}(K^{\epsilon})^{2}(k^{\epsilon}h/2)^{-1}\tan(k^{\epsilon}h/2)\right]}\right\}. \tag{11.7.28}$$

We may now use these to find the frequency dependence of the dielectric tensor in the piezoelectric resonance region with the use of the electric displacement boundary condition (11.4.13) with $\mathbf{E}^{o}$ (in the electrode) set equal to zero,

$$Q=A^{e}N_{i}\left(\epsilon_{0}\kappa_{ij}N_{j}\frac{\partial\Phi}{\partial z}-e_{ijk}N_{k}\frac{\partial u_{j}}{\partial z}\right)\qquad\left(z=\pm\frac{h}{2}\right) \tag{11.7.29}$$

where A^{e} is the electroded area and $Q=A^{e}\Sigma^{c}$ is the charge on it. Substituting Φ and $\mathbf{u}$ into this yields

$$Q=CV\cos\omega t \tag{11.7.30}$$

with the capacitance C given by

$$C=\frac{\epsilon_0 A\,{}^e\bar{\kappa}}{h} \tag{11.7.31}$$

in which the effective dielectric constant $\bar{\kappa}$ is

$$\bar{\kappa}=\frac{\mathbf{N}\cdot\boldsymbol{\kappa}\cdot\mathbf{N}}{1-\sum_\alpha (K^\alpha)^2(k^\alpha h/2)^{-1}\tan(k^\alpha h/2)}. \tag{11.7.32}$$

This has resonances whenever

$$1-\sum_\alpha (K^\alpha)^2\left(\frac{k^\alpha h}{2}\right)^{-1}\tan\left(\frac{k^\alpha h}{2}\right)=0. \tag{11.7.33}$$

If damping were included, the resonances would produce maxima and minima in $\bar{\kappa}$ rather than actual singularities. We can conclude that the denominator of (11.7.32) tends toward unity as $\omega\to\infty$ because of the $(k^\alpha)^{-1}\sim\omega^{-1}$ factor in the sum. Thus, the high frequency limit of (11.7.32) is

$$\bar{\kappa}=\mathbf{N}\cdot\boldsymbol{\kappa}\cdot\mathbf{N} \qquad (\omega\to\infty) \tag{11.7.34}$$

as expected. The low frequency limit is

$$\bar{\kappa}=\frac{\mathbf{N}\cdot\boldsymbol{\kappa}\cdot\mathbf{N}}{1-\sum_\alpha (K^\alpha)^2} \qquad (\omega\to 0), \tag{11.7.35}$$

which differs from the low frequency dielectric tensor in (11.5.6) for an unelectroded body having finite dimensions.

If a combination of crystal symmetry and orientation allows only one acoustic mode β of the three to be driven piezoelectrically, the sum in (11.7.33) contains only one term and the resonances are then given by

$$(K^\beta)^2\tan\left(\frac{k^\beta h}{2}\right)=\frac{k^\beta h}{2}. \tag{11.7.36}$$

The solutions of this equation are not spaced evenly in frequency. However, if the equation is rewritten as

$$\frac{(K^\beta)^2\sin(k^\beta h/2)}{(k^\beta h/2)}-\cos\left(\frac{k^\beta h}{2}\right)=0, \tag{11.7.37}$$

it is apparent that, for frequencies high compared to the lowest resonant frequency, the resonances are determined approximately by

$$\cos\left(\frac{k^{\beta}h}{2}\right)=0 \tag{11.7.38}$$

and thus approach a uniform spacing. This equation becomes a better approximation when the electromechanical coupling factor K^{β} is small. The solutions of (11.7.38) are

$$\frac{k^{\beta}h}{2}=\frac{\omega h}{2v_{\beta}}=\frac{(2m-1)\pi}{2} \qquad (m=1,2,3\ldots). \tag{11.7.39}$$

In terms of the acoustic wavelength

$$\lambda_{\beta}\equiv\frac{v_{\beta}}{\nu}=\frac{2\pi v_{\beta}}{\omega}, \tag{11.7.40}$$

where $\nu\equiv\omega/2\pi$ is the circular frequency, the resonance condition (11.7.39)

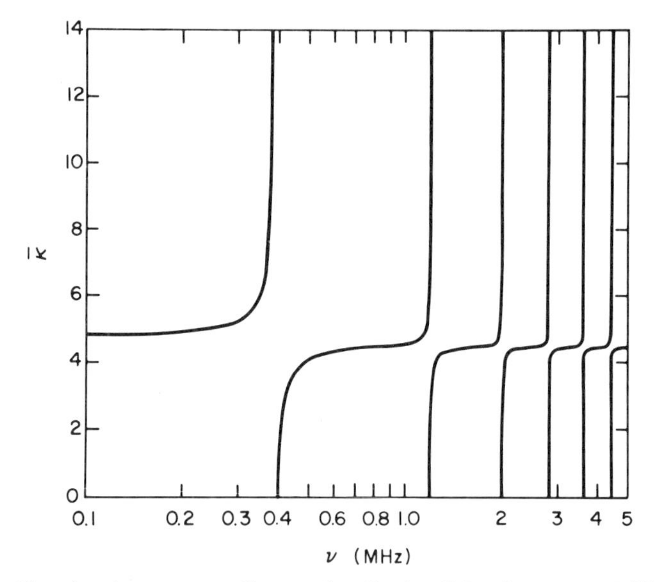

Fig. 11.3. Piezoelectric resonance effect on the effective dielectric constant $\bar{\kappa}$ (11.7.32) of a Y-cut quartz plate 0.5 mm thick versus frequency. The lowest six resonances governed by (11.7.36) are shown. No damping of them is included. For this orientation of quartz (point group 32) only the pure shear wave with the displacement in the X direction can be piezoelectrically driven. Its electromechanical coupling coefficient is -0.267 and its velocity is 4.023×10^{3} m/sec. The high frequency value for κ_{22} of 4.435 is used in the plot.

becomes

$$\frac{(2m-1)}{2}\lambda_\beta = h, \qquad (11.7.41)$$

which states that resonance occurs whenever an odd integral number of half acoustic wavelengths equals the plate thickness. Measurement of the resonant frequency determines the velocity v_β which is related to the stiffened elastic constants.

In order to better visualize the frequency dependence of the effective dielectric constant $\bar{\kappa}$ for a piezoelectric crystal plate we consider the special case of a Y-cut quartz plate, that is, with the Y crystallographic axis normal to the broad area surface of the plate. Since the electromechanical coupling constant vanishes for two of the three eigenmodes that propagate in the Y direction, the summation in (11.7.32) for $\bar{\kappa}$ contains only one term. That one represents a pure shear wave with the displacement eigenvector in the X direction. A plot of $\bar{\kappa}$ for a Y-cut quartz plate is given in Fig. 11.3. No damping of the resonances is included. Note that $\bar{\kappa}$ is lower in the almost flat region above each resonance compared to that below the same resonance. Note, however, that the major lowering of $\bar{\kappa}$ occurs on passage through the fundamental resonance.

11.8 Edge-Excited Thickness Vibrations

The thickness vibrations of a plate that are excited by a field between the major plate faces are all stiffened modes, since the exciting electric field is in the longitudinal direction of the traveling acoustic waves that compose the standing wave vibration. If the resonant frequencies and thus the velocities and elastic constants are to be measured for the unstiffened modes, a different method of excitation of those waves is needed. Some of them can be excited by placing the electrodes on two opposite edges of a plate. For this reason it is important to analyze edge-excited thickness vibrations of plates [Yamada and Niizeki, 1971].

Consider the plate shown on Fig. 11.4. We wish to find the form of the thickness vibrations between the broad area faces that are separated by a distance h. The equations (11.7.1) and (11.7.2) of the last section must again be solved. If the crystal considered is pyroelectric, we assume that $\mathbf{E}^S$ is canceled by extrinsic surface charge. Arbitrary crystal symmetry and arbitrary crystallographic orientation are again permitted.

The boundary conditions used on the broad area surfaces are the stress-free condition (11.7.3) used in the preceding section and the continuity of the normal component of electric displacement condition. For the

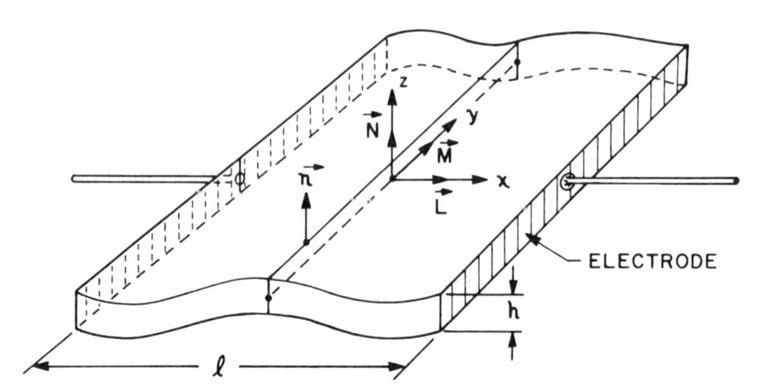

Fig. 11.4. Edge-excited plate geometry for exciting thickness vibrations.

latter we assume the plate to be essentially infinite. For such a plate the depolarizing field (see Section 8.5) causes the electric field outside the plate to vanish. Thus the electric displacement boundary condition (11.4.13) can be taken as

$$N_i(\epsilon_0\kappa_{ij}\Phi_{,j} - e_{ijk}u_{j,k}) = 0 \qquad \text{at } z = \pm\frac{h}{2}. \tag{11.8.1}$$

The meaning of this modification of the **D** field boundary condition is that the depolarizing field of a plate causes **D** (and not **E**) to be normal to the edge electrodes.

A relatively simple solution to this problem may be found provided we do not require the solution to obey either stress or potential boundary conditions on the electroded surfaces at $x = \pm l/2$. Thus the solution we find is a good approximation to the true solution from the center of the plate to a distance from the edges about equal to the plate thickness. To make up for the lack of the potential boundary condition at the electrodes we impose on the electric potential Φ a form consisting of three parts: (1) one (piezoelectrically driven) part independent of x and y, (2) a second (driving) part linear in x and independent of y and z, and (3) a third (depolarizing) part linear in z and independent of x and y. The displacement $\mathbf{u}$ can be taken independent of x and y. All solutions are taken as oscillatory in time at a single frequency $\omega = 2\pi\nu$.

In line with the discussion above we search for a displacement of the form

$$\mathbf{u}(z,t) = \mathbf{u}(z)\cos\omega t. \tag{11.8.2}$$

Since the potential can also depend on x, we express the gradient operator as

$$\nabla = \mathbf{L}\frac{\partial}{\partial x} + \mathbf{N}\frac{\partial}{\partial z} \tag{11.8.3}$$

where $\mathbf{L}, \mathbf{N}$ are unit vectors in the x, z directions respectively (see Fig. 11.4). Since the desired form of solution for Φ is

$$\Phi(x,z,t) = \left[\Phi(z) - E_0 x + Fz\right]\cos\omega t \tag{11.8.4}$$

where E_0 is the applied electric field and F the depolarizing field, we see that

$$\nabla\nabla\Phi = \mathbf{NN}\frac{\partial^2\Phi}{\partial z^2}. \tag{11.8.5}$$

Because of this the dynamical equations take the form of (11.7.6) and (11.7.7) of the preceding section.

Dropping the even vibrations, since they cannot be driven (see Section 11.7), we adopt solutions consistent with (11.8.4),

$$\mathbf{u} = \sum_{\alpha=1}^{3} \mathbf{b}^\alpha B^\alpha \sin k^\alpha z \cos\omega t, \tag{11.8.6}$$

$$\Phi = \left[\frac{N_i e_{ijk} N_k}{\epsilon_0 \mathbf{N}\cdot\boldsymbol{\kappa}\cdot\mathbf{N}} \sum_\alpha b_j^\alpha B^\alpha \sin k^\alpha z + Fz - E_0 x\right]\cos\omega t \tag{11.8.7}$$

where $\mathbf{b}^\alpha$ $(\alpha = 1,2,3)$ are displacement eigenvectors defined in (11.7.11) to (11.7.13) and discussed in Section 10.8. The presence of the stiffened stiffness tensor in those equations does not guarantee that every mode will be piezoelectrically stiffened. Symmetry sometimes does not permit it. As mentioned earlier, it is just such modes in which we are most interested in this section. The depolarizing field F can now be determined from the boundary condition (11.8.1) with the result

$$F = \frac{\mathbf{N}\cdot\boldsymbol{\kappa}\cdot\mathbf{L}}{\mathbf{N}\cdot\boldsymbol{\kappa}\cdot\mathbf{N}} E_0. \tag{11.8.8}$$

Substitution of the solutions into the stress boundary condition (11.7.3) yields a vector equation

$$N_j N_l \bar{c}_{ijkl} \sum_\alpha k^\alpha B^\alpha b_k^\alpha \cos\frac{k^\alpha h}{2} = N_j e_{kji}\left[L_k - N_k \frac{\mathbf{N}\cdot\boldsymbol{\kappa}\cdot\mathbf{L}}{\mathbf{N}\cdot\boldsymbol{\kappa}\cdot\mathbf{N}}\right] E_0, \tag{11.8.9}$$

which can be used to solve for the three amplitudes B^α ($\alpha=1,2,3$) of the odd vibrations. Substitution of the eigenvector equation (11.7.11) into this, forming a scalar product with $\mathbf{b}^\beta$, and use of the orthonormality of those vectors leads to

$$B^\beta = \frac{N_j e_{kji} b_i^\beta}{\rho^0 v_\beta^2 k^\beta \cos\dfrac{k^\beta h}{2}} \left[-N_k \frac{\mathbf{N}\cdot\boldsymbol{\kappa}\cdot\mathbf{L}}{\mathbf{N}\cdot\boldsymbol{\kappa}\cdot\mathbf{N}} + L_k \right] E_0. \tag{11.8.10}$$

Though the approximate solution we now have, consisting of (11.8.6) and (11.8.7) for **u** and Φ with the constants F and B^α given by (11.8.8) and (11.8.10), does not satisfy a constant potential condition over each electrode, we can nevertheless use it to determine the charge on the electrodes by way of the boundary condition on the normal component of the electric displacement. This boundary condition, (11.4.13), takes the form

$$Q = \int \Sigma^c \, dz = -\int \mathbf{D}\cdot\mathbf{L}\, dz \tag{11.8.11}$$

where Q is the charge per unit length in the y direction (see Fig. 11.4) and the electric displacement in the metal is set equal to zero. Substitution into this condition leads to

$$Q = C V_0 \cos\omega t, \tag{11.8.12}$$

where

$$V_0 \equiv -E_0 l \tag{11.8.13}$$

and where the capacitance C per unit length in the y direction is

$$C = \frac{\epsilon_0 h \bar{\kappa}}{l}. \tag{11.8.14}$$

Here $\bar{\kappa}$ is the effective dielectric constant,

$$\bar{\kappa} \equiv \kappa_P \left(1 + \sum_{\alpha=1}^{3} \frac{2(K^\alpha)^2}{hk^\alpha} \tan\frac{k^\alpha h}{2} \right), \tag{11.8.15}$$

κ_P is the effective dielectric constant of the edge-electroded plate in the absence of any piezoelectric effect,

$$\kappa_P \equiv \mathbf{L}\cdot\boldsymbol{\kappa}\cdot\mathbf{L} - \frac{(\mathbf{N}\cdot\boldsymbol{\kappa}\cdot\mathbf{L})^2}{\mathbf{N}\cdot\boldsymbol{\kappa}\cdot\mathbf{N}}, \tag{11.8.16}$$

K^α is the electromechanical coupling constant of the α mode,

$$K^\alpha \equiv \frac{\bar{e}^\alpha}{\left[\epsilon_0 \kappa_P \rho^0 v_\alpha^2\right]^{1/2}}, \tag{11.8.17}$$

and $\bar{e}^\alpha$ is the effective piezoelectric stress constant for the α mode,

$$\bar{e}^\alpha \equiv L_i e_{ijk} b_j^\alpha N_k - \left(\frac{\mathbf{N}\cdot\boldsymbol{\kappa}\cdot\mathbf{L}}{\mathbf{N}\cdot\boldsymbol{\kappa}\cdot\mathbf{N}}\right) N_i e_{ijk} b_j^\alpha N_k. \tag{11.8.18}$$

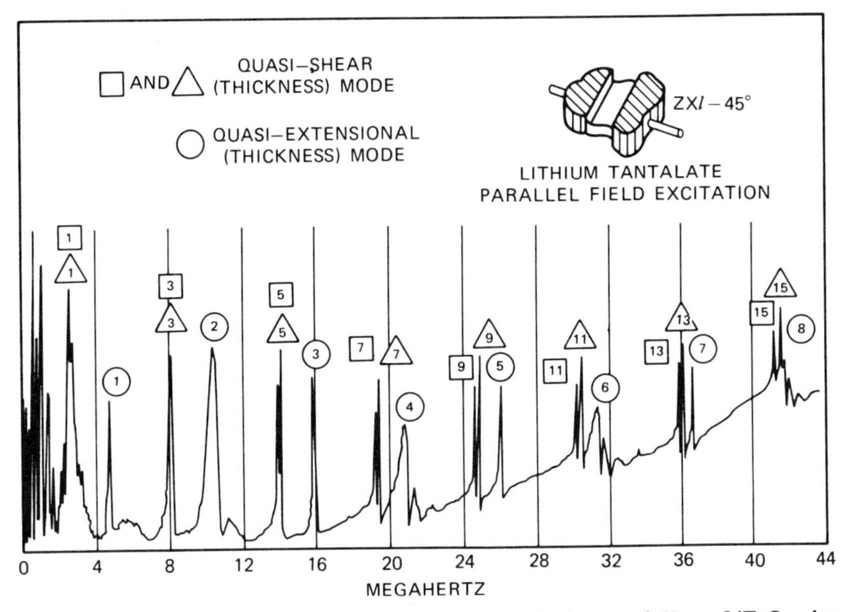

Fig. 11.5. Observed thickness resonances of an edge-excited, rotated Y-cut, $LiTaO_3$ plate [Warner et al., 1967, with permission]. The three series of eigenmodes are separately numbered. Only the odd numbered resonances of the two quasi-shear modes were excited, as expected from (11.8.21). However, both even and odd numbered resonances were found for the quasi-extensional mode. An explanation of how the even numbered resonances are excited requires a more sophisticated solution than that presented in Section 11.8. The many resonances in the 0 to 2 MHz range involve the larger lateral dimensions of the plate.

Vibrational resonances occur when

$$\bar{\kappa}=\infty, \tag{11.8.19}$$

which corresponds to

$$\frac{k^{\alpha}h}{2}=(2m-1)\frac{\pi}{2} \qquad (\alpha=1,2,3;\ m=1,2,3\ldots) \tag{11.8.20}$$

or

$$\nu=\frac{(2m-1)v_{\alpha}}{2h}. \tag{11.8.21}$$

Here we see that the resonant frequencies are evenly spaced even for low order resonances. Their measurement can be used to determine the acoustic velocities v_{α}. Figure 11.5 shows the resonances observed in an edge-excited $LiTaO_3$ plate. The measurement of antiresonances,

$$\bar{\kappa}=0, \tag{11.8.22}$$

leads to a transcendental relation which is much more complicated for experimental comparisons and so less useful.

11.9 Piezoelectrically Excited Extensional Vibrations of Rods

In this section we find an approximate one-dimensional solution for the piezoelectrically excited longitudinal or extensional vibrations of a rectangular cross-section rod. We take the rod to be of length l, width w, and thickness h where

$$l \gg w > h \tag{11.9.1}$$

and where the broad faces of area lw are electroded (see Fig. 11.6). A

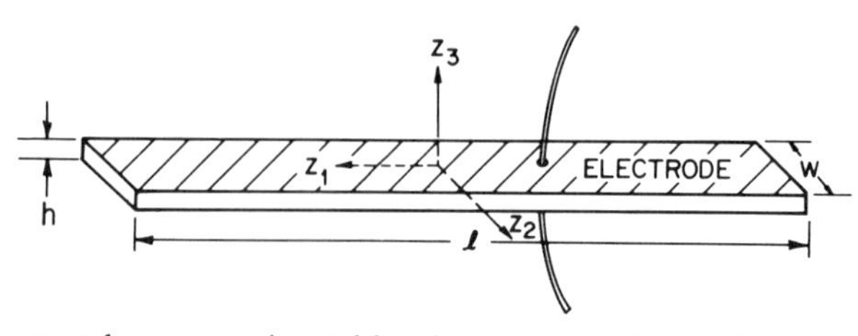

Fig. 11.6. Rectangular cross section rod for piezoelectric excitation of extensional vibrations.

geometric coordinate system with z_1, z_2, and z_3 measured in the length, width, and thickness directions respectively is used. An arbitrary orientation of crystal axes and an arbitrary symmetry crystal are permitted.

For extensional motion the controlling spatial variation is the lengthwise or z_1 direction. As we saw in Section 10.6 for an isotropic rod, the sides of the rod must move outward when the ends move inward and vice versa. Thus the displacement must depend on all three coordinates. The strain need not, however; it need depend only on the z_1 coordinate. Because of this an approximate solution to the problem can be found from a one-dimensional differential equation.

We consider first the stress boundary condition on the side surfaces. For stress-free sides we have

$$0 = t_{i2} = c_{i2kl}u_{k,l} - e_{mi2}E_m \qquad \text{at } z_2 = \pm\frac{w}{2} \tag{11.9.2}$$

and

$$0 = t_{i3} = c_{i3kl}u_{k,l} - e_{mi3}E_m \qquad \text{at } z_3 = \pm\frac{h}{2}. \tag{11.9.3}$$

Since the rod is thin in the z_2 and z_3 dimensions, these stresses may be taken as vanishing throughout the rod. Thus the only nonvanishing stress is t_{11} given by

$$t_{11} = c_{11kl}u_{k,l} - e_{m11}E_m. \tag{11.9.4}$$

The strain is given by

$$u_{(k,l)} = s_{kl11}t_{11} + d_{mkl}E_m \tag{11.9.5}$$

by (10.5.2). Since this equation holds for all k and l, it implies a dependence of the displacement on the lateral coordinates, z_2 and z_3, as mentioned earlier.

Next we show that t_{11} can be expressed in terms of a single displacement gradient $u_{1,1}$. Equation (11.9.4) may be rewritten as

$$t_{11} = c_{1111}u_{1,1} + \sum_{kl}{}' c_{11kl}u_{k,l} - e_{m11}E_m, \tag{11.9.6}$$

where the primed sum excludes the $k=1, l=1$ term. Equation (11.9.5) is now substituted, and the sum completed by adding and subtracting the missing term. The result is

$$\begin{aligned} t_{11} = c_{1111}u_{1,1} + c_{11kl}(s_{kl11}t_{11} + d_{mkl}E_m) \\ - c_{1111}(s_{1111}t_{11} + d_{m11}E_m) - e_{m11}E_m. \end{aligned} \tag{11.9.7}$$

Two pairs of terms cancel. Solving t_{11} gives

$$t_{11} = \frac{1}{s_{11}}u_{1,1} - \frac{d_{m1}}{s_{11}}E_m, \tag{11.9.8}$$

where we now use the matrix notation of Section 10.5. This shows that the effective stiffness coefficient of a rod is $1/s_{11}$ rather than c_{11} and that the effective piezoelectric stress tensor is d_{m1}/s_{11} rather than e_{m1}.

The electrical boundary condition that need be invoked is the vanishing of the tangential electric field at the electroded surfaces. This gives

$$E_1 = E_2 = 0 \qquad \text{at } z_3 = \pm\frac{h}{2}. \tag{11.9.9}$$

Because h is the thinnest dimension, these conditions may be taken as approximately true throughout the volume of the rod. This means that the acoustic waves traveling in the z_1 direction are unable to create piezoelectrically any appreciable longitudinal electric field that they normally would. To a good approximation then the only electric field is due to the applied potential V, which oscillates at an angular frequency ω and is independent of position. It is

$$E_3 = \frac{V}{h}\cos\omega t. \tag{11.9.10}$$

Since to this approximation no gradients of electric field exist, no piezoelectric coupling in the dynamic elasticity equation exists.

The only component of the elasticity equation that need be solved now is

$$\rho^0\frac{\partial^2 u_1}{\partial t^2} = t_{11,1} = \frac{1}{s_{11}}u_{1,11}. \tag{11.9.11}$$

This has a solution of the form

$$u_1(z_1,t) = (A\sin kz_1 + B\cos kz_1)\cos\omega t \tag{11.9.12}$$

with

$$k=(\rho^0 s_{11})^{1/2}\omega. \tag{11.9.13}$$

The vanishing of the stress at the rod ends,

$$0=\frac{1}{s_{11}}u_{1,1}\left(\pm\frac{l}{2}\right)-\frac{d_{31}}{s_{11}}E_3, \tag{11.9.14}$$

requires

$$A=\frac{d_{31}V}{kh\cos kl/2}, \tag{11.9.15}$$

$$B=0, \tag{11.9.16}$$

showing once again that only odd vibrations are excited. Thus the longitudinal displacement is given by

$$u_1(z_1,t)=\frac{d_{31}V}{kh}\frac{\sin kz_1}{\cos kl/2}\cos\omega t. \tag{11.9.17}$$

This shows that resonance occurs whenever

$$\frac{kl}{2}=(2n-1)\frac{\pi}{2} \qquad (n=1,2,3,\ldots) \tag{11.9.18}$$

or

$$\nu=\frac{(2n-1)}{2l(\rho^0 s_{11})^{1/2}}. \tag{11.9.19}$$

Observation of these resonant frequencies can be used to measure s_{11}, which, because we have employed a geometric coordinate system, can be any combination of the crystallographic compliance components.

PROBLEMS

11.1. If the elastic constants, the dielectric constant, and the density of a $\bar{4}3m$ crystal such as GaP are regarded as known, how can a plate resonance measurement be used to determine the piezoelectric constant?

11.2. If the stiffness constants are regarded as known, describe the measurement of the dielectric constant of a plate that is needed to determine e_{14} for a $\bar{4}3m$ crystal. Give an expression for e_{14} in terms of the measured quantities.

11.3. Derive an expression for the capacitance of the vibrating rod of Section 11.9. What are the high and low frequency limits of the effective dielectric constant?

11.4. An *electromechanical transducer* is a device that transforms electrical energy to acoustic energy or vice versa. One of the most commonly used transducers can be modeled as a layer of a piezoelectric crystal of uniform thickness l intimately bonded to a solid half-space. The outer surface of the layer is stress-free. Assume the half-space to be a piezoelectric crystal also. Assume arbitrary crystalline orientations of both the layer and half-space except that the one particular displacement vector eigenmode excited be common to both media, a condition normally used in practice. An oscillatory voltage $V \exp - i\omega t$ is applied across the layer. Ignore the material that forms the electrode layers. Find the displacement of the outgoing acoustic plane wave in the half-space. Show that the resonances of its amplitude agree exactly with those of thickness vibrations of the piezoelectric plate of Section 11.7 when the elastic stiffness of the half-space is negligibly small compared to that of the layer. Note that the exact resonances are damped even though no loss terms are present in the differential equations. Why?

11.5. Calculate the electrical impedance of the device of Prob. 11.4. Show that the zero frequency limit is a simple capacitance containing the low frequency dielectric constant.

11.6. Using the electrical impedance found in Prob. 11.5, calculate the input electrical power for the applied voltage $V \exp - i\omega t$ and show that it is equal to the acoustic power of the outgoing wave.

CHAPTER 12

Three-Field Optical Interactions

The linear interactions studied in the three preceding chapters may be termed two-field interactions because there is one input and one output field. Three-field interactions are the lowest order nonlinear interactions and include those with two input and one output field and those with one input and two output fields. Here we consider three-field interactions having an optical output. This implies that at least one of the input fields must be optical also. We allow the other input field to be either optical or acoustic.

The goal of this chapter is to obtain the constitutive relation for the nonlinear polarization in terms of two driving fields. The nonlinear polarization is the driving term in the wave equation for the output optical field. We develop the expression for the nonlinear polarization in a form that can be applied in later chapters to the direct and indirect elastooptic effects, the direct and indirect electrooptic effects, optical mixing, and optical parametric oscillation.

This theory makes a new prediction [Nelson and Lax, 1970, 1971b] for the form of the elastooptic susceptibility. It shows that besides the Pockels susceptibility, which couples the strain $u_{(i,j)}$ of an input acoustic wave to the electric field of an input optical wave to produce an output optical wave, there is an additional susceptibility term, which couples the rotation $u_{[i,j]}$ of an input acoustic wave with the electric field of an input optical wave to help produce the output optical wave. The latter term is present only in birefringent (uniaxial and biaxial) crystals and its size is related to the amount of birefringence. In strongly birefringent crystals such as rutile and calcite the new susceptibility term is comparable to the Pockels term.

12.1 Driven Internal Motion Equations

First, the general internal motion equation (7.3.3),

$$m^\nu \ddot{y}_i^\nu = q^\nu \mathscr{E}_i - \rho^0 \frac{\partial \Sigma}{\partial \Pi_C^\nu} X_{C,i}, \tag{12.1.1}$$

must be expanded to bilinear order. The material time derivative $\ddot{y}_i^\nu$ may be expanded in terms of spatial time derivatives with the transformation formula (3.11.15). The material time derivative $\dot{\mathbf{x}}=\dot{\mathbf{u}}$ appearing within $\mathscr{E}$ may be replaced by $\partial \mathbf{u}/\partial t$, since it multiplies another field. Thus we have

$$m^\nu\left(\frac{\partial^2 y_i^\nu}{\partial t^2}+2\frac{\partial y_{i,j}^\nu}{\partial t}\frac{\partial u_j}{\partial t}+y_{i,j}^\nu\frac{\partial^2 u_j}{\partial t^2}\right)=q^\nu\left[E_i+\left(\frac{\partial \mathbf{u}}{\partial t}\times\mathbf{B}\right)_i\right]$$
$$-\rho^0\frac{\partial \Sigma}{\partial \Pi_C^\nu}X_{C,i}. \tag{12.1.2}$$

The stored energy expansion must be carried to trilinear order,

$$\begin{aligned}\rho^0\Sigma={}&{}^{01}M_{AB}E_{AB}+{}^{02}M_{ABCD}E_{AB}E_{CD}+{}^{03}M_{ABCDEF}E_{AB}E_{CD}E_{EF}\\&+\sum_\nu{}^{10}M_C^\nu\Pi_C^\nu+\sum_{\nu\mu}{}^{20}M_{CD}^{\nu\mu}\Pi_C^\nu\Pi_D^\mu+\sum_{\nu\mu\lambda}{}^{30}M_{CDE}^{\nu\mu\lambda}\Pi_C^\nu\Pi_D^\mu\Pi_E^\lambda\\&+\sum_\nu{}^{11}M_{CAB}^\nu\Pi_C^\nu E_{AB}+\sum_{\nu\mu}{}^{21}M_{CDAB}^{\nu\mu}\Pi_C^\nu\Pi_D^\mu E_{AB}\\&+\sum_\nu{}^{12}M_{CABDE}^\nu\Pi_C^\nu E_{AB}E_{DE}.\end{aligned}\tag{12.1.3}$$

A number of simplifications can be made in the calculation at the beginning. Since we wish to consider an optical output field, at least one input field must be an optical field. Here "optical" refers to the frequency at which a particular field oscillates. In principle all fields entering the calculation can oscillate at optical frequencies. However, the displacement field is a center-of-mass variable and does not have any appreciable amplitude at an optical frequency. This is because optical frequencies are far above the resonant frequencies of the displacement field, which are the mechanical resonances of the body discussed in Chapter 11 (see also Prob. 10.8). Thus bilinear driving terms in (12.1.2) that involve two displacement fields are completely negligible for the problem considered, since at least one of them is at an optical frequency. This means that the terms involving ${}^{12}\mathbf{M}$ in (12.1.2) may be dropped at this point.

Other simplifications help reduce the problem. First, it is apparent that the terms in the stored energy involving ${}^{01}\mathbf{M}$, ${}^{02}\mathbf{M}$, and ${}^{03}\mathbf{M}$ cannot enter (12.1.2), since they possess no derivative with respect to $\mathbf{\Pi}^\nu$. Second, if the crystal considered is pyroelectric, we assume the condition that is most commonly met, namely, the spontaneous electric field being canceled by extrinsic surface charge as given in (11.3.2). The natural-state condition

(7.3.7) then gives

$$^{10}M_A^\nu = 0. \tag{12.1.4}$$

The stored energy derivative entering (12.1.2) can now be written as

$$\begin{aligned}\frac{\partial \rho^0 \Sigma}{\partial \Pi_C^\nu} &= 2\sum_\mu {}^{20}M_{CD}^{\nu\mu}\Pi_D^\mu + 3\sum_{\mu\lambda} {}^{30}M_{CDE}^{\nu\mu\lambda}\Pi_D^\mu \Pi_E^\lambda \\ &\quad + {}^{11}M_{CAB}^\nu E_{AB} + 2\sum_\mu {}^{21}M_{CDAB}^{\nu\mu}\Pi_D^\mu E_{AB}.\end{aligned} \tag{12.1.5}$$

Since we are dropping terms bilinear in the displacement, the Green finite strain tensor (3.10.5) may be taken as

$$E_{AB} \cong u_{(A,B)}. \tag{12.1.6}$$

Also, the rotationally invariant measure of the internal coordinates Π_C^ν of (6.5.6) is

$$\Pi_C^\nu = y_C^\nu - Y_i^\nu u_{C,i} - u_{C,i} y_i^\nu. \tag{12.1.7}$$

With these expressions the force term arising from the stored energy in (12.1.2) may be written as

$$\begin{aligned}\frac{\partial \rho^0 \Sigma}{\partial \Pi_C^\nu} X_{C,i} &= 2\sum_\mu {}^{20}M_{id}^{\nu\mu}\left[y_d^\mu - Y_j^\mu u_{d,j} - u_{d,j} y_j^\mu \right] \\ &\quad - 2\sum_\mu {}^{20}M_{cd}^{\nu\mu} y_d^\mu u_{c,i} + {}^{11}M_{iab}^\nu u_{a,b} \\ &\quad + 2\sum_\mu {}^{21}M_{idab}^{\nu\mu} y_d^\mu u_{a,b} \\ &\quad + 3\sum_{\mu\lambda} {}^{30}M_{ide}^{\nu\mu\lambda}\left[y_d^\mu - Y_j^\mu u_{d,j} \right]\left[y_e^\lambda - Y_k^\lambda u_{e,k} \right].\end{aligned} \tag{12.1.8}$$

Note that at this point only lowercase subscripts need be used, since the transformation from material to spatial derivatives is now accomplished to the order of accuracy desired. Gathering the linear terms on the left side and the bilinear terms on the right side, we write the internal motion equations as

$$m^\nu \frac{\partial^2 y_i^\nu}{\partial t^2} - q^\nu E_i + 2\sum_\mu {}^{20}M_{id}^{\nu\mu}\left[y_d^\mu - u_{d,j} Y_j^\mu \right] + {}^{11}M_{iab}^\nu u_{a,b} = F_i^\nu \tag{12.1.9}$$

where F_i^ν is the bilinear driving force acting on the ν internal coordinate and is defined by

$$F_i^\nu \equiv -2m^\nu \frac{\partial y_{i,j}^\nu}{\partial t}\frac{\partial u_j}{\partial t} - m^\nu y_{i,j}^\nu \frac{\partial^2 u_j}{\partial t^2} + q^\nu \epsilon_{ijk}\frac{\partial u_j}{\partial t}B_k$$
$$+2\sum_\mu {}^{20}M_{cd}^{\nu\mu}(u_{c,i}y_d^\mu + y_j^\mu u_{d,j}\delta_{ic})$$
$$-2\sum_\mu {}^{21}M_{idab}^{\nu\mu}y_d^\mu u_{a,b}$$
$$-3\sum_{\mu\lambda} {}^{30}M_{ide}^{\nu\mu\lambda}\left(y_d^\mu y_e^\lambda - 2y_d^\mu u_{e,k}Y_k^\lambda\right). \tag{12.1.10}$$

Further simplification of this requires the consideration of the frequency of each field.

12.2 Fourier Expansion of Fields

To determine which are the nonlinear terms of major importance we expand all the fields in Fourier series. The optical input frequency is denoted by ω_1 and the other input frequency, which we allow to be either another optical frequency or a low frequency in the acoustic frequency region, is denoted by ω_2. We consider an output field at the sum frequency ω_S given by

$$\omega_S = \omega_1 + \omega_2 \tag{12.2.1}$$

and regard $\omega_1 \neq \omega_2$. A development similar to that here could be done for the difference frequency ω_1,

$$\omega_1 = \omega_S - \omega_2. \tag{12.2.2}$$

Each field is expanded in a series of the form

$$\mathbf{f}(\mathbf{z},t) = \frac{1}{2}\sum_{m,n=-\infty}^{+\infty}(1+\delta_{m0}+\delta_{n0}-\delta_{m0}\delta_{n0})\mathbf{f}(\mathbf{z},t;m,n) \tag{12.2.3}$$

where

$$\mathbf{f}(\mathbf{z},t;m,n) \equiv \mathbf{f}(\mathbf{z};m,n)e^{-i(m\omega_1+n\omega_2)t}, \tag{12.2.4}$$

$$\mathbf{f}(\mathbf{z};-m,-n) \equiv \mathbf{f}^*(\mathbf{z};m,n). \tag{12.2.5}$$

In the last equation the star denotes the complex conjugate. That condition makes $\mathbf{f}(\mathbf{z},t)$ real. The somewhat complicated coefficient in the summation is merely to avoid any unnecessary factors of two if one or both input fields are static.

The Fourier expansion is now substituted for $\mathbf{u}$, $\mathbf{E}$, $\mathbf{B}$, and $\mathbf{y}^{\nu}$ in (12.1.9) and (12.1.10). As each group of terms at a given frequency is independent of the groups of terms at other frequencies, they each must vanish separately. Consider the group of terms at the sum frequency ω_S. Equation (12.1.9) yields

$$-\omega_S^2 m^{\nu} y_i^{\nu}(\mathbf{z};1,1) - q^{\nu} E_i(\mathbf{z};1,1) + 2\sum_{\mu} {}^{20}M_{id}^{\nu\mu}\left[y_d^{\mu}(\mathbf{z};1,1) - u_{d,j}(\mathbf{z};1,1)\,Y_j^{\mu}\right]$$
$$+\,{}^{11}M_{iab}^{\nu} u_{a,b}(\mathbf{z};1,1) = F_i^{\nu}(\mathbf{z};1,1) \qquad (12.2.6)$$

and (12.1.10) yields

$$\begin{aligned}
F_i^{\nu}(\mathbf{z};1,1) = \tfrac{1}{2}\{ & 2\omega_1\omega_2 m^{\nu} y_{i,j}^{\nu}(\mathbf{z};1,0)u_j(\mathbf{z};0,1) \\
& + \omega_2^2 y_{i,j}^{\nu}(\mathbf{z};1,0)u_j(\mathbf{z};0,1) - q^{\nu} i\omega_2 \epsilon_{ijk} u_j(\mathbf{z};0,1) B_k(\mathbf{z};1,0) \\
& + 2\sum_{\mu} {}^{20}M_{cd}^{\nu\mu}\left[u_{c,i}(\mathbf{z};0,1) y_d^{\mu}(\mathbf{z};1,0)\right. \\
& \left. + y_j^{\mu}(\mathbf{z};1,0) u_{d,j}(\mathbf{z};0,1)\delta_{ic}\right] \\
& - 2\sum_{\mu} {}^{21}M_{idab}^{\nu\mu} y_d^{\mu}(\mathbf{z};1,0) u_{a,b}(\mathbf{z};0,1) \\
& + 3\sum_{\mu\lambda} {}^{30}M_{ide}^{\nu\mu\lambda}\left[2u_{d,j}(\mathbf{z};0,1)\,Y_j^{\mu} y_e^{\lambda}(\mathbf{z};1,0)\right. \\
& \left. - y_d^{\mu}(\mathbf{z};1,0) y_e^{\lambda}(\mathbf{z};0,1)\right] \\
& + \text{interchange of } (0,1)\leftrightarrow(1,0) \text{ and } \omega_1\leftrightarrow\omega_2\}.
\end{aligned} \qquad (12.2.7)$$

A number of terms in the last two equations are negligible and may be dropped. As discussed before a displacement at an optical frequency is negligible in size. Thus two such terms on the left side in (12.2.6) and the $(0,1)\leftrightarrow(1,0)$ interchange term corresponding to every term explicitly listed in (12.2.7) except the last one can be dropped. If ω_2 is an optical frequency, all terms involving $\mathbf{u}$ may be dropped. Since we wish to allow ω_2 to be a low frequency also, we must keep the largest terms containing $\mathbf{u}(0,1)$. Among the negligible ones are the first two terms in (12.2.7) which arose from the time derivative transformation. This is most easily seen by a comparison of their magnitudes, M_1 and M_2, to the magnitude M of one of the largest terms such as that involving $^{21}\mathbf{M}$. This gives

$$\frac{M_1}{M} \sim \frac{\omega_1 \omega_2 m(\omega_1/v_O) y(1,0) u(0,1)}{^{21}M y(1,0)(\omega_2/v_A) u(0,1)} \tag{12.2.8}$$

where v_O and v_A are optical and acoustic velocities. We estimate [see (9.8.15)] that

$$^{21}M \sim {}^{20}M \sim m\omega_R^2 \tag{12.2.9}$$

where ω_R is the resonant frequency of an internal mode of motion. Typically such a frequency lies in the infrared for an ionic mode and in the ultraviolet for an electronic mode. Inserting (12.2.9) into the preceding equation gives

$$\frac{M_1}{M} \sim \frac{\omega_1 \omega_2 v_A}{\omega_R^2 v_O} \sim 10^{-5}. \tag{12.2.10}$$

A similar procedure for the second term gives

$$\frac{M_2}{M} \sim \frac{\omega_1^2 v_A}{\omega_R^2 v_O} \sim 10^{-5}. \tag{12.2.11}$$

Another term that can be dropped is the third one in (12.2.7) that involves the magnetic field. For it we get

$$\frac{M_3}{M} \sim \frac{q\omega_2 B(1,0) u(0,1)}{^{21}M y(1,0)(\omega_2/v_A) u(0,1)}. \tag{12.2.12}$$

From (7.1.11) we have

$$B(1,0) \sim \frac{k}{\omega_1} E(1,0) \sim \frac{E(1,0)}{v_O} \tag{12.2.13}$$

and from (9.1.5) we have

$$E(1,0) \sim \frac{^{20}M y(1,0)}{q}. \tag{12.2.14}$$

Substituting the last two equations and (12.2.9) into the expression for M_3 results in

$$\frac{M_3}{M} \sim \frac{v_A}{v_O} \sim 10^{-5}, \tag{12.2.15}$$

thus justifying the neglect of the magnetic field term.

The internal motion equation can now be solved for $\mathbf{y}^{\nu}$ with the use of $\boldsymbol{\Upsilon}$ defined in (9.1.8),

$$y_i^{\nu}(\mathbf{z};1,1)=\sum_{\mu}\Upsilon_{ij}^{\nu\mu}(\omega_S)\big(q^{\mu}E_j(\mathbf{z};1,1)+F_j^{\mu}(\mathbf{z};1,1)\big). \qquad (12.2.16)$$

The bilinear driving force $\mathbf{F}^{\mu}$ after dropping the small terms is

$$\begin{aligned}F_i^{\nu}(\mathbf{z};1,1)=&\sum_{\mu}{}^{20}M_{cd}^{\nu\mu}\big[u_{c,i}(\mathbf{z};0,1)y_d^{\mu}(\mathbf{z};1,0)\\&+y_j^{\mu}(\mathbf{z};1,0)u_{d,j}(\mathbf{z};0,1)\delta_{ic}\big]\\&-\sum_{\mu}{}^{21}M_{idab}^{\nu\mu}y_d^{\mu}(\mathbf{z};1,0)u_{a,b}(\mathbf{z};0,1)\\&+3\sum_{\mu\lambda}{}^{30}M_{ide}^{\nu\mu\lambda}\big[u_{d,j}(\mathbf{z};0,1)\,Y_j^{\mu}y_e^{\lambda}(\mathbf{z};1,0)\\&-y_d^{\mu}(\mathbf{z};1,0)y_e^{\lambda}(\mathbf{z};0,1)\big].\end{aligned} \qquad (12.2.17)$$

With these expressions we are ready to consider the driven wave equation.

12.3 Driven Wave Equation

The electric field wave equation,

$$\nabla\times(\nabla\times\mathbf{E})+\frac{1}{c^2}\frac{\partial^2\mathbf{E}}{\partial t^2}=-\mu_0\frac{\partial\mathbf{j}^D}{\partial t}, \qquad (12.3.1)$$

found in Section 9.2, produces the output optical wave we desire. It involves the dielectric current density,

$$\mathbf{j}^D=\frac{\partial\mathbf{P}}{\partial t}+\nabla\times(\mathbf{P}\times\dot{\mathbf{x}}), \qquad (12.3.2)$$

and the polarization,

$$\mathbf{P}=\frac{1}{J}\sum_{\nu}q^{\nu}\mathbf{y}^{T\nu}, \qquad (12.3.3)$$

which are discussed in Sections 4.6 and 4.7. We now wish to divide $\mathbf{j}^D$ into linear and bilinear parts in the manner used for the internal motion equations. Since no term should be more than linear in the displacement for the effects we are studying, we expand the Jacobian J only to first order by (10.2.3). The dielectric current density then becomes

$$j_i^D = \sum_\nu q^\nu \frac{\partial}{\partial t}\left[(1-u_{j,j})y_i^{T\nu}\right] + \sum_\nu q^\nu\left[(1-u_{k,k})\left(y_i^{T\nu}\frac{\partial u_j}{\partial t} - y_j^{T\nu}\frac{\partial u_i}{\partial t}\right)\right]_{,j}$$

$$= \sum_\nu q^\nu\left[\frac{\partial y_i^\nu}{\partial t} - Y_j^\nu \frac{\partial u_{i,j}}{\partial t}\right]$$

$$- \sum_\nu q^\nu\left[\frac{\partial}{\partial t}(u_{j,j}y_i^\nu) + \left(y_j^\nu \frac{\partial u_i}{\partial t} - y_i^\nu \frac{\partial u_j}{\partial t}\right)_{,j}\right]. \tag{12.3.4}$$

We now apply the Fourier expansion to $\mathbf{j}^D$, $\mathbf{u}$, and $\mathbf{y}^\nu$ in the manner done in the preceding section. The terms at the frequency ω_S in the last equation are

$$j_i^D(\mathbf{z};1,1) = -i\omega_S \sum_\nu q^\nu\left(y_i^\nu(\mathbf{z};1,1) - Y_j^\nu u_{i,j}(\mathbf{z};1,1)\right) + I_i(\mathbf{z};1,1) \tag{12.3.5}$$

where the bilinear current $\mathbf{I}$ is given by

$$I_i(\mathbf{z};1,1) = \frac{i}{2}\sum_\nu q^\nu\big[\omega_S u_{j,j}(\mathbf{z};0,1)y_i^\nu(\mathbf{z};1,0) - \omega_2 y_{i,j}^\nu(\mathbf{z};1,0)u_j(\mathbf{z};0,1)$$
$$+\omega_2 y_{j,j}^\nu(\mathbf{z};1,0)u_i(\mathbf{z};0,1) + \omega_2 y_j^\nu(\mathbf{z};1,0)u_{i,j}(\mathbf{z};0,1)$$
$$-\omega_2 y_i^\nu(\mathbf{z};1,0)u_{j,j}(\mathbf{z};0,1)$$
$$+ \text{ interchange of } (0,1)\leftrightarrow(1,0) \text{ and } \omega_1\leftrightarrow\omega_2\big]. \tag{12.3.6}$$

Terms in the last two equations with a displacement at an optical frequency can be dropped because of their smallness. If ω_2 is an optical frequency, then all the terms in $\mathbf{I}$ can be dropped. However, if ω_2 is an acoustic frequency, the first term must be retained while the remaining four are negligible. Their magnitudes M_2, M_3, M_4, M_5 in comparison to that of the first term M_1 are

$$\frac{M_2}{M_1} \sim \frac{\omega_2(\omega_1/v_O)y(1,0)u(0,1)}{\omega_S(\omega_2/v_A)y(1,0)u(0,1)} \sim \frac{v_A}{v_O} \sim 10^{-5}, \tag{12.3.7}$$

$$\frac{M_3}{M_1} \sim \frac{v_A}{v_O} \sim 10^{-5}, \tag{12.3.8}$$

$$\frac{M_4}{M_1} \sim \frac{\omega_2}{\omega_S} \lesssim 10^{-5}, \tag{12.3.9}$$

$$\frac{M_5}{M_1} \sim \frac{\omega_2}{\omega_S} \lesssim 10^{-5}. \tag{12.3.10}$$

Thus we have

$$\mathbf{j}^D(\mathbf{z};1,1) = -i\omega_S \sum_{\nu} q^{\nu}\mathbf{y}^{\nu}(\mathbf{z};1,1) + \mathbf{I}(\mathbf{z};1,1) \tag{12.3.11}$$

and

$$\mathbf{I}(\mathbf{z};1,1) = \frac{i\omega_S}{2}\nabla\cdot\mathbf{u}(\mathbf{z};0,1)\sum_{\nu} q^{\nu}\mathbf{y}^{\nu}(\mathbf{z};1,0) \tag{12.3.12}$$

for the dielectric current density and the bilinear driving current.

The wave equation now can be written as

$$\nabla\times[\nabla\times\mathbf{E}(\mathbf{z};1,1)] - \frac{\omega_S^2}{c^2}\mathbf{E}(\mathbf{z};1,1) = \frac{\omega_S^2}{\epsilon_0 c^2}\sum_{\nu} q^{\nu}\mathbf{y}^{\nu}(\mathbf{z};1,1) + \frac{i\omega_S}{\epsilon_0 c^2}\mathbf{I}(\mathbf{z};1,1). \tag{12.3.13}$$

Substitution for $\mathbf{y}^{\nu}$ from (12.2.16) yields

$$\left(\frac{c}{\omega_S}\right)^2\nabla\times[\nabla\times\mathbf{E}(\mathbf{z};1,1)] - \boldsymbol{\kappa}(\omega_S)\cdot\mathbf{E}(\mathbf{z};1,1) = \frac{\mathscr{P}(\mathbf{z};1,1)}{\epsilon_0} \tag{12.3.14}$$

where the definitions for the linear electric susceptibility (9.1.13) and the dielectric tensor (9.2.5) are used and the *nonlinear polarization* $\mathscr{P}$,

$$\mathscr{P}_i(\mathbf{z};1,1) \equiv \sum_{\nu\mu} q^{\nu}\Upsilon_{ij}^{\nu\mu}(\omega_S)F_j^{\mu}(\mathbf{z};1,1) + \frac{i}{\omega_S}I_i(\mathbf{z};1,1), \tag{12.3.15}$$

that drives the wave equation is introduced. It can be shown by estimates like those made above that the term in $\mathbf{I}$ and those in $\mathbf{F}^{\mu}$ retained in (12.2.17) and (12.3.12) all make comparable contributions to $\mathscr{P}$ with a magnitude of about $q\omega_2 u(0,1)y(1,0)/v_A$.

12.4 Nonlinear Polarization

We now wish to reexpress the equation for the nonlinear polarization $\mathscr{P}$ in terms having material tensors summed over the two macroscopic driving fields, the electric field and the acoustic displacement gradient. To do this we must eliminate the internal coordinates from $\mathbf{F}^{\mu}$ and $\mathbf{I}$ contained in $\mathscr{P}$.

The Fourier component at ω_1 of $\mathbf{y}^{\nu}$ can be obtained from (9.1.9) to be

$$y_i^{\nu}(\mathbf{z};1,0) = \sum_{\mu}\Upsilon_{ij}^{\nu\mu}(\omega_1)q^{\mu}E_j(\mathbf{z};1,0) \tag{12.4.1}$$

while the Fourier component at ω_2 of $\mathbf{y}^\nu$ can be found from (10.1.5) with the use of (10.1.4) and (12.1.7) to be

$$y_i^\nu(\mathbf{z};0,1)=\sum_\mu \Upsilon_{ij}^{\nu\mu}(\omega_2)\left[q^\mu E_j(\mathbf{z};0,1)-{}^{11}M_{jkl}^\mu u_{k,l}(\mathbf{z};0,1)\right] + Y_j^\nu u_{i,j}(\mathbf{z};0,1) \tag{12.4.2}$$

where as before we are allowing ω_2 to be either a low acoustic frequency or a second optical frequency. Note that the use of Fourier components allows an iterative procedure by which the constitutive relation of a physically nonlinear interaction is obtained only by the solution of mathematically linear equations.

Substitution of the last two equations into (12.2.17) for $\mathbf{F}^\mu$ and (12.3.12) for $\mathbf{I}$ and these quantities in turn into (12.3.15) for $\mathcal{P}$ leads to

$$\begin{aligned}\mathcal{P}_i(\mathbf{z};1,1)&=\epsilon_0\chi_{i\ j\ kl}^{\omega_S\omega_1\omega_2}E_j(\mathbf{z};1,0)u_{k,l}(\mathbf{z};0,1)\\&\quad+2\epsilon_0 b_{i\ j\ k}^{\omega_S\omega_1\omega_2}E_j(\mathbf{z};1,0)E_k(\mathbf{z};0,1)\\&=\epsilon_0\chi_{i\ j\ (kl)}^{\omega_S\omega_1\omega_2}E_j(\mathbf{z};1,0)u_{(k,l)}(\mathbf{z};0,1)\\&\quad+\epsilon_0\chi_{i\ j\ [kl]}^{\omega_S\omega_1\omega_2}E_j(\mathbf{z};1,0)u_{[k,l]}(\mathbf{z};0,1)\\&\quad+2\epsilon_0 b_{i\ j\ k}^{\omega_S\omega_1\omega_2}E_j(\mathbf{z};1,0)E_k(\mathbf{z};0,1)\end{aligned} \tag{12.4.3}$$

where the factor of 2 results from the assumption that $\omega_1 \neq \omega_2$ (see Section 15.1). Equation (12.4.3) introduces two new susceptibilities. The *elastooptic susceptibility* $\chi_{i\ j\ kl}^{\omega_S\omega_1\omega_2}$ is defined by

$$\chi_{i\ j\ kl}^{\omega_S\omega_1\omega_2}\equiv\chi_{i\ j\ (kl)}^{\omega_S\omega_1\omega_2}+\chi_{i\ j\ [kl]}^{\omega_S\omega_1\omega_2} \tag{12.4.4}$$

where

$$\begin{aligned}\epsilon_0\chi_{i\ j\ (kl)}^{\omega_S\omega_1\omega_2}\equiv&-\frac{\epsilon_0}{2}\chi_{ij}(\omega_1)\delta_{kl}-\sum_{\nu\mu\lambda\rho} q^\mu\Upsilon_{im}^{\mu\nu}(\omega_S)\,{}^{21}M_{mnkl}^{\nu\lambda}\Upsilon_{nj}^{\lambda\rho}(\omega_1)q^\rho\\&+3\sum_{\substack{\nu\mu\lambda\\\rho\zeta\theta}} q^\mu\Upsilon_{im}^{\mu\nu}(\omega_S)\,{}^{30}M_{mnp}^{\nu\lambda\rho}\Upsilon_{pj}^{\rho\zeta}(\omega_1)q^\zeta\Upsilon_{nq}^{\lambda\theta}(\omega_2)\,{}^{11}M_{qkl}^\theta\\&+\sum_{\nu\mu\lambda\rho} q^\mu\Upsilon_{im}^{\mu\nu}(\omega_S)\,{}^{20}M_{m(k}^{\nu\lambda}\Upsilon_{l)j}^{\lambda\rho}(\omega_1)q^\rho\\&+\sum_{\mu\nu\lambda\rho} q^\mu\Upsilon_{i(k}^{\mu\nu}(\omega_S)\,{}^{20}M_{l)m}^{\nu\lambda}\Upsilon_{mj}^{\lambda\rho}(\omega_1)q^\rho\end{aligned} \tag{12.4.5}$$

and

$$\epsilon_0 \chi_{i\ j\ [kl]}^{\omega_S \omega_1 \omega_2} \equiv - \sum_{\nu\mu\lambda\rho} q^{\mu} \Upsilon_{i[k}^{\mu\nu}(\omega_S)\, {}^{20}M_{l]m}^{\nu\lambda} \Upsilon_{mj}^{\lambda\rho}(\omega_1) q^{\rho}$$

$$+ \sum_{\nu\mu\lambda\rho} q^{\mu} \Upsilon_{im}^{\mu\nu}(\omega_S)\, {}^{20}M_{m\ [k}^{\nu\lambda} \Upsilon_{l]\ j}^{\lambda\rho}(\omega_1) q^{\rho}. \tag{12.4.6}$$

The *optical mixing tensor* or *susceptibility* is defined by

$$\epsilon_0 b_{i\ j\ k}^{\omega_S \omega_1 \omega_2} \equiv -\tfrac{3}{2} \sum_{\substack{\nu\mu\lambda \\ \rho\zeta\theta}} q^{\mu} \Upsilon_{il}^{\mu\nu}(\omega_S)\, {}^{30}M_{lmn}^{\nu\lambda\rho} \Upsilon_{nj}^{\rho\zeta}(\omega_1) q^{\zeta} \Upsilon_{mk}^{\lambda\theta}(\omega_2) q^{\theta}. \tag{12.4.7}$$

Often the optical mixing susceptibility is represented by **d** rather than **b**. We use **b** to avoid confusion with the piezoelectric strain tensor (10.5.4).

These equations are *derived constitutive relations* that include a predicted form of the frequency dispersion. Note that the frequency dispersion enters entirely from the $\Upsilon(\omega)$ factors. Note also that the susceptibilities are functions of all three frequencies ω_1, ω_2, and $\omega_S = \omega_1 + \omega_2$ and that there is an association between a tensor index of the susceptibility and a particular frequency. The numerical constants entering the susceptibilities are the various ${}^{mn}\mathbf{M}$ coefficients of the stored energy expansion and the effective charges q^{ν} and (within the $\Upsilon(\omega)$ factors) the effective masses m^{ν}. Note that the constants $\mathbf{Y}^{\nu}$ do not enter these susceptibilities.

The constitutive relations just derived apply to dielectrics, to piezoelectrics, and to pyroelectrics for which $\mathbf{E}^S = 0$. They also apply to any crystal symmetry group. The antisymmetric part, $\chi_{i\ j\ [kl]}^{\omega_S \omega_1 \omega_2}$, of the elastooptic susceptibility is a new prediction of this theory and is discussed in detail in the next chapter. The meaning and consequences of the nonlinear polarization (12.4.3) are examined in the next three chapters.

PROBLEMS

12.1. Derive the analogue of (12.4.3) for the output difference frequency given in (12.2.2).

12.2. Show that the optical mixing susceptibility and the elastooptic susceptibility may be reexpressed with the use of the normal coordinates defined in (9.P.7) to (9.P.10) as

$$\epsilon_0 b_{i\ j\ k}^{\omega_S\omega_1\omega_2} \equiv -\frac{3}{2}\sum_{\substack{\nu\mu\lambda\\abc}} \frac{c_i^a c_j^b c_k^c\, {}^{30}M_{pqr}^{\nu\mu\lambda} i_p^{a\nu} i_q^{b\mu} i_r^{c\lambda}}{(m^\nu m^\mu m^\lambda)^{1/2}(\Omega_a^2-\omega_S^2)(\Omega_b^2-\omega_1^2)(\Omega_c^2-\omega_2^2)}, \tag{12.P.1}$$

$$\begin{aligned}\epsilon_0 \chi_{i\ j\ kl}^{\omega_S\omega_1\omega_2} \equiv &-\sum_{\substack{\nu\mu\\ab}} \frac{c_i^a c_j^b\, {}^{21}M_{mnkl}^{\nu\mu} i_m^{a\nu} i_n^{b\mu}}{(m^\nu m^\mu)^{1/2}(\Omega_a^2-\omega_S^2)(\Omega_b^2-\omega_1^2)}\\ &+3\sum_{\substack{\nu\mu\lambda\rho\\abc}} \frac{c_i^a c_j^b\, {}^{30}M_{pqr}^{\nu\mu\lambda} i_p^{a\nu} i_q^{b\mu} i_r^{c\lambda} i_s^{c\rho}\, {}^{11}M_{skl}^{\rho}}{(m^\nu m^\mu m^\lambda m^\rho)^{1/2}(\Omega_a^2-\omega_S^2)(\Omega_b^2-\omega_1^2)(\Omega_c^2-\omega_2^2)}\\ &+\frac{1}{2}\sum_{\substack{\nu\\ab}} \frac{c_i^a c_j^b\left(i_k^{a\nu} i_l^{b\nu}\Omega_a^2 + i_k^{b\nu} i_l^{a\nu}\Omega_b^2\right)}{(\Omega_a^2-\omega_S^2)(\Omega_b^2-\omega_1^2)}\\ &-\frac{\epsilon_0}{2}\chi_{ij}(\omega_1)\delta_{kl}.\end{aligned} \tag{12.P.2}$$

CHAPTER 13

Elastooptic Effect

The *elastooptic effect* (also called the *photoelastic* or *piezooptic effect*) is the interaction of an elastic deformation with a light wave to produce an altered output light wave. It is the oldest known nonlinear optical interaction, having been discovered by Brewster in 1816 when he observed birefringence in a glass plate induced by bending. The elastooptic effect was studied for more than a century using static elastic deformations. Pockels gave a phenomenological formulation of the elastooptic effect in 1890 that was regarded as adequate until very recently.

The modern era of elastooptic studies began with the prediction by Brillouin [1914, 1922] of light scattering from thermal fluctuations, which is now called *Brillouin scattering*. This phenomenon was later related to the elastooptic effect by viewing the fluctuations as thermally excited acoustic waves. Gross [1930] was the first to observe Brillouin scattering. A short time later Debye and Sears [1932] and Lucas and Biquard [1932] observed light diffracted from coherently generated acoustic waves, an effect now called *acoustooptic diffraction*. This effect has grown in importance in recent years as a means of deflecting and modulating laser beams. Stimulated Brillouin scattering, which produces both an intense frequency shifted light wave and an intense hypersonic wave, was discovered by Chiao, Townes, and Stoicheff [1964].

A prediction [Nelson and Lax, 1970] was made in 1970 on the basis of the theory being presented here that the Pockels formulation of the elastooptic effect was inadequate when applied to shear deformations in birefringent crystals. It was shown that an interaction tensor that couples to rotation was needed in addition to the Pockels tensor, which couples to strain. The new part of the interaction may be called the rotational elastooptic effect or more simply the *rotooptic effect*. The new tensor was shown to be directly related to the anisotropy of the optical dielectric tensor and predicted to be comparable in size to the Pockels tensor in strongly birefringent crystals. Experimental verification was first found in rutile [Nelson and Lazay, 1970].

In this chapter we first give a full discussion of the elastooptic susceptibility including the direct part caused by strain, the direct part caused by rotation, and the indirect part caused by strain acting through a combination of the piezoelectric and electrooptic effects [Chapelle and Taurel, 1955]. We then calculate the characteristics of the optical output from acoustooptic diffraction for plane waves. This is done by making the physical approximation of no depletion of power in the input waves and solving the resulting mathematical problem exactly by the method of free-plus-forced waves. The acoustooptically diffracted light wave is shown to have a maximum strength when the interaction is *phase-matched*, by which is meant the wavevector of the output wave is equal to the sum of the wavevectors of the input waves. An examination of the simplified form of the solution in the region near phase matching reveals a number of characteristics that we use in deriving coupled mode equations for three-wave parametric interactions.

The availability of intense monochromatic lasers has made Brillouin scattering an important technique for the study of the elastooptic effect. A treatment of it requires determining the population of acoustic excitations (phonons) from the thermal content of the crystal in addition to the analysis that we present for acoustooptic diffraction. Since thermal phenomena are not included in our Lagrangian formulation, we do not treat Brillouin scattering. The interested person is referred to Cummins and Schoen [1972] and Nelson, Lazay, and Lax [1972].

13.1 Elastooptic Susceptibility

The nonlinear polarization (12.4.3) developed in the preceding chapter contains two terms with an electric field and a displacement gradient (the symmetric part being involved in one, the antisymmetric part in the other) acting as input fields and a term with two electric fields acting as input fields. The first two of these terms produce elastooptic interactions directly. The last term can also produce an elastooptic effect but by an indirect manner. The input displacement gradient field produces a piezoelectrically generated electric field in crystals of appropriate symmetry. This field then acts as the low frequency electric field in the nonlinear polarization (12.4.3). The result is a susceptibility term representing the *indirect elastooptic effect*.

The piezoelectrically generated electric field,

$$E_l(\mathbf{z};0,1) = -\frac{s_l s_i e_{ijk}}{\epsilon_0 \mathbf{s}\cdot\boldsymbol{\kappa}(\omega_2)\cdot\mathbf{s}}\, u_{j,k}(\mathbf{z};0,1), \tag{13.1.1}$$

is derived in Section 10.7. Here we drop the piezoelectric-like term depending on the spontaneous electric field of a pyroelectric, that is, we regard the field to be canceled by surface charge if a pyroelectric is considered. Substitution of this into the nonlinear polarization (12.4.3) yields

$$\mathcal{P}_i(\mathbf{z};1,1)=\epsilon_0\left[\chi_{i\,j\,(kl)}^{\omega_S\omega_1\omega_2}+\chi_{i\,j\,[kl]}^{\omega_S\omega_1\omega_2}\right.$$
$$\left.-\frac{2b_{i\,j\,m}^{\omega_S\omega_1\omega_2}s_m s_n e_{nkl}}{\epsilon_0\mathbf{s}\cdot\boldsymbol{\kappa}(\omega_2)\cdot\mathbf{s}}\right]E_j(\mathbf{z};1,0)u_{k,l}(\mathbf{z};0,1). \tag{13.1.2}$$

This is the complete elastooptic susceptibility. The terms within the brackets in order are the direct elastooptic susceptibility arising from strain, the direct elastooptic susceptibility arising from rotation (the rotooptic susceptibility), and the indirect elastooptic susceptibility arising from a combination of the piezoelectrically generated electric field and the electrooptic effect.

The derivation of (13.1.2) is based on the assumption that the displacement gradient field is a plane wave with a unit propagation vector $\mathbf{s}$. It is worth reconsidering the derivation for the case of a static deformation. If the treatment of the preceding chapter is repeated for the static case, we find that

$$\mathcal{P}_i(\mathbf{z};1,0)=2\epsilon_0\left[\chi_{i\,j\,(kl)}^{\omega_1\omega_1 0}+\chi_{i\,j\,[kl]}^{\omega_1\omega_1 0}\right]E_j(\mathbf{z};1,0)u_{k,l}(\mathbf{z};0,0)$$
$$+4\epsilon_0 b_{i\,j\,m}^{\omega_1\omega_1 0}E_j(\mathbf{z};1,0)E_m(\mathbf{z};0,0). \tag{13.1.3}$$

The extra factor of two in this equation compared to (12.4.3) results from the coefficient in the Fourier expansion (12.2.3) becoming unity when one input field is static. This factor is readily interpretable. When the elastic deformation oscillates at a finite frequency ω_2, two output frequencies or sidebands result (the $\omega_1-\omega_2$ output is called the *Stokes component* and the $\omega_1+\omega_2$ output is called the *anti-Stokes component*). The nonlinear polarization (12.4.3) represents only one of these. When the elastic deformation is static, the nonlinear polarization (13.1.3) represents, so to speak, both frequency components, since they are then degenerate in frequency. Hence the factor of 2 in (13.1.3) must enter.

The elimination of the static electric field from (13.1.3) in a piezoelectric crystal can be done only through a consideration of the electrical boundary conditions of the crystal studied. Analytically the simplest boundary conditions to consider are opposite crystal surfaces covered with metallic coatings and connected electrically together. Such boundary conditions prevent any piezoelectrically generated electric field; in this case $\mathbf{E}(\mathbf{z};0,0)$ may be set equal to zero in (13.1.3). The electrical boundary value problem of an uncoated crystal all of whose dimensions are finite and which possesses two opposite flat sides for the application of tractions is complicated and is not considered here. Instead we treat an idealized and somewhat unrealistic situation as an illustration of the static piezoelectrically created electric field.

Consider a broad area (effectively infinite) slab that is piezoelectric but nonpyroelectric and on whose surfaces uniform static tractions are applied. A homogeneous static strain and a homogeneous static electric field are created. They induce a polarization

$$P_i(0,0)=\epsilon_0\chi_{ij}(0)E_j(0,0)+e_{ijk}u_{j,k}(0,0). \tag{13.1.4}$$

For the infinite slab the boundary conditions yield (8.5.19),

$$\mathbf{E}(0,0)=-\frac{1}{\epsilon_0}\mathbf{n}\,\mathbf{n}\cdot\mathbf{P}(0,0), \tag{13.1.5}$$

where $\mathbf{n}$ is the normal to the broad area surfaces. Eliminating $\mathbf{P}(0,0)$ yields

$$\begin{aligned} E_i(0,0)&=-n_in_j\left[\chi_{jk}(0)E_k(0,0)+\frac{1}{\epsilon_0}e_{jkl}u_{k,l}(0,0)\right] \\ &=-n_in_j\left[\chi_{jk}(0)n_kn_lE_l(0,0)+\frac{1}{\epsilon_0}e_{jkl}u_{k,l}(0,0)\right] \end{aligned} \tag{13.1.6}$$

where the second form results because the first form states that $\mathbf{E}\|\mathbf{n}$. The scalar product of this equation with $\mathbf{n}$ can be used to solve for $\mathbf{n}\cdot\mathbf{E}$,

$$n_lE_l(0,0)=-\frac{n_je_{jkl}}{\epsilon_0\mathbf{n}\cdot\boldsymbol{\kappa}(0)\cdot\mathbf{n}}u_{k,l}(0,0). \tag{13.1.7}$$

Substitution of this expression back into (13.1.6) then gives the piezoelectrically created electric field,

$$E_i(0,0)=-\frac{n_in_j}{\epsilon_0\mathbf{n}\cdot\boldsymbol{\kappa}(0)\cdot\mathbf{n}}e_{jkl}u_{k,l}(0,0), \tag{13.1.8}$$

for the slab geometry.

From (13.1.3) the nonlinear polarization for the elastooptic effect from a static homogeneous elastic deformation in an infinite plate is thus

$$\mathscr{P}_i(\mathbf{z};1,0)=2\epsilon_0\left[\chi_{i\ j\ (kl)}^{\omega_1\omega_1 0}+\chi_{i\ j\ [kl]}^{\omega_1\omega_1 0}-\frac{2b_{i\ j\ m}^{\omega_1\omega_1 0}n_m n_n e_{nkl}}{\epsilon_0\mathbf{n}\cdot\boldsymbol{\kappa}(0)\cdot\mathbf{n}}\right]E_j(\mathbf{z};1,0)u_{k,l}(0,0). \tag{13.1.9}$$

Note that this nonlinear polarization is the same as that for the dynamic elastic deformation case (13.1.2) except for the factor of 2 discussed above and the replacement of the unit propagation vector **s** by the unit surface normal **n** in the indirect elastooptic effect susceptibility. The unrealistic aspect of this solution is the use of the boundary condition result (13.1.5) which assumes a vacuum in contact with the large area surfaces while at the same time we are assuming that tractions, which are produced in practice by contact with material bodies, are applied to these surfaces.

13.2 Direct Elastooptic Effect from Strain

The first susceptibility term in both (13.1.2) and (13.1.9) represents the direct interaction of the electric field of the input light and the strain portion of the elastic deformation. It is that part of the elastooptic effect embodied in the Pockels formulation.

As seen from (12.4.5) the direct elastooptic susceptibility from a *static* strain possesses the exact interchange symmetry

$$\chi_{i\ j\ (kl)}^{\omega_1\omega_1 0}=\chi_{(ij)\ (kl)}^{\omega_1\omega_1 0} \tag{13.2.1}$$

while that from a *dynamic* strain possesses the same symmetry

$$\chi_{i\ j\ (kl)}^{\omega_S\omega_1\omega_2}=\chi_{(ij)\ (kl)}^{\omega_S\ \omega_1\omega_2} \tag{13.2.2}$$

to the extent that the dispersion in the tensor between ω_1 and ω_S is negligible. Since normally 10^{13} Hz$\lesssim\omega_1/2\pi\lesssim10^{15}$ Hz and $0\leqslant\omega_2/2\pi\lesssim10^{10}$ Hz, this approximation is usually excellent. It should be noted that there is no interchange symmetry between the first and second *pairs* of tensor indices. This tensor thus has lower interchange symmetry than the elasticity tensor.

Pockels' formulation of the elastooptic effect began from a consideration of distortion of the index ellipsoid caused by the strain. The index ellipsoid is simply a slightly altered form of the eigenvector normalization condition (9.3.6),

$$\mathcal{E}^{\alpha}\cdot\boldsymbol{\kappa}\cdot\mathcal{E}^{\alpha}=1, \tag{13.2.3}$$

for $\alpha=1$ or 2 corresponding to the propagating waves. With the electric displacement eigenvector $\mathfrak{D}^{\alpha}$ (9.3.8) introduced this becomes

$$\mathfrak{D}^{\alpha}\cdot\boldsymbol{\kappa}^{-1}\cdot\mathfrak{D}^{\alpha}=1. \tag{13.2.4}$$

Next we introduce the unit electric displacement vector $\mathbf{d}^{\alpha}$ by (9.3.24),

$$\mathfrak{D}^{\alpha}=n^{\alpha}\mathbf{d}^{\alpha}. \tag{13.2.5}$$

The index ellipsoid now becomes

$$\frac{1}{(n^{\alpha})^2}=\mathbf{d}^{\alpha}\cdot\boldsymbol{\kappa}^{-1}\cdot\mathbf{d}^{\alpha}. \tag{13.2.6}$$

Pockels hypothesized that the strain portion of the elastic deformation perturbed the inverse dielectric tensor into $(\boldsymbol{\kappa}^{-1}+\boldsymbol{\Delta\kappa}^{-1})$ where

$$(\Delta\kappa^{-1})_{ij}=p_{ijkl}S_{kl}. \tag{13.2.7}$$

The tensor p_{ijkl} is called the *Pockels elastooptic tensor*. Because of the interchange symmetry of $\boldsymbol{\Delta\kappa}^{-1}$ (for a static deformation) and of $\mathbf{S}$, it obeys

$$p_{ijkl}=p_{(ij)(kl)}. \tag{13.2.8}$$

Since a change in the inverse dielectric tensor is related to a change in the dielectric tensor by

$$\boldsymbol{\Delta\kappa}^{-1}=-\boldsymbol{\kappa}^{-1}\cdot\boldsymbol{\Delta\kappa}\cdot\boldsymbol{\kappa}^{-1} \tag{13.2.9}$$

to first order, the Pockels tensor is related to the direct elastooptic susceptibility caused by a static strain (13.1.9) by

$$p_{ij(kl)}=-2(\kappa^{-1})_{im}(\kappa^{-1})_{jn}\chi_{mn(kl)}. \tag{13.2.10}$$

Though the frequency superscripts are dropped from $\chi_{mn(kl)}$, it refers to a single output Fourier component as indicated by the factor of 2. The relation inverse to (13.2.10) is

$$\chi_{ij(kl)}=-\tfrac{1}{2}\kappa_{im}\kappa_{in}p_{mnkl}. \tag{13.2.11}$$

The nonzero components of χ_{mn} or p_{mn} (contracted notation) allowed by crystal symmetry are given in the Appendix.

For later reference a reexpression of the elastooptic susceptibility coupling to strain (12.4.5) is needed. Substitution of the definition of Υ into the last two terms of that definition allows it to be written as

$$\begin{aligned}
\epsilon_0 \chi_{i\;j\;(kl)}^{\omega_S \omega_1 \omega_2} = & - \sum_{\nu\mu\lambda\rho} q^{\mu} \Upsilon_{im}^{\mu\nu}(\omega_S)\, {}^{21}M_{mnkl}^{\nu\lambda} \Upsilon_{nj}^{\lambda\rho}(\omega_1) q^{\rho} \\
& + 3 \sum_{\substack{\nu\mu\lambda \\ \rho\zeta\theta}} q^{\mu} \Upsilon_{im}^{\mu\nu}(\omega_S)\, {}^{30}M_{mnp}^{\nu\lambda\rho} \Upsilon_{pj}^{\rho\zeta}(\omega_1) q^{\zeta} \Upsilon_{nq}^{\lambda\theta}(\omega_2)\, {}^{11}M_{qkl}^{\theta} \\
& + \tfrac{1}{2}\left(\omega_1^2 + \omega_S^2\right) \sum_{\mu\sigma\rho} q^{\mu} \Upsilon_{i(k}^{\mu\sigma}(\omega_S) m^{\sigma} \Upsilon_{l)j}^{\sigma\rho}(\omega_1) q^{\rho} \\
& + \frac{\epsilon_0}{2} \delta_{j(k} \chi_{l)i}(\omega_S) + \frac{\epsilon_0}{2} \delta_{i(k} \chi_{l)j}(\omega_1) - \frac{\epsilon_0}{2} \chi_{ij}(\omega_1) \delta_{kl}.
\end{aligned} \tag{13.2.12}$$

Note that the form of the last three terms is describable as that which produces Maxwell stresses in a polarizable medium.

13.3 Rotooptic Effect

The rotooptic susceptibility couples to the antisymmetric combination of displacement gradients or, in other words, the rotation part of the deformation. It can be readily interpreted if (12.4.6) is first reexpressed in the manner just done for the elastooptic susceptibility that couples to strain. This yields

$$\begin{aligned}
\epsilon_0 \chi_{i\;j\;[kl]}^{\omega_S \omega_1 \omega_2} = & \tfrac{1}{2} \sum_{\mu\nu\rho} q^{\mu} \Upsilon_{i[k}^{\mu\nu}(\omega_S) \Upsilon_{l]j}^{\nu\rho}(\omega_1) m^{\nu} \left(\omega_S^2 - \omega_1^2\right) q^{\rho} \\
& - \frac{\epsilon_0}{2} \left(\chi_{i[k}(\omega_S) \delta_{l]j} + \chi_{j[k}(\omega_1) \delta_{l]i}\right).
\end{aligned} \tag{13.3.1}$$

Since $\omega_S \approx \omega_1$, the first term may be dropped to an excellent level of accuracy. Dropping the display of the frequency dependence, we then obtain

$$\begin{aligned}
\chi_{ij[kl]} &= -\tfrac{1}{2}\left(\chi_{i[k}\delta_{l]j} + \chi_{j[k}\delta_{l]i}\right) \\
&= -\tfrac{1}{2}\left(\kappa_{i[k}\delta_{l]j} + \kappa_{j[k}\delta_{l]i}\right).
\end{aligned} \tag{13.3.2}$$

If we define a Pockels-like tensor

$$p_{ij[kl]} \equiv -2(\kappa^{-1})_{im}(\kappa^{-1})_{jn}\chi_{mn[kl]} \tag{13.3.3}$$

analogous to that in (13.2.10), then

$$p_{ij[kl]} = \delta_{i[k}\kappa^{-1}_{l]j} + \delta_{j[k}\kappa^{-1}_{l]i}. \tag{13.3.4}$$

Thus we see that the direct rotooptic susceptibility is expressible in terms of the optical dielectric tensor, a linear property of the crystal. It is readily seen that if the dielectric tensor is isotropic as in cubic crystals and isotropic solids,

$$\kappa_{ij} = \kappa\delta_{ij}, \tag{13.3.5}$$

then

$$p_{ij[kl]} = 0. \tag{13.3.6}$$

Thus, this tensor is nonvanishing only in birefringent crystals. This includes uniaxial crystals of the trigonal, tetragonal, and hexagonal systems and biaxial crystals of the triclinic, monoclinic, and orthorhombic systems. For a uniaxial crystal the only nonvanishing components are

$$p_{(23)[23]} = p_{(13)[13]} = \tfrac{1}{2}\left(\kappa^{-1}_{33} - \kappa^{-1}_{22}\right) = \frac{1}{2}\left[\frac{1}{n_e^2} - \frac{1}{n_o^2}\right] \tag{13.3.7}$$

where n_o and n_e are the principal refractive indices. For a *positive uniaxial crystal* ($n_e > n_o$) $p_{(23)[23]}$ is negative and for a *negative uniaxial crystal* ($n_e < n_o$) $p_{(23)[23]}$ is positive. For an orthorhombic crystal the only nonvanishing components are

$$p_{(23)[23]} = \tfrac{1}{2}\left(\kappa^{-1}_{33} - \kappa^{-1}_{22}\right), \tag{13.3.8a}$$

$$p_{(13)[13]} = \tfrac{1}{2}\left(\kappa^{-1}_{33} - \kappa^{-1}_{11}\right), \tag{13.3.8b}$$

$$p_{(12)[12]} = \tfrac{1}{2}\left(\kappa^{-1}_{22} - \kappa^{-1}_{11}\right). \tag{13.3.8c}$$

For a monoclinic crystal $\kappa_{(13)}$ and the diagonal components of $\boldsymbol{\kappa}$ are nonvanishing if the Y crystallographic axis is chosen along the twofold symmetry axis. The values of the components in (13.3.8) also apply to a monoclinic crystal; in addition we have

$$p_{(11)[13]} = -p_{(33)[13]} = -2p_{(23)[12]} = 2p_{(12)[23]} = \kappa^{-1}_{13}. \tag{13.3.9}$$

For a triclinic crystal all components of $\boldsymbol{\kappa}$ may be nonzero and so all components of $p_{(ij)[kl]}$ are possible.

The physical interpretation of the rotooptic susceptibility is now apparent. It arises simply from the rotation of the linear optical anisotropy relative to the propagation direction of the light. Since the refractive indices depend on the orientation of the propagation vector relative to the principal axes of the optical dielectric tensor, this rotation affects the light propagation and so contributes to the elastooptic effect. Though this effect persists even when the rotation is homogeneous, it is then merely the effect of a rigid body rotation and of trivial interest, since it could be compensated for by experimental adjustments. However, if the rotation is inhomogeneous, the effect is entwined with the elastooptic effect from strain and cannot be removed from the observations by experimental adjustments. This is true, for instance, for acoustooptic diffraction or Brillouin scattering observations, since there the rotation varies throughout an acoustic wavelength. As discussed in Section 10.9 rotation occurs in any acoustic wave that possesses a shear component and for a pure shear wave the rotation $u_{[k,l]}$ is numerically equal to the strain $u_{(k,l)}$.

Though this effect was first discovered as a result of the general theory that is presented here, its simple interpretation indicates that it can be derived by a very elementary argument. Consider an infinitesimal body rotation that carries a body point at x_j into the new position

$$x_i' = x_j\left(\delta_{ij} + u_{[i,j]}\right). \tag{13.3.10}$$

The dielectric tensor κ_{ij} must transform on each index i,j under a body rotation exactly as the vector x_j transforms under the same rotation. Hence

$$\begin{aligned} \Delta\kappa_{ij} &= \left(\delta_{ik} + u_{[i,k]}\right)\left(\delta_{jl} + u_{[j,l]}\right)\kappa_{kl} - \kappa_{ij} \\ &= -\left(\kappa_{j[k}\delta_{l]i} + \kappa_{i[k}\delta_{l]j}\right)u_{k,l}. \end{aligned} \tag{13.3.11}$$

Since this is a static calculation, the coefficient of $u_{k,l}$ on the right is twice the susceptibility relating to a single output Fourier component. Thus we have

$$\chi_{ij[kl]} = -\tfrac{1}{2}\left(\kappa_{j[k}\delta_{l]i} + \kappa_{i[k}\delta_{l]j}\right) \tag{13.3.12}$$

in agreement with (13.3.2).

It is worth emphasizing at this point that the deficiency of the Pockels formulation of the elastooptic effect is in assuming that the change in the dielectric tensor resulting from an elastic deformation is caused only by the strain $u_{(i,j)}$. This derivation shows that the rotation $u_{[i,j]}$ can play an equally

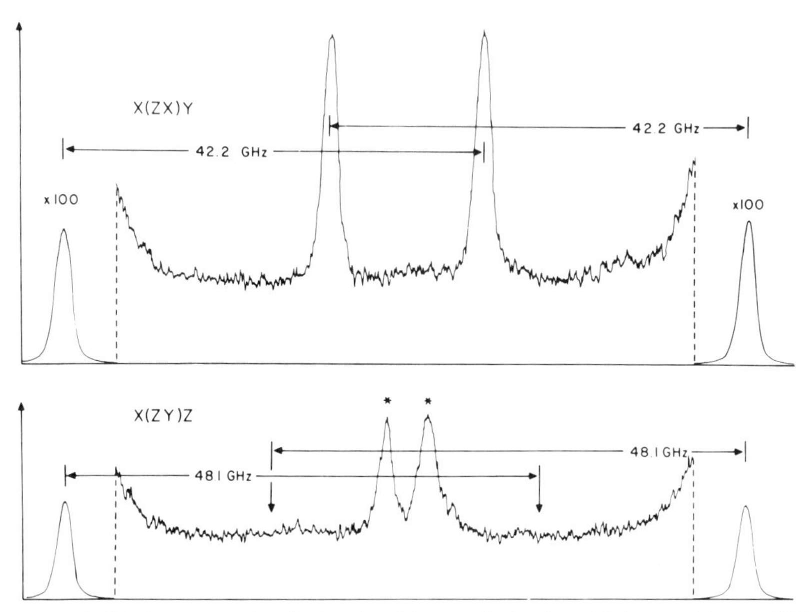

Fig. 13.1. Brillouin spectra of rutile [reprinted by permission from Nelson and Lazay, 1970]. In the top spectrum the incident light traveled along the X axis and was polarized along the Z axis; the scattered light traveled along the Y axis and was polarized along the X axis. This is indicated by the notation $X(ZX)Y$. $X(ZY)Z$ applies to the lower spectrum. A free spectral range of the interferometer that analyzes the scattered light is shown. The strong peaks at the ends of the free spectral range are caused by scattering of the laser light without a frequency shift by static imperfections. The two spectra are shown at identical gain settings. Scattering from a pure transverse mode which measures $|p_{1331}|=|p_{2332}|$ is observed in the upper trace while that which should measure $|p_{2323}|$ is not observed in the lower trace. The expected position of the unobserved mode (splitting of 48.1 GHz) is indicated. In each free spectral range the Stokes peak and the anti-Stokes peak which is associated with the adjacent order of the interferometer appear. The weak modes indicated by an asterisk are spurious.

important role in a birefringent medium. Stated another way, this derivation shows that the measure of elastic deformation relevant to the elastooptic effect is the *displacement gradient* which is the sum of strain $u_{(i,j)}$ and rotation $u_{[i,j]}$.

The rotooptic effect was first verified in rutile [Nelson and Lazay, 1970] by measuring the values of both $|p_{2332}|$ and $|p_{2323}|$ and comparing their difference to the prediction of (13.3.7). The measurements yielded $p_{23[23]}=\pm 0.0132$ compared with the prediction of -0.0138. The discrepancy, which is only 5%, was about equal to the experimental uncertainty. The observations differed dramatically from predictions based on the Pockels formulation because the rotooptic susceptibility happens by chance to be close to the value of the elastooptic susceptibility that couples to strain. This means that p_{2332} is sizable, since for it the susceptibilities add, while p_{2323} is quite small, since for it the susceptibilities subtract. Because the scattered light intensity is proportional to the square of the elastooptic coefficient, the observed light intensities shown in Fig. 13.1, which by the Pockels formulation were believed to be equal, differed by more than a factor of 100.

The rotooptic effect has now been verified in a number of crystals. For a compilation of data see Hearmon and Nelson [1978].

13.4 Indirect Elastooptic Effect

The indirect elastooptic effect occurs only in piezoelectric crystals and is a succession of the piezoelectric effect and the electrooptic effect. The form of the indirect elastooptic effect for an elastic deformation in the form of an acoustic wave is exhibited in (13.1.2) and for a static elastic deformation of a slab in (13.1.9). The indirect effect is experimentally distinguishable from the direct effect because of its dependence on a characteristic of the applied influence, the unit propagation vector of the acoustic wave in one case and the unit normal to the surface to which a traction is applied in the other case. Though the susceptibility of the indirect effect is simply a fourth rank tensor as far as coordinate transformations are concerned, it must be regarded as a *fourth rank tensor function of a vector* as far as crystal symmetry operations are concerned. This is necessitated because the unit propagation vector or the unit surface normal do not transform under crystal symmetry transformations. The result is that for a general direction of either unit vector the indirect elastooptic effect susceptibility may not vanish for a particular set of values of its four tensor indices for which the direct susceptibility is required to vanish by crystal symmetry.

If the *Pockels direct electrooptic effect tensor* r^S_{ijk},

$$r^S_{ijk} \equiv -4(\kappa^{-1})_{im}(\kappa^{-1})_{jn} b^{\omega_1 \omega_1 0}_{m\ n\ k}, \tag{13.4.1}$$

is introduced, the indirect elastooptic effect susceptibility can be converted to a Pockels-like elastooptic tensor. The effective Pockels-like tensor $\mathbf{p}^{\text{eff}}$ is then

$$\begin{aligned} p^{\text{eff}}_{ijkl} &\equiv p_{ijkl} - \frac{r^S_{ijm} s_m s_n e_{nkl}}{\epsilon_0 \mathbf{s} \cdot \boldsymbol{\kappa} \cdot \mathbf{s}} \\ &= p_{(ij)(kl)} + \delta_{i[k} \kappa^{-1}_{l]\,j} + \delta_{j[k} \kappa^{-1}_{l]\,i} - \frac{r^S_{ijm} s_m s_n e_{nkl}}{\epsilon_0 \mathbf{s} \cdot \boldsymbol{\kappa} \cdot \mathbf{s}} \end{aligned} \tag{13.4.2}$$

where **s** is the unit acoustic wave propagation vector. An analogous form holds for the statically deformed piezoelectric slab if **s** is replaced by **n**. Note that the r_{ijk}^{S} tensor is defined for a statically applied electric field. The relation inverse to (13.4.1) gives the *direct electrooptic susceptibility* $b_{m\ n\ k}^{\omega_1\omega_1 0}$ [which is a low frequency limit of the optical mixing tensor of (12.4.7)] in terms of the Pockels electrooptic tensor

$$b_{i\ j\ k}^{\omega_1\omega_1 0} = -\tfrac{1}{4}\kappa_{im}\kappa_{jn}r_{mnk}^{S}. \tag{13.4.3}$$

The indirect elastooptic effect in many crystals has negligible size but is numerically significant in crystals such as $LiNbO_3$ and $\alpha-HIO_3$.

13.5 Acoustooptic Diffraction

We wish to calculate the light intensity diffracted from a coherent acoustic wave using three simplifications—the use of plane waves, the neglect of input light wave depletion, and the near satisfaction of phase matching. The latter condition, more than a simplification, specifies the regime that is most useful and interesting. Our first solution, however, applies to any level of phase mismatching between the input and output waves. An approximate form of the solution close to phase matching is then found and the mathematical approximations applicable in the region are noted for later use. The method of exact solution that we use is patterned after that of Kleinman [1962b] who used it in studying optical harmonic generation (see also [Bloembergen and Pershan, 1962]). The acoustooptic interaction is shown schematically in Fig. 13.2.

Our task is to solve the wave equation (12.3.14),

$$\left(\frac{c}{\omega_S}\right)^2 \nabla\times(\nabla\times\mathbf{E}(\mathbf{z};1,1)) - \boldsymbol{\kappa}\cdot\mathbf{E}(\mathbf{z};1,1) = \frac{\mathscr{P}(\mathbf{z};1,1)}{\epsilon_0}, \tag{13.5.1}$$

driven by the nonlinear polarization $\mathscr{P}$ of (13.1.2). The plane wave input fields are denoted as

$$\mathbf{E}(\mathbf{z};1,0) = \mathbf{E}(1,0)e^{i\mathbf{k}_0\cdot\mathbf{z}}, \tag{13.5.2}$$

$$u_{i,j}(\mathbf{z};0,1) = u_{i,j}(0,1)e^{i\mathbf{k}_A\cdot\mathbf{z}} \tag{13.5.3}$$

and the nonlinear polarization as

$$\mathscr{P}(\mathbf{z};1,1) = \mathscr{P}(1,1)e^{i(\mathbf{k}_0+\mathbf{k}_A)\cdot\mathbf{z}}. \tag{13.5.4}$$

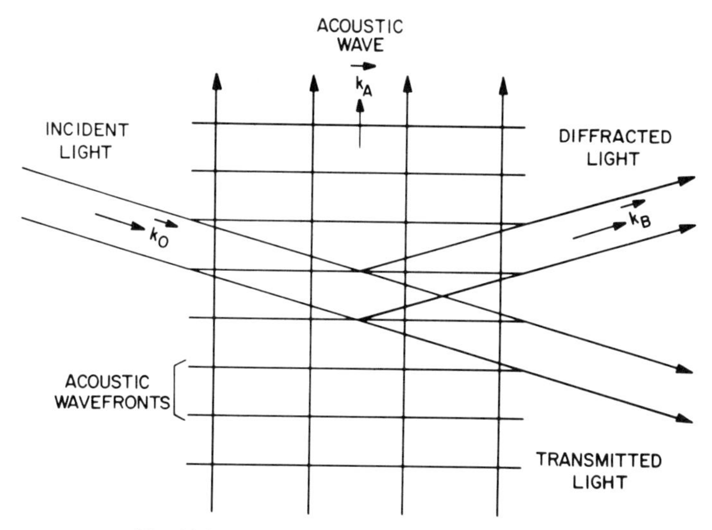

Fig. 13.2. Acoustooptic diffraction geometry.

We change notation here slightly by denoting the wavevector and frequency associated with (1,0) by $\mathbf{k}_O$ and ω_O (for optical), those associated with (0,1) by $\mathbf{k}_A$ and ω_A (for acoustic), and those associated with the (1,1) field by $\mathbf{k}_B$ and ω_B (for Brillouin) where

$$\omega_B = \omega_O + \omega_A. \tag{13.5.5}$$

The coefficients $\mathbf{E}(1,0)$, $u_{i,j}(0,1)$, and $\mathscr{P}(1,1)$ are taken as constants. This simplification excludes input wave depletion. Hence the results apply only to small power conversions into the diffracted field.

The solution to the driven wave equation (13.5.1) is the sum of the general homogeneous solution (the *free wave*) and a particular solution of the inhomogeneous equation (the *driven* or *forced wave*). The free wave must be proportional to one of the two propagating eigenvectors of Section 9.3 denoted by $\mathfrak{E}^{\xi}(\mathbf{s}_B, \omega_B)$ where

$$\mathbf{s}_B \equiv \frac{\mathbf{k}_B}{k_B}. \tag{13.5.6}$$

Since we do not consider the effects of an exit surface of the medium, the free-wave solution (denoted by a superscript f) may be taken as the sum of two free waves traveling in the $+\mathbf{z}$ direction only,

$$\mathbf{E}^f(\mathbf{z}; 1,1) = \sum_{\xi=1}^{2} C^{\xi} \mathfrak{E}^{\xi}(\mathbf{s}_B, \omega_B) e^{i\mathbf{k}_B^{\xi}\cdot\mathbf{z}}, \tag{13.5.7}$$

where C^{ξ} is an amplitude constant and $\mathbf{k}_B^{\xi}$ is the wavevector of the ξ eigenmode. [If the nonpropagating eigenmode were included in this equation (as a constant term), it would later be excluded by the boundary conditions.]

The forced or driven wave (denoted by a superscript d) has the space dependence,

$$\mathbf{E}^d(\mathbf{z};1,1)=\mathbf{E}^d(1,1)e^{i(\mathbf{k}_O+\mathbf{k}_A)\cdot\mathbf{z}}, \tag{13.5.8}$$

like that of the nonlinear polarization (13.5.4). Substitution into the driven wave equation leads to

$$-\left(\frac{c}{\omega_B}\right)^2(\mathbf{k}_O+\mathbf{k}_A)\times\left[(\mathbf{k}_O+\mathbf{k}_A)\times\mathbf{E}^d(1,1)\right]-\boldsymbol{\kappa}\cdot\mathbf{E}^d(1,1)=\frac{\mathscr{P}(1,1)}{\epsilon_0}. \tag{13.5.9}$$

By expanding the vector triple product and introducing

$$\mathbf{s}_D\equiv\frac{\mathbf{k}_O+\mathbf{k}_A}{|\mathbf{k}_O+\mathbf{k}_A|}, \tag{13.5.10}$$

$$n_D\equiv\frac{c}{\omega_B}|\mathbf{k}_O+\mathbf{k}_A| \tag{13.5.11}$$

(13.5.9) can be rewritten as

$$\left[n_D^2(\mathbf{1}-\mathbf{s}_D\mathbf{s}_D)-\boldsymbol{\kappa}(\omega_B)\right]\cdot\mathbf{E}^d(1,1)=\frac{\mathscr{P}(1,1)}{\epsilon_0}. \tag{13.5.12}$$

The bracketed coefficient on the left side is just the tensor $\boldsymbol{\alpha}$ of (10.7.7) whose inverse is given in (10.7.16). Using that result we have for the driven wave amplitude

$$\mathbf{E}^d(1,1)=\frac{1}{\epsilon_0}\sum_{\alpha=1}^{3}\frac{\mathfrak{E}^{\alpha}(\mathbf{s}_D,\omega_B)\mathfrak{E}^{\alpha}(\mathbf{s}_D,\omega_B)\cdot\mathscr{P}(1,1)}{\left[n_D/n^{\alpha}(\mathbf{s}_D)\right]^2-1}, \tag{13.5.13}$$

where n^{α} is the refractive index associated with the $\mathfrak{E}^{\alpha}(\mathbf{s}_D,\omega_B)$ eigenvector. We emphasize that the refractive index n_D of the driven wave does not, in general, equal either of the free wave refractive indices for the direction of propagation $\mathbf{s}_D$ and frequency ω_B of the driven wave. Phase matching, which is the condition that arranges this equality, is discussed shortly.

The complete solution inside the medium,

$$\begin{aligned}\mathbf{E}(\mathbf{z};1,1)&=\mathbf{E}^f(\mathbf{z};1,1)+\mathbf{E}^d(\mathbf{z};1,1)\\&=\sum_{\alpha=1}^{2}C^\alpha\boldsymbol{\mathcal{E}}^\alpha(\mathbf{s}_B,\omega_B)e^{i\mathbf{k}_B^\alpha\cdot\mathbf{z}}\\&\quad+\frac{1}{\epsilon_0}\sum_{\alpha=1}^{3}\frac{\boldsymbol{\mathcal{E}}^\alpha(\mathbf{s}_D,\omega_B)\boldsymbol{\mathcal{E}}^\alpha(\mathbf{s}_D,\omega_B)\cdot\boldsymbol{\mathcal{P}}(1,1)e^{i(\mathbf{k}_O+\mathbf{k}_A)\cdot\mathbf{z}}}{[n_D/n^\alpha(\mathbf{s}_D)]^2-1},\end{aligned}\tag{13.5.14}$$

must now be made to match input surface boundary conditions. These boundary conditions can be satisfied only if a backward wave at the output sum frequency ω_B is generated at the input surface. We denote this field,

$$\mathbf{E}^o(\mathbf{z};1,1)=\mathbf{E}^o e^{-i\mathbf{k}\cdot\mathbf{z}},\tag{13.5.15}$$

by a superscript o; its wavevector is simply $\mathbf{k}=\mathbf{s}\omega_B/c$.

The two components of $\mathbf{E}^o$ and the two constants C^ξ can be determined by the two tangential field boundary conditions (7.2.15) and (7.2.22). The moving medium terms in those equations as well as the comparable term in the constitutive relation (7.1.8) between the magnetic intensity $\mathbf{H}$ and the magnetic induction $\mathbf{B}$ have negligible size by estimates similar to those presented in Sections 12.2 and 12.3. Thus the boundary conditions are continuity of tangential electric field $\mathbf{E}$ and tangential magnetic intensity $\mathbf{H}$ obtained from

$$\mathbf{H}=\frac{1}{\mu_0}\mathbf{B}=\frac{1}{i\omega_B\mu_0}\nabla\times\mathbf{E}.\tag{13.5.16}$$

If we place the origin of the coordinate system in the plane of the input surface and denote the coordinate vector to any point in this plane by $\mathbf{z}_P$, then these boundary conditions become

$$\begin{aligned}\mathbf{n}\times\mathbf{E}^o e^{-i\mathbf{k}\cdot\mathbf{z}_P}=\mathbf{n}\times\Bigg(&\sum_{\alpha=1}^{2}C^\alpha\boldsymbol{\mathcal{E}}^\alpha(\mathbf{s}_B)e^{i\mathbf{k}_B^\alpha\cdot\mathbf{z}_P}\\&+\mathbf{E}^d(1,1)e^{i(\mathbf{k}_O+\mathbf{k}_A)\cdot\mathbf{z}_P}\Bigg),\end{aligned}\tag{13.5.17}$$

$$\begin{aligned}-\mathbf{n}\times(\mathbf{s}\times\mathbf{E}^o)e^{-i\mathbf{k}\cdot\mathbf{z}_P}=\mathbf{n}\times\Bigg[&\sum_{\alpha=1}^{2}C^\alpha n_B^\alpha(\mathbf{s}_B\times\boldsymbol{\mathcal{E}}^\alpha(\mathbf{s}_B))e^{i\mathbf{k}_B^\alpha\cdot\mathbf{z}_P}\\&+n_D\mathbf{s}_D\times\mathbf{E}^d(1,1)e^{i(\mathbf{k}_O+\mathbf{k}_A)\cdot\mathbf{z}_P}\Bigg],\end{aligned}\tag{13.5.18}$$

where $\mathbf{n}$ is the unit normal of the input plane surface and $\mathbf{E}^d(1,1)$ is the driven wave amplitude from (13.5.13). These equations can be satisfied at all points in the plane only if the components of the wavevectors tangent to the plane are equal,

$$-(\mathbf{k})_t = (\mathbf{k}_B^\alpha)_t = (\mathbf{k}_O + \mathbf{k}_A)_t. \qquad (13.5.19)$$

This double equation determines the tangential components of $\mathbf{k}$ and $\mathbf{k}_B^\alpha$ in terms of those of the driving (input) waves that are known. Since the magnitudes of $\mathbf{k}$ and $\mathbf{k}_B^\alpha$ are also known from the homogeneous wave equation, their normal components are also determined.

The condition (13.5.19) causes the exponential propagation factors to divide out of the two boundary conditions. They must now be solved for C^α and $\mathbf{E}^o$. If a scalar product is formed between $\mathfrak{E}^\beta$ and (13.5.17), the factor in the sum $\mathfrak{E}^\beta \cdot \mathbf{n} \times \mathfrak{E}^\alpha$ vanishes when $\alpha = \beta$. Thus we get

$$C^\alpha = \frac{\mathfrak{E}^\beta \cdot \mathbf{n} \times (\mathbf{E}^o - \mathbf{E}^d(1,1))}{\mathfrak{E}^\beta \cdot \mathbf{n} \times \mathfrak{E}^\alpha} \qquad (\alpha \neq \beta; \alpha, \beta = 1, 2). \qquad (13.5.20)$$

If next a scalar product is formed between $\mathbf{s}_B^\beta \times \mathfrak{E}^\beta$ and (13.5.18), the same reasoning as just employed plus the use of (13.5.20) for C^α leads to

$$\mathbf{Q}_{\beta\alpha} \cdot \mathbf{E}^o = p_{\beta\alpha} \qquad (\alpha \neq \beta) \qquad (13.5.21)$$

where

$$\mathbf{Q}_{\beta\alpha} \equiv \mathbf{s} \times \left[(\mathbf{s}_B \times \mathfrak{E}^\beta) \times \mathbf{n} \right] - \frac{n_B^\alpha (\mathbf{s}_B \times \mathfrak{E}^\beta) \cdot \mathbf{n} \times (\mathbf{s}_B \times \mathfrak{E}^\alpha)}{\mathfrak{E}^\beta \cdot \mathbf{n} \times \mathfrak{E}^\alpha} \mathfrak{E}^\beta \times \mathbf{n}$$

$$(\alpha \neq \beta), \qquad (13.5.22)$$

$$p_{\beta\alpha} \equiv n_D (\mathbf{s}_B \times \mathfrak{E}^\beta) \cdot \mathbf{n} \times (\mathbf{s}_D \times \mathbf{E}^d(1,1))$$
$$- \frac{n_B^\alpha (\mathbf{s}_B \times \mathfrak{E}^\beta) \cdot \mathbf{n} \times (\mathbf{s}_B \times \mathfrak{E}^\alpha)}{\mathfrak{E}^\beta \cdot \mathbf{n} \times \mathfrak{E}^\alpha} \mathfrak{E}^\beta \cdot \mathbf{n} \times \mathbf{E}^d(1,1) \qquad (\alpha \neq \beta).$$

$$(13.5.23)$$

Clearly α and β can be interchanged in (13.5.21) to produce

$$\mathbf{Q}_{\alpha\beta} \cdot \mathbf{E}^o = p_{\alpha\beta} \qquad (\alpha \neq \beta). \qquad (13.5.24)$$

Combining these two equations gives

$$(p_{\alpha\beta} \mathbf{Q}_{\beta\alpha} - p_{\beta\alpha} \mathbf{Q}_{\alpha\beta}) \cdot \mathbf{E}^o = 0 \qquad (\alpha \neq \beta), \qquad (13.5.25)$$

which states that $\mathbf{E}^o$ is orthogonal to the enclosed vector. $\mathbf{E}^o$ is also orthogonal to the unit wavevector $\mathbf{s}$,

$$\mathbf{s}\cdot\mathbf{E}^o = 0, \tag{13.5.26}$$

and so we may write

$$\mathbf{E}^o = A\mathbf{s}\times(p_{\alpha\beta}\mathbf{Q}_{\beta\alpha} - p_{\beta\alpha}\mathbf{Q}_{\alpha\beta}) \qquad (\alpha\neq\beta). \tag{13.5.27}$$

The constant A is determined by substituting this into either (13.5.21) or (13.5.24) with the result

$$\mathbf{E}^o = \frac{\mathbf{s}\times(p_{\alpha\beta}\mathbf{Q}_{\beta\alpha} - p_{\beta\alpha}\mathbf{Q}_{\alpha\beta})}{\mathbf{s}\cdot(\mathbf{Q}_{\beta\alpha}\times\mathbf{Q}_{\alpha\beta})} \qquad (\alpha\neq\beta). \tag{13.5.28}$$

Equations (13.5.14), (13.5.19), (13.5.20), and (13.5.28) along with the definitions of the various quantities constitute the formal solution of the problem.

We have presented this solution because it is mathematically straightforward and exact; its complexity in its general form, however, obscures the physical understanding that we wish to obtain from it. Our task now is to find an approximate form of this solution that clearly exhibits this understanding. We first note that every term in the solution is proportional to $\mathbf{E}^d(1,1)$ and that by (13.5.13) the denominator of one of the two free wave terms (never that of the nonpropagating solution) can approach zero as

$$n_D \to n^{\xi}(\mathbf{s}_D). \tag{13.5.29}$$

In a birefringent crystal this can happen for only one eigenmode at a time (except along optic axes, which we exclude); in a cubic crystal it could happen for both propagating eigenmodes. However, in the latter case the elastooptic interaction tensor contained in $\mathscr{P}$ is zero for one of the two modes. In this case, of course, we consider the nonzero term. The term for which $n_D \to n^{\xi}(\mathbf{s}_D)$ dominates $\mathbf{E}^d(1,1)$ and so the other terms may be dropped as being negligibly small. Thus we take

$$\mathbf{E}^d(1,1) = \frac{\boldsymbol{\mathcal{E}}^{\xi}(\mathbf{s}_D,\omega_B)\boldsymbol{\mathcal{E}}^{\xi}(\mathbf{s}_D,\omega_B)\cdot\mathscr{P}(1,1)}{\epsilon_0\left[\left(\dfrac{n_D}{n^{\xi}(\mathbf{s}_D)}\right)^2 - 1\right]}. \tag{13.5.30}$$

This means that we are, in effect, considering only the solution of a scalar wave equation equal to the scalar product of the vector wave equation with $\mathcal{E}^{\xi}(\mathbf{s}_D, \omega_B)$.

Since the limit (13.5.29) is a physically attainable situation, we can expect that the vanishing of the denominator of $\mathbf{E}^d(1,1)$ is compensated by a suitable vanishing of the numerators of the terms in the solution (13.5.14). In order to take the limit we must identify the basic quantity that is approaching zero and then express the various quantities in terms of it. Since the directions of waves as well as their speeds may need to be changed in order to approach the limit (13.5.29), it is clear that the difference of propagation vectors of the free and forced waves,

$$\Delta\mathbf{k} \equiv \mathbf{k}_B^{\xi} - \mathbf{k}_O - \mathbf{k}_A, \tag{13.5.31}$$

is the basic quantity approaching zero. We will speak of the vanishing of $\Delta\mathbf{k}$ as being caused by *phase matching* the forced wave to the ξ free wave. The result (13.5.19) of applying the boundary conditions has guaranteed that the component of $\Delta\mathbf{k}$ tangent to the plane input surface vanishes. Thus we may write

$$\Delta\mathbf{k} = \Delta k_n \mathbf{n}. \tag{13.5.32}$$

First, we reexpress the denominator of $\mathbf{E}^d(1,1)$ in terms of Δk_n. To do this we must expand $n^{\xi}(\mathbf{s}_D)$ to first order about $n^{\xi}(\mathbf{s}_B)$ by

$$n^{\xi}(\mathbf{s}_D) = n^{\xi}(\mathbf{s}_B) + (\mathbf{s}_D - \mathbf{s}_B)\cdot\nabla_s n^{\xi}(\mathbf{s}_B) \tag{13.5.33}$$

where ∇_s is the gradient with respect to the components of $\mathbf{s}$. Since $\mathbf{s}$ is a unit vector, ∇_s is perpendicular to $\mathbf{s}$ (see Section 9.10). Thus we may reexpress the first order term by

$$\begin{aligned} \mathbf{s}_D - \mathbf{s}_B &\cong \frac{1}{n^{\xi}(\mathbf{s}_B)}\left(n_D \mathbf{s}_D - n^{\xi}(\mathbf{s}_B)\mathbf{s}_B\right) \\ &= \frac{c}{\omega_B n^{\xi}(\mathbf{s}_B)}\left(\mathbf{k}_O + \mathbf{k}_A - \mathbf{k}_B^{\xi}\right) \\ &= -\frac{c\Delta k_n \mathbf{n}}{\omega_B n^{\xi}(\mathbf{s}_B)} \end{aligned} \tag{13.5.34}$$

since only the component of this normal to $\mathbf{s}_B$ enters the equation. Equation (13.5.33) then becomes

$$n^{\xi}(\mathbf{s}_D) = n^{\xi}(\mathbf{s}_B) - \frac{c\Delta k_n \mathbf{n}}{\omega_B n^{\xi}(\mathbf{s}_B)}\cdot\nabla_s n^{\xi}(\mathbf{s}_B). \tag{13.5.35}$$

Thus to first order we have

$$\left(\frac{n_D}{n^\xi(\mathbf{s}_D)}\right)^2-1=\frac{1}{\left[n^\xi(\mathbf{s}_B)\right]^2}\left\{n_D^2-\left[n^\xi(\mathbf{s}_B)\right]^2+\frac{2c\Delta k_n\mathbf{n}}{\omega_B}\cdot\nabla_s n^\xi(\mathbf{s}_B)\right\}$$

$$=\left(\frac{c}{\omega_B n^\xi}\right)^2\left\{k_D^2-(k_B^\xi)^2+\frac{2\omega_B}{c}\Delta k_n\mathbf{n}\cdot\nabla_s n^\xi\right\}$$

$$=\left(\frac{c}{\omega_B n^\xi}\right)^2\left\{\left[\mathbf{k}_O+\mathbf{k}_A-\mathbf{k}_B^\xi\right]\cdot\left[\mathbf{k}_O+\mathbf{k}_A+\mathbf{k}_B^\xi\right]+\frac{2\omega_B}{c}\Delta k_n\mathbf{n}\cdot\nabla_s n^\xi\right\}$$

$$=\left(\frac{c}{\omega_B n^\xi}\right)^2\left\{-\Delta k_n\mathbf{n}\cdot(2\mathbf{k}_B^\xi)+\frac{2\omega_B}{c}\Delta k_n\mathbf{n}\cdot\nabla_s n^\xi\right\}$$

$$=\frac{-2c\Delta k_n}{\omega_B n^\xi}\mathbf{n}\cdot\left\{\mathbf{s}_B-\frac{1}{n^\xi}\nabla_s n^\xi\right\}$$

$$=-\frac{2c\Delta k_n(\mathbf{n}\cdot\mathbf{t}_B^\xi)}{\omega_B n^\xi(\mathbf{s}_B\cdot\mathbf{t}_B^\xi)} \tag{13.5.36}$$

where the last step uses the expression (9.10.19) for the direction $\mathbf{t}_B^\xi$ of the group velocity.

Since the quantities $p_{\beta\xi}$ and $p_{\xi\beta}$ have vanishing denominators, we must examine their values in the limit of small Δk_n. Continuing to denote the phase-matchable free wave by ξ, we first reexpress $p_{\beta\xi}$ as

$$p_{\beta\xi}=\frac{c}{\epsilon_0\omega_B}\left\{(\mathbf{s}_B\times\mathcal{E}^\beta)\cdot\mathbf{n}\times\left[(\mathbf{k}_O+\mathbf{k}_A)\times\mathcal{E}^\xi\right]-(\mathbf{s}_B\times\mathcal{E}^\beta)\cdot\mathbf{n}\times\left[\mathbf{k}_B^\xi\times\mathcal{E}^\xi\right]\right\}\frac{\mathcal{E}^\xi\cdot\mathscr{P}(1,1)}{\left(\frac{n_D}{n^\xi(\mathbf{s}_D)}\right)^2-1}$$

$$=-\left\{\left[(\mathbf{s}_B\times\mathcal{E}^\beta)\times\mathbf{n}\right]\times\mathcal{E}^\xi\cdot\mathbf{n}\right\}\frac{\mathcal{E}^\xi\cdot\mathscr{P}(1,1)n^\xi(\mathbf{s}_B\cdot\mathbf{t}_B^\xi)}{2\epsilon_0(\mathbf{n}\cdot\mathbf{t}_B^\xi)}\qquad(\beta\neq\xi). \tag{13.5.37}$$

Reexpression of $p_{\xi\beta}$ leads to

$$p_{\xi\beta}=\frac{c}{\epsilon_0\omega_B}\{(\mathbf{s}_B\times\mathcal{E}^\xi)\cdot\mathbf{n}\times[(\mathbf{k}_O+\mathbf{k}_A)\times\mathcal{E}^\xi]\}\frac{\mathcal{E}^\xi\cdot\mathcal{P}(1,1)}{\left(\dfrac{n_D}{n^\xi(\mathbf{s}_D)}\right)^2-1}$$

$$=\frac{c}{\epsilon_0\omega_B}\{-[(\mathbf{s}_B\times\mathcal{E}^\xi)\times\mathbf{n}]\times\mathcal{E}^\xi\cdot(\mathbf{k}_O+\mathbf{k}_A)$$

$$+[(\mathbf{s}_B\times\mathcal{E}^\xi)\times\mathbf{n}]\times\mathcal{E}^\xi\cdot\mathbf{k}_B^\xi\}\frac{\mathcal{E}^\xi\cdot\mathcal{P}(1,1)}{\left(\dfrac{n_D}{n^\xi(\mathbf{s}_D)}\right)^2-1}$$

$$=-\{[(\mathbf{s}_B\times\mathcal{E}^\xi)\times\mathbf{n}]\times\mathcal{E}^\xi\cdot\mathbf{n}\}\frac{\mathcal{E}^\xi\cdot\mathcal{P}(1,1)n^\xi(\mathbf{s}_B\cdot\mathbf{t}_B^\xi)}{2\epsilon_0(\mathbf{n}\cdot\mathbf{t}_B^\xi)}\qquad(\beta\neq\xi).\tag{13.5.38}$$

The last two expressions can be used in (13.5.28) to obtain the expression for the electric field $\mathbf{E}^o$ of the backward wave from the input surface in the limit of small Δk_n. Since the Δk_n factors divide out of $\mathbf{E}^o$, it can be seen to be insensitive to this limit.

The integration constants (13.5.20) are now given by

$$C^\alpha=\frac{\mathcal{E}^\beta\cdot\mathbf{n}\times\mathbf{E}^o}{\mathcal{E}^\beta\cdot\mathbf{n}\times\mathcal{E}^\alpha}+\frac{\mathcal{E}^\xi\cdot\mathcal{P}(1,1)\omega_B n^\xi(\mathbf{s}_B\cdot\mathbf{t}_B^\xi)}{2\epsilon_0 c\Delta k_n(\mathbf{n}\cdot\mathbf{t}_B^\xi)}\qquad(\beta\neq\alpha=\xi)$$

$$=\frac{\mathcal{E}^\beta\cdot\mathbf{n}\times\mathbf{E}^o}{\mathcal{E}^\beta\cdot\mathbf{n}\times\mathcal{E}^\alpha}\qquad(\alpha\neq\beta=\xi).\tag{13.5.39}$$

The presence of Δk_n in one term here makes that term much larger than either of the terms involving $\mathbf{E}^o$; thus we may drop the latter terms. This gives us the final expression of the generated electric field near phase matching,

$$\mathbf{E}(\mathbf{z};1,1)=\frac{\omega_B n^\xi(\mathbf{s}_B\cdot\mathbf{t}_B^\xi)\mathcal{E}^\xi\,\mathcal{E}^\xi\cdot\mathcal{P}(1,1)}{2\epsilon_0 c\,\Delta k_n(\mathbf{n}\cdot\mathbf{t}_B^\xi)}\left(e^{i\mathbf{k}_B^\xi\cdot\mathbf{z}}-e^{i(\mathbf{k}_O+\mathbf{k}_A)\cdot\mathbf{z}}\right)$$

$$=\frac{i\omega_B z\mathbf{e}^\xi\mathbf{e}^\xi\cdot\mathcal{P}(1,1)\Phi^{1/2}(\Delta k_n z/2)}{2\epsilon_0 c n^\xi\cos\delta^\xi(\mathbf{n}\cdot\mathbf{t}_B^\xi)}e^{i\mathbf{k}_B^\xi\cdot\mathbf{z}}\tag{13.5.40}$$

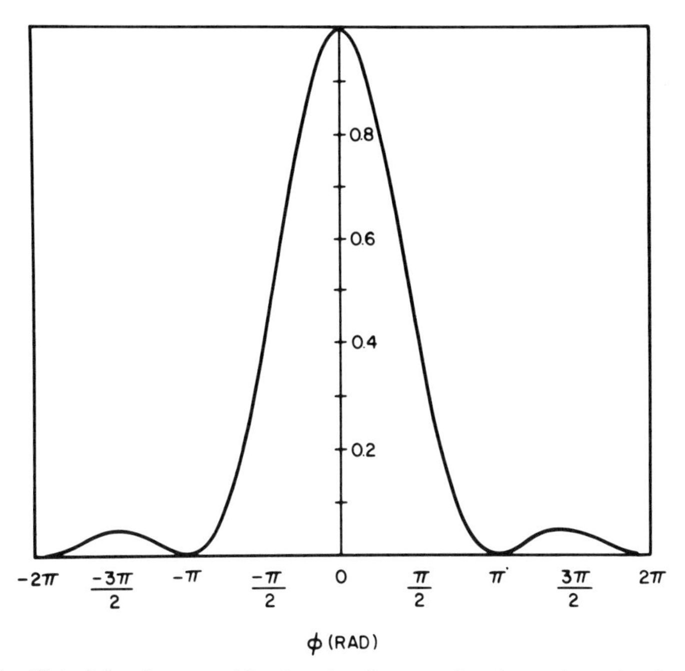

Fig. 13.3. Plot of the phase-matching function Φ versus the phase mismatch ϕ between the forced and free waves. Exact phase matching corresponds to $\phi=0$.

where

$$z \equiv \mathbf{z}\cdot\mathbf{n}, \tag{13.5.41}$$

$$\Phi(\phi) \equiv \frac{\sin^2\phi}{\phi^2}, \tag{13.5.42}$$

$$\cos\delta^{\xi} \equiv \mathbf{s}_B\cdot\mathbf{t}_B^{\xi} \tag{13.5.43}$$

and the eigenvectors $\mathcal{E}^{\xi}$ are converted to unit eigenvectors by (9.3.22). The Φ function is called the *phase-matching function* and has a maximum value of 1 when its argument is 0, that is, at exact phase matching. It is shown in Fig. 13.3.

Several aspects of the approximate solution (13.5.40) should be noted. First, the amplitude of $\mathbf{E}(\mathbf{z};1,1)$ depends only on the scalar coordinate z measured normal to the input surface. Second, near phase matching this dependence is slowly varying compared to the propagation factor $\exp i\mathbf{k}_B^{\xi}\cdot\mathbf{z}$ and becomes simply a linear dependence on z at exact phase matching. Third, $\mathbf{E}(z=z_P;1,1)=0$ for the approximate solution; thus this could be used as an approximate boundary condition for obtaining this solution.

Fourth, both the electric field eigenvector and the propagation vector of the nonlinearly generated light near phase matching are those of a free wave. Fifth, as pointed out earlier, the approximate solution obeys the scalar product of the wave equation with the electric field eigenvector of the phase matched solution and not the entire vector wave equation. Sixth, phase matching involves only the vanishing of the component $\Delta\mathbf{k}$ normal to the input surface, since the boundary conditions enforce the vanishing of the tangential components of $\Delta\mathbf{k}$. We make use of these facts in deriving coupled mode equations that describe parametric generation of light in Section 15.5.

The time-averaged Poynting vector of the generated light may now be obtained [see (9.P.4)] from

$$\langle \mathbf{S}(\mathbf{z})\rangle = \tfrac{1}{2}\mathcal{R}\{\mathbf{E}(\mathbf{z},t)\times\mathbf{H}^*(\mathbf{z},t)\}, \tag{13.5.44}$$

with $\mathbf{H}(\mathbf{z},t)$ given by

$$\mathbf{H}(\mathbf{z},t) = \frac{n^\xi}{\mu_0 c}\mathbf{s}_B\times\mathbf{E}(\mathbf{z},t). \tag{13.5.45}$$

In the last expression the derivative of the slowly varying amplitude is dropped in comparison to the derivative of the exponential propagation factor. The Poynting vector is now given by

$$\langle \mathbf{S}_B^\xi(z)\rangle = \frac{\omega_B^2 z^2|\mathbf{e}^\xi\cdot\mathscr{P}(1,1)|^2\Phi(\Delta k_n z/2)\mathbf{t}_B^\xi}{8\epsilon_0 c n^\xi\cos\delta^\xi(\mathbf{n}\cdot\mathbf{t}_B^\xi)^2}. \tag{13.5.46}$$

This formula shows that the generated light intensity at exact phase matching ($\Delta k_n = 0, \Phi = 1$) grows as the square of the coordinate z measured normal to the input surface (until input wave depletion becomes significant).

Equation (13.5.46) may be put into a more useful form by expressing

$$|\mathbf{e}^\xi\cdot\mathscr{P}(1,1)| = \epsilon_0\chi|A||B|k_A^\zeta \tag{13.5.47}$$

where

$$\chi \equiv e_i^\xi \chi_{ijkl}^{\text{eff}} e_j^\eta b_k^\zeta s_l^A. \tag{13.5.48}$$

Here $\boldsymbol{\chi}^{\text{eff}}$ is the effective susceptibility given by the expression inside the brackets of (13.1.2); $\mathbf{e}^\eta$ is the unit eigenvector of the input light; $|A|$ is its

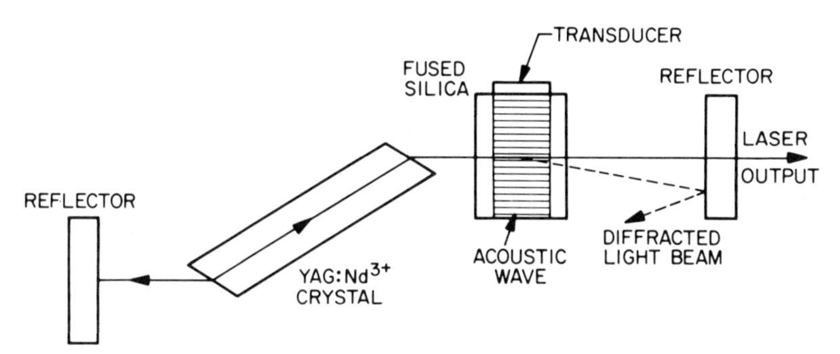

Fig. 13.4. Acoustooptic Q-switch for a laser. Acoustooptic diffraction in a fused silica block deflects any light attempting to travel along the laser cavity oscillation axis, thus preventing laser oscillation even though the neodymium doped yittrium-aluminum garnet (YAG:Nd) crystal is excited to an amplifying condition. Turning off the acoustic wave allows restoration of feedback in the oscillator and causes a large output light pulse to be generated from the excess excitation of the laser crystal.

amplitude given by (9.10.27) in terms of the magnitude of the time-averaged input light Poynting vector $\langle S_O^\eta \rangle$ by

$$|A|^2 = \frac{2\langle S_O^\eta \rangle}{\epsilon_0 c n^\eta \cos\delta^\eta}; \tag{13.5.49}$$

$\mathbf{b}^\zeta$ is the unit eigenvector of the input acoustic wave; $\mathbf{s}^A$ is its unit propagation vector; k_A^ζ is the magnitude of its propagation vector; $|B|$ is the amplitude of the acoustic wave obtained in terms of the magnitude of the time-averaged acoustic energy flux vector from (10.11.16) to be

$$|B|^2 = \frac{2\langle S_A^\zeta \rangle \cos\delta^\zeta}{\omega_A^2 \rho^0 v_\zeta} \tag{13.5.50}$$

where ω_A is the acoustic angular frequency, v_ζ is the acoustic velocity, and $\cos\delta^\zeta$ is defined in (10.11.15). Combining (13.5.46) to (13.5.50) yields

$$\langle \mathbf{S}_B^\xi(z) \rangle = \frac{(\omega_B z \chi)^2 \cos\delta^\zeta \langle S_O^\eta \rangle \langle S_A^\zeta \rangle \mathbf{t}_B^\xi}{2c^2 \rho^0 v_\zeta^3 n^\xi n^\eta \cos\delta^\xi \cos\delta^\eta (\mathbf{n}\cdot\mathbf{t}_B^\xi)^2}. \tag{13.5.51}$$

Alternately this may be expressed as

$$\langle \mathbf{S}_B^\xi(z) \rangle = \frac{(\omega_B z p)^2 (n^\xi n^\eta)^3 \cos\delta^\xi \cos\delta^\eta \cos\delta^\zeta \langle S_O^\eta \rangle \langle S_A^\zeta \rangle \mathbf{t}_B^\xi}{8c^2 \rho^0 v_\zeta^3 (\mathbf{n}\cdot\mathbf{t}_B^\xi)^2} \tag{13.5.52}$$

if we introduce a Pockels-like coefficient p given by

$$p \equiv d_m^{\xi} p_{mnkl}^{\text{eff}} d_n^{\eta} b_k^{\zeta} s_l^A \tag{13.5.53}$$

where $\mathbf{p}^{\text{eff}}$ is defined in (13.4.2). Note that when an effective susceptibility $\boldsymbol{\chi}^{\text{eff}}$ is used in the formula, unit electric field eigenvectors $\mathbf{e}$ occur, but when an effective Pockels-like tensor $\mathbf{p}^{\text{eff}}$ is used, unit electric displacement eigenvectors $\mathbf{d}$ occur. Both (13.5.51) and (13.5.52) express the output intensity measured inside the crystal.

Figure 13.4 illustrates one of the many uses of acoustooptic diffraction.

13.6 Phase Matching

The development of the preceding section shows that the output optical field from acoustooptic diffraction at the frequency

$$\omega_B = \omega_O + \omega_A \tag{13.6.1}$$

is greatly enhanced when

$$\mathbf{k}_B = \mathbf{k}_O + \mathbf{k}_A, \tag{13.6.2}$$

called the *phase-matching condition*, is met. The physical meaning of this can best be understood by referring to (13.5.40). The electric field consists of a driven wave term propagating with the wavevector $\mathbf{k}_O + \mathbf{k}_A$, a free wave term propagating with the wavevector $\mathbf{k}_B$, and a denominator that vanishes when these two wavevectors are equal. Thus we see that when the nonlinear driving polarization is *in phase* with the freely propagating wave of the same frequency and direction of propagation, the nonlinear driving polarization can most effectively couple energy into the radiated output wave. An enhanced or phase matched output wave results.

The conditions (13.6.1) and (13.6.2) can be met in a variety of ways. Consider an optically isotropic medium first. Since the acoustic frequency ω_A is much smaller than either of the optical frequencies ω_B or ω_O, an adequate approximation is to let $\omega \equiv \omega_B \cong \omega_O$. This means that $n \equiv n_B \cong n_O$ and hence $k_B \cong k_O$. The triangle of wavevectors shown in Fig. 13.5 is thus isoceles to an excellent level of approximation. The vector component of (13.6.2) parallel to $\mathbf{k}_A$ is then

$$\frac{n\omega}{c} \sin \Theta = -\frac{n\omega}{c} \sin \Theta + \frac{\omega_A}{v_A} \tag{13.6.3}$$

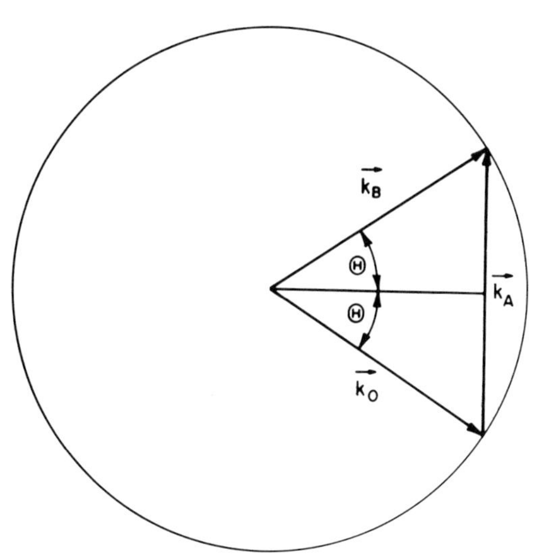

Fig. 13.5. Phase-matching diagram for acoustooptic diffraction in optically isotropic substances (cubic crystals and amorphous solids such as glasses).

where Θ is half of the apex angle. This yields immediately the Bragg law

$$\sin\Theta = \frac{c\omega_A}{2n\omega v_A} = \frac{\lambda}{2n\Lambda} \tag{13.6.4}$$

where

$$\lambda \equiv \frac{2\pi c}{\omega}, \tag{13.6.5}$$

$$\Lambda \equiv \frac{2\pi v_A}{\omega_A} \tag{13.6.6}$$

are the free-space optical wavelength and the acoustic wavelength respectively. Equation (13.6.4) shows that the angle between the incident and diffracted light beams is proportional to the acoustic frequency and inversely proportional to the optical frequency for small angles. As is evident from Fig. 13.5 the maximum acoustic frequency for a given optical frequency corresponds to $\Theta = \pi/2$ and has the value

$$\omega_A = \frac{2n\omega v_A}{c}. \tag{13.6.7}$$

Typical values of the parameters give this maximum acoustic frequency a

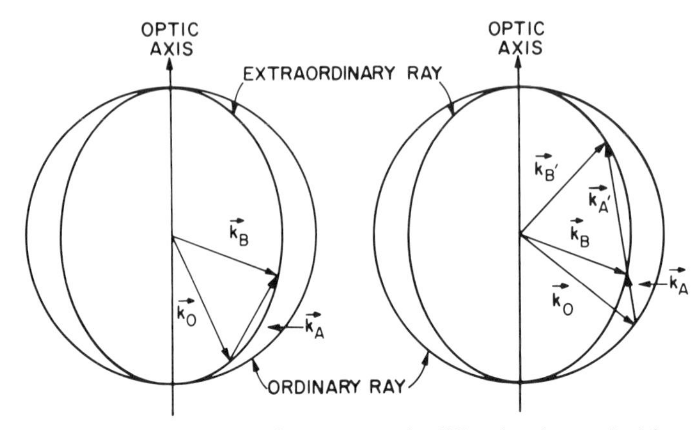

Fig. 13.6. Phase-matching diagram for acoustooptic diffraction in a uniaxial crystal in a plane containing the optic axis. The case where both input and output light waves are extraordinarily polarized is shown on the left. On the right the case where the input light wave has ordinary polarization and the output light wave has extraordinary polarization is shown. For fixed directions of the input light wave and acoustic wave two output light wave directions ($\mathbf{k}_B$ and $\mathbf{k}_B'$) corresponding to different acoustic frequencies exist.

value of $\nu_A = \omega_A/2\pi \sim 25$ GHz. In optically isotropic media the phase-matching condition is unaffected by whether the output light wave has the same or a different polarization than the input light wave.

The more complicated phase-matching geometries that can occur in optically anisotropic crystals were first studied by Dixon [1967]. Even when the input and output polarizations are the same the triangle of wavevectors need not be isoceles. Figure 13.6 shows such a case when both input and output light waves have extraordinary polarization in a negative uniaxial crystal. Figure 13.6 also shows a more complicated case where a polarization change from ordinary to extraordinary occurs. Note that two output light wave directions can occur for fixed directions of the input optical and acoustic waves but differing acoustic frequencies.

The algebraic conditions that replace the Bragg law in anisotropic media can be derived easily from the components of (13.6.2) taken parallel to $\mathbf{k}_A$,

$$k_B \sin\theta_d = -k_O \sin\theta_i + k_A, \tag{13.6.8}$$

and perpendicular to $\mathbf{k}_A$,

$$k_B \cos\theta_d = k_O \cos\theta_i. \tag{13.6.9}$$

The angles of incidence θ_i and diffraction θ_d are shown in Fig. 13.7. The

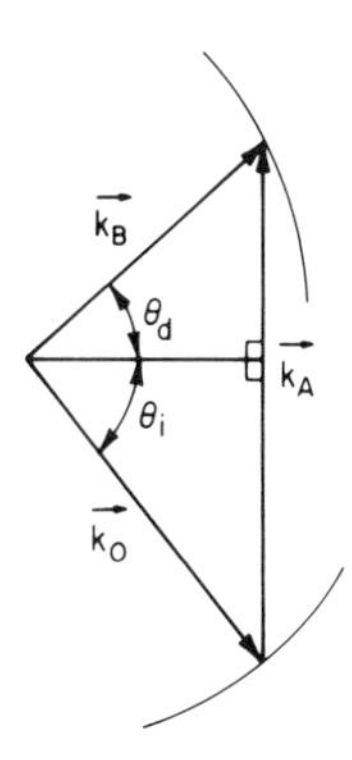

Fig. 13.7. Definition of angles for a general phase-matching diagram applying to acoustooptic diffraction in any material. The segments of the wavevector surfaces shown for the input and output light waves may be portions of the same surface or of different surfaces.

diagram of that figure as well as the last two equations apply to *any* material. Elimination of θ_d from the equations yields

$$\sin\theta_i = \frac{k_A^2 + k_O^2 - k_B^2}{2k_A k_O} \tag{13.6.10}$$

while elimination of θ_i yields

$$\sin\theta_d = \frac{k_A^2 - k_O^2 + k_B^2}{2k_A k_B}. \tag{13.6.11}$$

Making the approximation $\omega_O \approx \omega_B \equiv \omega$ leads to

$$\sin\theta_i = \frac{\lambda}{2n_O\Lambda}\left[1 + \frac{\Lambda^2}{\lambda^2}\left(n_O^2 - n_B^2\right)\right], \tag{13.6.12}$$

$$\sin\theta_d = \frac{\lambda}{2n_B\Lambda}\left[1 - \frac{\Lambda^2}{\lambda^2}\left(n_O^2 - n_B^2\right)\right]. \tag{13.6.13}$$

Note that when $n_O = n_B$ we have $\theta_d = \theta_i$ and the isoceles triangle of the isotropic case results. The first term on the right side of both of these equations is thus the Bragg law term while the difference of the second and third terms contains the effect of anisotropy. These formulas apply whether or not a polarization change occurs in the diffraction process and regardless of the type of optical anisotropy present.

For a fixed acoustic frequency and direction of propagation the formulas above fix not only the angle $\theta_i + \theta_d$ between the two optical wavevectors but the specific orientations of these wavevectors to the principal axes of the dielectric tensor. Light having a range of input angles, however, can be acoustooptically diffracted into light having a range of

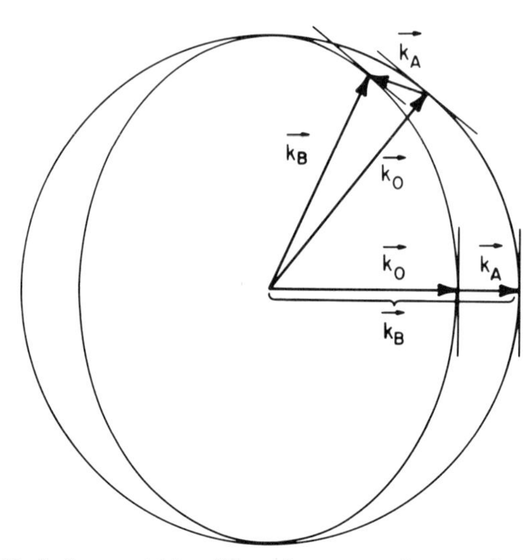

Fig. 13.8. Noncritical phase matching. When the tangent planes to the wavevector surfaces are parallel, the phase matching condition is approximately met over a range of angles of the input and output light waves.

output angles by a single acoustic wave provided the tangents to the two optical wavevector surfaces are parallel, a process called *noncritical phase matching*. This can be accomplished for a collinear acoustooptic interaction normal to the optic axis of a uniaxial crystal as illustrated in Fig. 13.8. Harris and Wallace [1969] used this geometry to obtain a sizable aperture in a tunable optical filter using the acoustooptic interaction (see Fig. 13.9). Chang [1974] showed that a sizable aperture can also be obtained from a noncollinear interaction provided that the same condition of parallel tangents is met. Figure 13.8 also illustrates the noncritical phase-matching scheme of Chang.

We close these remarks on phase matching of acoustooptic interactions by noting that if we consider difference frequency generation,

$$\omega_O = \omega_B - \omega_A, \qquad (13.6.14)$$

the reality condition (12.2.5) requires us to consider plane waves with $\exp i\mathbf{k}_B\cdot\mathbf{z}$ for the space dependence of the input optical field and $\exp - i\mathbf{k}_A\cdot\mathbf{z}$ for the space dependence of the input acoustic field. The subscript O here labels the output optical field. The result is an interaction which is phase matched when

$$\mathbf{k}_O = \mathbf{k}_B - \mathbf{k}_A. \qquad (13.6.15)$$

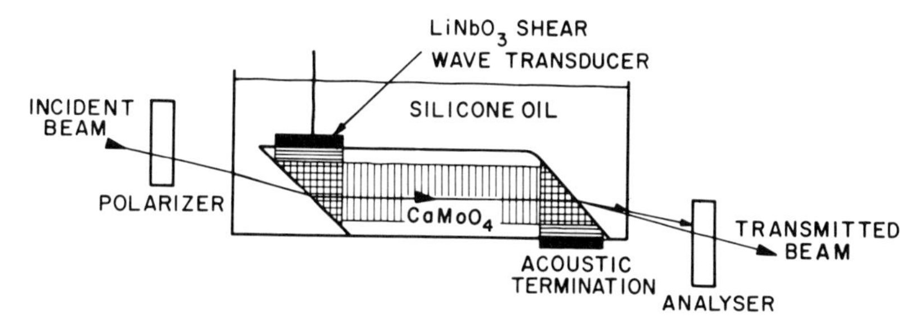

Fig. 13.9. Acoustooptic filter. Collinear propagation of an acoustic shear wave and an ordinary optical wave normal to the optic axis of $CaMoO_4$ can cause the production of a strong extraordinary optical wave in the same direction when phase matching is satisfied. The combination of (13.6.1) and (13.6.2) shows that the transmitted beam frequency is given by $c\nu_A/v_A(n^e - n^o)$ where ν_A and v_A are the frequency and velocity of the acoustic wave, c is the velocity of light, and n^e and n^o are the principal extraordinary and ordinary refractive indices. Tuning the acoustic frequency from 40 to 68 MHz tunes the transmitted beam wavelength from 670 to 510 nm with a 0.8 nm bandwidth [reprinted by permission from Harris et al., 1970]. The analyzer blocks the incident beam.

Note that these two equations are mathematically equivalent to the frequency and wavevector conservation equations for sum frequency generation. Thus sum and difference frequency generation are phase matched under identical conditions; only the input and output roles of the optical waves are reversed and the acoustic wave propagates in the reverse direction.

13.7 Static Elastooptic Effect

Section 13.1 presents the elastooptic susceptibility for a static deformation regarding the displacement gradient as an input field. This facilitates comparison with the elastooptic susceptibility for a traveling acoustic wave where the displacement gradient is a natural input field. It also facilitates comparison to the Pockels elastooptic tensor. In many respects, however, the static elastooptic effect is better formulated with the stress regarded as the input field. We now do this. In the process we wish to show that rotation can affect elastooptic measurements that employ static homogeneous deformations. Since we are not interested in electric field effects from piezoelectrics here, we assume that if the crystal is piezoelectric the appropriate surfaces of it are electroded and shorted together so as to prevent any piezoelectrically induced electric field.

Consider a large area (effectively infinite) slab of a crystal. A uniform normal compressive traction $-\mathbf{n}\mathfrak{T}$ is applied to the top surface, $\mathbf{n}$ being the outward surface normal and $\mathfrak{T}$ being the force per unit area. The bottom surface is regarded as held stationary. Reasoning as in Section 11.6, we find that a homogeneous stress $\mathbf{t}(0,0)$ given by

$$\mathbf{t}(0,0) = -\mathbf{nn}\mathfrak{T} \tag{13.7.1}$$

is induced in the slab. The (0,0) notation on the stress tensor is the Fourier amplitude notation introduced in Section 12.2. The stress is related to the strain by

$$t_{ij}(0,0) = c_{ijkl} u_{k,l}(0,0). \tag{13.7.2}$$

Inverting this equation gives the strain S_{mn}

$$S_{mn}(0,0) = s_{mnij} t_{ij}(0,0). \tag{13.7.3}$$

Figure 3.6 shows that rotation can occur when a slab is compressed against a stationary plane. The amount of rotation that occurs is related to the strain through the constraint imposed on the slab. If the normal to the slab is in the x_1 direction as in Fig. 3.6, it is apparent from the discussion of that figure that the stationary plane imposes

$$u_{1,2} = u_{(1,2)} + u_{[1,2]} = 0, \tag{13.7.4a}$$

$$u_{1,3} = u_{(1,3)} + u_{[1,3]} = 0. \tag{13.7.4b}$$

If $\mathbf{n} = [1,0,0]$ is a unit normal to the slab, then these equations may be written as

$$n_i u_{[i,k]} n_j - n_i u_{[i,j]} n_k + 2 n_i S_{i[k} n_{j]} = 0 \tag{13.7.5}$$

where $S_{jk} = u_{(j,k)}$ is the strain and Ω_{jk}, defined as the first two terms in (13.7.5)

$$\Omega_{jk} \equiv n_i u_{[i,k]} n_j - n_i u_{[i,j]} n_k, \tag{13.7.6}$$

is identified as the antisymmetric part of $u_{j,k}$ corresponding to rotation about two axes in the stationary plane. Thus

$$u_{j,k} = \Omega_{jk} + S_{jk} \tag{13.7.7}$$

where

$$\Omega_{jk} = 2 n_i S_{i[j} n_{k]} = 2 n_i n_{[k} s_{j]imn} t_{mn}(0,0). \tag{13.7.8}$$

The nonlinear polarization of (13.1.3) may now be expressed as

$$\mathscr{P}_i(\mathbf{z}; 1,0) = \epsilon_0 \Delta\kappa_{ij} E_j(\mathbf{z}; 1,0) \tag{13.7.9}$$

with $\Delta\kappa_{ij}$ given by

$$\Delta\kappa_{ij} \equiv 2\left[\chi_{i\ j\ (kl)}^{\omega_1\omega_1 0} s_{klmn} + 2\chi_{i\ j\ [kl]}^{\omega_1\omega_1 0} n_p n_{[l} s_{k]pmn}\right] t_{mn}(0,0). \tag{13.7.10}$$

This may be converted to the change in the inverse dielectric tensor

$$(\Delta\kappa^{-1})_{ij} = q_{ijmn}^{\text{eff}} t_{mn}(0,0), \tag{13.7.11}$$

$$q_{ijmn}^{\text{eff}} \equiv q_{ijmn} + 2n_l \kappa_{l(i}^{-1} s_{j)pmn} n_p - 2n_{(i} \kappa_{j)l}^{-1} s_{lpmn} n_p, \tag{13.7.12}$$

$$q_{ijmn} \equiv p_{ijkl} s_{klmn} \tag{13.7.13}$$

with the use of (13.2.10), (13.3.3), and (13.3.4). The quantity q_{ijmn} is the *Pockels piezooptic tensor*. It is apparent from either (13.7.10) or (13.7.12) that the rotation part of the deformation contributes to the static elastooptic effect. It arises through the action of the boundary constraint and can be present even though the independent variable, the stress, is a symmetric tensor.

Because the method of measuring the static elastooptic effect is the same as a standard method of measuring the electrooptic effect, we treat their measurement together in Section 14.10.

PROBLEMS

13.1. Figure 13.1 shows that both anti-Stokes and Stokes components (sum and difference frequencies) appear in Brillouin scattering. Show by means of the frequency and wavevector conditions how this can happen. Draw wavevector diagrams for the two cases. For each case show that the vector sum of the wavevectors representing each of the input waves equals the corresponding sum for the output waves.

13.2. Find an expression for the angular variation of p of (13.5.53) for phase matched collinear acoustooptic diffraction in the YZ crystallographic plane for calcite (negative uniaxial crystal, crystal class $\bar{3}m$).

CHAPTER 14

Electrooptic Effect

The *linear electrooptic effect* is the collection of phenomena that result from a change of the refractive index of a light wave that is linear in the electric field. This perturbing field may be either static or oscillating at a frequency as high as microwave frequencies. If the perturbing field is static, the linear electrooptic effect is detected by the resultant phase change of the light wave. If the perturbing field is oscillatory, the linear electrooptic effect may also be detected by the presence of sideband frequencies displaced from the input light wave frequency by the frequency of the perturbing electric field.

The linear electrooptic effect, though discovered independently by Roentgen and Kundt in 1883, has become known as the Pockels effect because of the latter's extensive studies of it. One of Pockels' main contributions was to show that a *direct electrooptic effect* existed in addition to an *indirect electrooptic effect*, that is, a combination of the piezoelectric and elastooptic effects. This showed that the electrooptic effect was a distinct effect, not just a combination of other known effects.

After the period of initial investigation near the end of the nineteenth century the linear electrooptic effect lay largely dormant until the discovery of the laser in 1960. Soon after that the effect became an important means to modulate the phase or intensity of a laser beam. The years since have seen a greatly expanded interest in the effect.

In this chapter we discuss the direct and indirect linear electrooptic effects. Only the direct effect is significant at frequencies well above the piezoelectric resonances of the crystal studied. Well below these resonances both effects contribute and usually contribute comparable amounts. For frequencies well below the piezoelectric resonances the indirect effect has the same crystal symmetry as the direct effect in contrast to the relation of the direct and indirect elastooptic effects. However, we show by an example that in the resonance region the indirect effect can lose the symmetry of the direct effect [Nelson and Turner, 1968].

Since in many applications of the electrooptic effect the light wave may be approximated as a plane wave and the electrooptic perturbation may be taken as homogeneous, we develop a perturbation theory of the plane wave eigenvector equation of crystal optics. We then use it to obtain a general solution for the electrooptic effect applicable to crystals of any symmetry, to light waves of any propagation direction and of either polarization, and to perturbing electric fields applied in any direction [Nelson, 1975]. We go on to discuss the solution of the electrooptic problem both from the viewpoint of Fourier components and phase matching as given in Section 13.5 and from a more conventional phase modulation viewpoint. We conclude the chapter with the development of a variational technique for calculating the electrooptic effect on optical waveguide modes.

14.1 Direct Electrooptic Susceptibility

The direct linear electrooptic effect is produced by the last term of the nonlinear polarization of (12.4.3)

$$\mathcal{P}_i(\mathbf{z};1,1)=2\epsilon_0 b_{i\,j\,k}^{\omega_S\omega_1\omega_2}E_j(\mathbf{z};1,0)E_k(\mathbf{z};0,1) \tag{14.1.1}$$

where $b_{i\,j\,k}^{\omega_S\omega_1\omega_2}$ is defined by (12.4.7) and referred to as the optical mixing susceptibility for arbitrary input frequencies ω_1 and ω_2. The direct effect exists for all frequencies ω_2 of the perturbing field $\mathbf{E}(\mathbf{z};0,1)$ and is the entire electrooptic effect for frequencies well above the piezoelectric resonances of the crystal. This is because no significant displacement gradient can be generated at the frequency ω_2 by the applied electric field and so the elastooptic terms in (12.4.3) drop out, leaving only the term in (14.1.1).

The definition of $b_{i\,j\,k}^{\omega_S\omega_1\omega_2}$ gives an interchange symmetry, called *permutation symmetry* [Armstrong et al., 1962],

$$b_{i\,j\,k}^{\omega_S\omega_1\omega_2}=b_{j\,k\,i}^{\omega_1\omega_2\omega_S}=b_{k\,i\,j}^{\omega_2\omega_S\omega_1}=b_{i\,k\,j}^{\omega_S\omega_2\omega_1}=b_{k\,j\,i}^{\omega_2\omega_1\omega_S}=b_{j\,i\,k}^{\omega_1\omega_S\omega_2}, \tag{14.1.2}$$

in which any pair of tensor indices and their associated frequencies may be interchanged. For the electrooptic effect it is assumed that one frequency, ω_2, is much smaller ($\omega_2/2\pi \lesssim 10^9$Hz) than the other two ($\omega_S/2\pi \approx \omega_1/2\pi \sim 3\times 10^{14}$ Hz). In the usual case where the dispersion in $b_{i\,j\,k}^{\omega_S\omega_1\omega_2}$ is small between ω_S and ω_1 we thus have the further interchange symmetry

$$b_{ijk}=b_{(ij)k} \tag{14.1.3}$$

where the frequency notation may be omitted provided that we remember that the low frequency field is associated with the third index. Since $\omega_1 \neq \omega_2$

for the electrooptic effect, the factor of 2 in (14.1.1) always is present. Thus we define the susceptibility for the direct electrooptic effect as that appearing in (14.1.1) when the frequencies are such that the symmetry expressed in (14.1.3) is a good approximation.

A linear electrooptic tensor for the direct effect may also be defined as the coefficient of the perturbation of the inverse dielectric tensor that appears in the index ellipsoid (13.2.6). Thus we let

$$(\Delta\kappa^{-1})_{ij} = r^S_{ijk} E_k \tag{14.1.4}$$

where $\mathbf{E}$ is a static electric field. We call r^S_{ijk} the *Pockels strain-free* (or *direct*) *electrooptic tensor* even though this definition differs from that of Pockels. Pockels expanded $\boldsymbol{\Delta\kappa}^{-1}$ in terms of the linear polarization. The electric field is more accessible as an experimental variable and so is now preferred as the expansion variable in (14.1.4). The Pockels tensor r^S_{ijk} is also called the *clamped* or *high frequency electrooptic tensor* to indicate that the indirect electrooptic effect does not contribute through the strain as an intermediary.

Equation (13.2.9) along with the expression (13.1.3) for the nonlinear polarization induced by a static electric field leads to

$$r^S_{ijk} = -4(\kappa^{-1})_{im}(\kappa^{-1})_{jn} b^{\omega_1\omega_1 0}_{m\ n\ k} \tag{14.1.5}$$

or

$$b^{\omega_1\omega_1 0}_{i\ j\ k} = -\tfrac{1}{4}\kappa_{im}\kappa_{jn} r^S_{mnk}. \tag{14.1.6}$$

When the perturbing electric field is at a finite frequency, this relation between the $\mathbf{r}^S$ and $\mathbf{b}$ tensors is still used, that is, the extra factor of $\frac{1}{2}$ is retained so that $\mathbf{r}^S$ is defined as if the perturbing field were static. The Pockels electrooptic tensor clearly has the same interchange symmetry,

$$r^S_{ijk} = r^S_{(ij)k}, \tag{14.1.7}$$

as the electrooptic susceptibility in (14.1.3). This allows $r^S_{(ij)k}$ to be contracted in the usual way (10.5.7) to r^S_{mk}. Since r^S_{ijk} has the same interchange symmetry as the piezoelectric tensors (10.2.2) and (10.5.4) the electrooptic effect is restricted to piezoelectric crystals. The components allowed for a given crystal class can be found from the tables in the Appendix.

14.2 Indirect Electrooptic Susceptibility

When the frequency of the perturbing electric field is below the frequencies of mechanical resonance, a full piezoelectric response of the strain to the electric field is possible and a full indirect contribution to the linear electrooptic effect exists. If no mechanical stress is imposed on the crystal by surface tractions, then the strain induced by the electric field is

$$u_{(m,n)} = d_{kmn} E_k \tag{14.2.1}$$

as found in (11.5.4). If there are no constraints on the crystal surfaces, then no rotation,

$$u_{[m,n]} = 0, \tag{14.2.2}$$

can be induced by the electric field. The nonlinear polarization (12.4.3) in this case becomes

$$\mathcal{P}_i(\mathbf{z}; 1,1) = \epsilon_0 \left[2b_{i\ j\ k}^{\omega_S \omega_1 \omega_2} + \chi_{i\ j\ (mn)}^{\omega_S \omega_1 \omega_2} d_{kmn} \right] E_j(\mathbf{z}; 1,0) E_k(\mathbf{z}; 0,1) \tag{14.2.3}$$

when $\omega_2 \neq 0$ and from (13.1.3) becomes

$$\mathcal{P}_i(\mathbf{z}; 1,0) = 2\epsilon_0 \left[2b_{i\ j\ k}^{\omega_1 \omega_1 0} + \chi_{i\ j\ (mn)}^{\omega_1 \omega_1 0} d_{kmn} \right] E_j(\mathbf{z}; 1,0) E_k(\mathbf{z}; 0,0) \tag{14.2.4}$$

when $\omega_2 = 0$. The second term in each equation arises from the *indirect electrooptic effect* and is seen to arise from a succession of the piezoelectric and elastooptic effects.

In this low frequency regime below the mechanical resonances of the crystal the electrooptic effect may be characterized by the *Pockels stress-free electrooptic tensor* r_{ijk}^T according to

$$(\Delta \kappa^{-1})_{ij} = r_{ijk}^T E_k \tag{14.2.5}$$

with

$$\begin{aligned} r_{ijk}^T &\equiv -2(\kappa^{-1})_{ip}(\kappa^{-1})_{jq} \left[2b_{p\ q\ k}^{\omega_1 \omega_1 0} + \chi_{p\ q\ (mn)}^{\omega_1 \omega_1 0} d_{kmn} \right] \\ &\equiv r_{ijk}^S + p_{ijmn} d_{kmn} \end{aligned} \tag{14.2.6}$$

with the use of (14.1.5) and (13.2.10). As always, the Pockels tensor is defined in terms of a static perturbation. The stress-free tensor is also called the *low frequency* or *unclamped electrooptic tensor*.

Since the indirect electrooptic effect is represented in the second term of (14.2.6) simply by a product of two material tensors, it has the same crystal symmetry as the direct electrooptic tensor. Hence the strain-free and stress-free tensors have the same crystal symmetry. However, we show in Section 14.11 that at frequencies between these two regimes, that is, at or near the piezoelectric resonant frequencies the effective electrooptic coefficient does not possess the crystal symmetry of the infinite crystal.

14.3 Plane Wave Solution for Homogeneous Perturbations

In many uses of the electrooptic effect the light wave may be adequately represented as a plane wave of a single frequency and the electrooptic perturbation of the dielectric tensor may be taken as spatially homogeneous. The latter is true because the wavelength of the perturbing electric field is typically large compared to the crystal dimension across which it is applied. For instance, if the perturbing electric field is applied across a 1 mm thick crystal, it is approximately homogeneous for frequencies as high as several gigahertz. Similar remarks can be made about elastooptic perturbations except there the maximum frequency is reduced by about 10^5, since the sound velocity is about 10^5 times smaller than the electromagnetic wave velocity.

Since the frequency of the perturbation of the dielectric tensor is minute compared to the frequency of the light wave (regardless of whether it arises from the electrooptic or elastooptic effects), it may be regarded as an adiabatic variation of the dielectric tensor and ignored in the wave equation. Thus the wave equation for the optical electric field at frequency ω_O is

$$\left(\frac{c}{\omega_O}\right)^2 \nabla\times[\nabla\times\mathbf{E}(\mathbf{z};1,0)]-\boldsymbol{\kappa}\cdot\mathbf{E}(\mathbf{z};1,0)=\frac{\mathcal{P}(\mathbf{z};1,0)}{\epsilon_0}$$
$$=\boldsymbol{\Delta\kappa}\cdot\mathbf{E}(\mathbf{z};1,0). \tag{14.3.1}$$

Taking the electric field to be a plane wave eigenmode of the perturbed crystal,

$$\mathbf{E}(\mathbf{z};1,0)=\boldsymbol{\mathcal{E}}^{\alpha}e^{ik_O\mathbf{s}\cdot\mathbf{z}}, \tag{14.3.2}$$

gives

$$(n^{\alpha})^2(\mathbf{1}-\mathbf{ss})\cdot\boldsymbol{\mathcal{E}}^{\alpha}=(\boldsymbol{\kappa}+\boldsymbol{\Delta\kappa})\cdot\boldsymbol{\mathcal{E}}^{\alpha} \tag{14.3.3}$$

for the perturbed wave equation with

$$n^{\alpha}\equiv\frac{k_O c}{\omega_O}. \tag{14.3.4}$$

One approach to determining the effect of $\boldsymbol{\Delta\kappa}$ on the light beam is to diagonalize $\boldsymbol{\kappa}+\boldsymbol{\Delta\kappa}$, finding its new principal axes and principal values. The eigenvectors $\mathcal{E}^{\alpha}$ are referred to the new axes. The perturbed values of the refractive indices can then be found from any of the several forms of the dispersion relation found in Section 9.4. A more common approach is to work from the index ellipsoid (13.2.6) as done by Pockels. Here $\boldsymbol{\kappa}^{-1}+\boldsymbol{\Delta\kappa}^{-1}$ must be diagonalized and its principal axes and principal values determined. The perturbed refractive indices referred to the new axes are then found.

Diagonalization of the perturbed dielectric tensor (or its inverse) involves the solution of a cubic equation in general since the dielectric tensor is a 3×3 array. However, both the electrooptic effect and the elastooptic effect are by definition effects linear in the applied field (electric field or strain) and are typically small changes of the dielectric tensor. Hence only the first order terms are relevant. Perturbation theory thus comes to mind as a way of solving the problem quite generally and simply. It could be applied to the eigenvector equation for the principal directions of the dielectric tensor (or its inverse). Such a procedure, however, leads to the uniaxial crystal case being doubly degenerate and the isotropic crystal case being triply degenerate.

These complications can be avoided if perturbation theory is applied to the wave equation. The solution will apply to any effect that perturbs the dielectric tensor slowly. The procedure to be used here has the advantage of making the uniaxial crystal case nondegenerate and the isotropic crystal case only doubly degenerate. It also has the advantage of giving the perturbed refractive indices without the need in practice of explicitly calculating perturbed principal directions of the dielectric tensor or perturbed eigenvectors of the electric or electric displacement fields.

We choose to work from the electric displacement eigenvector equation rather than from the electric field eigenvector equation (14.3.3). Define the perturbed electric displacement eigenvector by

$$\mathfrak{D}^{\alpha}\equiv(\boldsymbol{\kappa}+\boldsymbol{\Delta\kappa})\cdot\mathcal{E}^{\alpha}. \tag{14.3.5}$$

Therefore

$$\mathcal{E}^{\alpha}=(\boldsymbol{\kappa}+\boldsymbol{\Delta\kappa})^{-1}\cdot\mathfrak{D}^{\alpha}. \tag{14.3.6}$$

To first order

$$\begin{aligned}(\boldsymbol{\kappa}+\boldsymbol{\Delta\kappa})^{-1}&=\boldsymbol{\kappa}^{-1}-\boldsymbol{\kappa}^{-1}\cdot\boldsymbol{\Delta\kappa}\cdot\boldsymbol{\kappa}^{-1}\\&\equiv\boldsymbol{\kappa}^{-1}+\boldsymbol{\Delta\kappa}^{-1}.\end{aligned} \tag{14.3.7}$$

A biorthogonality condition for the perturbed eigenvectors,

$$\mathfrak{E}^{\alpha}\cdot\mathfrak{D}^{\beta}=\delta^{\alpha\beta}, \tag{14.3.8}$$

may be derived just as was done for (9.3.9). Equation (14.3.3) now becomes

$$(\mathbf{1}-\mathrm{ss})\cdot(\boldsymbol{\kappa}^{-1}+\Delta\boldsymbol{\kappa}^{-1})\cdot\mathfrak{D}^{\alpha}=\frac{1}{(n^{\alpha})^{2}}\mathfrak{D}^{\alpha}. \tag{14.3.9}$$

We now proceed to solve this equation by perturbation techniques.

14.4 Nondegenerate Perturbation Theory of the Wave Equation

We wish to apply first order perturbation theory to the eigenvector equation (14.3.9). First, we consider the nondegenerate case, that is, all unperturbed values of n^{α} different. The ordinary perturbation formulas must be altered in two respects. We deal here with a biorthogonal set of eigenvectors (see Section 9.3) for one thing. Also, the first order term in the definition (14.3.5) leads to extra terms in the first order perturbed eigenvectors $\mathfrak{E}_1^{\alpha}$. Denote

$$\boldsymbol{\Gamma}\equiv(\mathbf{1}-\mathrm{ss})\cdot\boldsymbol{\kappa}^{-1}, \tag{14.4.1}$$

$$\Delta\boldsymbol{\Gamma}\equiv(\mathbf{1}-\mathrm{ss})\cdot\Delta\boldsymbol{\kappa}^{-1} \tag{14.4.2}$$

and represent the perturbed operator on the left side of (14.3.9) by $\boldsymbol{\Gamma}+\lambda\Delta\boldsymbol{\Gamma}$, λ being an indicator of smallness. Also denote

$$G^{\alpha}\equiv\left(\frac{1}{n^{\alpha}}\right)^{2}. \tag{14.4.3}$$

$\mathfrak{D}^{\alpha}$, $\mathfrak{E}^{\alpha}$, and G^{α} are now expanded in power series in the smallness indicator such as

$$\mathfrak{D}^{\alpha}=\sum_{n=0}^{\infty}\lambda^{n}\mathfrak{D}_{n}^{\alpha}. \tag{14.4.4}$$

These give the eigenvector equation the form

$$\sum_{n}\lambda^{n}(\boldsymbol{\Gamma}+\lambda\Delta\boldsymbol{\Gamma})\cdot\mathfrak{D}_{n}^{\alpha}=\sum_{n,m}\lambda^{n+m}G_{m}^{\alpha}\mathfrak{D}_{n}^{\alpha}. \tag{14.4.5}$$

The expansions in λ such as (14.4.4) assume that the perturbed eigenvectors $\mathfrak{D}^\alpha$ approach the unperturbed eigenvector $\mathfrak{D}_0^\alpha$ as $\lambda \to 0$. This is always true when the eigenvalues are nondegenerate as considered in this section.

The coefficients of each power of λ in (14.4.5) may be equated separately. The zeroth order terms that are independent of the perturbation give

$$\mathbf{\Gamma}\cdot\mathfrak{D}_0^\alpha = G_0^\alpha \mathfrak{D}_0^\alpha. \tag{14.4.6}$$

From Eq. (14.3.6) we have

$$\mathfrak{E}_0^\alpha = \boldsymbol{\kappa}^{-1}\cdot\mathfrak{D}_0^\alpha. \tag{14.4.7}$$

We also have the biorthogonality condition (9.3.9) for the unperturbed eigenvectors

$$\mathfrak{E}_0^\alpha\cdot\mathfrak{D}_0^\beta = \delta^{\alpha\beta}. \tag{14.4.8}$$

The terms first order in λ in (14.4.5) yield

$$\mathbf{\Gamma}\cdot\mathfrak{D}_1^\alpha + \Delta\mathbf{\Gamma}\cdot\mathfrak{D}_0^\alpha = G_0^\alpha \mathfrak{D}_1^\alpha + G_1^\alpha \mathfrak{D}_0^\alpha. \tag{14.4.9}$$

The first order eigenvector correction can be expanded in terms of the unperturbed eigenvectors by

$$\mathfrak{D}_1^\alpha = \sum_{\gamma=1}^{3} a_{\gamma\alpha}\mathfrak{D}_0^\gamma, \tag{14.4.10}$$

giving (14.4.9) the form

$$\Delta\mathbf{\Gamma}\cdot\mathfrak{D}_0^\alpha = \sum_\gamma a_{\gamma\alpha}(G_0^\alpha - G_0^\gamma)\mathfrak{D}_0^\gamma + G_1^\alpha\mathfrak{D}_0^\alpha \tag{14.4.11}$$

with the use of (14.4.6). The scalar product of this equation with $\mathfrak{E}_0^\beta$ yields

$$\mathfrak{E}_0^\beta\cdot\Delta\mathbf{\Gamma}\cdot\mathfrak{D}_0^\alpha = a_{\beta\alpha}\left(G_0^\alpha - G_0^\beta\right) + G_1^\alpha\delta^{\alpha\beta} \tag{14.4.12}$$

with the use of (14.4.8). If $\alpha=\beta$, this gives the first order correction to the eigenvalues

$$G_1^\alpha = \mathfrak{E}_0^\alpha\cdot\Delta\mathbf{\Gamma}\cdot\mathfrak{D}_0^\alpha = \mathfrak{E}_0^\alpha\cdot(\mathbf{1}-\mathbf{ss})\cdot\Delta\boldsymbol{\kappa}^{-1}\cdot\mathfrak{D}_0^\alpha = G_0^\alpha\mathfrak{D}_0^\alpha\cdot\Delta\boldsymbol{\kappa}^{-1}\cdot\mathfrak{D}_0^\alpha \tag{14.4.13}$$

with the use of the definition of $\Delta\mathbf{\Gamma}$ (14.4.2) and the unperturbed eigenvector equation (9.3.1). If $\alpha\neq\beta$, it gives an expression for the expansion coefficients $a_{\alpha\beta}$ of the perturbed eigenvectors. The biorthogonality condition can be used to obtain the remaining $a_{\alpha\alpha}$ coefficients. Since we have no need for the perturbed eigenvectors, we do not need to obtain these coefficients.

The first order perturbed eigenvalues may now be written as

$$\frac{1}{(n^\alpha)^2} = \frac{1}{(n_0^\alpha)^2}\left(1 + \mathfrak{D}_0^\alpha \cdot \Delta\boldsymbol{\kappa}^{-1} \cdot \mathfrak{D}_0^\alpha\right). \tag{14.4.14}$$

Since $n_0^\alpha = \infty$ for the longitudinal nonpropagating mode in any crystal, we see from this equation that the perturbation of the eigenvalue of this mode is zero. Thus $n^\alpha = n_0^\alpha = \infty$, that is, the nonpropagating mode remains nonpropagating under the perturbation. If for the propagating modes we set

$$n^\alpha = n_0^\alpha + \Delta n^\alpha, \tag{14.4.15}$$

then (14.4.14) can be expanded to first order with the use of the unit electric displacement vectors (9.3.24) to give

$$\begin{aligned}\Delta n^\alpha &= -\frac{n_0^\alpha}{2} \mathfrak{D}_0^\alpha \cdot \Delta\boldsymbol{\kappa}^{-1} \cdot \mathfrak{D}_0^\alpha \\ &= -\frac{(n_0^\alpha)^3}{2} \mathbf{d}_0^\alpha \cdot \Delta\boldsymbol{\kappa}^{-1} \cdot \mathbf{d}_0^\alpha \qquad (\alpha = 1, 2)\end{aligned} \tag{14.4.16}$$

for the change in the refractive index for either propagating mode whenever those modes are nondegenerate.

14.5 Degenerate Perturbation Theory of the Wave Equation

When the eigenvalues of the two propagating eigenvectors are degenerate, the derivation of the preceding section needs modification. In that derivation the expansion of the perturbed eigenvectors in the smallness indicator λ assumes that the limit of a perturbed eigenvector as $\lambda \to 0$ is the unperturbed eigenvector used. For nondegenerate modes this is always true. But when the two propagating modes are degenerate, there is an infinite number of linear combinations of eigenvectors of these modes that can equally well serve as unperturbed eigenvectors. However, for a given perturbation that lifts the degeneracy only one certain combination of the unperturbed eigenvectors will join on continuously in λ to each perturbed eigenvector. This particular combination must be found in order for perturbation theory to be applied to a degenerate set of eigenvectors.

Consider the two unperturbed eigenvectors $\mathfrak{D}_0^\alpha(\alpha=1,2)$ which have the same eigenvalue G_0^α. The linear combinations that join on continuously with the perturbed eigenvectors are denoted by $\mathfrak{D}_0^a$ where a Latin letter is used for the mode index. They may be expressed as

$$\mathfrak{D}_0^a \equiv \sum_{\gamma=1}^{2} S_{\gamma a}\mathfrak{D}_0^\gamma \qquad (a=1,2). \tag{14.5.1}$$

The first order correction to the eigenvector, $\mathfrak{D}_1^a$, is expanded in terms of the new eigenvectors,

$$\mathfrak{D}_1^a = \sum_{b=1}^{2} a_{ba}\mathfrak{D}_0^b. \tag{14.5.2}$$

This equation is now substituted into the equation of first order terms (14.4.9) for a mode to give

$$\boldsymbol{\Gamma}\cdot\sum_b a_{ba}\mathfrak{D}_0^b + \boldsymbol{\Delta\Gamma}\cdot\mathfrak{D}_0^a = G_0^a\sum_b a_{ba}\mathfrak{D}_0^b + G_1^a\mathfrak{D}_0^a. \tag{14.5.3}$$

Taking a scalar product with $\mathfrak{E}_0^\beta$ and using

$$\boldsymbol{\Gamma}\cdot\mathfrak{D}_0^a = G_0^a\mathfrak{D}_0^a, \tag{14.5.4}$$

$$\mathfrak{E}_0^\beta\cdot\mathfrak{D}_0^a = S_{\beta a} \tag{14.5.5}$$

yields

$$\sum_b a_{ba}S_{\beta b}G_0^b + \mathfrak{E}_0^\beta\cdot\boldsymbol{\Delta\Gamma}\cdot\mathfrak{D}_0^a = G_0^a\sum_b a_{ba}S_{\beta b} + G_1^aS_{\beta a}. \tag{14.5.6}$$

Since $G_0^a = G_0^b$ for all a and b considered in this equation, the first terms of the two sides cancel, leaving after reexpression of the matrix element in the manner done in (14.4.13)

$$\sum_{\gamma=1}^{2} S_{\gamma a}\left[G_0^\beta\mathfrak{D}_0^\beta\cdot\boldsymbol{\Delta\kappa}^{-1}\cdot\mathfrak{D}_0^\gamma - G_1^a\delta^{\beta\gamma}\right]=0 \tag{14.5.7}$$

to determine $S_{\gamma a}$ and G_1^a. A nontrivial solution exists provided that

$$\det\left[G_0^\beta\mathfrak{D}_0^\beta\cdot\boldsymbol{\Delta\kappa}^{-1}\cdot\mathfrak{D}_0^\gamma - G_1^a\delta^{\beta\gamma}\right]=0, \tag{14.5.8}$$

a condition that determines the two values of G_1^a $(a=1,2)$. For each G_1^a value (14.5.7) yields all but one $S_{\gamma a}$ coefficient. The last one is determined by requiring the normalization of the new unperturbed eigenvectors,

$$\mathfrak{D}_0^a \cdot \mathfrak{E}_0^a = 1. \tag{14.5.9}$$

With the new unperturbed eigenvectors the nondegenerate perturbation theory of Section 14.4 may be applied. In particular the expression for G_1^α (14.4.13) may be used to obtain the first order correction to the eigenvalues. When α corresponds to one of the two new unperturbed eigenvectors, that equation yields the solutions found from (14.5.8). Since we have no use for the first order perturbed eigenvectors, we do not obtain them here.

The method just presented is a straightforward procedure for determining the perturbed eigenvalues and the appropriate unperturbed eigenvectors. However, it is sometimes more lengthy than an alternate and equivalent procedure which is suggested by (14.5.7) and (14.5.8). Those equations are used to obtain the particular unperturbed eigenvectors that diagonalize the perturbation, namely, the eigenvectors for which

$$\mathfrak{D}^a \cdot \boldsymbol{\Delta\kappa}^{-1} \cdot \mathfrak{D}^b = 0 \qquad (a \neq b). \tag{14.5.10}$$

It is apparent from those equations that the eigenvectors which satisfy this last condition can be used to obtain the eigenvalue perturbations,

$$G_1^a = G_0^a \mathfrak{D}^a \cdot \boldsymbol{\Delta\kappa}^{-1} \cdot \mathfrak{D}^a \qquad (a=1,2), \tag{14.5.11}$$

and thus the refractive index changes,

$$\begin{aligned} \Delta n^a &= -\frac{n_0}{2} \mathfrak{D}_0^a \cdot \boldsymbol{\Delta\kappa}^{-1} \cdot \mathfrak{D}_0^a \\ &= -\frac{n_0^3}{2} \mathbf{d}_0^a \cdot \boldsymbol{\Delta\kappa}^{-1} \cdot \mathbf{d}_0^a \qquad (a=1,2). \end{aligned} \tag{14.5.12}$$

This method is convenient when the set of all degenerate eigenvectors can be readily parameterized; then the condition (14.5.10) is used to fix the value of the parameter.

14.6 Application to Optically Isotropic Crystals

Cubic crystals, being optically isotropic or anaxial for linear propagation, have the same refractive index for the two polarizations of light for any direction of propagation. Hence they must be treated by the degenerate perturbation theory of the preceding section.

The unperturbed eigenvectors of the two propagating modes were found in Section 9.5 to be

$$\mathcal{E}_0^1 = \frac{\mathbf{e}}{n_0}, \qquad \mathcal{D}_0^1 = n_0 \mathbf{e} \tag{14.6.1}$$

and

$$\mathcal{E}_0^2 = \frac{\mathbf{s}\times\mathbf{e}}{n_0}, \qquad \mathcal{D}_0^2 = n_0 \mathbf{s}\times\mathbf{e} \tag{14.6.2}$$

where $\mathbf{e}$ is a unit vector normal to the direction of propagation $\mathbf{s}$,

$$\mathbf{e}\cdot\mathbf{s} = 0. \tag{14.6.3}$$

The unperturbed refractive index of each of these modes is

$$n_0 = \kappa^{1/2}. \tag{14.6.4}$$

Since $\mathbf{e}$ is a unit vector whose direction is only partially determined by (14.6.3), it may be parameterized by an azimuthal angle about $\mathbf{s}$ and (14.5.10) of the second procedure of the preceding section used to determine the angle relative to the particular perturbation $\boldsymbol{\Delta\kappa}^{-1}$ considered. Thus the particular $\mathbf{e}$ vector determined in this manner may be denoted by $\boldsymbol{\epsilon}$ and must satisfy

$$\boldsymbol{\epsilon}\cdot\boldsymbol{\Delta\kappa}^{-1}\cdot(\mathbf{s}\times\boldsymbol{\epsilon}) = 0. \tag{14.6.5}$$

The refractive index changes associated with the eigenvectors of (14.6.1) and (14.6.2) having $\mathbf{e} = \boldsymbol{\epsilon}$ are

$$\Delta n^{(1)} = -\frac{n_0^3}{2}\boldsymbol{\epsilon}\cdot\boldsymbol{\Delta\kappa}^{-1}\cdot\boldsymbol{\epsilon}, \tag{14.6.6a}$$

$$\Delta n^{(2)} = -\frac{n_0^3}{2}(\mathbf{s}\times\boldsymbol{\epsilon})\cdot\boldsymbol{\Delta\kappa}^{-1}\cdot(\mathbf{s}\times\boldsymbol{\epsilon}). \tag{14.6.6b}$$

The result (14.6.6) applies to either orientation of polarization, to any direction of propagation, and to an arbitrary dielectric perturbation such as given in (14.1.4) and (14.2.5) for the high and low frequency electrooptic effects and in (13.7.11) for the elastooptic effect.

14.7 Application to Uniaxial Crystals

Next we obtain the change in refractive index induced by the perturbation $\boldsymbol{\Delta\kappa}^{-1}$ for the two propagating modes in a uniaxial crystal. For all propagation directions except along the optic axis the ordinary ray and extraordinary ray have nondegenerate refractive indices. For these directions we may apply the nondegenerate perturbation theory. We also treat propagation along the optic axis by degenerate perturbation theory in this section.

Consider the ordinary ray first. The unperturbed refractive index is

$$n_0^o = \kappa_{11}^{1/2} \tag{14.7.1}$$

and the unperturbed eigenvectors of the ordinary ray found in Section 9.6 are

$$\mathcal{E}^o = \frac{\mathbf{s}\times\mathbf{c}}{N^o}, \tag{14.7.2}$$

$$\mathcal{D}^o = (n_0^o)^2 \frac{\mathbf{s}\times\mathbf{c}}{N^o} \tag{14.7.3}$$

with

$$N^o \equiv n_0^o\left[1-(\mathbf{s}\cdot\mathbf{c})^2\right]^{1/2}. \tag{14.7.4}$$

Substitution of these quantities into (14.4.16) yields

$$\Delta n^o = -\frac{(n_0^o)^3}{2}\left[\frac{s_2^2(\Delta\kappa^{-1})_{11}+s_1^2(\Delta\kappa^{-1})_{22}-2s_1s_2(\Delta\kappa^{-1})_{12}}{s_1^2+s_2^2}\right]. \tag{14.7.5}$$

If the unit propagation vector $\mathbf{s}$ is expressed in terms of the spherical angles θ, φ by

$$\mathbf{s} = \left[\sin\theta\cos\varphi, \sin\theta\sin\varphi, \cos\theta\right], \tag{14.7.6}$$

then an alternate form of the result,

$$\Delta n^o = -\frac{(n_0^o)^3}{2}\left[(\Delta\kappa^{-1})_{11}\sin^2\varphi+(\Delta\kappa^{-1})_{22}\cos^2\varphi-(\Delta\kappa^{-1})_{12}\sin 2\varphi\right], \tag{14.7.7}$$

is obtained. This is the general result for the ordinary ray in a uniaxial crystal.

Consider next the extraordinary ray. The unperturbed refractive index is given by

$$\frac{1}{\left[n_0^e(\mathbf{s})\right]^2}=\frac{s_3^2}{\kappa_{11}}+\frac{s_1^2+s_2^2}{\kappa_{33}} \tag{14.7.8}$$

in terms of the unit propagation vector $\mathbf{s}$, or

$$n_0^e(\theta)=\left[\frac{\kappa_{11}\kappa_{33}}{\kappa_{11}\sin^2\theta+\kappa_{33}\cos^2\theta}\right]^{1/2} \tag{14.7.9}$$

in terms of the spherical angles. The unperturbed eigenvectors are

$$\mathcal{E}_0^e=\frac{\boldsymbol{\kappa}^{-1}\cdot[\mathbf{s}\times(\mathbf{s}\times\mathbf{c})]}{N^e}, \tag{14.7.10}$$

$$\mathfrak{D}_0^e=\frac{\mathbf{s}\times(\mathbf{s}\times\mathbf{c})}{N^e} \tag{14.7.11}$$

with

$$N^e\equiv\frac{1}{n_0^e(\mathbf{s})}\left[1-(\mathbf{s}\cdot\mathbf{c})^2\right]^{1/2}. \tag{14.7.12}$$

Substitution into (14.4.16) leads to

$$\begin{aligned}\Delta n^e = & -\frac{\left[n_0^e(\mathbf{s})\right]^3}{2}\frac{1}{s_1^2+s_2^2}\Big[s_1^2s_3^2(\Delta\kappa^{-1})_{11}+s_2^2s_3^2(\Delta\kappa^{-1})_{22}+\left(s_1^2+s_2^2\right)^2(\Delta\kappa^{-1})_{33}\\ & +2s_1s_2s_3^2(\Delta\kappa^{-1})_{12}-2s_2s_3\left(s_1^2+s_2^2\right)(\Delta\kappa^{-1})_{23}\\ & -2s_1s_3\left(s_1^2+s_2^2\right)(\Delta\kappa^{-1})_{13}\Big].\end{aligned} \tag{14.7.13}$$

If $\mathbf{s}$ is expressed in terms of the spherical angles, then this becomes

$$\begin{aligned}\Delta n^e = & -\frac{\left[n_0^e(\theta)\right]^3}{2}\Big[(\Delta\kappa^{-1})_{11}\cos^2\theta\cos^2\varphi+(\Delta\kappa^{-1})_{22}\cos^2\theta\sin^2\varphi\\ & +(\Delta\kappa^{-1})_{33}\sin^2\theta+(\Delta\kappa^{-1})_{12}\cos^2\theta\sin 2\varphi\\ & -(\Delta\kappa^{-1})_{23}\sin 2\theta\sin\varphi-(\Delta\kappa^{-1})_{13}\sin 2\theta\cos\varphi\Big].\end{aligned} \tag{14.7.14}$$

This is the general result for the extraordinary ray in a uniaxial crystal. Both this result and the one for the ordinary ray in (14.7.7) apply to any direction of propagation in a uniaxial crystal except along the optic axis where the two modes become degenerate. To handle this case we now apply degenerate perturbation theory.

First we need to find the combination of unperturbed eigenvectors that are the limits of the perturbed eigenvectors as the perturbation vanishes. The unperturbed electric displacement eigenvectors of (14.7.3) and (14.7.11) become

$$\mathfrak{D}_0^o = (n_0^o)^2 \mathcal{E}_0^o = n_0^o[\sin\varphi, -\cos\varphi, 0], \tag{14.7.15}$$

$$\mathfrak{D}_0^e = (n_0^o)^2 \mathcal{E}_0^e = n_0^o[\cos\varphi, \sin\varphi, 0] \tag{14.7.16}$$

for propagation along the optic axis ($\theta = 0$). Since φ is undetermined in this direction, we may use the reasoning of (14.5.10) to determine φ in terms of the perturbation. We thus require

$$\mathfrak{D}_0^o \cdot \boldsymbol{\Delta\kappa}^{-1} \cdot \mathfrak{D}_0^e = 0. \tag{14.7.17}$$

Substitution of the previous two equations into this gives

$$(\Delta\kappa^{-1})_{11}\sin\varphi\cos\varphi + (\Delta\kappa^{-1})_{12}(\sin^2\varphi - \cos^2\varphi) \\ -(\Delta\kappa^{-1})_{22}\sin\varphi\cos\varphi = 0. \tag{14.7.18}$$

Solution of this for φ,

$$\tan 2\varphi = \frac{2(\Delta\kappa^{-1})_{12}}{(\Delta\kappa^{-1})_{11} - (\Delta\kappa^{-1})_{22}}, \tag{14.7.19}$$

determines the appropriate unperturbed eigenvectors.

Because the refractive index change formulas (14.7.7) and (14.7.14) apply to any combination of θ, φ, they must apply to $\theta = 0$ and φ determined by (14.7.19). Those formulas thus give

$$\Delta n\binom{e}{o} = \tfrac{1}{4}(n_0^o)^3\Big\{(\Delta\kappa^{-1})_{11} + (\Delta\kappa^{-1})_{22} \\ \pm\Big[\big((\Delta\kappa^{-1})_{11} - (\Delta\kappa^{-1})_{22}\big)^2 + 4\big((\Delta\kappa^{-1})_{12}\big)^2\Big]^{1/2}\Big\}, \tag{14.7.20}$$

where the + and e and the − and o are associated, for the refractive index changes when the light propagates along the optic axis of a uniaxial crystal. Alternatively, this result follows from (14.5.12).

14.8 Application to Biaxial Crystals

Next we find the change in refractive index induced by the perturbation $\Delta\boldsymbol{\kappa}^{-1}$ for the two propagating modes in a biaxial crystal. We denote these two modes by $\alpha = +,-$ associated with the sign occurring in the unperturbed refractive index,

$$n_0^{\pm} = \left[\frac{\kappa_{11}\kappa_{33}}{\kappa_{11}\sin^2\frac{1}{2}(\theta_1 \pm \theta_2) + \kappa_{33}\cos^2\frac{1}{2}(\theta_1 \pm \theta_2)} \right]^{1/2}, \tag{14.8.1}$$

obtained from (9.7.13). The unperturbed eigenvectors of these modes expressed in the principal coordinate system are

$$\mathcal{E}_{0i}^{\alpha} = \frac{s_i}{\left[(n_0^{\alpha})^2 - \kappa_{ii} \right] N^{\alpha}} \qquad (\alpha = +,-), \tag{14.8.2}$$

$$\mathfrak{D}_{0i}^{\alpha} = \frac{\kappa_{ii} s_i}{\left[(n_0^{\alpha})^2 - \kappa_{ii} \right] N^{\alpha}} \qquad (\alpha = +,-) \tag{14.8.3}$$

with

$$N^{\alpha} \equiv \left[\sum_{i=1}^{3} \frac{s_i^2 \kappa_{ii}}{\left[(n_0^{\alpha})^2 - \kappa_{ii} \right]^2} \right]^{1/2}, \tag{14.8.4}$$

as found in Section 9.7.

Straightforward substitution of the eigenvectors into (14.4.16) leads to

$$\Delta n^{\pm} = -\frac{n_0^{\pm}}{2(N^{\pm})^2} \sum_{i,j=1}^{3} \frac{s_i \kappa_{ii} (\Delta\kappa^{-1})_{ij} s_j \kappa_{jj}}{\left[(n_0^{\pm})^2 - \kappa_{ii} \right]\left[(n_0^{\pm})^2 - \kappa_{jj} \right]}. \tag{14.8.5}$$

This represents a general solution for the refractive index change in any biaxial crystal for a light wave having either ($\pm$) polarization and an arbitrary direction of propagation $\mathbf{s}$. The refractive index change is produced by any change in the inverse dielectric tensor such as given in (13.7.11) for the elastooptic effect or in (14.1.4) and (14.2.5) for the high and low frequency electrooptic effects.

Formulas for the index change could be obtained for propagation along either of the optic axes using degenerate perturbation theory. It does not seem that the likelihood of their use warrants recording the expressions here.

14.9 Measurement of Electrooptic Frequency Shifts

When the perturbing electric field is applied at an angular frequency ω_2, the input light wave of frequency ω_1 acquires frequency sidebands at $\omega_1 \pm \omega_2$. Such sidebands can be detected by mixing the output beam on a photomultiplier tube with another coherent beam having a nearby frequency ω_M. Since coherent electric fields must be added before squaring to produce the detected light intensity, difference frequencies of $|\omega_M - \omega_1 \pm \omega_2|$ and $|\omega_M - \omega_1|$ are created in the radio frequency region. If the bandwidth of the detection system is small compared to the separation of these difference frequencies, only one is detected at a time. If ω_M is then swept through this difference frequency range at a frequency still much less, measurements of the strength of the sideband components compared to the unmodulated component can be made.

We may obtain equivalent expressions for this sideband strength by two quite different procedures. The first is to adapt the expression of the output light intensity for acoustooptic diffraction developed in Section 13.5. In an electrooptic experiment the low frequency electric field typically has a frequency less than 1 GHz and is applied across electrodes on the crystal spaced by a centimeter or less. This field is thus substantially homogeneous in space. The nonlinear polarization then is a plane wave with a wavevector $\mathbf{k}_1$ of the input light wave and a frequency which is $\omega_1 + \omega_2$, the sum of the input light wave frequency and the perturbing field frequency. The output wavevector $\mathbf{k}_S$ must thus be set equal to $\mathbf{k}_1$ and the output frequency ω_S set equal to $\omega_1 + \omega_2$,

$$\mathbf{k}_S = \mathbf{k}_1, \tag{14.9.1}$$

$$\omega_S = \omega_1 + \omega_2. \tag{14.9.2}$$

This is equivalent to viewing the output as a phase matched output wave of the sum frequency field when $\mathbf{k}_2 \ll \mathbf{k}_1$ and thus negligible.

For these conditions the real part of the electric field of (13.5.40) with the time dependence included becomes

$$\mathbf{E}(z,t) = -\frac{n^{\xi}\omega_S z}{2\epsilon_0 c}\,\mathcal{E}^{\xi}\,\mathcal{E}^{\xi}\cdot\mathcal{P}(1,1)\sin\frac{\omega_S}{c}(n^{\xi}z - ct) \tag{14.9.3}$$

for propagation normal to the input surface ($\mathbf{n}\cdot\mathbf{t}_S^{\xi} = \cos\delta^{\xi}$). Here $\mathcal{P}(1,1)$ is given by either (14.1.1) or (14.2.3) depending on whether ω_2 is above the piezoelectric resonance region or below it.

The time-averaged Poynting vector of the sum frequency $(\omega_1 + \omega_2)$ sideband may be found from (13.5.46) by the insertion of the nonlinear polarization (14.1.1) (which assumes that ω_2 is well above the piezoelectric resonances), with the use of the strain-free electrooptic tensor in (14.1.5), and by taking the input and output optical modes the same ($\xi=\eta$). The result is

$$\langle S^{\xi}(1,1)\rangle = \frac{\omega_S^2 z^2 (n^{\xi})^6 (r^S)^2 |E^{\zeta}(0,1)|^2 \langle S^{\eta}(1,0)\rangle}{16c^2}. \tag{14.9.4}$$

Here the effective strain-free (high frequency) electrooptic coefficient r^S, which appears squared, is defined by

$$r^S \equiv d_m^{\xi} r_{mnk}^S d_n^{\eta} e_k^{\zeta} \tag{14.9.5}$$

where $\mathbf{d}$ is a unit electric displacement field, $\mathbf{e}$ is a unit electric field, and ξ, $\eta=\xi$, ζ refer to the output light wave, the input light wave, and the perturbing electric field. An analogous formula could be written for the difference frequency sideband.

An alternate treatment of the electrooptic effect regards the dielectric tensor perturbation $\boldsymbol{\Delta\kappa}$ as slowly varying (even at 1 GHz frequency) compared to the optical field. This time dependence is then ignored in solving the wave equation. This was done in finding the index change resulting from $\boldsymbol{\Delta\kappa}$ in Sections 14.6, 14.7, and 14.8 for cubic, uniaxial, and biaxial crystals. The time dependence of $\boldsymbol{\Delta\kappa}$ leads to a time dependence of Δn^{ξ} which then appears in the solution as

$$\mathbf{E}(z,t) = \mathcal{E}^{\xi} \sin\left(\frac{\omega_1 n_0^{\xi} z}{c} - \omega_1 t + \frac{\omega_1 \Delta n^{\xi} z}{c} \sin \omega_2 t \right). \tag{14.9.6}$$

In this form the electrooptic effect is seen to modulate the phase of the wave. By a mathematical identity this may be reexpressed as

$$\mathbf{E}(z,t) = \mathcal{E}^{\xi} \sum_{n=-\infty}^{\infty} J_n\left(\frac{\omega_1 \Delta n^{\xi} z}{c} \right) \sin\left(\frac{\omega_1 n_0^{\xi} z}{c} - \omega_1 t + n\omega_2 t \right) \tag{14.9.7}$$

where J_n is the Bessel function of order n. This expression shows that many sidebands, all separated by ω_2, occur when the perturbation, which is the argument of the Bessel function, is sufficiently large. When the argument x of the Bessel functions $J_n(x)$ is small, the only significant terms involve $J_0(x) \cong 1, J_1(x) = -J_{-1}(x) \cong x$. Thus we obtain

$$\mathbf{E}(z,t) = \mathcal{E}^{\xi} \left[\sin\left(\frac{\omega_1 n_0^{\xi} z}{c} - \omega_1 t \right) + \frac{\omega_1 \Delta n^{\xi} z}{c} \sin\left[\frac{\omega_1 n_0^{\xi} z}{c} - (\omega_1 - \omega_2) t \right] \right.$$
$$\left. - \frac{\omega_1 \Delta n^{\xi} z}{c} \sin\left[\frac{\omega_1 n_0^{\xi} z}{c} - (\omega_1 + \omega_2) t \right] \right]. \tag{14.9.8}$$

This gives us the unmodulated component and the two sidebands (the sum and difference frequencies). If the expression (14.4.16) is reexpressed as

$$\Delta n^{\xi} = -\frac{n_0^{\xi}}{2}\,\mathfrak{D}_0^{\xi}\cdot\Delta\kappa^{-1}\cdot\mathfrak{D}_0^{\xi}$$

$$= \frac{n_0^{\xi}}{2}\,\mathcal{E}_0^{\xi}\cdot\Delta\kappa\cdot\mathcal{E}_0^{\xi}$$

$$= \frac{n_0^{\xi}}{2\epsilon_0}\,\mathcal{E}_0^{\xi}\cdot\mathcal{P}, \tag{14.9.9}$$

the amplitude of the sum frequency term in (14.9.8) is seen to agree with that of (14.9.3) to within the $\omega_S \approx \omega_1$ approximation. Thus we see that the second approach agrees with the first approach which considers the phase matching of Fourier components. The second approach has the advantage of obtaining the strengths of all the sidebands at once.

14.10 Measurement of Electrooptic Phase Shifts

The electrooptic effect resulting from either a static or time varying electric field [see (14.9.6)] may be regarded as producing a phase shift of the light wave through a refractive index change. Thus the electrooptic effect may be measured by optical phase measurements. Such measurements are done relative to a second portion of the light wave which may pass through the crystal under study in the other eigenmode, may pass around the crystal under study, or be generated separately if both light waves are coherent. The former case is spoken of as a relative phase measurement and the latter two cases are spoken of as absolute phase measurements. We will consider briefly the relative phase measurement case. The treatment also applies to the measurement of phase shifts caused by the elastooptic effect.

A polarizer in front of the crystal under study allows a plane polarized light beam to impinge on the crystal with its electric field $\mathbf{E}_P$ at an angle η to one of the eigenmodes of the crystal (see Fig. 14.1). This field is projected onto the two eigenmodes of the crystal by the requirement of continuous tangential components of the electric field. We ignore the existence of a reflected wave, that is, we take the surface transmission coefficient to be unity. The electric fields of the propagating eigenmodes (denoted by + and −) are

$$E^- = E_P \cos\eta\, e^{i(\omega/c)(n^-\mathbf{s}\cdot\mathbf{z}-ct)}, \tag{14.10.1}$$

$$E^+ = E_P \sin\eta\, e^{i(\omega/c)(n^+\mathbf{s}\cdot\mathbf{z}-ct)}. \tag{14.10.2}$$

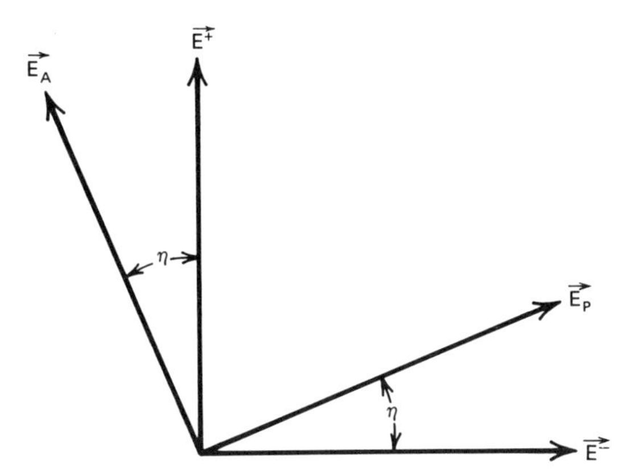

Fig. 14.1. Orientation of the electric fields in a relative phase measurement. $\mathbf{E}_P$ and $\mathbf{E}_A$ are directions of the transmitted electric fields of the polarizer and analyzer, respectively. $\mathbf{E}^+$ and $\mathbf{E}^-$ are the electric fields of the + and − eigenmodes of the crystal.

The length of the crystal in the direction of propagation **s** is l. The perturbation applied to the crystal (either an electric field or a stress) causes a change in this length,

$$l = l_0 + \Delta l, \tag{14.10.3}$$

as well as a change in refractive indices,

$$n^{\pm} = n_0^{\pm} + \Delta n^{\pm}. \tag{14.10.4}$$

After passing through the crystal the light from both modes encounters a second polarizer called an analyzer which passes light whose electric field vector is orthogonal to that passing the first polarizer. Setting the transmission coefficient of the output surface to unity and projecting the electric field of each mode onto the transmitting direction of the analyzer, we find an output electric field E_A of the light to be

$$E_A = E_P \cos\eta \sin\eta \left(e^{i(\varphi^+ - \omega t)} - e^{i(\varphi^- - \omega t)}\right). \tag{14.10.5}$$

Here we use

$$\varphi^{\pm} \equiv \frac{2\pi}{\lambda}\left(n_0^{\pm} + \Delta n^{\pm}\right)\left(l_0 + \Delta l\right) \tag{14.10.6}$$

where λ is the free-space wavelength of the light. It is apparent from (14.10.5) that E_A is largest for $\eta = 45°$, which we now assume.

The intensity of the output light is given by

$$\begin{aligned}\langle S_A\rangle &= \tfrac{1}{2}|\Re\{\mathbf{E}_A\times\mathbf{H}_A^*\}|\\ &= \frac{\epsilon_0 c}{2}|E_A|^2\\ &= \langle S_P\rangle \sin^2\frac{(\varphi^+-\varphi^-)}{2}\end{aligned} \tag{14.10.7}$$

where $\langle S_P\rangle$ is the intensity of the input light. This shows that the intensity of the light after the analyzer varies as the square of the sine of half the phase difference between the eigenmodes. Measurement of this intensity thus yields the phase difference

$$\varphi^+-\varphi^- = \frac{2\pi}{\lambda}\left[(n_0^+-n_0^-)l_0+(n_0^+-n_0^-)\Delta l+(\Delta n^+-\Delta n^-)l_0\right]. \tag{14.10.8}$$

The first pair of terms is the phase difference in the absence of any perturbation and may be taken as a reference phase. The second pair of terms is the difference of optical paths between the two eigenmodes caused by a change of crystal length as a result of the perturbation. In an electrooptic measurement Δl is given by a product of the original crystal length l_0, the appropriate piezoelectric stress tensor component, and the applied electric field. In the elastooptic measurement Δl is given by a product of the original crystal length l_0, the appropriate compliance component, and the stress. These factors can be separately determined so that this pair of terms is known. Note that this effect exists only in birefringent crystals. The third pair of terms is due to the electrooptic (or elastooptic) effect. The appropriate refractive index changes from Sections 14.6, 14.7, and 14.8 are to be used. Measurement of the phase difference $\varphi^+-\varphi^-$ versus the applied electric field (or stress) then yields the appropriate component of the interaction tensor. The phase difference is usually measured by introducing an additional phase difference with a device called a compensator before the light reaches the analyzer. When this additional phase difference is equal and opposite to that introduced electrooptically, no light passes through the analyzer ($\langle S_A\rangle=0$).

14.11 Symmetry Breaking Electrooptic Effect

As is shown in Section 14.2 the indirect electrooptic effect for frequencies well below the piezoelectric resonances has the same crystal symmetry as the direct electrooptic effect. The total electrooptic effect in this low frequency regime consists of a sum of the direct and indirect electrooptic effects and so has the same crystal symmetry as the direct effect. The direct effect produces the entire effect for frequencies well above the resonances. It is thus tempting to believe that the electrooptic effect possesses the same crystal symmetry for all frequencies. However, this is not true. The reason is very simple. At or near a piezoelectric resonance the strain induced by an applied electric field is strongly affected by the size and shape of the crystal and need not be the strain induced by the electric field in a crystal of infinite extent. Thus the indirect electrooptic effect need not have crystal symmetry in the resonance region. Demonstration of this fact offers a good example of various topics discussed previously.

Consider the geometry illustrated in Fig. 14.2. A long rectangular plate of a cubic crystal of the piezoelectric class $\bar{4}3m$ is oriented so that an electric field may be applied across its thickness in the [001] crystallographic direction, a light beam may propagate along its length in the [$1\bar{1}0$]

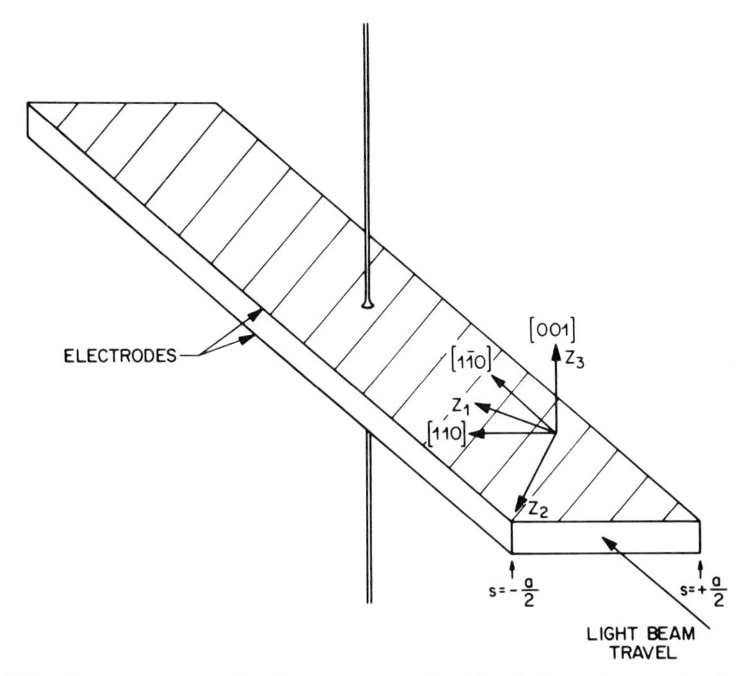

Fig. 14.2. Arrangement for observing a symmetry breaking indirect electrooptic effect in a $\bar{4}3m$ crystal. Resonant lateral vibrations are piezoelectrically excited between the $s = \pm a/2$ surfaces.

direction, and resonant vibrations may be excited across its width in the [110] direction. The thickness is chosen very small compared to the width which is chosen very small compared to the length.

Since the thickness is small compared to the electromagnetic wavelength at the frequency of the applied electric field, we may take this field as homogeneous throughout the crystal. We also consider a crystal whose piezoelectric constants are sufficiently small that the electric field term in the acoustic propagation equation,

$$\rho^0 \frac{\partial^2 u_i}{\partial t^2} = t_{ij,j}, \qquad (14.11.1)$$

$$t_{ij} = c_{ijkl} u_{k,l} - e_{kij} E_k \qquad (14.11.2)$$

may be neglected. Thus the electric field is taken as

$$E_1 = E_2 = 0, \qquad E_3 = E \sin \omega t \qquad (14.11.3)$$

throughout the crystal, E being a constant and ω the angular frequency of this perturbing field.

We consider all the crystal surfaces to be traction free,

$$t_{ij} N_j = 0 \qquad \text{(on surfaces)}, \qquad (14.11.4)$$

where **N** refers to the normal to any surface. On the two electroded surfaces this states that t_{i3} vanishes. Since these two surfaces are very close together, it is an adequate approximation to take

$$t_{i3} = 0 \qquad (14.11.5)$$

throughout the volume of the crystal. Under this condition the crystal is said to be in a state of *plane stress*. Because of this condition and the symmetry of the stress tensor we may drop the $i=3$ component of (14.11.1). It may also be used to alter the coefficients in (14.11.2). If that equation is solved for the strain,

$$S_{ij} = s_{ijkl} t_{kl} + d_{kij} E_k, \qquad (14.11.6)$$

and (14.11.3) and (14.11.5) used to simplify this equation, we find

$$S_1 = s_{11} t_1 + s_{12} t_2, \qquad (14.11.7a)$$

$$S_2 = s_{12} t_1 + s_{11} t_2, \qquad (14.11.7b)$$

$$S_6 = s_{44} t_6 + d_{14} E_3 \qquad (14.11.7c)$$

where the symmetry restrictions on the **s** and **e** tensors for a $\bar{4}3m$ crystal and contracted notation are used. Solving these for the stresses gives

$$t_1 = \gamma_{11} S_1 + \gamma_{12} S_2, \tag{14.11.8a}$$

$$t_2 = \gamma_{12} S_1 + \gamma_{11} S_2, \tag{14.11.8b}$$

$$t_6 = \gamma_{44} S_6 - e_{14} E_3 \tag{14.11.8c}$$

where

$$\gamma_{11} \equiv \frac{s_{11}}{s_{11}^2 - s_{12}^2}, \tag{14.11.9a}$$

$$\gamma_{12} \equiv \frac{-s_{12}}{s_{11}^2 - s_{12}^2}, \tag{14.11.9b}$$

$$\gamma_{44} \equiv \frac{1}{s_{44}} \tag{14.11.9c}$$

are the *reduced stiffness constants* of the plate. Equations (14.11.8) in conjunction with (14.11.5) eliminate the z_3 coordinate dependence from (14.11.1), leaving

$$\rho^0 \frac{\partial^2 u_1}{\partial t^2} = \gamma_{11} u_{1,11} + \gamma_{12} u_{2,21} + \gamma_{44} u_{1,22} + \gamma_{44} u_{2,12}, \tag{14.11.10a}$$

$$\rho^0 \frac{\partial^2 u_2}{\partial t^2} = \gamma_{11} u_{2,22} + \gamma_{12} u_{1,12} + \gamma_{44} u_{1,22} + \gamma_{44} u_{2,12}. \tag{14.11.10b}$$

At the frequencies near the resonances of the width dimension, which we wish to consider, the length of the plate acts as if it were clamped, that is, the applied electric field frequency is far above the resonant frequencies of the length dimension. Thus no coordinate dependence in the length dimension need be considered. This is best accomplished by first introducing coordinates measured parallel to the length (r) and to the width (s) in terms of the coordinates z_1 and z_2 referred to crystallographic axes used until now. Thus we let

$$r \equiv \frac{1}{\sqrt{2}}(z_1 - z_2), \tag{14.11.11}$$

$$s \equiv \frac{1}{\sqrt{2}}(z_1 + z_2). \tag{14.11.12}$$

We also introduce displacement components parallel to the length (w_1) and width (w_2) by

$$w_1 \equiv \frac{1}{\sqrt{2}}(u_1 - u_2), \tag{14.11.13}$$

$$w_2 \equiv \frac{1}{\sqrt{2}}(u_1 + u_2). \tag{14.11.14}$$

Substitution of these definitions into (14.11.10) leads to

$$\rho^0 \frac{\partial^2 w_1}{\partial t^2} = \tfrac{1}{2}(\gamma_{11} - \gamma_{12}) \frac{\partial^2 w_1}{\partial s^2}, \tag{14.11.15}$$

$$\rho^0 \frac{\partial^2 w_2}{\partial t^2} = \tfrac{1}{2}(2\gamma_{44} + \gamma_{11} + \gamma_{12}) \frac{\partial^2 w_2}{\partial s^2} \tag{14.11.16}$$

where the lack of r dependence eliminates derivatives with respect to r. The first of these is the propagation equation for a purely transverse wave traveling in the width direction and the second is for a purely longitudinal wave traveling in this direction. The boundary condition (14.11.4) applied to the surfaces limiting the width dimension gives

$$\frac{\partial w_1}{\partial s} = 0 \qquad \left(s = \pm \frac{a}{2}\right), \tag{14.11.17}$$

$$\frac{\partial w_2}{\partial s} = \frac{2e_{14}E_3}{2\gamma_{44} + \gamma_{11} + \gamma_{12}} \qquad \left(s = \pm \frac{a}{2}\right) \tag{14.11.18}$$

where a is the width. We now see from (14.11.17) that w_1 is undriven. We may thus ignore it. This leaves us with (14.11.16) and (14.11.18) to solve.

It can be seen at this point that the length direction and the width direction, which are crystallographically equivalent for the orientation assumed, affect the vibration problem quite differently. The lack of standing vibrations in the length direction in the range of frequencies considered, which results from the large length-to-width ratio and the consequent neglect of dependence on the coordinate r, causes the strains involved to lack the symmetry expected of them from the symmetry of the infinite crystal plus the applied electric field. As we now show, this leads to a time and space varying indirect electrooptic effect which differs in symmetry from the direct electrooptic effect.

We now choose a standing wave solution to (14.11.16) of the form

$$w_2 = \left(A \sin \frac{\omega s}{v} + B \cos \frac{\omega s}{v}\right) \sin \omega t \tag{14.11.19}$$

where v is the velocity of the longitudinal wave,

$$v \equiv \left[\frac{2\gamma_{44} + \gamma_{11} + \gamma_{12}}{2\rho^0} \right]^{1/2}. \tag{14.11.20}$$

Applying the boundary condition (14.11.18) leads to

$$A = \frac{e_{14}E}{\omega\rho^0 v \cos(\omega a/2v)}, \tag{14.11.21}$$

$$B = 0, \tag{14.11.22}$$

which shows that only odd vibrations of w_2 are driven, a result found also in Sections 11.7, 11.8, and 11.9. Equation (14.11.21) shows that a resonance occurs for every frequency satisfying

$$\frac{\omega a}{2v} = (2m-1)\frac{\pi}{2} \qquad (m = 1, 2, \ldots), \tag{14.11.23}$$

which corresponds to an odd integral multiple of the acoustic half-wavelength equaling the width a. The solution is now

$$w_2 = \frac{e_{14}E \sin(\omega s/v) \sin \omega t}{\omega\rho^0 v \cos(\omega a/2v)} \tag{14.11.24}$$

and the only nonvanishing strain component is $\partial w_2/\partial s$.

The dielectric perturbation is

$$(\Delta\kappa^{-1})_{ij} = r_{ijk}^{S} E_k(\mathbf{z}, t; 0, 1) + p_{ijmn} u_{m,n}(\mathbf{z}, t; 0, 1). \tag{14.11.25}$$

Since

$$u_{1,1} = u_{1,2} = u_{2,1} = u_{2,2} = \tfrac{1}{2} \frac{\partial w_2}{\partial s}, \tag{14.11.26}$$

we have

$$(\Delta\kappa^{-1})_{11} = (\Delta\kappa^{-1})_{22} = \tfrac{1}{2}(p_{11} + p_{12}) \frac{\partial w_2}{\partial s}, \tag{14.11.27}$$

$$(\Delta\kappa^{-1})_{33} = p_{12} \frac{\partial w_2}{\partial s}, \tag{14.11.28}$$

$$(\Delta\kappa^{-1})_{12} = (\Delta\kappa^{-1})_{21} = r_{41} E_3 + p_{44} \frac{\partial w_2}{\partial s} \tag{14.11.29}$$

for the only nonzero components of $\boldsymbol{\Delta\kappa}^{-1}$ in the crystallographic coordinate system. The symmetries of the $\mathbf{r}$ and $\mathbf{p}$ tensors for a $\bar{4}3m$ crystal are used here.

We now wish to obtain the refractive index changes for the eigenmodes of the perturbed crystal. Since the crystal is cubic, the procedure of Section 14.6 must be employed. The unit propagation vector for the light **s** (not to be confused with the scalar coordinate s) is

$$\mathbf{s}=\frac{1}{\sqrt{2}}[1,-1,0] \tag{14.11.30}$$

and the unit electric field vector **e** of the light may be expressed as

$$\mathbf{e}=\left[\frac{1}{\sqrt{2}}\sin\theta,\frac{1}{\sqrt{2}}\sin\theta,\cos\theta\right] \tag{14.11.31}$$

where θ is the angle between **e** and the [001] axis. Hence

$$\mathbf{s}\times\mathbf{e}=\left[\frac{-\cos\theta}{\sqrt{2}},\frac{-\cos\theta}{\sqrt{2}},\sin\theta\right]. \tag{14.11.32}$$

Substitution of these into (14.6.5),

$$\mathbf{e}\cdot\boldsymbol{\Delta\kappa}^{-1}\cdot(\mathbf{s}\times\mathbf{e})=0, \tag{14.11.33}$$

yields immediately

$$\sin\theta\cos\theta=0. \tag{14.11.34}$$

Of the two solutions we choose $\theta=0$ (the other works equally well). Thus the eigenmodes of the perturbed crystal have their electric fields in the [001] and $[\bar{1}\bar{1}0]$ directions as seen from (14.6.1) and (14.6.2). The refractive index perturbations of these modes according to (14.6.6) are

$$\begin{aligned}\Delta n^{[001]}&=-\frac{n_0^3}{2}\mathbf{e}\cdot\boldsymbol{\Delta\kappa}^{-1}\cdot\mathbf{e}\\&=-\frac{n_0^3}{2}p_{12}\frac{\partial w_2}{\partial s}\\&=-\frac{n_0^3p_{12}e_{14}E\sin(\omega s/v)\sin\omega t}{\rho^0v^2\cos(\omega a/2v)},\end{aligned} \tag{14.11.35}$$

$$\begin{aligned}\Delta n^{[\bar{1}\bar{1}0]}&=-\frac{n_0^3}{2}(\mathbf{s}\times\mathbf{e})\cdot\boldsymbol{\Delta\kappa}^{-1}\cdot(\mathbf{s}\times\mathbf{e})\\&=-\frac{n_0^3}{2}\left[r_{41}E_3+\tfrac{1}{2}(2p_{44}+p_{11}+p_{12})\frac{\partial w_2}{\partial s}\right]\\&=-\frac{n_0^3}{2}\left[r_{41}+\frac{(2p_{44}+p_{11}+p_{12})e_{14}\sin(\omega s/v)}{2\rho^0v^2\cos(\omega a/2v)}\right]E\sin\omega t.\end{aligned} \tag{14.11.36}$$

It is now quite apparent that the indirect electrooptic effect terms possess a different symmetry than the one direct electrooptic effect term. In particular we see that the index change $\Delta n^{[001]}$ is entirely due to the indirect electrooptic effect. Through the factor $\partial w_2/\partial s$ the indirect terms are both space and time dependent while the direct term is time dependent only.

Equations (14.11.35) and (14.11.36) give the phase modulation of the light beam when substituted into (14.9.6). With the use of polarizers as described in Section 14.10 an intensity modulation can be obtained. The phase difference,

$$\varphi^{[001]} - \varphi^{[\overline{1}\overline{1}0]} = \frac{2\pi l_0}{\lambda}\left(\Delta n^{[001]} - \Delta n^{[\overline{1}\overline{1}0]}\right), \qquad (14.11.37)$$

where l_0 is the crystal length and λ is the free-space optical wavelength, may be used in (14.10.7) for the present case.

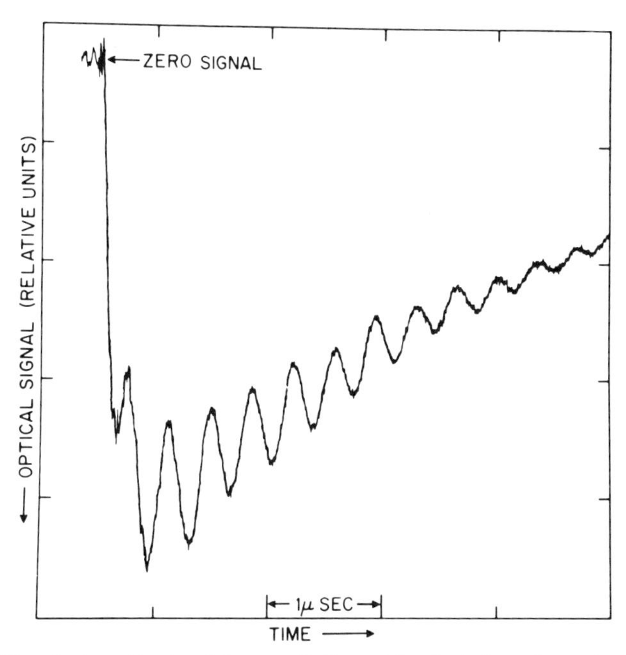

Fig. 14.3. The electrooptic signal from a GaP crystal oriented as shown in Fig. 14.2. A step-function electric field was applied. The direct electrooptic effect was then also a step function in time. However, the conductivity of the crystal led to a dielectric relaxation time of a few microseconds, thus causing the signal decay. Extrapolation to the pulse turn-on time eliminated the effects of initial electrical transients. The oscillatory part of the signal is caused by the indirect electrooptic effect. The step-function field excited the fundamental width dimension resonance and its odd overtones. The observed fundamental resonant frequency was 2.81 MHz; that predicted from (14.11.23) was 2.85 MHz [reprinted by permission from Nelson and Turner, 1968].

The geometry just analyzed was used in a measurement of the direct and indirect electrooptic effects in GaP, a $\bar{4}3m$ crystal [Nelson and Turner, 1968]. In that experiment a step-function electric field was applied. The direct electrooptic effect then had a step function time dependence; the indirect electrooptic effect had a nearly triangular wave dependence on time because the step-function electric field piezoelectrically excited the fundamental longitudinal resonance of the width dimension and a large number of its odd overtones. The direct and indirect effects were thus easily distinguished. Figure 14.3 shows the time dependent electrooptic effect observed in that experiment.

14.12 Variational Principle for Waveguide Calculations

The perturbation approach of Sections 14.4 to 14.8 for obtaining a general solution to the electrooptic effect is very useful whenever the light beam can be adequately approximated as a plane wave. However, the direction of much optical research is toward manipulating light beams within minute dielectric optical waveguides and the electrooptic effect is an important means of this manipulation. Since waveguide modes have a rapid variation of intensity with coordinates measured normal to the direction of propagation, they cannot be considered as plane waves. Thus some different approximation method is needed to handle the electrooptic effect on waveguide modes; the variational technique offers this. In this section we develop a variational expression for the propagation constant of a mode of a dielectric waveguide. The development is patterned after Berk [1956].

An optical dielectric waveguide can exist only if there is a central region having a higher value of the optical dielectric constant κ (and thus of the refractive index) when compared to the surrounding medium. A dielectric waveguide that confines the light in one of the two directions normal to the direction of propagation is illustrated in Fig. 14.4.

The two Maxwell equations governing propagation of optical frequency fields in a dielectric are

$$\nabla \times \mathbf{E} + \mu_0 \frac{\partial \mathbf{H}}{\partial t} = 0, \tag{14.12.1}$$

$$\nabla \times \mathbf{H} - \epsilon_0 \boldsymbol{\kappa} \cdot \frac{\partial \mathbf{E}}{\partial t} = 0. \tag{14.12.2}$$

We consider solutions of these with a form appropriate to waveguide

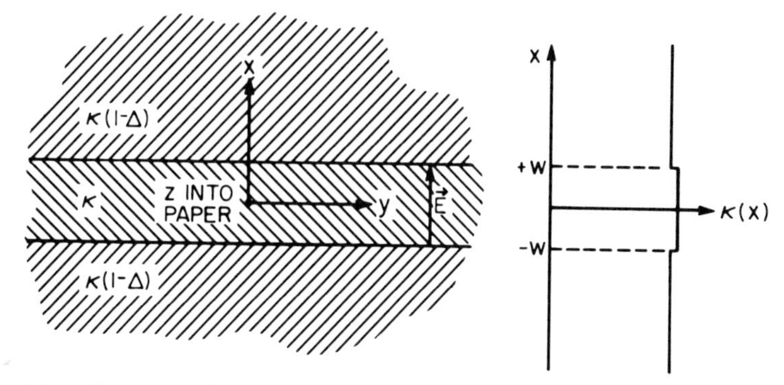

Fig. 14.4. Geometry of planar dielectric waveguide and its dielectric constant profile.

modes

$$\mathbf{E}=\mathbf{e}(x,y)e^{i(\beta\mathbf{s}\cdot\mathbf{z}-\omega t)}, \tag{14.12.3}$$

$$\mathbf{H}=\mathbf{h}(x,y)e^{i(\beta\mathbf{s}\cdot\mathbf{z}-\omega t)} \tag{14.12.4}$$

where β is the propagation constant, $\mathbf{s}$ is a unit vector in the z direction, and the amplitude functions $\mathbf{e}$ and $\mathbf{h}$ depend only on the lateral coordinates x and y. Substitution of the latter pair of equations into the former gives

$$\nabla\times\mathbf{e}+i\beta\mathbf{s}\times\mathbf{e}-i\omega\mu_0\mathbf{h}=0, \tag{14.12.5}$$

$$\nabla\times\mathbf{h}+i\beta\mathbf{s}\times\mathbf{h}+i\omega\epsilon_0\boldsymbol{\kappa}\cdot\mathbf{e}=0. \tag{14.12.6}$$

Next we form a scalar product of the first of these equations with $\mathbf{h}$ and of the second of these with $\mathbf{e}$, subtract the resulting equations, integrate the two lateral coordinates over the entire region that the fields exist, and solve for β. The result is

$$\beta=\frac{\omega\int\{\mu_0\mathbf{h}\cdot\mathbf{h}^*+\epsilon_0\mathbf{e}\cdot\boldsymbol{\kappa}\cdot\mathbf{e}^*\}\,da+i\int\{\mathbf{h}^*\cdot\nabla\times\mathbf{e}-\mathbf{e}^*\cdot\nabla\times\mathbf{h}\}\,da}{\int(\mathbf{e}\times\mathbf{h}^*+\mathbf{e}^*\times\mathbf{h})\cdot\mathbf{s}\,da}. \tag{14.12.7}$$

This is a variational formula for β because (1) if the exact solutions $\mathbf{e}(x,y)$ and $\mathbf{h}(x,y)$ of a dielectric waveguide problem are substituted into the right side, the exact expression for β results, and (2) the variation of the propagation constant $\delta\beta$ with respect to variations of the fields $\delta\mathbf{e}$ and $\delta\mathbf{h}$

is zero (*stationary*) provided the fields **e** and **h** satisfy the Maxwell equations (14.12.5) and (14.12.6). The correctness of the first statement is obvious, since the formula for β is derived from the Maxwell equations and the proof of the second statement is left for an exercise. Since the formula for β is stationary for variations about the true value, approximate forms of the fields may be used to obtain rather good approximations for β. The approximate fields may contain adjustable parameters $A^{(i)}$ ($i = 1,2,\ldots$) which may then be determined by

$$\frac{\partial \beta}{\partial A^{(i)}} = 0, \tag{14.12.8}$$

which expresses the *stationarity* of β. Also, if the fields **e** and **h** of some waveguide problem are known, then the change in the propagation constant brought about by some perturbation of the geometry of problem, such as a change of the dielectric tensor $\boldsymbol{\kappa}$ resulting from the electrooptic effect, can be found from (14.12.7).

14.13 Optical Waveguides in Cubic Crystals

The first study of the electrooptic effect in an optical dielectric waveguide was done in GaP, an acentric cubic crystal of class $\bar{4}3m$ [Nelson and Reinhart, 1964]. The waveguide was found to occur naturally in the plane of a *pn* junction. The dielectric constant was found to be higher within the width ($\sim 0.6\ \mu m$) of the *pn* junction by about one part in a thousand over the bulk crystal. Light passing through this waveguide was modulated in phase (and thus in intensity, see Section 14.10) by applying voltage in the reverse bias polarity. The applied electric field existed only in the central higher dielectric constant region; that region became birefringent in a manner determined by the orientation of the crystallographic axes relative to the electric field.

In order to apply the variational principle to the electrooptic effect in the waveguide just described, we must first find the **e** and **h** fields for waves of both polarizations in the absence of any electrooptic effect. We denote the isotropic dielectric constant inside the waveguide by $\kappa = n^2$ and that in the bulk crystal by $\kappa(1-\Delta)$ where $\Delta \approx 10^{-3}$. The electric field wave equation may be written as

$$\nabla(\nabla \cdot \mathbf{E}) - (\nabla \cdot \nabla)\mathbf{E} + \frac{\kappa(x)}{c^2}\frac{\partial^2 \mathbf{E}}{\partial t^2} = 0 \tag{14.13.1}$$

and the electric field **E** taken in the form (14.12.3) except that no dependence on y is present for the one-dimensional waveguide. For this form of field the first term of the wave equation can be seen to couple only the x and z components of **E**; the y component is left uncoupled to the others by the equation. The solution having only the y component of **E** is called the *TE* (transverse electric) solution. Its wave equation becomes

$$\frac{\partial^2 e_y}{\partial x^2} - \beta^2 e_y = -\frac{\kappa(x)\omega^2}{c^2} e_y. \tag{14.13.2}$$

In each of the regions $\kappa(x)$ may be taken as a constant. It is thus a simple matter to solve this equation in each region. The required boundary conditions are that e_y vanish for $x = \pm\infty$ and be continuous for $x = \pm w$. Because of the reflection symmetry of the geometry in the x direction, *TE* solutions both even and odd in this coordinate result. They are

$$e_y(x) = A\left\{\begin{array}{l}\cos bx\\ \sin bx\end{array}\right. \qquad (|x| \leqslant w) \tag{14.13.3a}$$

$$= A\left\{\begin{array}{c}\cos bw\\ \pm\sin bw\end{array}\right\} e^{p(w-|x|)} \qquad \left(\begin{array}{c}x \geqslant w\\ x \leqslant w\end{array}\right), \tag{14.13.3b}$$

$$e_x(x) = e_z(x) = 0 \qquad (\text{all } x) \tag{14.13.4}$$

where the parameters p, b are related to the propagation constant β through

$$\beta^2 = \left(\frac{\omega n}{c}\right)^2 (1-\Delta) + p^2, \tag{14.13.5}$$

$$\beta^2 = \left(\frac{\omega n}{c}\right)^2 - b^2, \tag{14.13.6}$$

$$b \tan bw = p \qquad (\text{even solutions}), \tag{14.13.7}$$

$$b \cot bw = -p \qquad (\text{odd solutions}). \tag{14.13.8}$$

The latter two equations result from the boundary conditions at $x = \pm w$. The magnetic intensity amplitude **h** of the *TE* modes is

$$h_x = -\frac{\beta}{\mu_0 \omega} e_y, \tag{14.13.9}$$

$$h_y = 0, \tag{14.13.10}$$

$$h_z = \frac{1}{i\mu_0 \omega} \frac{\partial e_y}{\partial x}. \tag{14.13.11}$$

The solution involving e_x and e_z can be seen to have only an h_y component nonzero and so may be identified as the *TM* (transverse magnetic) solution. It may be found by a completely analogous procedure to that for the *TE* solution if we first form the magnetic field wave equation,

$$\nabla\times(\boldsymbol{\kappa}^{-1}\cdot\nabla\times\mathbf{H})+\frac{1}{c^2}\frac{\partial^2\mathbf{H}}{\partial t^2}=0, \tag{14.13.12}$$

by eliminating $\mathbf{E}$ between (14.12.1) and (14.12.2). In a cubic crystal $\boldsymbol{\kappa}$ is isotropic and so in regions where it is constant this equation may be written as

$$\nabla\times(\nabla\times\mathbf{H})+\frac{\kappa}{c^2}\frac{\partial^2\mathbf{H}}{\partial t^2}=0, \tag{14.13.13}$$

which has the form of (14.13.1). Following an analogous procedure to that used for the *TE* solution now yields the *TM* solution

$$h_y(x)=B\begin{cases}\cos bx\\ \sin bx\end{cases} \qquad (|x|\leqslant w) \tag{14.13.14a}$$

$$=B\begin{Bmatrix}\cos bw\\ \pm\sin bw\end{Bmatrix}e^{p(w-|x|)} \qquad \begin{pmatrix}x\geqslant w\\ x\leqslant w\end{pmatrix}, \tag{14.13.14b}$$

$$h_x(x)=h_z(x)=0 \qquad (\text{all } x) \tag{14.13.15}$$

where the parameters p,b are here related to the propagation constant β through (14.13.5) and (14.13.6) and

$$b(1-\Delta)\tan bw=p \qquad (\text{even solutions}), \tag{14.13.16}$$

$$b(1-\Delta)\cot bw=-p \qquad (\text{odd solutions}). \tag{14.13.17}$$

The electric field amplitude $\mathbf{e}$ of the *TM* modes is

$$e_x=\frac{\beta h_y}{\epsilon_0\omega\kappa(x)}, \tag{14.13.18}$$

$$e_y=0, \tag{14.13.19}$$

$$e_z=\frac{i}{\epsilon_0\omega\kappa(x)}\frac{\partial h_y}{\partial x}. \tag{14.13.20}$$

Let us consider the solution of the *TE* mode equations (14.13.5) to (14.13.8) for the eigenvalues of p,b, and β. The first two of these equations can be subtracted to eliminate β and so yield a circle in p,b space,

$$b^2+p^2=\left(\frac{\omega n\sqrt{\Delta}}{c}\right)^2. \tag{14.13.21}$$

The intersections of this circle with the tangent curve (even solutions) and the cotangent curve (odd solutions) then determine the eigenvalues of p,b and hence of β. Figure 14.5 illustrates the intersection for the case applicable to the original *pn* junction experiment for which the radius of the circle is small enough that it intersects only the curve for the lowest order even solution. For this case the tangent may be expanded to first order with the result

$$b^2w \cong p. \tag{14.13.22}$$

Equation (14.13.21) can then be solved for p to first order,

$$p=\frac{1}{2w}\left[\left(1+4\frac{\omega^2w^2n^2\Delta}{c^2}\right)^{1/2}-1\right]\cong\frac{\omega^2n^2w\Delta}{c^2}, \tag{14.13.23}$$

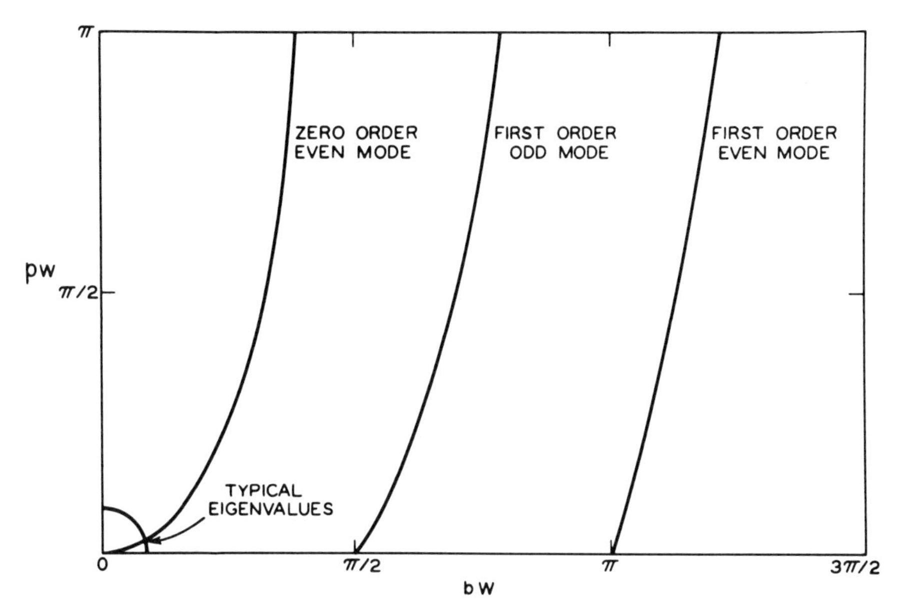

Fig. 14.5. Graphical solution of eigenvalue equations for *TE* modes. Parameter values used are typical of the *pn* junction problem. The eigenvalue equations of *TM* modes are only minutely different [reprinted with permission from Nelson and McKenna, 1967].

since $4(\omega wn/c)^2\Delta \cong 0.4$ when $\Delta \approx 10^{-3}$, $w \approx 0.3\ \mu m$, $\lambda = 2\pi c/\omega = 0.6\ \mu m$ and $n \approx 3.3$. The constant b is then given approximately by

$$b \cong \frac{\omega n}{c}\sqrt{\Delta} \tag{14.13.24}$$

and the propagation constant by

$$\beta \cong \frac{\omega n}{c}\left(1 - \frac{\Delta}{2}\right). \tag{14.13.25}$$

The presence in this geometry of only a single *TE* mode is caused by the smallness of the dimensionless parameters pw and bw which results from the small size of the waveguide half-width w and the fractional increase of dielectric constant Δ in the waveguide. For larger values of either or both of these parameters many solutions (modes) exist. Because the equations for p, b, and β for the *TM* solutions are so similar to the equations for the *TE* solutions, (14.13.23) to (14.13.25) can also be used for the lowest order even *TM* mode.

14.14 Electrooptic Effect in Optical Waveguides

We now wish to use the variational principle derived in Section 14.12 to evaluate the electrooptic phase modulation of the modes of the optical dielectric waveguide considered in Section 14.13. We work with the lowest order even *TE* and *TM* solutions appropriate to the *pn* junction problem.

As a first example consider the normal to the plane of the waveguide to be the [111] crystallographic direction and the direction of propagation to be the $[\bar{1}10]$ crystallographic direction. This direction of propagation is frequently used, since it is normal to one of the {110} cleavage planes that are used as crystal surfaces. In the crystallographic coordinate system the change in the dielectric tensor is

$$\Delta\kappa_{kl}^{c} = -\kappa_{km}\kappa_{ln}(\Delta\kappa^{-1})_{mn} = -\kappa_{km}\kappa_{ln}r_{mnp}^{S}E_p \tag{14.14.1}$$

where we have assumed that the clamping of the thin waveguide layer by the surrounding crystal prevents any contribution from the indirect electrooptic effect. For a class $\bar{4}3m$ crystal such as GaP the only nonzero electrooptic tensor components are $r_{41} = r_{52} = r_{63}$. Thus we have

$$\mathbf{\Delta\kappa^{c}} = \kappa\begin{bmatrix} 0 & \delta & \delta \\ \delta & 0 & \delta \\ \delta & \delta & 0 \end{bmatrix} \tag{14.14.2}$$

where

$$\delta \equiv -\frac{\kappa r_{41}^S E}{\sqrt{3}}. \tag{14.14.3}$$

The geometric coordinate system x,y,z of the waveguide (see Fig. 14.4) is related to the crystallographic coordinate system X,Y,Z by

$$\begin{bmatrix} x \\ y \\ z \end{bmatrix} = \frac{1}{\sqrt{6}} \begin{bmatrix} \sqrt{2} & \sqrt{2} & \sqrt{2} \\ -1 & -1 & 2 \\ -\sqrt{3} & \sqrt{3} & 0 \end{bmatrix} \begin{bmatrix} X \\ Y \\ Z \end{bmatrix}, \tag{14.14.4}$$

which has the form

$$x_i = T_{ij} X_j, \qquad T_{ij} \equiv \frac{\partial x_i}{\partial X_j}. \tag{14.14.5}$$

Under this coordinate transformation the dielectric perturbation $\mathbf{\Delta\kappa}$ in the geometric coordinate system is

$$\Delta\kappa_{ij} = T_{ik} T_{jl} \Delta\kappa_{kl}^c, \tag{14.14.6}$$

or in matrix notation ($\sim$ indicates transpose) is

$$\begin{aligned} \mathbf{\Delta\kappa} &= \mathbf{T} \cdot \mathbf{\Delta\kappa}^c \cdot \tilde{\mathbf{T}} \\ &= \kappa \begin{bmatrix} 2\delta & 0 & 0 \\ 0 & -\delta & 0 \\ 0 & 0 & -\delta \end{bmatrix}. \end{aligned} \tag{14.14.7}$$

Thus in this first example the perturbed dielectric tensor is diagonal in the geometric coordinate system that we use.

The electrooptically perturbed propagation constant $\beta' \equiv \beta + \Delta\beta$ may now be calculated from the variational formula (14.12.7) by inserting the perturbed dielectric tensor $\boldsymbol{\kappa}' \equiv \boldsymbol{\kappa} + \mathbf{\Delta\kappa}$ and the field components of the unperturbed mode. Since the dielectric perturbation (14.14.7) is diagonal in the geometric coordinate system, no mixing between the *TE* and *TM* modes occurs. Thus we may consider the modes individually in the variational formula. For any *TE* mode it gives

$$\beta' = \frac{2\beta^2 \int e_y^2\, dx + (\omega/c)^2 \int (\kappa_{yy}' - \kappa) e_y^2\, dx}{2\beta \int e_y^2\, dx} \tag{14.14.8}$$

or

$$\Delta\beta = \frac{\omega^2}{2c^2\beta}\,\frac{\int \Delta\kappa_{yy} e_y^2\,dx}{\int e_y^2\,dx}. \tag{14.14.9}$$

Note that this formula holds for $\Delta\kappa_{yy}$ depending on x, a situation that would make the wave equation very difficult to solve exactly. If $\Delta\kappa_{yy}$ is independent of x, then

$$\Delta\beta = -\frac{\omega^2 n^2 \delta}{2c^2\beta}F \tag{14.14.10}$$

where F is the fraction of the mode intensity residing in the electrooptically perturbed region and is given by

$$F \equiv \frac{\int_{-w}^{w} e_y^2\,dx}{\int_{-\infty}^{\infty} e_y^2\,dx}. \tag{14.14.11}$$

For even TE modes this may be calculated to be

$$F = \frac{pbw + p\sin bw \cos bw}{pbw + p\sin bw\cos bw + b\cos^2 bw}. \tag{14.14.12}$$

For the lowest order even TE mode this has the approximate value of $2pw$. Equation (14.14.10) for the propagation constant change of this mode (now labeled TE) becomes

$$\Delta\beta_{TE} = -\left(\frac{\omega n}{c}\right)^3 w^2\delta\Delta. \tag{14.14.13}$$

A similar procedure can be followed for the TM modes with the result

$$\Delta\beta_{TM} = \frac{\beta^2 \int \frac{\Delta\kappa_{xx}}{\kappa^2} h_y^2\,dx + \int \frac{\Delta\kappa_{zz}}{\kappa^2}\left(\frac{\partial h_y}{\partial x}\right)^2 dx}{2\beta\int \frac{h_y^2}{\kappa}\,dx}. \tag{14.14.14}$$

The first term in the numerator in ratio to the denominator has the form of (14.14.12) with p replaced by $p(1-\Delta)$. Thus its approximate value for the lowest order even mode is also $2pw$. The second term in the numerator in ratio to the denominator has an approximate value of $pb^4w^3/3$, which is negligibly small, for the lowest order even mode. The smallness results because this term represents the electrooptic coupling to the longitudinal component of the mode, a very small component. Thus we get

$$\Delta\beta_{TM} = 2\left(\frac{\omega n}{c}\right)^3 w^2 \delta\Delta \tag{14.14.15}$$

for the lowest order even TM mode.

The phase difference occurring in a length l between the two lowest order even modes that would be observed in an intensity modulation measurement (see Section 14.10) is

$$\Delta\varphi \equiv (\Delta\beta_{TE} - \Delta\beta_{TM})l = \frac{\sqrt{3}\; n^5\omega^3 w^2 \Delta r_{41}^S El}{c^3}. \tag{14.14.16}$$

The much different functional dependence of $\Delta\varphi$ for waveguide modes compared to plane wave modes results from the fractional coupling (14.14.11) of the electrooptic effect to the waveguide mode in the geometry studied.

Phase differences of the order of several radians can be obtained for a *pn* junction length $l \approx 1$ mm even though w and Δ are so small. The reasons are the large refractive index, which enters as the fifth power, and the very large electric fields ($\sim 4\times 10^5$ V/cm) obtainable in full reverse bias.

As a further example of the variational technique let us consider the electrooptic modulation of the lowest order even modes of the waveguide discussed in the preceding section when the normal to the waveguide plane (the electric field direction) is [110] and the propagation direction is $[\bar{1}10]$ [McKenna and Reinhart, 1976]. The electrooptic perturbation of the dielectric tensor in this case is

$$\boldsymbol{\Delta\kappa}^c = \kappa \begin{bmatrix} 0 & 0 & \delta \\ 0 & 0 & \delta \\ \delta & \delta & 0 \end{bmatrix} \tag{14.14.17}$$

in the crystallographic coordinate system with

$$\delta \equiv -\frac{\kappa r_{41}^S E}{\sqrt{2}}. \tag{14.14.18}$$

The coordinate transformation matrix for transforming $\boldsymbol{\Delta\kappa}^c$ to the geometric coordinate system of the waveguide [see (14.14.5)–(14.14.7)] is

$$\mathbf{T} = \frac{1}{\sqrt{2}} \begin{bmatrix} 1 & 1 & 0 \\ 0 & 0 & -\sqrt{2} \\ -1 & 1 & 0 \end{bmatrix} \tag{14.14.19}$$

and the dielectric perturbation in the geometric coordinate system is

$$\boldsymbol{\Delta\kappa}=\kappa\begin{bmatrix} 0 & -\sqrt{2}\,\delta & 0 \\ -\sqrt{2}\,\delta & 0 & 0 \\ 0 & 0 & 0 \end{bmatrix}. \tag{14.14.20}$$

Because this is nondiagonal it cannot perturb either the *TE* or *TM* mode if either is used *alone* in the variational formula. Thus the effect of this perturbation is to couple the *TE* and *TM* modes of the unperturbed waveguide together to form two new lowest order even modes of the perturbed waveguide.

The trial fields in the variational formula may be taken as a sum of the *TE* and *TM* fields (see Section 14.13) with the ratio of their arbitrary constant multipliers (A for *TE* and B for *TM*) to be determined from the variational principle. For the lowest order even modes, as seen from the solutions of the last section, the values of p, b, and β are very close for the *TE* and *TM* modes. We now regard these parameters as the same in both the *TE* and *TM* parts of the new modes and to have the values found in (14.13.23) to (14.13.25). In other words we are ignoring the minute waveguide birefringence, because in practice it is small compared to the electrooptically induced birefringence, and regarding the original *TE* and *TM* modes as degenerate. Substituting the trial fields into the variational formula (14.12.7) leads to

$$\beta'=\beta+\frac{\omega\int e_y h_y \dfrac{\Delta\kappa_{xy}}{\kappa(x)}\,dx}{c^2\int\left\{\epsilon_0 e_y^2+\dfrac{\mu_0 h_y^2}{\kappa(x)}\right\}dx}. \tag{14.14.21}$$

The ratio B/A may now be found from the stationarity of β',

$$\frac{\partial\beta'}{\partial B}=0. \tag{14.14.22}$$

If we denote $\epsilon_y \equiv e_y/A$ and $\eta_y \equiv h_y/B$, this condition leads quickly to the relation

$$\frac{B}{A}=\pm\left[\frac{\epsilon_0\int \epsilon_y^2\,dx}{\mu_0\int \eta_y^2\,dx}\right]^{1/2}, \tag{14.14.23}$$

which means that the new modes have electric fields lying in a plane roughly midway between the $+x$ and $+y$ directions $(+)$ and between the $-x$ and $+y$ directions $(-)$. Evaluating the equation yields

$$\frac{B}{A}=\pm\frac{n}{\mu_0 c}. \tag{14.14.24}$$

The variational formula (14.14.21) may now be evaluated. Denoting $\beta'\equiv\beta+\Delta\beta$ and assuming $\Delta\kappa_{xy}$ homogeneous in the central layer, we obtain

$$\Delta\beta_{\pm}=\mp\sqrt{2}\,\delta\left(\frac{\omega n}{c}\right)^3 w^2\Delta \tag{14.14.25}$$

for the electrooptically induced perturbations of the propagation constants of the new modes designated + and −. This leads to a phase difference between these modes in a travel distance l to be

$$\Delta\varphi=(\Delta\beta_+-\Delta\beta_-)l=\frac{2n^5\omega^3w^2\Delta r_{41}^S El}{c^3}. \tag{14.14.26}$$

In this second example one result of the small electrooptic perturbation is to cause the two lowest order even modes to reorient their electric field vectors by about 45°. This occurs because we approximate the lowest order even *TE* and *TM* modes as being degenerate. The orientation of the electric field of the perturbed modes then is determined by the electrooptically induced anisotropy even for a minute applied electric field.

The results of the two examples above are in agreement with the results of an expansion of the exact solution of the perturbed waveguides [Nelson and McKenna, 1967]. The present method, however, is substantially shorter, particularly in the case of the second example.

PROBLEMS

14.1. Consider an arbitrarily oriented crystal slab with electrodes on its two large area surfaces and a light wave propagating parallel to these surfaces. One of the surfaces is attached to a massive rigid surface. What is the form of the low frequency electrooptic susceptibility of the slab?

14.2. Find specific expressions for the refractive index changes from the electrooptic effect in $LiNbO_3$ (point group $3m$) for beams of both ordinary and extraordinary polarization traveling in the $\mathbf{s}=[0, 1/\sqrt{2}, 1/\sqrt{2}]$ direction for a perturbing electric field applied in the $[0, 1/\sqrt{2}, -1/\sqrt{2}]$ direction.

14.3. Find general expressions for the refractive index changes in terms of dielectric tensor perturbations for the two eigenmodes propagating along an optic axis in a biaxial crystal.

14.4. How can the arrangement of Section 14.10 be used to make an intensity modulator linear in the applied electric field and giving the maximum modulation for small changes in this field? Give an explicit expression for the intensity versus the applied electric field.

14.5. Using the first procedure of Section 14.5, find the refractive index changes for a cubic crystal. Show their equivalence to (14.6.6).

14.6. Show that the variation of the propagation constant β in (14.12.7) vanishes for variations of the field amplitudes $\mathbf{e}$ and $\mathbf{h}$ provided that the amplitudes are solutions of the Maxwell equations.

14.7. Find the electrooptically induced change in the propagation constants of the two lowest order even modes of the dielectric waveguide described in Section 14.13 with the geometric coordinate axes x,y,z aligned along [100], $[01\bar{1}]$, and [011].

CHAPTER 15

Optical Mixing

Though both the elastooptic and electrooptic effects discussed in Chapters 13 and 14 are nonlinear optical interactions and both were discovered in the nineteenth century, the field of nonlinear optics is generally regarded as beginning with the discovery of *optical second harmonic generation* by Franken and co-workers [1961]. Their experiment was made possible by the discovery a year earlier of the ruby laser [Maiman, 1960; see also Collins et al., 1960]. In the Franken experiment a small amount of near ultraviolet light at 347 nm was detected emanating from a quartz crystal irradiated by a millisecond pulse of ruby laser light at 694 nm. Second harmonic generation is a three-wave optical interaction in which the input beam interacts with itself to create an optical output beam at twice the input or fundamental frequency. As such its interaction strength is characterized by a specialization of the susceptibility (12.4.3) to the case of optical input fields.

The same susceptibility that causes optical second harmonic generation can give rise to sum and difference frequency generation when the two optical input beams have different frequencies [Bass et al., 1962a]. The generation of a zero frequency signal as the difference frequency of two waves having the same frequency is called *optical rectification* [Bass et al., 1962b]. We refer to all these processes—optical harmonic generation, sum and difference frequency generation, and optical rectification—by the generic term *optical mixing*. When the interaction is reversed, it is called *optical parametric amplification.* In it one intense input beam of coherent light generates in an appropriate medium two output light beams whose frequencies add up to the frequency of the input beam. If this interaction is carried on in an optical cavity that provides feedback, the optical amplification may exceed the optical losses and produce an *optical parametric oscillator*. The generation of two intense coherent beams in this manner was first accomplished by Giordmaine and Miller [1965].

15.1 Optical Mixing Susceptibility

Equation (12.4.3) is an expression for the nonlinear polarization $\mathscr{P}$ generated by either two electric fields or one electric field and one acoustic displacement gradient. Here we wish to use that expression with two optical electric fields regarded as input fields. As is shown in Section 9.1, the displacement gradient created piezoelectrically by an optical electric field is wholly negligible. Thus the terms involving the displacement gradient may be dropped from (12.4.3) for the present application. If the Fourier expansion used in obtaining that equation is examined, it can be seen that the factor of 2 in the remaining term arises only when the two input frequencies ω_1 and ω_2 are different and is replaced by a factor of 1 when they are the same. The latter possibility is of considerable interest since it corresponds to second harmonic generation. We thus write (12.4.3) for a three-wave, entirely optical interaction that produces the sum frequency as

$$\mathscr{P}_i(\mathbf{z};1,1) = D\epsilon_0 b_{i\ j\ k}^{\omega_S\omega_1\omega_2} E_j(\mathbf{z};1,0)\,E_k(\mathbf{z};0,1) \tag{15.1.1}$$

where

$$\begin{aligned} D &= 1 \quad \text{if} \quad \omega_1 = \omega_2, \\ &= 2 \quad \text{if} \quad \omega_1 \neq \omega_2, \end{aligned} \tag{15.1.2}$$

$$\omega_S = \omega_1 + \omega_2, \tag{15.1.3}$$

and $b_{i\ j\ k}^{\omega_S\omega_1\omega_2}$ is the *optical mixing tensor*. It is an odd rank tensor and so can exist only in acentric materials (see the Appendix). If the derivation of (12.4.3) were carried through for difference frequency generation and (12.2.5) were used to introduce the complex conjugate electric field amplitude, the nonlinear polarization would be

$$\mathscr{P}_i(\mathbf{z};1,0) = D\epsilon_0 b_{i\ j\ k}^{\omega_1\omega_S\omega_2} E_j(\mathbf{z};1,1)\,E_k^*(\mathbf{z};0,1) \tag{15.1.4}$$

where

$$\omega_1 = \omega_S - \omega_2, \tag{15.1.5}$$

$$\begin{aligned} D &= 2 \quad \text{if} \quad \omega_S = \omega_2, \\ &= 2 \quad \text{if} \quad \omega_S \neq \omega_2. \end{aligned} \tag{15.1.6}$$

The $D = 2$ value for $\omega_S = \omega_2$ results from the nonlinear polarization being a static quantity in this case.

The derivation in Section 12.4 leads to the optical mixing susceptibility being defined as

$$\epsilon_0 b_{i\ j\ k}^{\omega_S\omega_1\omega_2} \equiv -\tfrac{3}{2} \sum_{\substack{\nu\mu\lambda \\ \rho\xi\theta}} q^\mu \Upsilon_{il}^{\mu\nu}(\omega_S)\, {}^{30}M_{lmn}^{\nu\lambda\rho} \Upsilon_{nj}^{\rho\xi}(\omega_1) q^\xi \Upsilon_{mk}^{\lambda\theta}(\omega_2) q^\theta \tag{15.1.7}$$

in terms of the material descriptor ${}^{30}\mathbf{M}$ and the mechanical admittances $\boldsymbol{\Upsilon}(\omega)$ evaluated at the three frequencies ω_1, ω_2, and ω_S. The interchange symmetry possessed by $\boldsymbol{\Upsilon}(\omega)$ and ${}^{30}\mathbf{M}$ leads to $b_{i\ j\ k}^{\omega_S\omega_1\omega_2}$ having permutation symmetry given in (14.1.2). It was first pointed out by Kleinman [1962a] that the optical mixing tensor has complete interchange symmetry on its tensor indices *alone*,

$$b_{i\ j\ k}^{\omega_S\omega_1\omega_2} = b_{i\ k\ j}^{\omega_S\omega_1\omega_2} = b_{j\ k\ i}^{\omega_S\omega_1\omega_2} = b_{j\ i\ k}^{\omega_S\omega_1\omega_2} = b_{k\ i\ j}^{\omega_S\omega_1\omega_2} = b_{k\ j\ i}^{\omega_S\omega_1\omega_2}, \tag{15.1.8}$$

if $b_{i\ j\ k}^{\omega_S\omega_1\omega_2}$ has no significant frequency dispersion in the region of the spectrum containing ω_1, ω_2, and ω_S. This symmetry, called *Kleinman symmetry*, is only approximate. However, it is obeyed to a sufficient degree by a wide variety of crystals in the visible region that it is a useful relationship. The additional restrictions that Kleinman symmetry places on the optical mixing susceptibility are shown in the Appendix. In some crystal classes otherwise allowed tensor components are required to vanish. These are called *Kleinman forbidden components.*

The definition of the optical mixing susceptibility (15.1.7) can be usefully reexpressed as follows. Define a *partial susceptibility* $\chi_{ia}^{\lambda}(\omega)$ by

$$\epsilon_0 \chi_{ia}^{\lambda}(\omega) \equiv \sum_{\xi} q^\xi \Upsilon_{ia}^{\xi\lambda}(\omega) q^\lambda \tag{15.1.9}$$

and a quantity $\delta_{abc}^{\lambda\mu\nu}$, which we call the *generalized Miller delta*, by

$$\delta_{abc}^{\lambda\mu\nu} \equiv -\frac{3\epsilon_0^3\, {}^{30}M_{abc}^{\lambda\mu\nu}}{2q^\lambda q^\mu q^\nu}. \tag{15.1.10}$$

The name partial susceptibility results from its relation to the linear electric susceptibility $\chi_{ia}(\omega)$ of (9.1.13),

$$\chi_{ia}(\omega) = \sum_{\lambda} \chi_{ia}^{\lambda}(\omega). \tag{15.1.11}$$

Note that the generalized Miller delta has permutation symmetry,

$$\delta_{abc}^{\lambda\mu\nu} = \delta_{acb}^{\lambda\nu\mu} = \delta_{bca}^{\mu\nu\lambda} = \delta_{bac}^{\mu\lambda\nu} = \delta_{cab}^{\nu\lambda\mu} = \delta_{cba}^{\nu\mu\lambda}. \tag{15.1.12}$$

The optical mixing tensor now becomes

$$\epsilon_0 b_{i\ j\ k}^{\omega_S\omega_1\omega_2} = \sum_{\lambda\mu\nu} \chi_{ia}^{\lambda}(\omega_S)\chi_{jb}^{\mu}(\omega_1)\chi_{kc}^{\nu}(\omega_2)\delta_{abc}^{\lambda\mu\nu}. \tag{15.1.13}$$

This relation is a generalization of one originally obtained by Miller [1964]. His derivation in effect assumed only a single internal coordinate representing an electronic mode of motion so that the summation was not needed in his equation and each partial susceptibility became the full linear electric susceptibility. Miller wrote the equation in the principal coordinate system of the susceptibility tensor so that the principal susceptibility components could be divided from the right side. He then found that the numerical values of δ_{abc} components allowed by Kleinman symmetry in a number of crystals varied by about a factor of 2 even though the components of b_{ijk} varied by a factor of 600. This observation, called *Miller's rule*, is very useful, since it shows that the material nonlinearity constants (15.1.10) do not vary greatly from one crystal to another and that large optical mixing tensor components occur in crystals having large linear electric susceptibilities. The variation of the δ_{abc} components between crystals has been related successfully to the characteristics of bond charges in the various crystals by Levine [1973].

In order to account for the frequency dispersion of material tensors of a real crystal even approximately, at least two internal modes of motion, one electronic with its resonance in the ultraviolet and one ionic with its resonance in the infrared, must be included. Let us consider a crystal having just these two. Further let us consider crystals of orthorhombic or higher symmetry so that the principal coordinate system of the linear susceptibility coincides with the crystallographic coordinate system. We also drop the coupling terms between the ionic and electronic modes because of the large separation of their resonances, that is, $^{20}\mathbf{M}$ is taken as diagonal with respect to its superscripts. With these simplifications the mechanical admittance $\boldsymbol{\Upsilon}(\omega)$ (9.1.8) takes the form

$$\Upsilon_{ij}^{\nu\mu}(\omega) = \frac{\delta_{ij}\delta^{\nu\mu}}{2\,{}^{20}M_{jj}^{\mu\mu} - m^{\mu}\omega^2} \tag{15.1.14}$$

and the partial susceptibility (15.1.9) becomes

$$\begin{aligned} \chi_{ij}^{\lambda}(\omega) &= \frac{\delta_{ij}(q^{\lambda})^2}{\epsilon_0\left(2\,{}^{20}M_{jj}^{\lambda\lambda} - m^{\lambda}\omega^2\right)} \\ &= \frac{\delta_{ij}\chi_{jj}^{\lambda}(0)\left(\omega_{Tj}^{\lambda}\right)^2}{\left(\omega_{Tj}^{\lambda}\right)^2 - \omega^2} \end{aligned} \tag{15.1.15}$$

where

$$\left(\omega_{Tj}^{\lambda}\right)^2 \equiv \frac{2\,{}^{20}M_{jj}^{\lambda\lambda}}{m^{\lambda}}, \tag{15.1.16}$$

$$\chi_{jj}^{\lambda}(0)\left(\omega_{Tj}^{\lambda}\right)^2 \equiv \frac{(q^{\lambda})^2}{\epsilon_0 m^{\lambda}}. \tag{15.1.17}$$

From the expression for the partial susceptibility it can be seen that

$$\chi_{ij}^{\lambda}(\omega) \cong 0 \qquad \text{for} \quad \omega \gg \omega_{Tj}^{\lambda}, \tag{15.1.18}$$

which expresses the fact that an oscillator cannot respond at frequencies well above its resonant frequency.

The form of the optical mixing tensor (15.1.13) can be written for the crystal with only one (anisotropic) ionic and one (anisotropic) electronic resonance as

$$\begin{aligned}\epsilon_0 b_{i\ j\ k}^{\omega_S\omega_1\omega_2} = {} & \chi_{ia}^{e}(\omega_S)\chi_{jb}^{e}(\omega_1)\chi_{kc}^{e}(\omega_2)\delta_{abc}^{eee} \\ & + \left[\chi_{ia}^{e}(\omega_S)\chi_{jb}^{e}(\omega_1)\chi_{kc}^{i}(\omega_2)\right. \\ & + \chi_{ia}^{e}(\omega_S)\chi_{jc}^{i}(\omega_1)\chi_{kb}^{e}(\omega_2) \\ & \left. + \chi_{ic}^{i}(\omega_S)\chi_{jb}^{e}(\omega_1)\chi_{ka}^{e}(\omega_2)\right]\delta_{abc}^{eei} \\ & + \text{interchange of } e \leftrightarrow i\end{aligned} \tag{15.1.19}$$

where the permutation symmetry (15.1.12) is used. For such a crystal four Miller delta coefficients, δ_{abc}^{eee}, δ_{abc}^{eei}, δ_{abc}^{eii}, and δ_{abc}^{iii}, need be determined by experiment. If optical harmonic generation or optical sum frequency generation is measured, all three frequencies $\omega_1, \omega_2, \omega_S = \omega_1 + \omega_2$ are large compared to ionic resonant frequencies and so application of (15.1.18) shows that

$$\epsilon_0 b_{i\ j\ k}^{\omega_S\omega_1\omega_2} = \chi_{ia}^{e}(\omega_S)\chi_{jb}^{e}(\omega_1)\chi_{kc}^{e}(\omega_2)\delta_{abc}^{eee} \tag{15.1.20}$$

is measured and so can determine δ_{abc}^{eee}. Similarly, an electrooptic experiment ($\omega_2 \ll \omega_1, \omega_S$) measures

$$\epsilon_0 b_{i\ j\ k}^{\omega_S\omega_1\omega_2} = \chi_{ia}^{e}(\omega_S)\chi_{jb}^{e}(\omega_1)\left[\chi_{kc}^{e}(\omega_2)\delta_{abc}^{eee} + \chi_{kc}^{i}(\omega_2)\delta_{abc}^{eei}\right], \tag{15.1.21}$$

which in conjunction with optical sum frequency generation measurements determines δ_{abc}^{eei}. Microwave mixing experiments [Boyd and Pollack, 1973] measure the entire expression (15.1.19) and so in conjunction with the other measurements just mentioned determine a susceptibility weighted combination of δ_{abc}^{eii} and δ_{abc}^{iii}. Another measurement near the ionic resonance is needed to determine these two coefficients separately.

15.2 Sum and Difference Frequency Generation

The original second harmonic generation experiment of Franken and co-workers [1961] obtained a very small conversion of optical power from the fundamental frequency to the harmonic frequency. It was soon realized by Giordmaine [1962] and Maker et al. [1962] that a dramatic increase in second harmonic power could be obtained by phase matching the interaction through the use of the birefringence of the crystal. Before discussing the technique of phase matching in detail we wish to present a solution for the output optical power for conditions close to phase matching.

We can exhibit many—but not all—of the important characteristics of optical mixing by considering all three waves involved to be plane waves and by ignoring depletion of the input waves. A similar problem is solved in Section 13.5 in the study of acoustooptic diffraction and we only need modify the meaning of some of the quantities in order to apply that solution here. If we consider sum frequency generation,

$$\omega_S = \omega_1 + \omega_2, \tag{15.2.1}$$

between two input optical waves of angular frequencies ω_1 and ω_2, we must solve the wave equation (13.5.1),

$$\left(\frac{c}{\omega_S}\right)^2 \nabla \times (\nabla \times \mathbf{E}(\mathbf{z};1,1)) - \boldsymbol{\kappa}\cdot\mathbf{E}(\mathbf{z};1,1) = \frac{\mathscr{P}(\mathbf{z};1,1)}{\epsilon_0}, \tag{15.2.2}$$

driven by the nonlinear polarization $\mathscr{P}$, for the output field $\mathbf{E}(\mathbf{z};1,1)$ at the sum frequency ω_S. When the input fields are plane waves,

$$\mathbf{E}(\mathbf{z};1,0) = \mathbf{e}^{\eta} E(1,0) e^{i\mathbf{k}_1^{\eta}\cdot\mathbf{z}}, \tag{15.2.3}$$

$$\mathbf{E}(\mathbf{z};0,1) = \mathbf{e}^{\zeta} E(0,1) e^{i\mathbf{k}_2^{\zeta}\cdot\mathbf{z}} \tag{15.2.4}$$

where $\mathbf{e}^{\eta}$ and $\mathbf{e}^{\zeta}$ are unit electric field eigenvectors, the nonlinear polarization (15.1.1) becomes

$$\begin{aligned}\mathscr{P}_i(\mathbf{z};1,1) &= D\epsilon_0 b_{i\ j\ k}^{\omega_S\omega_1\omega_2} e_j^{\eta} e_k^{\zeta} E(1,0)E(0,1)e^{i(\mathbf{k}_1^{\eta}+\mathbf{k}_2^{\zeta})\cdot\mathbf{z}} \\ &= \mathscr{P}_i(1,1)e^{i(\mathbf{k}_1^{\eta}+\mathbf{k}_2^{\zeta})\cdot\mathbf{z}}.\end{aligned} \tag{15.2.5}$$

The solution of this problem near the phase-matching condition,

$$\mathbf{k}_S^\xi = \mathbf{k}_1^\eta + \mathbf{k}_2^\zeta, \tag{15.2.6}$$

is given by the time-averaged Poynting vector $\langle \mathbf{S}^\xi \rangle$ of (13.5.46),

$$\langle \mathbf{S}^\xi(z;1,1) \rangle = \frac{\omega_S^2 z^2 |\mathbf{e}^\xi \cdot \boldsymbol{\mathscr{P}}(1,1)|^2 \Phi(\Delta k_n z/2) \mathbf{t}_S^\xi}{8\epsilon_0 c n^\xi \cos\delta^\xi (\mathbf{n}\cdot\mathbf{t}_S^\xi)^2}. \tag{15.2.7}$$

Here ξ refers to the output eigenmode that can be phase-matched, $\mathbf{e}^\xi$ is the unit electric field eigenvector, $\mathbf{t}_S^\xi$ is the unit vector in the direction of energy propagation, $z = \mathbf{z}\cdot\mathbf{n}$ is a scalar coordinate, $\mathbf{n}$ is the normal to the input surface, Φ is the phase-matching function (13.5.42), Δk_n is the component of the wavevector mismatch,

$$\Delta \mathbf{k} \equiv \mathbf{k}_S^\xi - \mathbf{k}_1^\eta - \mathbf{k}_2^\zeta, \tag{15.2.8}$$

normal to the input surface, and δ^ξ is the angle between the electric field vector and the electric displacement vector. As is shown in Section 13.5, only the normal component of the wavevector mismatch enters (15.2.7) because the tangential component is required to vanish by the input surface boundary condition. A plot of the phase-matching function versus wavevector mismatch is given in Fig. 13.3 and shows that $\langle \mathbf{S}^\xi \rangle$ reaches a maximum when the phase-matching condition (15.2.6) is met. A measurement of the phase-matching curve for collinear second harmonic generation with an infrared gas laser beam in KH_2PO_4 (called KDP) is shown in Fig. 15.1.

The phase-matching function arises from the interference of the free wave and forced wave electric field solutions of the wave equation (15.2.2). Initially they are in phase leading to an increasing sum frequency power; as they get out of phase a maximum is reached; finally they become sufficiently out of phase that sum frequency power is converted back into the power of the input waves and so decreases to zero. The distance (measured normal to the surface in the case of plane waves) taken for the output power to rise from zero to its maximum is called the *coherence length* and is defined by

$$l_c \equiv \frac{\pi}{\Delta k_n}. \tag{15.2.9}$$

Figure 15.2 illustrates the dependence of the sum frequency power on the distance in from the surface for various coherence lengths.

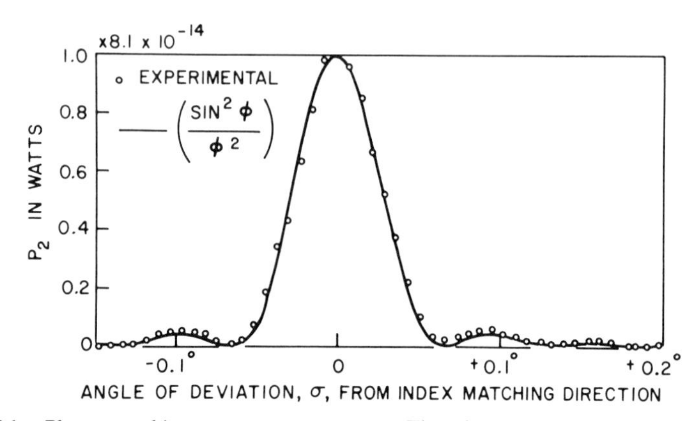

Fig. 15.1. Phase matching curve measurements. The phase matching function $\sin^2\phi/\phi^2$ where $\phi=\Delta k_n l/2$ was measured by varying the angle σ from the phase matching direction in KDP for second harmonic generation from an unfocused 1.1526 μm laser beam from a HeNe gas laser [reprinted by permission from Ashkin, Boyd, and Dziedzic, 1963].

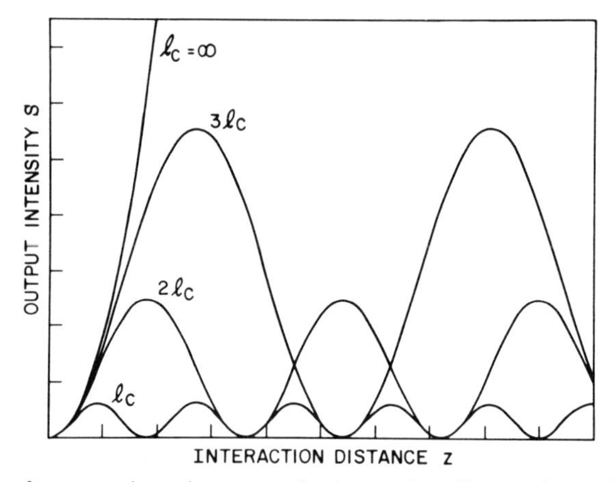

Fig. 15.2. Sum frequency intensity versus the interaction distance for various coherence lengths. The sum frequency formula (15.2.7) is plotted versus the distance into the crystal for various coherence lengths. Infinite coherence length corresponds to exact phase matching in which the sum frequency grows as the square of the distance [adapted with permission from Kleinman, 1972].

When a wavevector mismatch is present, the free wave propagation vector $\mathbf{k}_S^\xi$ and the driven wave propagation vector $\mathbf{k}_1^\eta+\mathbf{k}_2^\zeta$ are not parallel. However, the solution of Section 13.5 shows that the free wave excited is the one having the sum frequency ω_S and a wavevector component tangential to the input surface equal to that of the driven wave,

$$\left(\mathbf{k}_S^\xi\right)_t=\left(\mathbf{k}_1^\eta+\mathbf{k}_2^\zeta\right)_t. \tag{15.2.10}$$

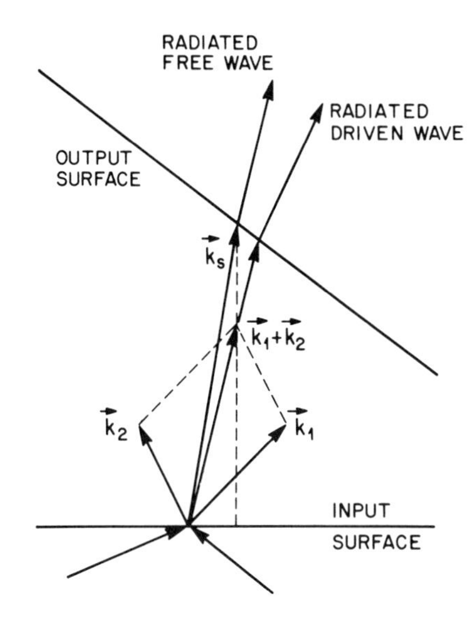

Fig. 15.3. Geometry for observation of free and forced waves separately. The two input waves are represented by wavevectors $\mathbf{k}_1$ and $\mathbf{k}_2$. They produce a driven wave with wavevector $\mathbf{k}_1+\mathbf{k}_2$ and a free wave with wavevector $\mathbf{k}_S$. The projections of the wavevectors of the latter two waves are the same on the input crystal surface but differ on the output surface because it is not parallel to the input surface. The free and forced waves are thus refracted out at different angles.

If the output crystal surface is parallel to the input surface, this condition guarantees that both the free and driven waves refract at the same exterior angle at the output surface. Their separate identity is then not apparent. If the crystal is made wedge-shaped as shown in Fig. 15.3, the free and driven waves refract at different exterior angles since their propagation vectors have different components tangential to the output surface. The free and driven waves can then be separately observed as done in an experiment by Giordmaine and Rentzepis [1967] shown in Fig. 15.4.

Equation (15.2.7) can be better expressed if we introduce the effective optical mixing susceptibility b, defined as

$$b \equiv e_i^{\xi} b_{i\ j\ k}^{\omega_S \omega_1 \omega_2} e_j^{\eta} e_k^{\zeta}, \tag{15.2.11}$$

and the magnitudes of the time-averaged Poynting vectors of the two input waves. The result is

$$\langle \mathbf{S}^{\xi}(1,1) \rangle = \frac{(\omega_S D z b)^2 \langle S^{\eta}(1,0) \rangle \langle S^{\zeta}(0,1) \rangle \Phi(\Delta k_n z/2) \mathbf{t}_S^{\xi}}{2\epsilon_0 c^3 n^{\xi} \cos\delta^{\xi} n^{\eta} \cos\delta^{\eta} n^{\zeta} \cos\delta^{\zeta} (\mathbf{n}\cdot\mathbf{t}_S^{\xi})^2} \tag{15.2.12}$$

where $D=2$ if the two input frequencies are different. For second harmonic generation ($D=1$) from a single input wave ($\eta=\zeta$) this equation becomes

$$\langle \mathbf{S}^{\xi}(2,0) \rangle = \frac{(\omega_H z b)^2 \langle S^{\eta}(1,0) \rangle^2 \Phi(\Delta k_n z/2) \mathbf{t}_H^{\xi}}{2\epsilon_0 c^3 n^{\xi} \cos\delta^{\xi} (n^{\eta})^2 \cos^2\delta^{\eta} (\mathbf{n}\cdot\mathbf{t}_H^{\xi})^2} \tag{15.2.13}$$

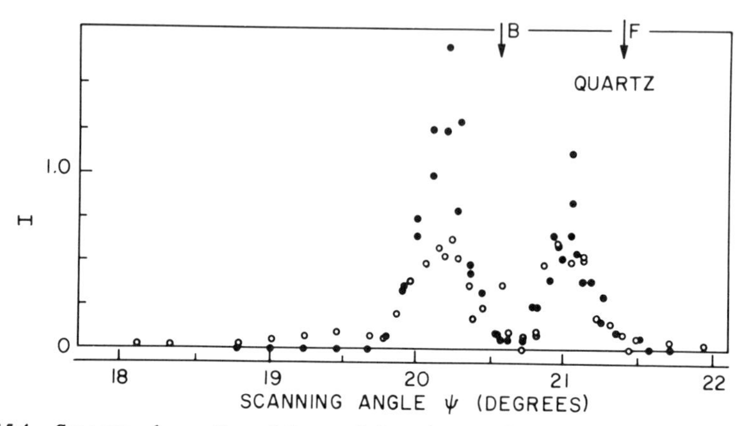

Fig. 15.4. Separate observation of free and forced waves [reprinted by permission from Giordmaine and Rentzepis, 1967]. The intensity of ultraviolet sum frequency light at 231.4 nm produced by mixing 347.2 nm and 694.3 nm light (see Fig. 15.3) is plotted versus the angle ψ measured relative to the transmitted 694.3 nm beam. The expected angles for the driven or bound wave (B) and the free wave (F) are shown. The slight angular discrepancies between expected and observed positions of the intensity peaks were probably caused by crystal misorientation.

where

$$\omega_H = 2\omega_F, \tag{15.2.14}$$

phase matching means

$$\mathbf{k}_H^\xi = 2\mathbf{k}_F^\eta, \tag{15.2.15}$$

and H and F refer to second harmonic and fundamental respectively. The presence of the square of the input *intensity* in the equation shows that for a given *power* in the fundamental wave the harmonic wave intensity increases if the fundamental wave is intensified by focusing.

The output intensity from difference frequency generation can be found by an analogous procedure. In the notation above the output frequency is

$$\omega_1 = \omega_S - \omega_2 \tag{15.2.16}$$

and phase matching is given by

$$\mathbf{k}_1 = \mathbf{k}_S - \mathbf{k}_2. \tag{15.2.17}$$

The output intensity is then given by

$$\langle \mathbf{S}^{\eta}(1,0)\rangle = \frac{2(\omega_1 zb)^2 \langle S^{\xi}(1,1)\rangle \langle S^{\zeta}(0,1)\rangle \Phi(\Delta k_n z/2)\mathbf{t}_1^{\eta}}{\epsilon_0 c^3 n^{\eta}\cos\delta^{\eta} n^{\xi}\cos\delta^{\xi} n^{\zeta}\cos\delta^{\zeta}(\mathbf{n}\cdot\mathbf{t}_1^{\eta})^2}. \quad (15.2.18)$$

In the plane wave geometry assumed here the phase matched output given by either (15.2.12), (15.2.13), or (15.2.18) increases as the square of the interaction length. This dependence is shown as the envelope curve in Fig. 15.2 labeled by $l_c = \infty$. The square law behavior can persist only for as long as input wave depletion remains negligible as was assumed in this calculation. Armstrong and co-workers [1962] have obtained the solution that includes input wave depletion.

With intense input waves a sizable fraction of the incident energy has been converted to the second harmonic in various experiments. Nath and co-workers [1970] obtained 40% power conversion of a 694 nm ruby laser beam into a 347 nm ultraviolet beam in $LiIO_3$. Schinke [1972] converted 54% of the power of a 532 nm light beam, itself a second harmonic of a 1064 nm beam from a YAG:Nd laser, into a 266 nm ultraviolet beam in $(NH_4)H_2PO_4$ (called ADP). Geusic and co-workers [1968] placed a $Ba_2NaNb_5O_{15}$ crystal within the cavity of a YAG:Nd laser to resonate the fundamental wave at 1064 nm. They obtained a second harmonic power equal to the fundamental power present in the absence of the $Ba_2NaNb_5O_{15}$ crystal and in that sense obtained 100% conversion. Figure 15.5 shows the experimental arrangement used by Geusic and co-workers.

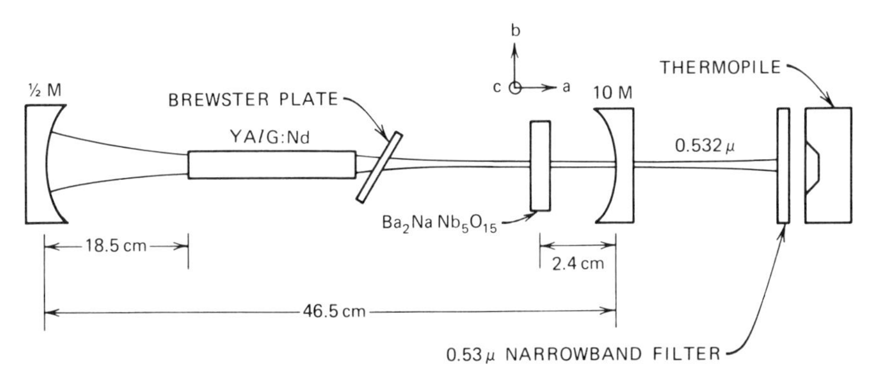

Fig. 15.5. Harmonic generation within a laser cavity. A second harmonic power of 1.1 W was obtained in a 532 nm wavelength beam generated by a $Ba_2NaNb_5O_{15}$ crystal placed inside the cavity of a YAG:Nd laser. Separate measurements with the $Ba_2NaNb_5O_{15}$ crystal removed found 1.1 W of infrared power at 1064 nm generated by the YAG:Nd laser. Thus 100% conversion of power to the second harmonic was obtained [reprinted by permission from Geusic et al., 1968].

15.3 Phase Matching

We saw in the preceding section that sum frequency generation,

$$\omega_S = \omega_1 + \omega_2, \tag{15.3.1}$$

had its maximum output when the interaction is phase matched,

$$\mathbf{k}_S^\xi = \mathbf{k}_1^\eta + \mathbf{k}_2^\zeta. \tag{15.3.2}$$

Though this condition is formally the same as the phase-matching condition for acoustooptic diffraction, discussed at length in Section 13.6, its implementation in an entirely optical interaction is quite different. In this section we describe various methods that are used. It should be remembered that any physical situation that satisfies the frequency condition (15.3.1) and the wavevector condition (15.3.2) for sum frequency generation can also be used for difference frequency generation, since (15.2.16) and (15.2.17) are equivalent to these equations.

The phase-matching condition (15.3.2) is a vector equation and so applies to interactions in which the input and output waves propagate noncollinearly as well as collinearly. Since in practice the input waves have a finite, and indeed rather small, lateral extent ($\lesssim 1\,\mathrm{mm}$) noncollinear interactions have a short interaction distance and so have a small conversion efficiency. For this reason optical mixing experiments are almost always done with all the waves propagating collinearly. We thus restrict the discussion to collinear phase matching. We also specialize to second harmonic generation for simplicity of illustration.

For collinear second harmonic generation we have

$$\omega_H = 2\omega_F, \tag{15.3.3}$$

$$k_H^\xi = 2k_F^\eta. \tag{15.3.4}$$

Combining these equations yields a condition on the refractive indices,

$$n^\xi(\omega_H) = n^\eta(\omega_F). \tag{15.3.5}$$

This condition cannot ordinarily be met if the eigenmode ξ of the second harmonic and the eigenmode η of the fundamental are the same type. This is because in the optical region of transparent crystals the refractive index increases with increasing frequency (*normal dispersion*). The solution to this problem, found almost simultaneously by Giordmaine [1962] and Maker and co-workers [1962], is to have the second harmonic and the fundamental propagate in different eigenmodes, that is, different polarizations, in a birefringent crystal. Thus a solution to (15.3.5) is obtained by using birefringence to compensate for dispersion. The method cannot work in cubic crystals since they possess no birefringence.

Two cases of birefringence phase matching arise in uniaxial crystals depending on whether the crystal is positive or negative. A positive uniaxial crystal has $\kappa_{33} > \kappa_{11}$; a negative uniaxial crystal has $\kappa_{33} < \kappa_{11}$. If we assume a region of normal dispersion, phase matching of collinear second harmonic generation in a negative crystal can occur only if the fundamental is an ordinary wave and the harmonic is an extraordinary wave. This gives (15.3.5) the form

$$n_H^e(\theta) = n_F^o \qquad (-\text{crystal}). \tag{15.3.6}$$

For a positive crystal the reverse is true,

$$n_H^o = n_F^e(\theta) \qquad (+\text{crystal}). \tag{15.3.7}$$

Figure 15.6 illustrates these two cases. If the expression (9.6.8) for the extraordinary refractive index as a function of θ is substituted into (15.3.6), the phase-matching angle θ_m can be found to satisfy

$$\sin^2\theta_m = \frac{(n_H^e)^2}{(n_F^o)^2} \cdot \frac{(n_H^o)^2 - (n_F^o)^2}{(n_H^o)^2 - (n_H^e)^2} \qquad (-\text{crystal}) \tag{15.3.8}$$

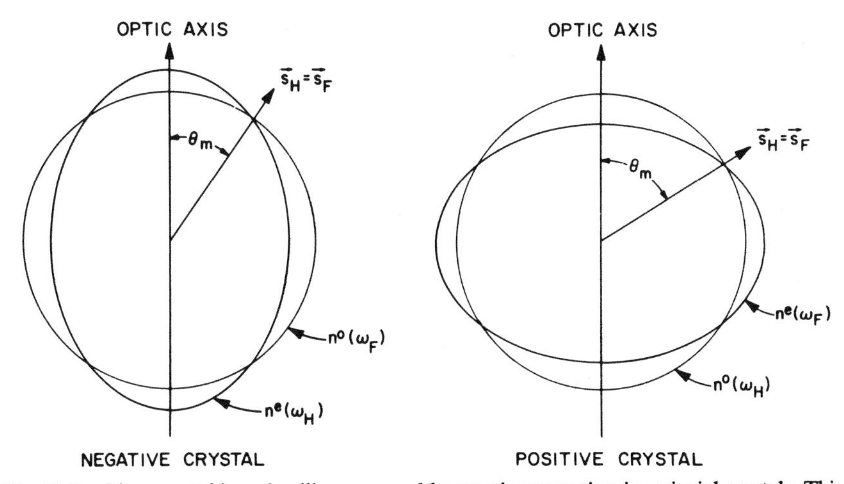

Fig. 15.6. Phase matching of collinear second harmonic generation in uniaxial crystals. This requires the equality of the refractive indices of the fundamental and the harmonic waves. The equality occurs in two cones of angles about the optic axis of half apex angle θ_m.

where the refractive indices are the principal refractive indices $[(n^o)^2=\kappa_{11},(n^e)^2=\kappa_{33}]$. Solution of (15.3.7) for the phase-matching angle in a positive crystal yields (15.3.8) if we interchange F and H.

Phase matching relates only the wavevectors of the input and output waves. As is shown in Section 9.10 the energy of a light wave in an anisotropic medium travels in a different direction from the wavevector. Since the input and output waves are in eigenmodes of different type for birefringence phase matching, phase matching cannot guarantee that the energy propagation directions of the fundamental and harmonic in a collinear second harmonic generation process are collinear. If the waves have infinite lateral extent, as is assumed in the plane wave analysis of the last section, this noncollinearity produces no problem, since there is no loss of overlap of the waves. However, for light beams of finite lateral extent the noncollinearity of energy propagation directions of the fundamental and harmonic lead to a loss of spatial overlap, called *walk-off*. The coherence length at perfect phase matching is then not the crystal length as would be concluded from the calculation of Section 15.2.

The calculation of harmonic generation with finite diameter beams including the effects of walk-off was carried out by Boyd and co-workers [1965]. Instead of presenting this complicated calculation we present a simple yet semiquantitative treatment of walk-off. Consider phase matched collinear second harmonic generation in a negative uniaxial crystal shown in Fig. 15.7. The energy propagation direction of the second harmonic extraordinary wave makes the angle δ with its wavevector which is aligned with that of the fundamental ordinary wave. This angle is found in (9.P.2) for the extraordinary ray in a uniaxial crystal to be given by

$$\tan\delta=\tfrac{1}{2}[n^e(\theta)]^2\left[\frac{1}{(n^e)^2}-\frac{1}{(n^o)^2}\right]\sin 2\theta. \tag{15.3.9}$$

The distance that the second harmonic typically overlaps the fundamental is called the *aperture length* l_a and can be seen from the figure to be given by

$$l_a=\frac{a}{\delta}, \tag{15.3.10}$$

when, as usual, δ is a small angle. Since the second harmonic is generated throughout the crystal length l, but has only an overlap length l_a, it is reasonable to conjecture that the total second harmonic *power* will be proportional to ll_a rather than l^2 as found for the infinitely wide waves. Note from the figure that the intensity would be lower still by the factor

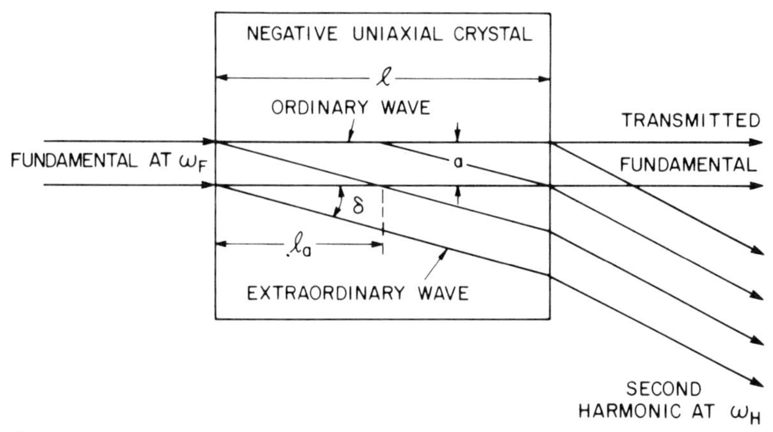

Fig. 15.7. Energy walk-off in collinear second harmonic generation. For the ordinary wave fundamental the energy propagates parallel to the wavevector which is aligned with the wavevector of the extraordinary wave harmonic. The energy of the harmonic wave, however, "walks off" from the collinear interaction at an angle δ, thus lessening the effective interaction distance.

$a/(a+l\delta)=l_a/(l+l_a)$. The theory of Boyd and coworkers [1965] confirms that this conjecture is approximately true. Clearly the same considerations apply for a positive uniaxial crystal in which the fundamental is an extraordinary wave and the second harmonic is an ordinary wave and similar ones also apply for a biaxial crystal.

It is apparent that walk-off seriously limits conversion efficiency in a second harmonic generation process. However, if for the frequencies of interest a crystal can be found that allows phase matching in a collinear process normal to the optic axis ($\theta=\pi/2$), (15.3.9) shows that the walk-off angle can be made zero. Phase matching normal to the optic axis has another major advantage analogous to one found for phase matching of acoustooptic diffraction in Section 13.6. Phase matching normal to the optic axis is maintained to first order in the angular deviation $\Delta\theta$. This can be seen by expanding the phase mismatch $\Delta k_H(\theta)$ about $\Delta k_H(\pi/2)=0$ for a collinear harmonic generation process in a negative uniaxial crystal,

$$\begin{aligned}\Delta k_H\left(\frac{\pi}{2}+\Delta\theta\right)&=k_H^e-2k_F^o\\&=\frac{\omega_H}{c}\left[n_H^e\left(\frac{\pi}{2}+\Delta\theta\right)-n_F^o\right]\\&=\frac{\omega_H n_H^e}{2c(n_H^o)^2}\left[(n_H^o)^2-(n_H^e)^2\right](\Delta\theta)^2.\end{aligned}\tag{15.3.11}$$

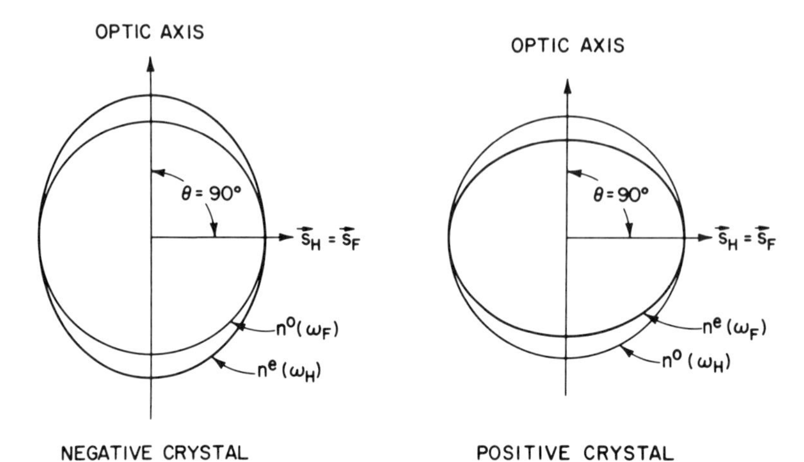

Fig. 15.8. Noncritical phase matching for second harmonic generation. Phase matching occurs only in the plane normal to the optic axis. Because the index curves are tangent to one another, phase matching is maintained to first order about this direction. Also, energy walk-off is zero.

This situation is called *noncritical phase matching* and results from the two refractive index curves being tangent for $\theta = \pi/2$ (see Fig. 15.8). Miller, Boyd, and Savage [1965] were the first to demonstrate noncritical phase matching. They used $LiNbO_3$ heated to adjust its refractive indices to permit noncritical phase matching for the 1064 nm wavelength of a $CaWO_4$: Nd laser used as the fundamental.

Plane wave treatments of harmonic generation can only suggest what occurs when the input light wave is focused. Because (15.2.13) shows that the second harmonic intensity increases as the square of the fundamental intensity, some focusing of the fundamental can be expected to improve conversion efficiency. Too strong focusing, however, should lead to a decrease in conversion efficiency caused by the phase mismatch from the increased spread of wavevector directions. Studies by Kleinman et al. [1966] and Boyd and Kleinman [1968] have determined the optimum focusing.

Another method of phase matched second harmonic generation which prevents walk-off uses a dielectric waveguide as the propagation medium. If the fundamental and second harmonic are to propagate in individual waveguide modes, the guide dimensions and the amount of refractive index increase inside the guide (needed to produce a waveguide) must be adjusted to produce modes that can be phase matched. The presence of birefringence is not essential and so acentric cubic crystals may be used. Typically the field distributions of the fundamental and harmonic will differ substantially. This leads to a reduced coupling per unit length of waveguide.

15.4 Measurement of Optical Mixing Tensor Dispersion

The $\Upsilon(\omega)$ factors present in the definition of the optical mixing tensor (15.1.7) give rise to strong frequency dispersion in the neighborhood of ionic and electronic resonances. Faust and Henry [1968] were the first to measure this dispersion in the vicinity of an ionic resonance. They chose GaP, a diatomic cubic crystal, for study, since it has only a single ionic resonant frequency and the parameters characterizing it were already known. Also, laser lines from a water vapor laser were available in the wavelength range of 20 to 60 μm and so covered the resonance region.

The experiment mixed a visible HeNe laser line at 632.8 nm with the various H_2O laser lines to produce seven sum frequencies and five difference frequencies. These combination frequencies appear as anti-Stokes and Stokes components spread less than 20 nm from the HeNe laser line. Since two of the three frequencies involved are visible and one infrared, (15.1.21) for the mixing tensor of a crystal with one ionic resonance and one (effective) electronic resonance,

$$\epsilon_0 b_{i\ j\ k}^{\omega_S\omega_1\omega_2} = \chi_{ia}^{e}(\omega_S)\chi_{jb}^{e}(\omega_1)\left[\chi_{kc}^{e}(\omega_2)\delta_{abc}^{eee} + \chi_{kc}^{i}(\omega_2)\delta_{abc}^{eei}\right], \qquad (15.4.1)$$

where ω_1 is the HeNe laser frequency and ω_2 is a H_2O laser frequency, is applicable. For a difference frequency, $\omega_D = \omega_1 - \omega_2$, the sum frequency ω_S in (15.4.1) may be simply replaced by ω_D; the slightly altered value of the $\chi_{ia}^{e}(\omega_S)$ factor is negligible in this experiment. For a cubic crystal the partial susceptibility $\chi_{ij}^{\mu}(\omega)$ is given from (15.1.15) to be

$$\chi_{ij}^{\mu}(\omega) = \frac{\delta_{ij}\chi^{\mu}(0)(\omega_T^{\mu})^2}{(\omega_T^{\mu})^2 - \omega^2}. \qquad (15.4.2)$$

With this expression (15.4.1) may be rewritten as

$$b_{i\ j\ k}^{\omega_S\omega_1\omega_2} = B_{ijk}\left[1 + \frac{C}{1-(\omega_2/\omega_T)^2}\right] \qquad (15.4.3)$$

where

$$\epsilon_0 B_{ijk} \equiv \chi^e(\omega_S)\chi^e(\omega_1)\chi^e(\omega_2)\delta_{ijk}^{eee}, \tag{15.4.4}$$

$$C \equiv \frac{\chi^i(0)\delta_{ijk}^{eei}}{\chi^e(\omega_2)\delta_{ijk}^{eee}}. \tag{15.4.5}$$

For the conditions of the experiment both B_{ijk} and C are substantially constant with respect to frequency. The superscript i is dropped from the transverse optic resonant frequency ω_T of the ionic mode.

When studying the dispersion of an interaction tensor close to a resonance, it is necessary to include the effects of absorptive loss in the ionic oscillator. Since we do not include absorptive loss in this treatment, we simply point out that its inclusion alters the form of the resonant denominator of (15.4.3) in a standard way to be

$$b_{i\ j\ k}^{\omega_S\omega_1\omega_2} = B_{ijk}\left[1 + \frac{C}{1 - \omega_2^2/\omega_T^2 - i\omega_2\Gamma/\omega_T^2}\right] \tag{15.4.6}$$

where Γ is the damping constant of the oscillator.

The Faust-Henry experiment measured the modulus of the mixing tensor, $|b_{i\ j\ k}^{\omega_S\omega_1\omega_2}|$, on a relative scale as a function of ω_2, the H_2O laser frequency. For GaP (crystal class $\bar{4}3m$) there is only one independent tensor element $d_{14} = d_{25} = d_{36}$ (see the Appendix). Because GaP is cubic, birefringence phase matching is not possible. However, the H_2O laser frequencies fall in a region of strong dispersion and so some of them could have been phase matched in a noncollinear interaction geometry. This was not possible for other frequencies used. Since good relative values of the optical mixing tensor were desired, all frequencies were studied in a single non-phase-matched geometry.

When a non-phase-matched geometry is studied, a slight tilt of the crystal can be used to put the phase-matching function on one of the maxima in Fig. 15.2 for which $\sin(\Delta k_n z/2) = 1$. In this case $z^2\Phi(\Delta k_n z/2) = (2l_c/\pi)^2$ and so the generated output power is proportional to the square of the coherence length which requires a knowledge of both visible and infrared refractive indices. The output power equation (15.2.7) needs modification to account for absorption at the ω_2 frequency, which, however, we do not consider here. The values of the absorption coefficient at the various frequencies along with ω_T and Γ are known for GaP from infrared reflectivity studies. The constant C was previously found from Raman scattering experiments. Thus the data from the Faust-Henry experiment could be compared with a predicted frequency dependence from

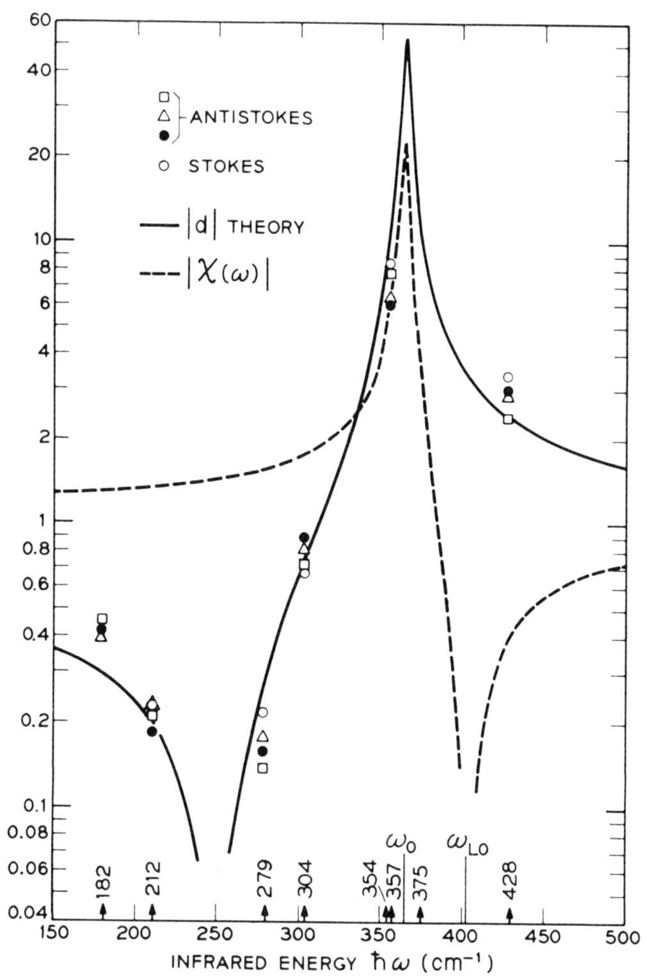

Fig. 15.9. Optical mixing tensor dispersion in the ionic resonance region of GaP. The measured points [adapted with permission from Faust et al., 1968] agree well with the predicted dispersion of $|d| \equiv |b_{14}|$. The linear susceptibility $|\chi|$, as well as the transverse optic frequency ω_O and the longitudinal optic frequency ω_{LO}, are also shown.

(15.4.6). This is shown in Fig. 15.9. The agreement with the strong frequency dispersion is seen to be very good.

Since optical mixing experiments measure the square of the effective optical mixing coefficient (15.4.6), they cannot be used to determine the algebraic sign of optical mixing tensor coefficients. However, the Faust-Henry experiment showed that the b_{14} optical mixing component in GaP had the same algebraic sign above the ionic resonance as below it. Since there is no further dispersion in the tensor at lower frequencies until the piezoelectric resonances occur, a measurement of the algebraic sign of the high frequency electrooptic tensor component r_{14} by determining the sign of the phase difference (14.10.8) [Nelson and Turner, 1968] thus determines the sign (positive) of the optical mixing tensor of GaP in the visible frequency range.

15.5 Three-Wave Parametric Interaction

The *parametric interaction* of three waves is one in which the highest frequency wave, called the *pump wave*, at frequency ω_P is an input wave and gives rise to gain at two lower frequencies. The generated waves at the two lower frequencies are called the *signal wave* and the *idler wave* and have frequencies ω_S and ω_I satisfying

$$\omega_P = \omega_S + \omega_I. \tag{15.5.1}$$

The three waves have propagation vectors that satisfy, at least approximately, phase matching,

$$\mathbf{k}_P = \mathbf{k}_S + \mathbf{k}_I, \tag{15.5.2}$$

where P, S, and I denote pump, signal, and idler waves. The parametric generation of the signal and idler waves is the reverse process of optical mixing of those waves to produce the pump wave.

The latter fact suggests that the technique of solution used for the optical mixing problem, which is developed in Section 13.5 for acoustooptic diffraction, could be used for the parametric generation problem. However, we wish instead to make use of the understanding gained from that solution near phase matching to approximate the wave equation and so obtain a simplified differential equation governing each of the three parametrically interacting waves. Such equations are called *coupled mode equations*.

Several simplifications occur near phase matching: all three waves have electric field eigenvectors and propagation vectors belonging to freely propagating waves; the weak coupling of the optical mixing tensor causes only a small change in the amplitude of each wave within its wavelength; this change depends only on the scalar coordinate,

$$z \equiv \mathbf{n} \cdot \mathbf{z}, \tag{15.5.3}$$

measured along the inward normal $\mathbf{n}$ to the input surface; and only the scalar component of the wave equation with the respective electric field eigenvector need be considered for each wave. The dependence of the amplitudes on the scalar coordinate z means that planes of *constant amplitude* for each wave are parallel to the slab surfaces even though the planes of *constant phase* are perpendicular to the propagation vector of each wave. Because of the high velocity of light transient effects in parametric generation processes are very rapid and not commonly measured. For this reason we develop only a steady state formulation for the interacting waves. Thus the amplitudes are regarded as time independent. For the time dependent generalization see Prob. 15.2.

We consider the three waves as plane waves interacting in a crystalline slab of arbitrary crystallographic orientation shown in Fig. 15.10. The three waves may be noncollinear as long as their propagation vectors satisfy exactly or approximately the phase-matching condition (15.5.2). Waves whose propagation vector or Poynting vector are parallel to the input surface are excluded; such waves could not have entered the medium by way of the input surface. We use the Fourier expansion given in (12.2.3) to (12.2.5) and associate the amplitude notation (1, 1), (1, 0), and (0, 1) with

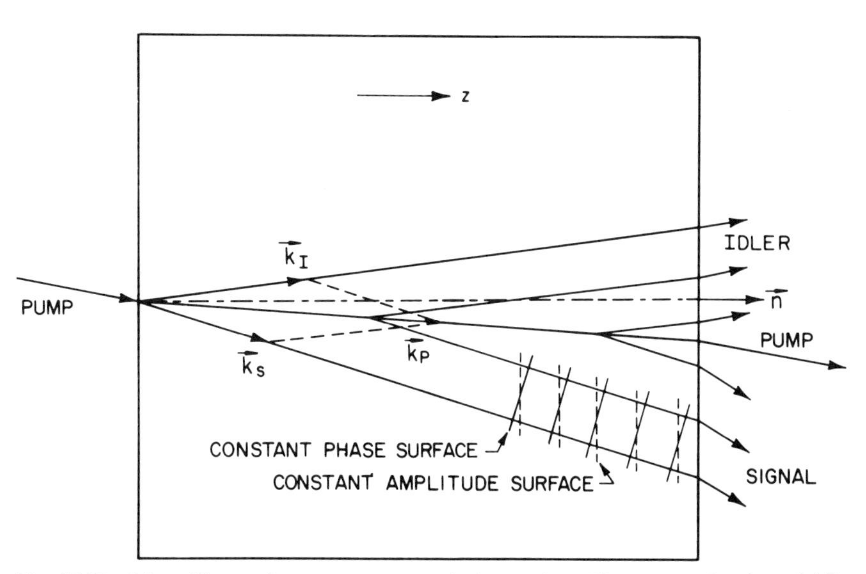

Fig. 15.10. Noncollinear three-wave parametric interaction. The pump, signal, and idler waves propagate in the directions $\mathbf{k}_P$, $\mathbf{k}_S$, and $\mathbf{k}_I$ which nearly satisfy phase matching. Wave amplitude variations can depend only on z, making constant phase and constant amplitude surfaces different.

the waves having frequencies ω_P, ω_S, and ω_I respectively. The unit vectors in the directions of propagation of the three waves are denoted by $\mathbf{s}_P$, $\mathbf{s}_S$, $\mathbf{s}_I$. The development is patterned after Kleinman [1968].

The Fourier expansion of the general driven wave equation (9.2.3) produces a wave equation of the form of (12.3.14),

$$\left(\frac{c}{\omega}\right)^2 \nabla\times(\nabla\times\mathbf{E}) - \boldsymbol{\kappa}\cdot\mathbf{E} = \frac{\mathscr{P}}{\epsilon_0}, \tag{15.5.4}$$

for each frequency, $\omega=\omega_P, \omega_S, \omega_I$, with the nonlinear polarization $\mathscr{P}$ given by (15.1.1). Consider the wave equation for the pump wave first. As pointed out above, we need solve only the scalar product of this wave equation with the unit electric field eigenvector of the pump wave $\mathbf{e}_P$,

$$\left(\frac{c}{\omega_P}\right)^2 \mathbf{e}_P\cdot\nabla\times(\nabla\times\mathbf{E}) - \mathbf{e}_P\cdot\boldsymbol{\kappa}\cdot\mathbf{E} = \frac{\mathbf{e}_P\cdot\mathscr{P}}{\epsilon_0}. \tag{15.5.5}$$

The time independent electric field of the pump wave is

$$\mathbf{E} = \mathbf{e}_P E(z\,;1,1)e^{i\mathbf{k}_P\cdot\mathbf{z}} \tag{15.5.6}$$

where $\mathbf{k}_P$ is the wavevector of a free wave. Corresponding expressions are used for the signal and idler waves. The assumption that the nonlinear coupling of the waves produces slowly varying plane wave amplitudes means that

$$k_P E(z\,;1,1) \gg \frac{\partial E(z\,;1,1)}{\partial z}. \tag{15.5.7}$$

This allows us to neglect second derivatives of the amplitude in the wave equation. Equation (15.5.5) for the pump wave is then

$$\left\{\left(\frac{c}{\omega_P}\right)^2\left[2i\mathbf{e}_P\cdot\mathbf{k}_P\mathbf{e}_P\cdot\nabla - (\mathbf{e}_P\cdot\mathbf{k}_P)^2 + k_P^2 - 2i\mathbf{k}_P\cdot\nabla\right] - \mathbf{e}_P\cdot\boldsymbol{\kappa}\cdot\mathbf{e}_P\right\} E(z\,;1,1) = \frac{\mathbf{e}_P\cdot\mathscr{P}}{\epsilon_0}e^{-i\mathbf{k}_P\cdot\mathbf{z}}. \tag{15.5.8}$$

Since the propagation constant is essentially unperturbed, we may subtract from this equation terms of the undriven wave equation,

$$\mathbf{e}_P \cdot \left[-\frac{c^2}{\omega_P^2} (\mathbf{k}_P \mathbf{k}_P - k_P^2 \mathbf{1}) - \boldsymbol{\kappa} \right] \cdot \mathbf{e}_P E(z;1,1) e^{i\mathbf{k}_P \cdot \mathbf{z}} = 0. \qquad (15.5.9)$$

The result is

$$\left[(\mathbf{e}_P \cdot \mathbf{s}_P)\mathbf{e}_P - \mathbf{s}_P \right] \cdot \nabla E(z;1,1) = -\frac{i\omega_P}{2\epsilon_0 c n_P} \mathbf{e}_P \cdot \mathscr{P} e^{-i\mathbf{k}_P \cdot \mathbf{z}}. \qquad (15.5.10)$$

From (9.10.28) we see that the bracketed quantity is $-\mathbf{t}_P \cos\delta_P$ where $\mathbf{t}_P$ is the unit vector in the direction of energy propagation and δ_P is the angle between that direction and the propagation direction $\mathbf{s}_P$. We also may write the gradient operator as

$$\nabla = \mathbf{n} \frac{\partial}{\partial z}, \qquad (15.5.11)$$

because it now acts only on the electric field amplitude which is only a function of the scalar coordinate z.

Equation (15.5.10) now becomes

$$\frac{\partial E(z;1,1)}{\partial z} = \frac{i\omega_P \mathbf{e}_P \cdot \mathscr{P} e^{-i\mathbf{k}_P \cdot \mathbf{z}}}{2\epsilon_0 c n_P \cos\delta_P \mathbf{n} \cdot \mathbf{t}_P}. \qquad (15.5.12)$$

If the nonlinear polarization $\mathscr{P}$ from (15.1.1) is introduced with $D=2$, since $\omega_S \neq \omega_I$, it is convenient to define the effective optical mixing susceptibility as

$$b \equiv (e_P)_i b_{i\ j\ k}^{\omega_P \omega_S \omega_I} (e_S)_j (e_I)_k. \qquad (15.5.13)$$

We take the signal and idler waves to be plane waves of the form (15.5.6). Thus the phase mismatch,

$$\Delta\mathbf{k} \equiv \mathbf{k}_P - \mathbf{k}_S - \mathbf{k}_I, \qquad (15.5.14)$$

may be introduced. Since the input surface boundary conditions force the component of $\Delta\mathbf{k}$ tangential to the surface to vanish, $\Delta\mathbf{k}$ may be replaced by $\Delta k_n \mathbf{n}$. The equation for the pump wave amplitude then takes its final form,

$$\frac{\partial E(z;1,1)}{\partial z}=\frac{i\omega_P bE(z;1,0)E(z;0,1)e^{-i\Delta k_n z}}{cn_P\cos\delta_P\mathbf{n}\cdot\mathbf{t}_P}. \tag{15.5.15}$$

The development of the equation for the signal wave amplitude proceeds analogously except that the Fourier expansion gives us the complex conjugate of the idler field $E^*(z;0,1)$ in the coupling term in order that the term oscillate at the signal frequency $\omega_S=\omega_P-\omega_I$. The result is

$$\frac{\partial E(z;1,0)}{\partial z}=\frac{i\omega_S bE(z;1,1)E^*(z;0,1)e^{i\Delta k_n z}}{cn_S\cos\delta_S\mathbf{n}\cdot\mathbf{t}_S}. \tag{15.5.16}$$

The permutation symmetry of the optical mixing tensor, whereby tensor indices and their associated frequencies may be permuted in any order [see (14.1.2)], must be used in the derivation so that the coefficient b entering (15.5.15) is the same as that in (15.5.16). The equation for the idler wave amplitude,

$$\frac{\partial E(z;0,1)}{\partial z}=\frac{i\omega_I bE(z;1,1)E^*(z;1,0)e^{i\Delta k_n z}}{cn_I\cos\delta_I\mathbf{n}\cdot\mathbf{t}_I}, \tag{15.5.17}$$

is found analogously. The last three equations are *coupled mode equations* that are first order differential equations for the scalar amplitudes of the three interacting waves.

15.6 Manley-Rowe Relations

The similarity of the right sides of the last three equations suggests an important relationship between the left sides. If we multiply (15.5.15) by $E^*(z;1,1)$ and add the complex conjugate of the result to itself, we have

$$\frac{\partial}{\partial z}\left[\frac{\epsilon_0 cn_P\cos\delta_P\mathbf{n}\cdot\mathbf{t}_P}{2\omega_P}|E(z;1,1)|^2\right]=\Gamma \tag{15.6.1}$$

where

$$\Gamma\equiv\Re\left\{i\epsilon_0 bE^*(z;1,1)E(z;1,0)E(z;0,1)e^{-i\Delta k_n z}\right\}. \tag{15.6.2}$$

Since the time-averaged Poynting vector of the pump beam is

$$\langle \mathbf{S}_P(z) \rangle = \frac{\epsilon_0 c}{2} n_P \cos\delta_P |E(z;1,1)|^2 \mathbf{t}_P, \tag{15.6.3}$$

(15.6.1) can be written as

$$\frac{\partial}{\partial z}\left(\frac{\langle \mathbf{S}_P(z) \cdot \mathbf{n} \rangle}{\omega_P} \right) = \Gamma. \tag{15.6.4}$$

Similar handling of the signal and idler equations gives equations analogous to this (except for a minus sign) and leads to

$$\frac{\partial}{\partial z}\left(\frac{\langle \mathbf{S}_S(z) \cdot \mathbf{n} \rangle}{\omega_S} \right) = \frac{\partial}{\partial z}\left(\frac{\langle \mathbf{S}_I(z) \cdot \mathbf{n} \rangle}{\omega_I} \right) = -\frac{\partial}{\partial z}\left(\frac{\langle \mathbf{S}_P(z) \cdot \mathbf{n} \rangle}{\omega_P} \right), \tag{15.6.5}$$

which are called the *Manley-Rowe relations* [Manley and Rowe, 1956].

Though the derivation of the Manley-Rowe relations is based on continuum physics, they are most easily interpreted from the viewpoint of quantum physics. Since a quantum of light energy, a *photon*, is $\hbar\omega$, where $\hbar$ is Planck's constant divided by 2π, dividing the Manley-Rowe relations by $\hbar$ makes them relations between the changes in the number of photons in the signal, idler, and pump waves. They then state that the *increase* in the number of signal wave photons per unit distance normal to the input surface equals the *increase* in the number of idler wave photons which in turn equals the *decrease* in the number of pump wave photons. Note that this growth and decay occur normal to the input surface even though the propagation vectors and Poynting vectors of the three waves are in other directions.

This interpretation is in agreement with a quantum interpretation of the frequency condition (15.5.1). If that equation is multiplied by $\hbar$ and the equation interpreted as an energy conservation statement of quanta, it states that the loss of one pump beam quantum causes the gain of one signal beam quantum and one idler beam quantum.

The close connection of the Manley-Rowe relations to energy conservation can also be seen by multiplying (15.6.4) by ω_P and adding to the result the analogous equations for the signal and idler waves. The result is

$$\frac{\partial}{\partial z}\left[\langle \mathbf{S}_P(z) + \mathbf{S}_S(z) + \mathbf{S}_I(z) \rangle \cdot \mathbf{n} \right] = \Gamma(\omega_P - \omega_S - \omega_I) = 0 \tag{15.6.6}$$

where the frequency condition (15.5.1) is used to show the vanishing of the right side. This equation can be recognized as the energy conservation equation of the three-wave interaction in the steady state $(\partial/\partial t \to 0)$.

15.7 Optical Parametric Amplification

We now wish to find a solution to the coupled mode equations governing the amplitudes of the waves in a steady state optical parametric interaction. We consider the case of no pump wave depletion for simplicity. Thus (15.5.15) for the pump wave amplitude need not be considered. Our treatment is patterned after Smith [1972].

Consider first the equation for the signal wave amplitude (15.5.16). By transferring the exponential factor to the left side, differentiating with respect to z, and introducing the idler equation we find

$$\frac{\partial^2 E(z;1,0)}{\partial z^2} e^{-i\Delta k_n z} - i\Delta k_n \frac{\partial E(z;1,0)}{\partial z} e^{-i\Delta k_n z} = ig_S \frac{\partial E^*(z;0,1)}{\partial z}$$
$$= g_S g_I^* E(z;1,0) e^{-i\Delta k_n z} \tag{15.7.1}$$

where

$$g_S \equiv \frac{\omega_S b E(1,1)}{c n_S \cos\delta_S \mathbf{n}\cdot\mathbf{t}_S}, \tag{15.7.2}$$

$$g_I \equiv \frac{\omega_I b E(1,1)}{c n_I \cos\delta_I \mathbf{n}\cdot\mathbf{t}_I}. \tag{15.7.3}$$

After dividing out the exponential factor we have a linear homogeneous differential equation having the solution

$$E(z;1,0) = Ae^{\eta^+ z} + Be^{\eta^- z}, \tag{15.7.4}$$

with

$$\eta^\pm \equiv \frac{i\Delta k_n}{2} \pm \gamma, \tag{15.7.5}$$

$$\gamma^2 \equiv \gamma_0^2 - \frac{\Delta k_n^2}{4}, \tag{15.7.6}$$

$$\gamma_0^2 \equiv g_S g_I^* = g_S^* g_I. \tag{15.7.7}$$

We note for use in manipulating later equations that

$$\eta^{+}-\eta^{-}=\eta^{+*}-\eta^{-*}=2\gamma, \tag{15.7.8}$$

$$\eta^{-*}\eta^{+*}=\eta^{-}\eta^{+}=-\gamma_0^2, \tag{15.7.9}$$

$$\eta^{+}\eta^{+*}=\eta^{-}\eta^{-*}=\gamma_0^2. \tag{15.7.10}$$

An analogous procedure yields a similar solution for the idler wave amplitude

$$E(z;0,1)=Ce^{\eta^{+}z}+De^{\eta^{-}z}. \tag{15.7.11}$$

Substitution of this solution and that for the signal wave into either of the original differential equations, (15.5.16) or (15.5.17), leads to two relations,

$$C=\frac{ig_I}{\eta^{+}}A^*, \tag{15.7.12}$$

$$D=\frac{ig_I}{\eta^{-}}B^*, \tag{15.7.13}$$

between the integration constants. If the signal and idler electric field amplitudes at the input surface ($z=0$) are denoted by $E(1,0)$ and $E(0,1)$ respectively, then A and B may be expressed in terms of them by

$$A=\frac{\eta^{+*}}{\eta^{+*}-\eta^{-*}}\left|E(1,0)-\frac{i\eta^{-*}E^*(0,1)}{g_I^*}\right|, \tag{15.7.14}$$

$$B=\frac{\eta^{-*}}{\eta^{+*}-\eta^{-*}}\left|-E(1,0)+\frac{i\eta^{+*}E^*(0,1)}{g_I^*}\right|. \tag{15.7.15}$$

Substitution of these constants into (15.7.4) and use of (15.7.6) to (15.7.9) lead to

$$E(z;1,0)=\left[\left(\gamma\cosh\gamma z-\frac{i\Delta k_n}{2}\sinh\gamma z\right)E(1,0)\right.$$
$$\left.+ig_S E^*(0,1)\sinh\gamma z\right]\frac{e^{i\Delta k_n z/2}}{\gamma} \tag{15.7.16}$$

for the electric field amplitude of the signal wave. It is convenient to denote the ratio of input photon intensities of the idler and signal waves by

$$r^2\equiv\frac{\omega_S\langle\mathbf{S}_I(0)\cdot\mathbf{n}\rangle}{\omega_1\langle\mathbf{S}_S(0)\cdot\mathbf{n}\rangle}=\frac{\omega_S n_I\cos\delta_I\mathbf{n}\cdot\mathbf{t}_I|E(0,1)|^2}{\omega_I n_S\cos\delta_S\mathbf{n}\cdot\mathbf{t}_S|E(1,0)|^2} \tag{15.7.17}$$

and to introduce the initial pump phase φ_P by

$$E(1,1)=|E(1,1)|e^{-i\varphi_P} \tag{15.7.18}$$

and the initial signal and idler phases φ_S and φ_I similarly. Letting φ denote

$$\varphi \equiv \varphi_P - \varphi_S - \varphi_I \tag{15.7.19}$$

we may rewrite (15.7.16) as

$$E(z;1,0)=\frac{E(1,0)e^{i\Delta k_n z/2}}{\gamma}\left[\gamma\cosh\gamma z + i\left(\gamma_0 r e^{-i\varphi}-\frac{\Delta k_n}{2}\right)\sinh\gamma z\right]. \tag{15.7.20}$$

Equations (15.7.11) to (15.7.15) lead to an analogous solution

$$E(z;0,1)=\frac{E(0,1)e^{i\Delta k_n z/2}}{\gamma}\left[\gamma\cosh\gamma z + i\left(\frac{\gamma_0 e^{-i\varphi}}{r}-\frac{\Delta k_n}{2}\right)\sinh\gamma z\right] \tag{15.7.21}$$

for the idler wave.

The magnetic intensity $\mathbf{H}$ of the signal wave is simply

$$\mathbf{H}=\frac{1}{\omega_S\mu_0}\mathbf{k}_S\times\mathbf{E} \tag{15.7.22}$$

since the contribution from the slowly varying amplitude is negligible by (15.5.7). The time-averaged signal wave Poynting vector is then

$$\langle\mathbf{S}_S(z)\rangle=\frac{\epsilon_0}{2}cn_S\cos\delta_S|E(z;1,0)|^2\mathbf{t}_S \tag{15.7.23}$$

where $\mathbf{t}_S$ is given in (9.10.28). Insertion of (15.7.20) for $E(z;1,0)$ into this gives

$$\langle\mathbf{S}_S(z)\cdot\mathbf{n}\rangle=\frac{\langle\mathbf{S}_S(0)\cdot\mathbf{n}\rangle}{\gamma^2}\left[(\gamma\cosh\gamma z+\gamma_0 r\sin\varphi\sinh\gamma z)^2 + \left(\gamma_0 r\cos\varphi-\frac{\Delta k_n}{2}\right)^2\sinh^2\gamma z\right]. \tag{15.7.24}$$

A similar procedure and the additional use of (15.7.17) gives

$$\langle \mathbf{S}_I(z)\cdot\mathbf{n}\rangle = \frac{\omega_I \langle \mathbf{S}_S(0)\cdot\mathbf{n}\rangle}{\omega_S \gamma^2}\left[(r\gamma\cosh\gamma z + \gamma_0 \sin\varphi \sinh\gamma z)^2 + \left(\gamma_0\cos\varphi - \frac{r\Delta k_n}{2}\right)^2 \sinh^2\gamma z\right] \tag{15.7.25}$$

for the idler wave Poynting vector. These formulas are quite general, since they allow for all three waves to be injected at the input surface. Note that the intensities of the signal and idler waves at a position z into the crystal depend on the relative phase φ of the three injected waves.

We now apply these equations to an *optical parametric amplifier*. By this we mean a device in which only the pump and signal waves are injected and amplification of the signal wave is desired. By setting $r=0$ (no injected idler wave) and $z=l$, the thickness of the slab, we find

$$\langle \mathbf{S}_S(l)\cdot\mathbf{n}\rangle = \langle \mathbf{S}_S(0)\cdot\mathbf{n}\rangle\left[1 + \frac{\gamma_0^2}{\gamma^2}\sinh^2\gamma l\right], \tag{15.7.26}$$

$$\langle \mathbf{S}_I(l)\cdot\mathbf{n}\rangle = \langle \mathbf{S}_S(0)\cdot\mathbf{n}\rangle \frac{\omega_I\gamma_0^2}{\omega_S\gamma^2}\sinh^2\gamma l. \tag{15.7.27}$$

Note that the growth of the idler wave in relation to the growth of the signal wave is just that expected from the Manley-Rowe relation. We see that the single pass power gain of the signal wave is

$$G_S \equiv \frac{\langle \mathbf{S}_S(l)\cdot\mathbf{n}\rangle - \langle \mathbf{S}_S(0)\cdot\mathbf{n}\rangle}{\langle \mathbf{S}_S(0)\cdot\mathbf{n}\rangle} = \frac{\gamma_0^2\sinh^2\gamma l}{\gamma^2}. \tag{15.7.28}$$

If the phase mismatch is large compared to the gain, $(\Delta k_n l/2)^2 \gg (\gamma_0 l)^2$, then $\gamma = i\Delta k_n/2$ and

$$G_S = (\gamma_0 l)^2\left[\frac{\sin^2(\Delta k_n l/2)}{(\Delta k_n l/2)^2}\right] = (\gamma_0 l)^2 \Phi\left(\frac{\Delta k_n z}{2}\right), \tag{15.7.29}$$

which shows the gain profile is that of the phase-matching function Φ.

When $\gamma l \ll 1$ corresponding to small gain and small or zero phase mismatch, we have

$$G_S = (\gamma_0 l)^2, \qquad \text{(small gain)}. \tag{15.7.30}$$

Use of (15.7.2), (15.7.3), and (15.7.7) gives

$$\begin{aligned} G_S &= \frac{\omega_I \omega_S b^2 l^2 |E(1,1)|^2}{c^2 n_I n_S \cos\delta_I \cos\delta_S (\mathbf{n}\cdot\mathbf{t}_I)(\mathbf{n}\cdot\mathbf{t}_S)} \\ &= \frac{2}{\epsilon_0} f_S f_I f_P b^2 l^2 \frac{\langle \mathbf{S}_P(0)\cdot\mathbf{n}\rangle}{\omega_P} \end{aligned} \tag{15.7.31}$$

where

$$f_A \equiv \frac{\omega_A}{c n_A \cos\delta_A (\mathbf{n}\cdot\mathbf{t}_A)} \qquad (A = S, I, P). \tag{15.7.32}$$

If the signal and idler frequencies are expressed in terms of their degenerate frequency $\omega_P/2$ and a relative deviation $\delta\omega$ from this frequency,

$$\omega_S = \frac{\omega_P}{2}(1+\delta\omega), \tag{15.7.33}$$

$$\omega_I = \frac{\omega_P}{2}(1-\delta\omega), \tag{15.7.34}$$

then the major dependence of the gain on frequency is

$$\omega_I \omega_S = \frac{\omega_P^2}{4}\left[1-(\delta\omega)^2\right]. \tag{15.7.35}$$

Thus maximum gain exists at the degenerate operating point, $\omega_S = \omega_I = \omega_P/2$, and decreases as $[1-(\delta\omega)^2]$ for a frequency deviation $\delta\omega$ from degeneracy. Lastly we point out, for example, that the value of the gain G_S for a collinear interaction normal to the optic axis in a 1 cm long crystal of $LiNbO_3$ with a pump intensity of 1 MW/cm^2 and signal and idler wavelengths of 1 μm is 10%.

15.8 Optical Parametric Oscillation

If the output of an amplifier is fed back to the input in phase with the input signal, oscillation occurs provided that the gain exceeds the loss. In optical parametric oscillators, as in lasers, the feedback is provided by two mirrors facing each other. The requirement that the fed back light wave be in phase with the initiating light wave is satisfied for any electromagnetic mode of the optical cavity. Since the bandwidth over which gain occurs is typically large compared to the frequency spacing of cavity modes, a mode is usually available for optical oscillation. For a cavity formed by two plane mirrors the spacing of the modes with the simplest field configuration is $c/2l$, where l is the optical path length between the mirrors. For $l = 1.5$ cm the mode spacing thus is 10 GHz.

We saw in (15.7.24) and (15.7.25) of the preceding section that the power in the signal and idler waves depends on the relative phase φ of the pump, signal, and idler waves. When the signal and idler waves are created from noise by the pump wave, they obtain phases corresponding to maximum power transfer from the pump. Thus the relative phase φ can be obtained by maximizing the signal wave intensity with respect to φ for $z = l$,

$$\frac{\partial\langle \mathbf{S}_S(l)\cdot\mathbf{n}\rangle}{\partial\varphi} = 0. \tag{15.8.1}$$

This yields

$$\cot\varphi = -\frac{\Delta k_n}{2\gamma}\tanh\gamma l. \tag{15.8.2}$$

To obtain the maximum signal wave intensity we must choose φ in the second quadrant, that is, $\cos\varphi < 0$ and $\sin\varphi > 0$.

Note that for a phase matched interaction ($\Delta k_n = 0$) the relative phase is

$$\varphi \equiv \varphi_P - \varphi_S - \varphi_I = \frac{\pi}{2}. \tag{15.8.3}$$

The same result for φ would be obtained if the idler wave intensity were maximized since the Manley-Rowe relations show the photon intensities of the signal and idler waves have equal growth. Elimination of φ from

(15.7.24) and (15.7.25) with the additional use of (15.7.17) yields the signal and idler intensities for the optimum phase,

$$\langle \mathbf{S}_S(l)\cdot\mathbf{n}\rangle = \langle \mathbf{S}_S(0)\cdot\mathbf{n}\rangle \left\{ \frac{r\gamma_0}{\gamma}\sinh\gamma l + \left[1+\left(\frac{\gamma_0}{\gamma}\right)^2\sinh^2\gamma l\right]^{1/2}\right\}^2, \tag{15.8.4}$$

$$\langle \mathbf{S}_I(l)\cdot\mathbf{n}\rangle = \langle \mathbf{S}_I(0)\cdot\mathbf{n}\rangle \left\{ \frac{\gamma_0}{r\gamma}\sinh\gamma l + \left[1+\left(\frac{\gamma_0}{\gamma}\right)^2\sinh^2\gamma l\right]^{1/2}\right\}^2. \tag{15.8.5}$$

Optical parametric oscillators are of two types, doubly resonant oscillators and singly resonant oscillators. In doubly resonant oscillators low loss cavities are provided for both the signal and idler waves while in a singly resonant oscillator a low loss cavity is provided for the signal wave only. It is important to realize that signal and idler gain exists for only one direction of travel in the cavity because of the phase-matching condition. In both directions of travel, however, there is loss.

Oscillation threshold for a doubly resonant oscillator is determined by

$$\langle \mathbf{S}_S(l)\cdot\mathbf{n}\rangle(1-\alpha_S) = \langle \mathbf{S}_S(0)\cdot\mathbf{n}\rangle, \tag{15.8.6}$$

$$\langle \mathbf{S}_I(l)\cdot\mathbf{n}\rangle(1-\alpha_I) = \langle \mathbf{S}_I(0)\cdot\mathbf{n}\rangle, \tag{15.8.7}$$

where l is the one-way length of the cavity and α_S and α_I are the round-trip power losses of the signal and idler waves. Substitution of the expressions (15.8.4) and (15.8.5) into these equations leads to the threshold conditions

$$\left(\frac{\gamma_0}{\gamma}\right)^2\sinh^2\gamma l = \frac{\alpha_S\alpha_I}{\left[(1-\alpha_S)^{1/2}+(1-\alpha_I)^{1/2}\right]^2}, \tag{15.8.8}$$

$$r^2 \equiv \frac{\omega_S\langle \mathbf{S}_I(0)\cdot\mathbf{n}\rangle}{\omega_I\langle \mathbf{S}_S(0)\cdot\mathbf{n}\rangle} = \frac{\alpha_S(1-\alpha_I)}{\alpha_I(1-\alpha_S)}. \tag{15.8.9}$$

Note that the left side of the first of these is just what is defined as the single pass power gain of the signal wave in (15.7.28). The threshold condition does not assume phase matching, since often there are no cavity modes for the signal and idler waves that permit *exact* phase matching. When this happens, oscillation occurs in those modes satisfying the frequency condition (15.5.1) and having the smallest phase mismatch (and thus the highest gain). When the losses are small compared to unity and the phase mismatch is negligible, the condition reduces to

$$(\gamma_0 l)^2 = \frac{\alpha_S \alpha_I}{4}. \tag{15.8.10}$$

Use of (15.7.30) and (15.7.31) converts this to an expression for the pump intensity required to reach threshold,

$$\langle \mathbf{S}_P(0)\cdot\mathbf{n}\rangle_{\mathrm{THR}} = \frac{\epsilon_0 \omega_P \alpha_S \alpha_I}{8 f_P f_S f_I b^2 l^2}. \tag{15.8.11}$$

If the round-trip power losses are held to 2% for both the signal and idler waves (each having a wavelength of 1 μm) in a collinear interaction normal to the optic axis in a 1 cm long $LiNbO_3$ crystal, the pump intensity must be about 1 kW/cm^2.

The threshold condition for a singly resonant oscillator can be found either by solving (15.8.6) alone with $r=0$ or by simply setting $\alpha_I = 1$ in the doubly resonant oscillator equation. The result is

$$\left(\frac{\gamma_0}{\gamma}\right)^2 \sinh^2 \gamma l = \frac{\alpha_S}{1-\alpha_S}. \tag{15.8.12}$$

For small losses and negligible phase mismatch this becomes

$$(\gamma_0 l)^2 = \alpha_S \tag{15.8.13}$$

which may be rewritten as

$$\langle \mathbf{S}_P(0)\cdot\mathbf{n}\rangle_{\mathrm{THR}} = \frac{\epsilon_0 \omega_P \alpha_S}{2 f_P f_S f_I b^2 l^2}. \tag{15.8.14}$$

Comparison of this equation with (15.8.11) shows that the singly resonant oscillator must be pumped $4/\alpha_I$ times harder than the doubly resonant oscillator in order to reach threshold. If $\alpha_I = 0.02$, this factor is 200.

Though the power of the pump wave needed to reach oscillation is strongly in favor of the doubly resonant oscillator, the singly resonant oscillator has a compensating advantage in tunability, which is one of the most important properties of optical parametric oscillators. Since only one wave is resonant, that is, is an eigenmode of the optical cavity, it can be tuned continuously when the optical path length in the cavity is varied continuously. In the doubly resonant oscillator both signal and idler waves [and the pump wave also by virtue of (15.5.2)] must be in cavity modes. What mode actually oscillates is a complicated function of the mode widths and spacings of the signal and idler waves, the gain profile, and the frequency width of the pump wave. This leads to the tuning of the output frequencies of a doubly resonant oscillator being discontinuous because of discrete mode jumps.

Phase matching of a parametric interaction is formally identical to that for optical mixing. Solutions of the phase matching condition for collinear interactions in uniaxial crystals analogous to those presented for harmonic generation in Section 15.3 are

$$(\omega_S+\omega_I)n^e(\omega_S+\omega_I,\theta)=\omega_S n^o(\omega_S)+\omega_I n^o(\omega_I) \qquad (-\text{crystal}), \quad (15.8.15)$$

$$(\omega_S+\omega_I)n^o(\omega_S+\omega_I)=\omega_S n^e(\omega_S,\theta)+\omega_I n^e(\omega_I,\theta) \qquad (+\text{crystal}). \quad (15.8.16)$$

This type of birefringence phase matching is called type I, and gives rise to optical parametric oscillator operating points illustrated in Fig. 15.11 for $LiNbO_3$. A second kind of phase matching, called type II, is also possible when the birefringence is large enough in comparison to the dispersion. In type II phase matching,

$$(\omega_S+\omega_I)n^e(\omega_S+\omega_I,\theta)=\omega_S n^o(\omega_S)+\omega_I n^e(\omega_I,\theta) \qquad (-\text{crystal}), \quad (15.8.17)$$

$$(\omega_S+\omega_I)n^o(\omega_S+\omega_I)=\omega_S n^o(\omega_S)+\omega_I n^e(\omega_I,\theta) \qquad (+\text{crystal}), \quad (15.8.18)$$

the signal and idler waves propagate in opposite polarizations. In general, either polarization can be used for either the signal or idler wave. A plot of optical parametric oscillator operating points using type II phase matching in $AgGaS_2$ is shown in Fig. 15.12.

The first optical parametric oscillator was made by Giordmaine and Miller [1965] who used an optical resonator containing a $LiNbO_3$ crystal as shown in Fig. 15.13. An extraordinary wave pump beam at 529 nm propagated normal to the optic axis to produce a collinear noncritically phase-matched interaction with the signal and idler waves both ordinary (type I phase matching in a negative uniaxial crystal). Coupling of the three waves occurred through the $b_{3\,2\,2}^{\omega_P\omega_S\omega_I}$ coefficient. At a pump power of

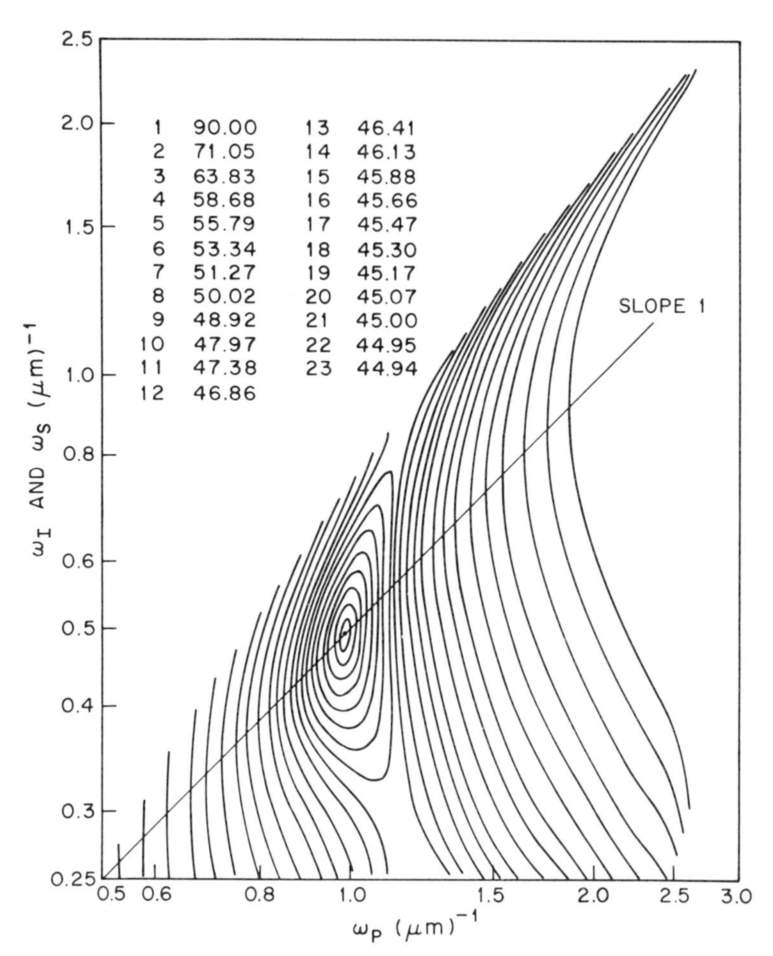

Fig. 15.11. Plot of type I optical parametric oscillator operating points for congruent $LiNbO_3$ [reprinted by permission from Nelson and Mikulyak, 1975), a negative uniaxial crystal. The signal and idler frequencies are found from the intersection of a vertical line representing the pump frequency with a curve parametrized by the angle θ measured from the optic axis. The values of θ in degrees are listed in order for the curves beginning from the right and ending at the point to which the curves shrink. The line of unit slope includes all degenerate operating points which also represent phase matched second harmonic generation. Solutions along the $\theta = 90°$ curve represent noncritical phase matching.

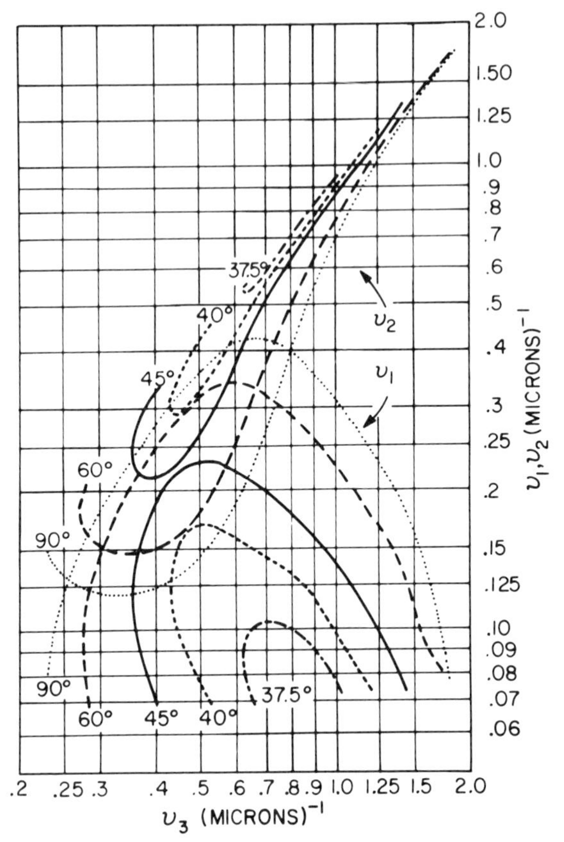

Fig. 15.12. Plot of type II optical parametric oscillator operating points for $AgGaS_2$, a negative uniaxial crystal [adapted with permission from Boyd et al., 1971]. The signal and idler frequencies are found from the intersection of a vertical line representing the pump frequency with curves parametrized by the angle θ measured from the optic axis. Solutions for $\theta = 90°$ represent noncritical phase matching. The wave of circular frequency ν_1 is extraordinary, that of frequency ν_2 is ordinary, and the pump wave of frequency ν_3 is extraordinary.

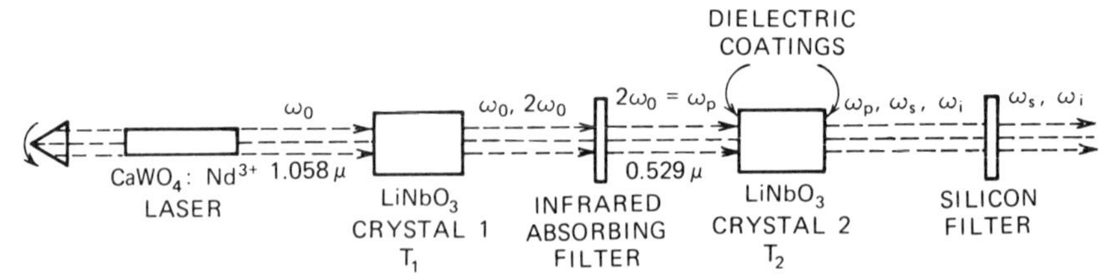

Fig. 15.13. Optical parametric oscillator apparatus [reprinted by permission from Giordmaine and Miller, 1965]. A collinear interaction of the pump (ω_P), signal (ω_S), and idler (ω_I) frequencies occurred in crystal 2 under noncritical phase-matching conditions.

4×10^5 W/cm^2 a distinct threshold for signal and idler waves was observed. These waves were highly collimated (divergence less than 3×10^{-3} rad) and highly monochromatic (line width less than 0.03 nm).

One of the most important characteristics of optical parametric oscillation is its tunability in frequency. This is caused by altering the conditions of phase matching. Giordmaine and Miller used a change in temperature of the oscillator crystal to do this, since the extraordinary refractive index in $LiNbO_3$ is particularly temperature sensitive. Their measurements of signal and idler tuning are shown in Fig. 15.14. Frequency tuning by rotating the oscillator crystal in its resonator was first done by Akhmanov

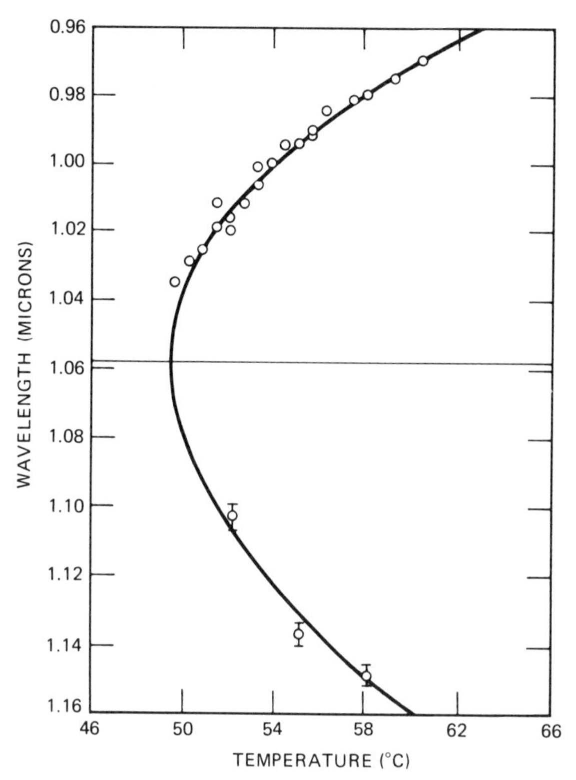

Fig. 15.14. Temperature tuning of signal and idler wavelengths in the optical parametric oscillator of Fig. 15.13 [reprinted by permission from Giordmaine and Miller, 1965]. Changing the temperature of the oscillator crystal changes its refractive indices and thus the wavelengths of the signal and idler waves that can be phase matched for a given pump wavelength and propagation direction ($\theta = 90°$).

and co-workers [1966]. A tuning range of 957.5 to 1177.5 nm for the signal and idler waves from a KDP crystal pumped at 530 nm was obtained. Kreuzer [1967] produced a 5.0 nm tuning of the signal wave from a $LiNbO_3$ crystal by varying the refractive indices by an applied electric field through the electrooptic effect.

Continuous operation of an optical parametric oscillator was first attained by Smith and co-workers [1968]. They used a $Ba_2NaNb_5O_{12}$ crystal pumped by 532 nm wavelength beam from the device shown in Fig. 15.5. Noncritical phase matching was used and the signal and idler waves were tuned from 980 nm to 1160 nm by changing the temperature of the oscillator crystal. An output of 3 mW was measured when the pump wave had 300 mW power. The output was unstable, consisting of pulsations of a few microseconds in duration.

The optical parametric oscillators discussed above used collinear interactions and doubly resonant cavities in order to obtain low thresholds. An optical parametric oscillator that used both a noncollinear interaction and a singly resonant cavity was constructed by Falk and Murray [1969]. It is shown in Fig. 15.15.

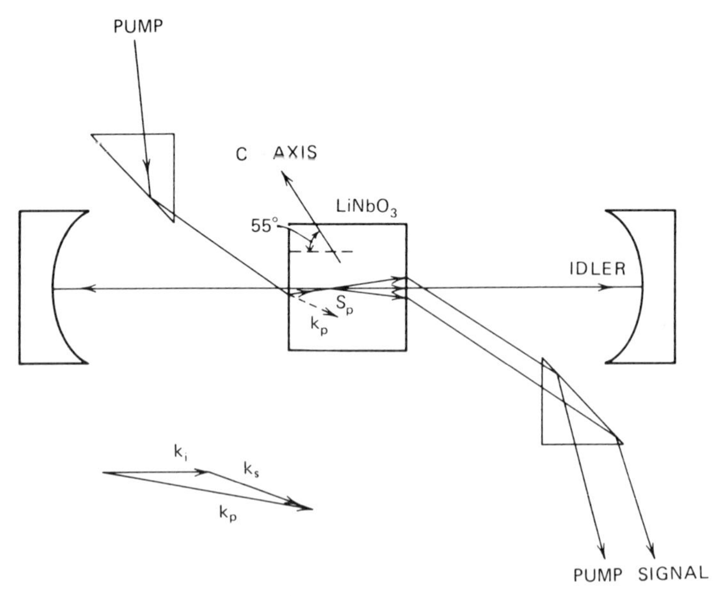

Fig. 15.15. Optical parametric oscillator with a noncollinear interaction and a singly resonant cavity [reprinted by permission from Falk and Murray, 1969]. Improved frequency stability compared to a doubly resonant oscillator was obtained. A signal power of 340 kW was measured for a 750 kW pump wave.

PROBLEMS

15.1. Beginning from (15.5.15) find the generated sum frequency flux in the absence of input wave depletion and show that it agrees with (15.2.12).

15.2. Show that the coupled mode equations for three waves whose amplitudes E_P, E_S, E_I are slowly varying functions of both space and time are

$$\frac{\partial E_P}{\partial z} + \frac{1}{\mathbf{v}_g^P \cdot \mathbf{n}} \frac{\partial E_P}{\partial t} = i\beta_P E_S E_I e^{-i\Delta k_n z}, \tag{15.P.1}$$

$$\frac{\partial E_S}{\partial z} + \frac{1}{\mathbf{v}_g^S \cdot \mathbf{n}} \frac{\partial E_S}{\partial t} = i\beta_S E_P E_I^* e^{i\Delta k_n z}, \tag{15.P.2}$$

$$\frac{\partial E_I}{\partial z} + \frac{1}{\mathbf{v}_g^I \cdot \mathbf{n}} \frac{\partial E_I}{\partial t} = i\beta_I E_P E_S^* e^{i\Delta k_n z} \tag{15.P.3}$$

where

$$\beta_i \equiv \frac{\omega_i b}{c n_i \cos \delta_i \mathbf{t}_i \cdot \mathbf{n}}, \tag{15.P.4}$$

$\mathbf{v}_g^i$ is the group velocity of the i wave, and all other quantities have the meanings used in Section 15.5. *Hint*: Begin from the wave equation in the form (9.2.3) and the internal motion equation and allow both the internal coordinates and the electric field to have amplitudes slowly varying in time as well as in space.

15.3. Some nonlinear equations have propagating pulse solutions in which the pulse retains its shape indefinitely (is *stationary*). The stationary shape results from an exact compensation of the broadening effect of frequency dispersion and the narrowing effect of nonlinearity. Such pulse solutions are known as *solitary waves* or, when they exhibit great stability to perturbations as frequently occurs, as *solitons*. When such a solution exists for the slowly varying amplitude function of a wave, the solution is called an *envelope soliton*. Show that a set of envelope soliton solutions exists for (15.P.1) to (15.P.3) at exact phase matching in which the pump wave amplitude is proportional to $\tanh k(z - vt)$ and the signal and idler wave amplitudes are each proportional to $\operatorname{sech} k(z - vt)$. Find the conditions on the velocity v and the associated conditions on the phases of the three waves. Is there any condition on k?

15.4. Find the energy conservation law from (15.P.1) to (15.P.3). Calculate the energy flux function that occurs in this law for each of the three envelope solitons found in Prob. 15.3. Note that the pump wave may be described as a "dark envelope soliton" and the signal and idler waves as "bright envelope solitons."

15.5. Find envelope soliton solutions of (15.P.1) to (15.P.3) at exact phase matching in which the signal wave is proportional to $\tanh k(z-vt)$ and the pump and idler waves are each proportional to $\operatorname{sech} k(z-vt)$. Find the conditions on the velocity v and the associated conditions on the phases of the three waves.

15.6. The interaction between two light waves and an optic mode of a medium is called the *Raman interaction*. It can be observed by *spontaneous Raman scattering* (first observed in crystals by Landsberg and Mandelstam [1928]), by *stimulated Raman scattering* [Woodbury and Ng, 1962; Eckhardt et al., 1962], and by certain sum and difference frequency experiments [Giordmaine and Kaiser, 1966]. In both scattering processes a single light wave at frequency ω_L is sent through the medium and a light wave at a frequency ω_R lower than ω_L by the optic mode frequency ω_Q,

$$\omega_L = \omega_R + \omega_Q, \tag{15.P.5}$$

is observed. In the stimulated process both the output light wave and the optic mode excitation are coherent and intense. The Raman interaction with the m optic mode can be characterized by a nonlinear polarization

$$\mathcal{P}_i^{\;\omega_L} = \epsilon_0 \chi_{i\;j}^{\omega_L\omega_R m} E_j^{\;\omega_R} \eta_{\omega_Q}^{m} \tag{15.P.6}$$

where $\chi_{i\;j}^{\omega_L\omega_R m}$ is the *Raman susceptibility* of the m optic mode and η^m is the optic mode normal coordinate given in (9.P.7) to (9.P.10). If the Raman susceptibility is defined for every optic mode (ionic and electronic), show that the Raman susceptibility is given by

$$\chi_{i\;j}^{\omega_L\omega_R m} \equiv -\frac{3}{\epsilon_0} \sum_{\substack{\nu\mu\lambda \\ kl}} \frac{c_i^{\;k} c_j^{\;l}\; {}^{30}M_{pqr}^{\nu\mu\lambda}\, i_p^{\;k\nu}\, i_q^{\;l\mu}\, i_r^{\;m\lambda}}{(m^\nu m^\mu m^\lambda)^{1/2}(\Omega_k^2-\omega_L^2)(\Omega_l^2-\omega_R^2)}. \tag{15.P.7}$$

What interchange symmetry does $\chi_{i\;j}^{\omega_L\omega_R m}$ have?

15.7. Derive the relation

$$2b_{i\,j\,k}^{\omega_L\omega_R\omega_Q}=\sum_m \frac{\chi_{i\,j}^{\omega_L\omega_R m}c_k^m}{\Omega_m^2-\omega_Q^2} \tag{15.P.8}$$

between the optical mixing tensor and the Raman susceptibility of the m optic mode. This relation is a generalization of one derived by Kaminow [1967]. The relation states that optic modes that contribute to optical mixing must be both infrared active and Raman active (the m optic mode is *Raman active* if crystal symmetry allows $\chi_{i\,j}^{\omega_L\omega_R m}\neq 0$; it is *Raman inactive* if crystal symmetry requires $\chi_{i\,j}^{\omega_L\omega_R m}=0$).

15.8. The Raman interaction can be characterized by three coupled mode equations. If the optic mode is infrared active, then its coupled mode equation at frequency ω_Q (see Prob. 15.6) is found from the electric field wave equation with the optic mode coordinate eliminated (just as for the other two frequencies). The process is then an optical mixing interaction with one wave in the polariton region [Kurtz and Giordmaine, 1969]. It is thus included in the analysis of Sections 15.2 and 15.5. If the optic mode is infrared inactive, then its coupled mode equation at ω_Q is found from the internal motion equation expressed in the normal coordinates of (9.P.7) to (9.P.10). Show that for the latter case the coupled mode equations are

$$\frac{\partial E_L}{\partial z}+\frac{1}{\mathbf{v}_g^L\cdot\mathbf{n}}\frac{\partial E_L}{\partial t}=\frac{i\omega_L\chi E_R\eta_Q^1 e^{-i\Delta k_n z}}{2n_L c\cos\delta_L(\mathbf{t}_L\cdot\mathbf{n})}, \tag{15.P.9}$$

$$\frac{\partial E_R}{\partial z}+\frac{1}{\mathbf{v}_g^R\cdot\mathbf{n}}\frac{\partial E_R}{\partial t}=\frac{i\omega_R\chi E_L\eta_Q^{1*} e^{i\Delta k_n z}}{2n_R c\cos\delta_R(\mathbf{t}_R\cdot\mathbf{n})}, \tag{15.P.10}$$

$$\frac{\partial \eta_Q^1}{\partial t}=\frac{i\chi E_L E_R^* e^{i\Delta k_n z}}{\omega_Q}, \tag{15.P.11}$$

where $\chi\equiv e_i^L\chi_{i\,j}^{\omega_L\omega_R 1}e_j^R$ is the effective Raman susceptibility of the $m=1$ optic mode, $\Delta\mathbf{k}\equiv\mathbf{k}_L-\mathbf{k}_R$, and the remaining notation parallels that of Section 15.5.

15.9. Loss in the optic mode can be included phenomenologically as a term $\eta_Q^1/2\tau$, where τ is the optic mode lifetime, on the left side of

(15.P.11). Since this loss is typically large ($\tau \lesssim 10^{-11}$ sec), we can assume $\partial\eta_Q^1/\partial t \ll \eta_Q^1/2\tau$. Derive under this condition and with no depletion of power at the frequency ω_L that the normal component of the Poynting vector of the wave at frequency ω_R is given by

$$\langle \mathbf{S}_R \cdot \mathbf{n} \rangle = F\left(z - \mathbf{v}_g^R \cdot \mathbf{n} t\right) e^{gz} \tag{15.P.12}$$

with the *Raman gain* g given by

$$g = \frac{4\tau f_L f_R \chi^2}{\epsilon_0 \omega_Q} \cdot \frac{\langle \mathbf{S}_L \cdot \mathbf{n} \rangle}{\omega_L} \tag{15.P.13}$$

and with F an arbitrary envelope function. The quantities f_L and f_R are defined in (15.7.32).

CHAPTER 16

Material Form of Electromagnetism

In Chapter 4 both a spatial and material form for the matter Lagrangian density and for the interaction Lagrangian density are given but the electromagnetic field Lagrangian density is given only in the spatial description. Thus only the spatial form of the electromagnetic equations can be obtained from it. This, of course, is the conventional handling of the electromagnetic equations in either the Maxwell or the Lorentz form. The reason simply is that electromagnetic fields can exist in a vacuum and so it is natural to refer them to a coordinate system that does not depend in any way on the existence of matter in the volume of space considered.

What then can be the usefulness of a material form of the electromagnetic equations? By *material form* we mean referring all vector components to the material coordinate axes and regarding all fields as functions of the material coordinate $\mathbf{X}$ and time. The usefulness is mainly in the consideration of electromagnetic interactions within deformable bodies in the low frequency region. In Chapter 11 it is shown that in the quasi-electrostatic regime, which applies to normal size laboratory samples ($\sim$1 cm dimensions) for frequencies up to 1 GHz, a truncated form of the electromagnetic equations is adequate. These equations do not allow propagating (radiating) electromagnetic fields, only time dependent stationary fields. These fields are often created by electrodes attached to the body. If the body is deforming, the electrical boundary conditions are easily known only in the material coordinate system. It is then particularly convenient to have the electromagnetic equations referred to the material coordinate system [Walker et al., 1965; Thurston, 1974].

We, therefore, consider in this chapter the transformation of the Maxwell equations for an arbitrary medium, not just a nonferromagnetic dielectric with which the remainder of this treatment is concerned. We require the Maxwell form of the electromagnetic equations to remain *form invariant* under the transformation to material form. We then transform the Lagrangian with material form electromagnetic fields and obtain the Lorentz form of the electromagnetic equations from it. The latter is found *not* to be form invariant under the transformation to material form. The treatment is based on Lax and Nelson [1976b].

Application of the totally material form Lagrangian density is made to linear and nonlinear electroacoustic interactions in the next chapter.

16.1 Material Form of Maxwell Equations

In this section we find the transformations of the four fields, **E**, **B**, **D**, and **H**, the free charge density q^f, and the conduction current density $\mathbf{j}^c$ which retain the form invariance of the Maxwell equations under the deformation transformation (3.1.2). The conventional spatial frame form of the Maxwell equations is

$$\nabla\times\mathbf{H}=\frac{\partial\mathbf{D}}{\partial t}+\mathbf{j}^f, \tag{16.1.1}$$

$$\nabla\times\mathbf{E}=-\frac{\partial\mathbf{B}}{\partial t}, \tag{16.1.2}$$

$$\nabla\cdot\mathbf{D}=q^f, \tag{16.1.3}$$

$$\nabla\cdot\mathbf{B}=0 \tag{16.1.4}$$

where the total free charge current density is

$$\mathbf{j}^f\equiv\mathbf{j}^c+q^f\dot{\mathbf{x}}. \tag{16.1.5}$$

We consider these equations in a volume of space occupied by matter.

In order to determine the transformation of the fields to material form we consider the integral forms of the equations. Consider (16.1.1) first. Integrating over a surface area fixed in the matter yields

$$\int(\nabla\times\mathbf{H})\cdot\mathbf{da}=\int\left(\frac{\partial\mathbf{D}}{\partial t}+\mathbf{j}^f\right)\cdot\mathbf{da}. \tag{16.1.6}$$

Use of (3.13.6) to reexpress $\partial\mathbf{D}/\partial t$ and Stokes' theorem to reexpress the curl term leads to

$$\int(\mathbf{H}-\dot{\mathbf{x}}\times\mathbf{D})\cdot\mathbf{dx}=\frac{d}{dt}\int\mathbf{D}\cdot\mathbf{da}+\int(\mathbf{j}^f-\nabla\cdot\mathbf{D}\dot{\mathbf{x}})\,\mathbf{da}. \tag{16.1.7}$$

The integration variables may be transformed to the material frame by

(3.3.1) and (3.3.11). Use of (16.1.3) to eliminate $\nabla \cdot \mathbf{D}$ then yields

$$\int (\mathbf{H} - \dot{\mathbf{x}} \times \mathbf{D})_i x_{i,J} \, dX_J = \frac{d}{dt} \int D_i J X_{J,i} \, dA_J + \int (j_i^f - q^f \dot{x}_i) J X_{J,i} \, dA_J. \tag{16.1.8}$$

Note that the integrand of the last integral on the right side may be rewritten as the conduction current through (16.1.5). The desired form invariance of the equation leads us to define the material form of the magnetic intensity $\mathcal{H}$, the electric displacement $\mathfrak{D}$, and the conduction current density $\mathcal{J}^c$ by

$$\mathcal{H}_J(\mathbf{X}, t) \equiv \{\mathbf{H}(\mathbf{x}(\mathbf{X}, t), t) - \dot{\mathbf{x}}(\mathbf{X}, t) \times \mathbf{D}(\mathbf{x}(\mathbf{X}, t), t)\}_i x_{i,J}(\mathbf{X}, t), \tag{16.1.9}$$

$$\mathfrak{D}_J \equiv J X_{J,i} D_i, \tag{16.1.10}$$

$$\mathcal{J}_K^c \equiv J X_{K,i} j_i^c \tag{16.1.11}$$

where the functional dependence typical of all the field transformation equations is shown in the first one. With these definitions (16.1.8) becomes

$$\int (\nabla_X \times \mathcal{H}) \cdot \mathbf{dA} = \int (\dot{\mathfrak{D}} + \mathcal{J}^c) \cdot \mathbf{dA}. \tag{16.1.12}$$

Here Stokes' theorem is used on the left side and the gradient operator with respect to $\mathbf{X}$ is denoted by ∇_X. On the right side the commutativity of material time differentiation and material space integration over a material area is used. Since the area of integration is arbitrary, we must have

$$\nabla_X \times \mathcal{H} = \dot{\mathfrak{D}} + \mathcal{J}^c \tag{16.1.13}$$

for the first transformed Maxwell equation.

Next consider (16.1.2). Following a procedure similar to that just used we integrate over a surface area fixed in the matter, use Stokes' theorem and (3.13.6), and so obtain

$$\int \mathbf{E} \cdot \mathbf{dx} = -\frac{d}{dt} \int \mathbf{B} \cdot \mathbf{da} + \int (\mathbf{B} \times \dot{\mathbf{x}}) \cdot \mathbf{dx} + \int \nabla \cdot \mathbf{B} \dot{\mathbf{x}} \cdot \mathbf{da}. \tag{16.1.14}$$

The last term vanishes because of (16.1.4). Transformation of the integration variables to the material frame now yields

$$\int (\mathbf{E} + \dot{\mathbf{x}} \times \mathbf{B})_i x_{i,J} \, dX_J = -\frac{d}{dt} \int B_i J X_{J,i} \, dA_J. \tag{16.1.15}$$

This leads us to define

$$\mathcal{E}_J \equiv (\mathbf{E} + \dot{\mathbf{x}} \times \mathbf{B})_i x_{i,J}, \tag{16.1.16}$$

$$\mathcal{B}_J \equiv J X_{J,i} B_i \tag{16.1.17}$$

for the material forms of the electric field and the magnetic induction. Reuse of Stokes' theorem and the commutativity of the differentiation and integration on the right side gives

$$\int (\nabla_X \times \mathcal{E}) \cdot \mathbf{dA} = \int -\dot{\mathcal{B}} \cdot \mathbf{dA}, \tag{16.1.18}$$

which in turn gives

$$\nabla_X \times \mathcal{E} + \dot{\mathcal{B}} = 0 \tag{16.1.19}$$

for the second transformed Maxwell equation.

Next we integrate (16.1.3) over a certain volume of matter in the spatial frame and use Gauss' theorem,

$$\int \nabla \cdot \mathbf{D}\, dv = \int \mathbf{D} \cdot \mathbf{da} = \int q^f\, dv. \tag{16.1.20}$$

Transformation to material frame integration variables leads to

$$\int \mathcal{D} \cdot \mathbf{dA} = \int \nabla_X \cdot \mathcal{D}\, dV = \int \mathcal{Q}^f\, dV \tag{16.1.21}$$

where the definitions of $\mathcal{D}$ (16.1.10) and of the material form of the free charge density

$$\mathcal{Q}^f \equiv q^f J \tag{16.1.22}$$

are used. Since the integration volume is arbitrary, we obtain

$$\nabla_X \cdot \mathcal{D} = \mathcal{Q}^f \tag{16.1.23}$$

for the third transformed Maxwell equation. By a completely analogous procedure we also obtain

$$\nabla_X \cdot \mathcal{B} = 0 \tag{16.1.24}$$

for the last transformed Maxwell equation.

The field transformation equations may be readily inverted. First, the inverse relations for the electric displacement and magnetic induction are

$$D_i = J^{-1} x_{i,K} \mathfrak{D}_K, \tag{16.1.25}$$

$$B_i = J^{-1} x_{i,K} \mathfrak{B}_K. \tag{16.1.26}$$

The inverse relation for the magnetic intensity is found from (16.1.9),

$$\begin{aligned} H_j &= \mathfrak{H}_J X_{J,j} + \epsilon_{jkl}\left(-x_{k,K}\frac{\partial X_K}{\partial t}\right)\left(J^{-1}x_{l,L}\mathfrak{D}_L\right) \\ &= \mathfrak{H}_J X_{J,j} - X_{J,j}\left(J^{-1}\epsilon_{ikl}x_{i,J}x_{k,K}x_{l,L}\right)\frac{\partial X_K}{\partial t}\mathfrak{D}_L \\ &= X_{J,j}\left(\mathfrak{H}_J - \epsilon_{JKL}\frac{\partial X_K}{\partial t}\mathfrak{D}_L\right) \end{aligned} \tag{16.1.27}$$

with the use of

$$\epsilon_{JKL} = J^{-1}\epsilon_{jkl}x_{j,J}x_{k,K}x_{l,L}, \tag{16.1.28}$$

which is obtained from (3.3.8). By an analogous procedure we obtain

$$E_j = X_{J,j}\left(\mathcal{E}_J + \epsilon_{JKL}\frac{\partial X_K}{\partial t}\mathfrak{B}_L\right). \tag{16.1.29}$$

The inverse conduction current transformation is

$$j_i^c = J^{-1} x_{i,J} \mathcal{J}_J^c. \tag{16.1.30}$$

16.2 Material Frame Boundary Conditions

Since the Maxwell equations are now transformed to the material coordinate system, the boundaries of the moving deforming body appear at rest. Because of this the derivations of the boundary conditions at a body surface may be carried through in the manner described in Section 7.2 except that none of the complications caused by moving boundaries, that are treated there, arise here. Because of this we simply quote the boundary conditions in this frame. They are

$$\mathbf{N}\times(\mathcal{H}^o-\mathcal{H}^i)=\mathbf{K}^c, \tag{16.2.1}$$

$$\mathbf{N}\times(\mathcal{E}^o-\mathcal{E}^i)=0, \tag{16.2.2}$$

$$\mathbf{N}\cdot(\mathcal{D}^o-\mathcal{D}^i)=\Sigma^f, \tag{16.2.3}$$

$$\mathbf{N}\cdot(\mathcal{B}^o-\mathcal{B}^i)=0 \tag{16.2.4}$$

where $\mathbf{N}$ is the unit outward (undeformed) surface normal and o and i refer to outside and inside the surface. The surface conduction current density $\mathbf{K}^c$ and the surface charge density Σ^f which consists of free charge and extrinsic immobile charge are defined by

$$\mathcal{J}^c\equiv\mathbf{K}^c\,\delta(S), \tag{16.2.5}$$

$$\mathcal{Q}^f\equiv\Sigma^f\,\delta(S) \tag{16.2.6}$$

where $\delta(S)$ is the one-dimensional Dirac delta function and S is a material coordinate measured normal to the surface. For a plane boundary surface S is defined by

$$S=\mathbf{N}\cdot(\mathbf{X}-\mathbf{X}^S). \tag{16.2.7}$$

Thus $S=0$ is the equation of the surface. If the surface is not planar, (16.2.7) is valid locally in the vicinity of any point $\mathbf{X}^S$ on the surface.

16.3 Equivalence of Material Frame and Spatial Frame Boundary Conditions

Demonstrating the equivalence of the material frame boundary conditions with those found in the spatial frame in Section 7.2 is not a trivial exercise. It requires finding the transformation relations between the spatial and material frame surface conduction currents, $\mathbf{k}^c$ and $\mathbf{K}^c$, and between the spatial and material frame surface charge densities, σ^f and Σ^f. It also requires understanding whether various material fields such as the deformation gradient and the material velocity are continuous or not at a material surface.

Let us consider the transformation of the surface conduction current first. Substituting the definitions in the material frame (16.2.5) and in the spatial frame (7.2.17) into the transformation (16.1.11) yields

$$K_K^c\delta(S)=JX_{K,i}k_i^c\delta(s). \tag{16.3.1}$$

Since the Dirac delta function transforms as a density, in this case a one-dimensional density, we have

$$\delta(S)=\delta(s)\frac{ds}{dS}. \tag{16.3.2}$$

From the definition of s (7.2.7) we obtain

$$\begin{aligned} ds &= \mathbf{n}\cdot\mathbf{dx} \\ &= \left(JX_{J,i}N_J\frac{dA}{da}\right)(x_{i,K}dX_K) \\ &= J\frac{dA}{da}\mathbf{N}\cdot\mathbf{dX} \\ &= J\frac{dA}{da}\,dS \end{aligned} \tag{16.3.3}$$

where (3.3.14) is used in the second equality. Here dA and da are scalar elements of a material surface expressed in the material and spatial descriptions. Combining the last three equations yields

$$K_K^c = X_{K,i}k_i^c\frac{da}{dA} \tag{16.3.4}$$

for the transformation law of surface conduction currents.

The transformation of surface charge densities is found similarly. Substitution of the definitions (7.2.6) and (16.2.6) into the charge density transformation (16.1.22) leads to

$$\Sigma^f\delta(S)=J\sigma^f\delta(s). \tag{16.3.5}$$

Use of (16.3.2) and (16.3.3) then gives

$$\Sigma^f=\sigma^f\frac{da}{dA}. \tag{16.3.6}$$

In order to understand what material fields are continuous at a material boundary, consider two different material media in adhesive contact. It is apparent that the position vector $\mathbf{x}(\mathbf{X},t)$ of a matter point is continuous across such a surface provided no fracture or slippage occurs. Imagine all the material properties of one medium such as the mass density, the polarization, and the stiffness to approach zero, that is, the value characteristic of a vacuum. Since throughout the limiting process continuity of $\mathbf{x}(\mathbf{X},t)$ holds, we conclude that $\mathbf{x}(\mathbf{X},t)$ must be continuous in the limit when one medium becomes a vacuum. This leads immediately to a second conclusion that gradients of the position *tangential* to the surface must also be continuous even when one medium is a vacuum. Gradients of the position normal to the surface, however, need not be continuous. The foregoing reasoning also indicates that the velocity $\dot{\mathbf{x}}$ is continuous across a body surface even when the second medium is a vacuum. Lastly, we remark that the surface normal and scalar element of area of a body surface have the same values when viewed from either side of the surface. This is true whether they are expressed in the material or spatial frames.

With these understandings we proceed to transform the boundary conditions on $\mathcal{H}$ to the spatial frame. Inserting the definition of $\mathcal{H}$ (16.1.9) into (16.2.1), we obtain

$$\epsilon_{ABC}N_B\left\{\left[(\mathbf{H}+\dot{\mathbf{x}}\times\mathbf{D})_j x_{j,C}\right]^o-\left[(\mathbf{H}+\dot{\mathbf{x}}\times\mathbf{D})_j x_{j,C}\right]^i\right\}=K_A^c. \quad (16.3.7)$$

Since $\mathbf{N}$ is normal to the surface, the gradient indicated by the subscript C must be tangential to the surface. Since such a deformation gradient is continuous at the surface, $x_{j,C}$ may be removed from each of the brackets. Equation (3.3.14) relating $\mathbf{N}$ to $\mathbf{n}$ may now be inserted into the equation and a scalar product of it formed with $x_{i,A}\,dA/da$ with the result

$$J^{-1}\epsilon_{ABC}x_{i,A}x_{j,B}x_{k,C}n_j\left[(\mathbf{H}-\dot{\mathbf{x}}\times\mathbf{D})_k^o-(\mathbf{H}-\dot{\mathbf{x}}\times\mathbf{D})_k^i\right]=\frac{dA}{da}x_{i,A}K_A^c. \quad (16.3.8)$$

Use of the transformation of the permutation symbol (3.3.9), the surface current transformation (16.3.4), and the continuity of $\dot{\mathbf{x}}$ at the surface then yields

$$\mathbf{n}\times\left\{\mathbf{H}^o-\mathbf{H}^i-\dot{\mathbf{x}}\times\left[\mathbf{D}^o-\mathbf{D}^i\right]\right\}=\mathbf{k}^c \quad (16.3.9)$$

for the boundary condition in the spatial frame in agreement with (7.2.22).

We leave the proof that the remaining material frame boundary conditions transform to the corresponding spatial frame boundary conditions of the Section 7.2 for exercises.

16.4 Material Form of Electromagnetic Lagrangian Density

The spatial form of the electromagnetic Lagrangian density $\mathcal{L}_S$ is introduced in Section 2.4. It consists of the field Lagrangian density $\mathcal{L}_{FS}$ and the interaction Lagrangian density $\mathcal{L}_{IS}$,

$$\mathcal{L}_S=\mathcal{L}_{FS}+\mathcal{L}_{IS}, \quad (16.4.1)$$

where

$$\mathcal{L}_{FS} = \frac{\epsilon_0}{2}(\mathbf{E}^2 - c^2\mathbf{B}^2), \tag{16.4.2}$$

$$\mathcal{L}_{IS} = \mathbf{j}\cdot\mathbf{A} - q\Phi \tag{16.4.3}$$

and $\mathbf{j}$ and q are the total current and charge density containing both free and bound charge. The electric field $\mathbf{E}$ and the magnetic induction $\mathbf{B}$ are given by

$$\mathbf{E} = -\nabla\Phi - \frac{\partial \mathbf{A}}{\partial t}, \tag{16.4.4}$$

$$\mathbf{B} = \nabla \times \mathbf{A} \tag{16.4.5}$$

where the vector potential $\mathbf{A}$ and the scalar potential Φ are regarded as the independent fields of the Lagrangian density.

The first task of transforming to a material form of the electromagnetic Lagrangian density is to find the material form of $\mathbf{A}$ and Φ. We begin by substituting the $\mathbf{B}$ field transformation (16.1.26) into the last equation,

$$J^{-1}x_{i,K}\mathcal{B}_K = \epsilon_{ijk}A_{k,j}. \tag{16.4.6}$$

Forming a scalar product with $X_{I,i}$ and using the chain rule to change the derivative on $\mathbf{A}$ leads to

$$\mathcal{B}_I = (JX_{I,i}X_{J,j}X_{K,k}\epsilon_{ijk})x_{l,K}A_{l,J}. \tag{16.4.7}$$

Equation (3.3.9) may be used to express the quantity in parentheses as ϵ_{IJK}. We thus obtain

$$\mathcal{B}_I = \epsilon_{IJK}[(x_{l,K}A_l)_{,J} - x_{l,KJ}A_l]. \tag{16.4.8}$$

The second term here vanishes, since it is a double scalar product between symmetric and antisymmetric quantities. The result is

$$\boldsymbol{\mathcal{B}} = \nabla_X \times \boldsymbol{\mathcal{A}}, \tag{16.4.9}$$

where

$$\mathcal{A}_K \equiv x_{l,K}A_l, \qquad A_j = X_{K,j}\mathcal{A}_K; \tag{16.4.10}$$

$\boldsymbol{\mathcal{A}}$ is the material form of the vector potential. It is a function of the material coordinate $\mathbf{X}$ (and t) and has its components referred to the material coordinate system.

Finding the material form of the scalar potential requires considerably more manipulation. We begin by substituting the **E** field transformation (16.1.29) into (16.4.4)

$$X_{J,j}\left(\mathcal{E}_J+\epsilon_{JKL}\frac{\partial X_K}{\partial t}\mathcal{B}_L\right)=-\Phi_{,j}-\frac{\partial A_j}{\partial t}$$

$$=-\Phi_{,M}X_{M,j}-\frac{\partial}{\partial t}(X_{K,j}\mathcal{A}_K) \quad (16.4.11)$$

and using (16.4.10). By forming the scalar product with $x_{j,I}$, expanding the time derivative, commuting the spatial time and space derivatives, and substituting (16.4.9), we obtain

$$\mathcal{E}_I=-\epsilon_{IKL}\epsilon_{LMN}\frac{\partial X_K}{\partial t}\mathcal{A}_{N,M}-\Phi_{,I}-x_{j,I}\left(\mathcal{A}_K\frac{\partial}{\partial x_j}\frac{\partial X_K}{\partial t}+X_{K,j}\frac{\partial \mathcal{A}_K}{\partial t}\right). \quad (16.4.12)$$

This may be expressed as

$$\mathcal{E}_I=-\frac{\partial X_K}{\partial t}\mathcal{A}_{K,I}+\frac{\partial X_K}{\partial t}\mathcal{A}_{I,K}-\Phi_{,I}-\mathcal{A}_K\frac{\partial}{\partial X_I}\frac{\partial X_K}{\partial t}-\frac{\partial \mathcal{A}_I}{\partial t}$$

$$=-\Phi_{,I}-\left(\frac{\partial X_K}{\partial t}\mathcal{A}_K\right)_{,I}+\frac{\partial X_K}{\partial t}\mathcal{A}_{I,K}-\frac{d\mathcal{A}_I}{dt}+\dot{x}_j\mathcal{A}_{I,j} \quad (16.4.13)$$

with the use of $\epsilon-\delta$ identity and the definition of the material time derivative. Substitution of (16.4.10) and the time derivative relation (3.11.5) yields

$$\mathcal{E}_I=-\Phi_{,I}+(X_{K,j}\dot{x}_j x_{l,K}A_l)_{,I}-\frac{d\mathcal{A}_I}{dt}+\frac{\partial X_K}{\partial t}\mathcal{A}_{I,K}-x_{j,K}\frac{\partial X_K}{\partial t}X_{L,j}\mathcal{A}_{I,L}$$

$$=-\varphi_{,I}-\dot{\mathcal{A}}_I \quad (16.4.14)$$

where

$$\varphi\equiv\Phi-\dot{\mathbf{x}}\cdot\mathbf{A}, \qquad \Phi=\varphi-\frac{\partial \mathbf{X}}{\partial t}\cdot\boldsymbol{\mathcal{A}}; \quad (16.4.15)$$

φ is the material form of the scalar potential and is a function of **X** and t. Note that its relationship to Φ and **A** involves the material velocity $\dot{\mathbf{x}}$.

We now wish to transform the Lagrangian density to a function of $\boldsymbol{\mathcal{A}}$ and φ. First, since it is a scalar density, the material form of the electromagnetic Lagrangian density $\mathfrak{L}_M$ must transform as

$$\mathfrak{L}_M = J\mathfrak{L}_S. \tag{16.4.16}$$

Substitution of the transformations for **E** and **B** into the field Lagrangian leads directly to

$$\mathfrak{L}_{FM} = \frac{\epsilon_0}{2}\left\{J\left(\mathcal{E} + \frac{\partial \mathbf{X}}{\partial t}\times \mathfrak{B}\right)\cdot \mathbf{C}^{-1}\cdot\left(\mathcal{E} + \frac{\partial \mathbf{X}}{\partial t}\times\mathfrak{B}\right) - c^2 J^{-1}\mathfrak{B}\cdot\mathbf{C}\cdot\mathfrak{B}\right\} \tag{16.4.17}$$

for the material form of the field Lagrangian density. The $\mathbf{C}$ and $\mathbf{C}^{-1}$ tensors are the Green deformation tensor and its inverse,

$$C_{AB} \equiv x_{i,A}x_{i,B}, \tag{16.4.18}$$

$$(C^{-1})_{AB} = X_{A,i}X_{B,i}, \tag{16.4.19}$$

originally introduced in (3.4.2) and (3.P.5). The field Lagrangian $\mathfrak{L}_{FM}$ is a function of $\mathcal{Q}$ and φ through (16.4.9) and (16.4.14).

The material form of the interaction Lagrangian density $\mathfrak{L}_{IM}$ can now be found to be

$$\begin{aligned}
\mathfrak{L}_{IM} &= J(\mathbf{j}\cdot\mathbf{A} - q\Phi) \\
&= Jj_k X_{K,k}\mathcal{Q}_K - Jq\varphi + Jq\frac{\partial X_K}{\partial t}\mathcal{Q}_K \\
&= JX_{K,k}(j_k - \dot{x}_k q)\mathcal{Q}_K - Jq\varphi \\
&= \mathcal{J}\cdot\mathcal{Q} - \mathcal{Q}\varphi
\end{aligned} \tag{16.4.20}$$

where

$$\mathcal{Q} \equiv Jq, \tag{16.4.21}$$

$$\mathcal{J}_K \equiv JX_{K,k}(j_k - \dot{x}_k q), \qquad j_i = J^{-1}x_{i,K}\left(\mathcal{J}_K - \frac{\partial X_K}{\partial t}\mathcal{Q}\right). \tag{16.4.22}$$

Here $\mathcal{Q}$ is the material measure of the total charge density including both bound and free charge and $\mathcal{J}$ is the material measure of the total current density arising from both bound and free charge. Note that the current arising from the transport of a charge density (both bound and free) at the material velocity $\dot{\mathbf{x}}$ is subtracted out of $\mathcal{J}$.

16.5 Material Form of Maxwell–Lorentz Equations

The material form of the electromagnetic Lagrangian density found in the preceding section is

$$\mathfrak{L}_M = \frac{\epsilon_0}{2}\left\{ J\left(\mathfrak{E} + \frac{\partial \mathbf{X}}{\partial t} \times \mathfrak{B}\right) \cdot \mathbf{C}^{-1} \cdot \left(\mathfrak{E} + \frac{\partial \mathbf{X}}{\partial t} \times \mathfrak{B}\right) - c^2 J^{-1} \mathfrak{B} \cdot \mathbf{C} \cdot \mathfrak{B} \right\} + \mathfrak{J} \cdot \mathfrak{A} - \mathfrak{Q}\varphi \tag{16.5.1}$$

where $\mathfrak{E}$ and $\mathfrak{B}$ are functions of $\mathfrak{A}$ and φ and $\mathbf{X}$ and t are the independent variables. The Lorentz form of the electromagnetic equations can now be found from $\mathfrak{L}_M$.

The Lagrange equation for $\mathfrak{A}$,

$$\frac{d}{dt}\frac{\partial \mathfrak{L}_M}{\partial \dot{\mathfrak{A}}_K} = \frac{\partial \mathfrak{L}_M}{\partial \mathfrak{A}_K} - \frac{\partial}{\partial X_L}\frac{\partial \mathfrak{L}_M}{\partial \mathfrak{A}_{K,L}}, \tag{16.5.2}$$

leads directly to

$$\epsilon_{KLM}\frac{\partial}{\partial X_L}\left\{ \frac{1}{\mu_0 J} C_{MN}\mathfrak{B}_N + \epsilon_{MNP}\frac{\partial X_N}{\partial t}\epsilon_0 J (C^{-1})_{PQ}\left(\mathfrak{E} + \frac{\partial \mathbf{X}}{\partial t} \times \mathfrak{B}\right)_Q \right\} = \frac{d}{dt}\left\{ \epsilon_0 J (C^{-1})_{KJ}\left(\mathfrak{E} + \frac{\partial \mathbf{X}}{\partial t} \times \mathfrak{B}\right)_J \right\} + \mathfrak{J}_K, \tag{16.5.3}$$

which is the material frame analog of (2.4.1). The Lagrange equation for φ,

$$\frac{d}{dt}\frac{\partial \mathfrak{L}_M}{\partial \dot{\varphi}} = \frac{\partial \mathfrak{L}_M}{\partial \varphi} - \frac{\partial}{\partial X_L}\frac{\partial \mathfrak{L}_M}{\partial \varphi_{,L}}, \tag{16.5.4}$$

leads to

$$\frac{\partial}{\partial X_L}\left\{ \epsilon_0 J (C^{-1})_{LM}\left(\mathfrak{E} + \frac{\partial \mathbf{X}}{\partial t} \times \mathfrak{B}\right)_M \right\} = \mathfrak{Q}, \tag{16.5.5}$$

which is the material frame analog of (2.4.2). The two remaining electromagnetic equations,

$$\nabla_X \times \mathfrak{E} + \dot{\mathfrak{B}} = 0, \tag{16.5.6}$$

$$\nabla_X \cdot \mathfrak{B} = 0, \tag{16.5.7}$$

follow directly from (16.4.14) and (16.4.9). While the latter two equations retain the form of their spatial frame analogs (16.1.2) and (16.1.4), (16.5.3) and (16.5.5) do not retain the form of their spatial frame analogs (16.1.1) and (16.1.3). Thus the form invariance under a deformation transformation required in Section 16.1 for the Maxwell form of the electromagnetic equations produces field transformations that do not allow form invariance of the Lorentz form of the electromagnetic equations.

16.6 Polarization and Magnetization

The Maxwell form of the electromagnetic equations is obtained from the Lorentz form in the spatial frame by expanding the charge and current densities in terms of multipole moments as done in Sections 4.4 and 4.5. If we retain for the time being free charge and current densities, then the charge density is

$$q = q^f - \nabla \cdot \mathbf{P} \tag{16.6.1}$$

and the current density is

$$\mathbf{j} = \mathbf{j}^f + \frac{\partial \mathbf{P}}{\partial t} + \nabla \times (\mathbf{M} + \mathbf{P} \times \dot{\mathbf{x}}), \tag{16.6.2}$$

where $\mathbf{j}^f$ is defined in (16.1.5). The polarization $\mathbf{P}$ and the magnetization $\mathbf{M}$ are the electric dipole and magnetic dipole contributions to the multipole expansions developed in Sections 4.4 and 4.5. The electric quadrupolarization terms are dropped from the two equations. We now wish to find the material frame transforms of $\mathbf{P}$ and $\mathbf{M}$.

We begin by substituting the spatial frame charge expression (16.6.1) into (16.4.21),

$$\begin{aligned}
\mathcal{Q} &= J(q^f - \nabla \cdot \mathbf{P}) \\
&= Jq^f - JX_{K,j}P_{j,K} \\
&= Jq^f - (JX_{K,j}P_j)_{,K} \\
&= \mathcal{Q}^f - \mathcal{P}_{K,K}
\end{aligned} \tag{16.6.3}$$

where use of the Euler–Jacobi–Piola identity (3.2.13) and of the definition of the material frame measure of the free charge density is made. Equation

(16.6.3) leads us to define the material frame measure of polarization $\mathcal{P}$ by

$$\mathcal{P}_K \equiv JX_{K,j}P_j, \qquad P_i = J^{-1}x_{i,K}\mathcal{P}_K. \tag{16.6.4}$$

Next we substitute both the spatial frame charge and current expressions (16.6.1) and (16.6.2) into (16.4.22),

$$\begin{aligned}
\mathcal{J}_K &= JX_{K,i}\left[j_i^f + \frac{\partial P_i}{\partial t} + \epsilon_{ijk}(M_k + \epsilon_{klm}P_l\dot{x}_m)_{,j} - \dot{x}_i(q^f - P_{j,j})\right] \\
&= JX_{K,i}j_i^c + JX_{K,i}\overset{*}{P}_i + (JX_{K,i}X_{L,j}X_{M,k}\epsilon_{ijk})x_{l,M}M_{l,L} \\
&= \mathcal{J}_K^c + JX_{K,i}J^{-1}x_{i,L}\frac{d}{dt}(JX_{L,j}P_j) + \epsilon_{KLM}x_{l,M}M_{l,L} \\
&= \mathcal{J}_K^c + \frac{d}{dt}(JX_{K,j}P_j) + \epsilon_{KLM}\left[(x_{l,M}M_l)_{,L} - x_{l,ML}M_l\right] \\
&= \mathcal{J}_K^c + \dot{\mathcal{P}}_K + \epsilon_{KLM}(\mathcal{M}_M)_{,L}.
\end{aligned} \tag{16.6.5}$$

Here $\mathcal{M}$ is the material frame measure of magnetization defined by

$$\mathcal{M}_K \equiv x_{l,K}M_l, \qquad M_j = X_{K,j}\,\mathcal{M}_K \tag{16.6.6}$$

and two expressions for the convected time derivative, (3.13.5) and (3.13.2), are used. In vector notation (16.6.5) is

$$\mathcal{J} = \mathcal{J}^c + \dot{\mathcal{P}} + \nabla_X \times \mathcal{M}. \tag{16.6.7}$$

We now turn to obtaining the material form of the relations of the magnetic intensity and the electric displacement to the other fields. If the charge and current expansions (16.6.1) and (16.6.2) are substituted into the general form of the Maxwell–Lorentz equations (2.4.1) to (2.4.4), it is necessary to define in the spatial frame the magnetic intensity $\mathbf{H}$ as

$$\mathbf{H} \equiv \frac{1}{\mu_0}\mathbf{B} - \mathbf{M} - \mathbf{P}\times\dot{\mathbf{x}} \tag{16.6.8}$$

and the electric displacement as

$$\mathbf{D} \equiv \epsilon_0\mathbf{E} + \mathbf{P}. \tag{16.6.9}$$

The first of these represents a generalization of the definition of $\mathbf{H}$ introduced previously for a nonmagnetic dielectric in (7.1.8). We leave as exercises the demonstration that these two equations in conjunction with the transformations for **B**, **H**, **M**, **E**, **D**, and **P** lead to

$$\mathcal{H}_A \equiv \frac{1}{\mu_0 J} C_{AB}\mathcal{B}_B + \epsilon_{ABC}\frac{\partial X_B}{\partial t}\epsilon_0 J(C^{-1})_{CD}\left(\mathcal{E} + \frac{\partial \mathbf{X}}{\partial t}\times\mathcal{B}\right)_D - \mathcal{M}_A \tag{16.6.10}$$

and

$$\mathcal{D}_K \equiv \epsilon_0 J(C^{-1})_{KL}\left(\mathcal{E} + \frac{\partial \mathbf{X}}{\partial t}\times\mathcal{B}\right)_L + \mathcal{P}_K. \tag{16.6.11}$$

These relations for $\mathcal{H}$ and $\mathcal{D}$ provide the connection between the material Lorentz-form equations and the material Maxwell-form equations. It is readily seen that substitution of these relations along with the expressions (16.6.3) and (16.6.5) for $\mathcal{Q}$ and $\mathcal{J}$ into the material Lorentz-form equations (16.5.3) and (16.5.5) leads to the corresponding material Maxwell-form equations (16.1.13) and (16.1.23).

16.7 Electric Dipole Form of Interaction Lagrangian

Just as is done in the spatial frame in Section 4.7, the interaction Lagrangian for a dielectric in the electric dipole approximation may be given an alternate form that is convenient to use. If the free charge and free current densities and the magnetization are dropped, the material frame charge and current densities become

$$\mathcal{J} = \dot{\mathcal{P}} \equiv \mathcal{J}^D, \tag{16.7.1}$$

$$\mathcal{Q} = -\nabla_X \cdot \mathcal{P} \equiv \mathcal{Q}^D. \tag{16.7.2}$$

The interaction Lagrangian may then be written as

$$\begin{aligned} L_I &= \int [\mathcal{J}\cdot\mathcal{A} - \mathcal{Q}\varphi]\,dV \\ &= \int [\dot{\mathcal{P}}\cdot\mathcal{A} + \nabla_X\cdot\mathcal{P}\varphi]\,dV \\ &= \int \left[\frac{d}{dt}(\mathcal{P}\cdot\mathcal{A}) - \mathcal{P}\cdot\dot{\mathcal{A}} + \nabla_X\cdot(\mathcal{P}\varphi) - \mathcal{P}\cdot\nabla_X\varphi\right] dV. \end{aligned} \tag{16.7.3}$$

Since the addition of total space or time derivatives to the Lagrangian density does not affect the equations of motion (see Section 2.1), the two total derivatives may be dropped from the integrand of (16.7.3) with the result

$$L_I=\int \mathcal{P}\cdot(-\nabla_X\varphi-\dot{\mathcal{Q}})\,dV$$
$$=\int \mathcal{P}\cdot\mathcal{E}\,dV. \tag{16.7.4}$$

Thus the material frame interaction Lagrangian density of a dielectric may be written as

$$\mathcal{L}_{IM}=\mathcal{P}\cdot\mathcal{E} \tag{16.7.5}$$

in the electric dipole approximation.

PROBLEMS

16.1. Show that the material frame boundary condition (16.2.3) transforms to the spatial frame boundary condition (7.2.9).

16.2. Derive (16.6.10) from (16.6.8) and the various field transformation equations.

16.3. Derive (16.6.11) from (16.6.9) and the various field transformation equations.

16.4. Transform the electromagnetic momentum continuity equation (8.1.7) of a nonmagnetic dielectric to

$$\frac{d}{dt}\{X_{I,i}[(\mathcal{D}-\mathcal{P})\times\mathcal{B}]_I\}-M_{iK,K}$$
$$=-X_{I,i}\left\{\left(\mathcal{E}+\frac{\partial\mathbf{X}}{\partial t}\times\mathcal{B}\right)_I\mathcal{Q}^D+\left[\left(\mathcal{J}^D+\mathcal{Q}^D\frac{\partial\mathbf{X}}{\partial t}\right)\times\mathcal{B}\right]_I\right\} \tag{16.P.1}$$

where

$$M_{iK}\equiv X_{I,i}\left\{\left(\mathcal{E}+\frac{\partial\mathbf{X}}{\partial t}\times\mathcal{B}\right)_I(\mathcal{D}-\mathcal{P})_K+\left[\mathcal{H}-\frac{\partial\mathbf{X}}{\partial t}\times(\mathcal{D}-\mathcal{P})\right]_I\mathcal{B}_K\right\}$$
$$-\tfrac{1}{2}X_{K,i}\left\{\left(\mathcal{E}+\frac{\partial\mathbf{X}}{\partial t}\times\mathcal{B}\right)_I(\mathcal{D}-\mathcal{P})_I+\left[\mathcal{H}-\frac{\partial\mathbf{X}}{\partial t}\times(\mathcal{D}-\mathcal{P})\right]_I\mathcal{B}_I\right\}$$
$$-X_{I,i}[(\mathcal{D}-\mathcal{P})\times\mathcal{B}]_I\frac{\partial X_K}{\partial t} \tag{16.P.2}$$

is the mixed frame Maxwell stress tensor expressed in terms of material frame electromagnetic fields. Note that these equations may be expressed in terms of $\mathfrak{E}$ and $\mathfrak{B}$ by using

$$\mathfrak{D} - \mathfrak{P} = \epsilon_0 J \mathbf{C}^{-1} \cdot \left(\mathfrak{E} + \frac{\partial \mathbf{X}}{\partial t} \times \mathfrak{B} \right) \tag{16.P.3}$$

and

$$\mathfrak{H} - \frac{\partial \mathbf{X}}{\partial t} \times (\mathfrak{D} - \mathfrak{P}) = \frac{1}{\mu_0 J} \mathbf{C} \cdot \mathfrak{B}. \tag{16.P.4}$$

16.5. Transform the electromagnetic energy continuity equation (8.7.6) for a nonmagnetic dielectric to

$$\begin{aligned}
&\frac{d}{dt}\left[\frac{\epsilon_0 J}{2} \left(\mathfrak{E} + \frac{\partial \mathbf{X}}{\partial t} \times \mathfrak{B} \right) \cdot (\mathbf{C}^{-1}) \cdot \left(\mathfrak{E} + \frac{\partial \mathbf{X}}{\partial t} \times \mathfrak{B} \right) \right. \\
&\quad \left. + \frac{1}{2\mu_0 J} \mathfrak{B} \cdot \mathbf{C} \cdot \mathfrak{B} \right] + \nabla_X \cdot \left[\frac{1}{\mu_0 J} \left(\mathfrak{E} + \frac{\partial \mathbf{X}}{\partial t} \times \mathfrak{B} \right) \times (\mathbf{C} \cdot \mathfrak{B}) \right. \\
&\quad + \frac{\partial \mathbf{X}}{\partial t} \frac{\epsilon_0 J}{2} \left(\mathfrak{E} + \frac{\partial \mathbf{X}}{\partial t} \times \mathfrak{B} \right) \cdot (\mathbf{C}^{-1}) \cdot \left(\mathfrak{E} + \frac{\partial \mathbf{X}}{\partial t} \times \mathfrak{B} \right) \\
&\quad \left. + \frac{\partial \mathbf{X}}{\partial t} \frac{1}{2\mu_0 J} \mathfrak{B} \cdot \mathbf{C} \cdot \mathfrak{B} \right] = -\left(\dot{\mathfrak{P}} + \frac{\partial \mathbf{X}}{\partial t} \nabla_X \cdot \mathfrak{P} \right) \cdot \left(\mathfrak{E} + \frac{\partial \mathbf{X}}{\partial t} \times \mathfrak{B} \right).
\end{aligned} \tag{16.P.5}$$

CHAPTER 17

Nonlinear Electroacoustics

We now turn to a consideration of nonlinearities of dielectric, piezoelectric, and pyroelectric crystals in the interaction of acoustic waves and electromagnetic fields in the low frequency region. Though the formulation can be applied to nonlinearities of any order, attention is placed on three-field interactions in this chapter. The formulation is initially fully electrodynamic but a truncated set of dynamic equations valid in the quasi-electrostatic regime is also obtained because most present day experiments in this field can be so characterized.

In the quasi-electrostatic regime boundary conditions are usually known only in the material coordinate system since electrodes and transducers are attached to body surfaces. This means that a material coordinate system formulation of a problem is most easily solved and also means that measurements determine material tensors referred to the material coordinates.

The three-field nonlinearities studied in this chapter include the third order stiffness tensor, the nonlinear piezoelectric tensor, the electrostriction tensor, and the electric field mixing tensor. The latter is found to be just the low frequency limit of the optical mixing tensor considered in Chapter 15. Interestingly, we find that the electrostriction tensor is not just the low frequency limit of the elastooptic tensor as long believed. The difference arises because the measured elastooptic tensor is a spatial frame tensor while the electrostriction tensor is a material frame tensor. The electrostriction tensor has no part coupling to rotations and differs by Maxwell stress terms from the low frequency limit of the elastooptic tensor that couples to strain.

Two applications of the theory are presented. The first is three-wave acoustic mixing which is considered from both the free-plus-forced-wave viewpoint and from the slowly-varying-amplitude viewpoint. This is the first correct treatment of this interaction in piezoelectric crystals. The second application is the interaction of a homogeneous oscillating electric field with two counterpropagating bulk acoustic waves, an interaction first studied by Thompson and Quate [1971]. The treatment here obtains the material interaction coefficient for this process for the first time.

The treatment in this chapter is based on [Nelson, 1978a, 1978b, 1978c] which places emphasis on electric and electroelastic nonlinearities. Purely elastic nonlinearities were described earlier by Toupin and Bernstein [1961], Thurston and Brugger [1964], and Thurston and Shapiro [1967]. The first fully correct formulation of electroelastic nonlinearities (in the quasi-electrostatic approximation) is in the work of Baumhauer and Tiersten [1973]; application to surface acoustic waves was made by them [Tiersten and Baumhauer, 1974].

17.1 Matter Equations with Material Frame Electromagnetic Fields

In this section we obtain the general matter equations, that is, the center-of-mass and internal motion equations, expressed in terms of the material frame electromagnetic fields. This may be done either by reexpressing the forms of those equations found in Sections 7.3 and 7.4 with the field transformations found in the preceding chapter or by obtaining them from a transformed Lagrangian. We choose the latter approach.

The total material frame Lagrangian density $\mathfrak{L}_M$ of a dielectric crystal consists of parts representing the electromagnetic field (16.4.17), the electromagnetic field-matter interaction in the electric dipole approximation (16.7.5), and the matter (4.7.12) and (6.5.8). Thus we have

$$\begin{aligned}\mathfrak{L}_M = &\frac{\epsilon_0}{2} J\left(\mathfrak{E} + \frac{\partial \mathbf{X}}{\partial t} \times \mathfrak{B}\right)\cdot \mathbf{C}^{-1} \cdot \left(\mathfrak{E} + \frac{\partial \mathbf{X}}{\partial t} \times \mathfrak{B}\right) \\ &- \frac{1}{2\mu_0 J} \mathfrak{B} \cdot \mathbf{C} \cdot \mathfrak{B} + \mathfrak{P} \cdot \mathfrak{E} + \frac{\rho^0}{2}(\dot{\mathbf{x}})^2 + \tfrac{1}{2}\sum_\nu m^\nu (\dot{\mathbf{y}}^{T\nu})^2 \\ &- \rho^0 \Sigma(E_{AB}, \Pi_C^\mu)\end{aligned} \qquad (17.1.1)$$

where

$$\mathfrak{P}_K = J X_{K,j} P_j = X_{K,j} \sum_\nu q^\nu y_j^{T\nu} = \sum_\nu q^\nu (Y_K^\nu + \Pi_K^\nu) \qquad (17.1.2)$$

and $\mathfrak{E}$ and $\mathfrak{B}$ are regarded as functions of the material frame vector and scalar potentials (16.4.9) and (16.4.14).

The internal motion equations are found from

$$\frac{d}{dt} \frac{\partial \mathfrak{L}_M}{\partial \dot{y}_i^{T\nu}} = \frac{\partial \mathfrak{L}_M}{\partial y_i^{T\nu}} \qquad (17.1.3)$$

to be

$$m^{\nu}\ddot{y}_i^{T\nu} = q^{\nu} X_{K,i} \mathcal{E}_K - \frac{\partial \rho^0 \Sigma}{\partial \Pi_K^{\nu}} X_{K,i}. \tag{17.1.4}$$

In this chapter we are interested in fields at frequencies in the sonic and ultrasonic regions and thus well below internal (optic mode) resonances. Thus in solving the last equation for $y_i^{T\nu}(x_{j,A}, \mathcal{E}_K)$ we drop the inertial force term, since it is negligible in size. The internal coordinate $\mathbf{y}^{T\nu}$ may then be substituted back into the Lagrangian density before the center-of-mass equation is found. This procedure of adiabatic elimination of internal coordinates is equivalent to eliminating $\mathbf{y}^{T\nu}$ from the center-of-mass equation of motion as done in Section 10.3. By using adiabatic elimination we exclude the case where two optical electric fields produce an acoustic wave output field at the difference frequency of the input fields. This case is easily handled as a slight generalization of the results of this section.

Next we evaluate the Lagrange equation for $\mathbf{x}$,

$$\frac{d}{dt} \frac{\partial \mathcal{L}_M}{\partial \dot{x}_i} = \frac{\partial \mathcal{L}_M}{\partial x_i} - \frac{\partial}{\partial X_A} \frac{\partial \mathcal{L}_M}{\partial x_{i,A}}, \tag{17.1.5}$$

regarding the adiabatic elimination to have given the dependence $\mathcal{P}(x_{i,A}, \mathcal{E}_K)$ to the polarization. Since $\mathcal{L}_M$ is not an explicit function of $\mathbf{x}$, this equation gives directly

$$\frac{d}{dt}\left[\rho^0 \dot{x}_i + \epsilon_0 J X_{K,i} \left\{ \left[\mathbf{C}^{-1} \cdot \left(\mathcal{E} + \frac{\partial \mathbf{X}}{\partial t} \times \mathcal{B} \right) \right] \times \mathcal{B} \right\}_K \right] = \frac{\partial}{\partial X_A}\left[T_{iA} - \mathcal{E}_K \frac{\partial \mathcal{P}_K}{\partial x_{i,A}} + M_{iA} \right] \tag{17.1.6}$$

where M_{iA} is the mixed frame Maxwell stress tensor expressed in terms of material frame fields and given in (16.P.2) and where the Piola-Kirchoff stress tensor is given by

$$T_{iA} \equiv \frac{\partial \rho^0 \Sigma}{\partial x_{i,A}}. \tag{17.1.7}$$

Equation (17.1.6) has the form of a conservation law, since all terms are either material time derivatives or material frame divergences. It can be shown, in fact, that this equation is just a reexpression of the momentum conservation law (8.2.3). We leave this as an exercise.

The development of the momentum conservation law showed that it is a sum of the center-of-mass equation and the electromagnetic momentum continuity equation. Thus the center-of-mass equation may be obtained from (17.1.6) by subtracting the electromagnetic continuity equation expressed in the material frame. The latter is given in (16.P.1). Performing this subtraction gives

$$\rho^0 \ddot{x}_i = \left(T_{iJ} - \mathcal{E}_K \frac{\partial \mathcal{P}_K}{\partial x_{i,J}} \right)_{,J} + X_{I,i} \left\{ \left(\mathfrak{E} + \frac{\partial \mathbf{X}}{\partial t} \times \mathfrak{B} \right)_I \mathcal{Q}^D + \left[\left(\mathcal{J}^D + \mathcal{Q}^D \frac{\partial \mathbf{X}}{\partial t} \right) \times \mathfrak{B} \right]_I \right\} \tag{17.1.8}$$

for the center-of-mass equation of motion expressed in material frame electromagnetic fields. $\mathcal{Q}^D$ and $\mathcal{J}^D$ are the dielectric charge and current of (16.7.1) and (16.7.2).

17.2 Linear Piezoelectricity with Material Frame Electric Field

Before proceeding to the exploration of nonlinear acoustic and electroacoustic phenomena it is worth considering the equations of linear piezoelectricity expressed in terms of the material frame electric field in order to reexamine the piezoelectric-like terms involving the spontaneous polarization of a pyroelectric crystal that are shown to be immeasurable and yet lead to so much complication in the spatial frame expression of these equations in Chapters 10 and 11. We show that such terms make no explicit appearance in a completely material frame formulation of linear piezoelectricity. Thus the material frame expression of linear piezoelectricity is preferable to the spatial frame expression when dealing with pyroelectrics. When dealing with nonpyroelectric piezoelectrics, there is no preference between the spatial and material frame expressions of linear piezoelectricity, since the equations in the two frames are identical in form.

Since our interest here is aimed at the linear piezoelectric-like terms involving the spontaneous polarization and not at those involving the spontaneous electric field, we set the latter to zero,

$$\mathbf{E}^S = 0, \tag{17.2.1}$$

that is, we regard it as canceled by collected extrinsic surface charge as discussed in Section 11.3. We also set the spontaneous stress equal to zero,

$$\mathbf{t}^S = 0, \tag{17.2.2}$$

for simplicity. These two simplifications permit us to drop the linear terms from the stored energy as can be seen from (7.3.7) and (10.3.7). The stored energy is then simply

$$\rho^0\Sigma=\sum_{\nu\mu}{}^{20}M_{KL}^{\nu\mu}\Pi_K^\nu\Pi_L^\mu+\sum_\nu{}^{11}M_{KAB}^\nu\Pi_K^\nu E_{AB}$$

$$+{}^{02}M_{ABCD}E_{AB}E_{CD}. \tag{17.2.3}$$

The internal motion equations (17.1.4) in the adiabatic regime now become

$$0=q^\nu X_{K,i}\,\mathcal{E}_K-2\sum_\mu{}^{20}M_{KL}^{\nu\mu}\Pi_L^\mu X_{K,i}-{}^{11}M_{KAB}^\nu E_{AB}X_{K,i}. \tag{17.2.4}$$

After forming a scalar product of this equation with $x_{i,I}$ this equation can be solved for Π_I^λ with the use of the zero frequency mechanical admittance $\Upsilon_{IK}^{\lambda\nu}$ referred to the material coordinate system [see (9.1.8)],

$$\Pi_I^\lambda=\sum_\nu\Upsilon_{IK}^{\lambda\nu}\left(q^\nu\,\mathcal{E}_K-{}^{11}M_{KAB}^\nu E_{AB}\right). \tag{17.2.5}$$

The equation for Π_I^λ may now be used to express the stored energy in the adiabatic regime as

$$\rho^0\Sigma=\tfrac{1}{2}c_{ABCD}E_{AB}E_{CD}+\frac{\epsilon_0}{2}\chi_{KL}\mathcal{E}_K\mathcal{E}_L \tag{17.2.6}$$

where the stiffness tensor c_{ABCD} is defined in (10.3.6) and the linear electric susceptibility χ_{KL} is defined by (9.1.13). The material frame polarization (17.1.2) may be reexpressed as

$$\mathcal{P}_K=P_K^S+\epsilon_0\chi_{KN}\mathcal{E}_N+e_{KAB}u_{A,B} \tag{17.2.7}$$

where the definition of the piezoelectric stress tensor e_{KAB} (10.2.2) is used. Note that the piezoelectric-like term involving $\mathbf{P}^S$ that enters the spatial frame polarization (10.2.7) does not appear in (17.2.7) for the material frame polarization. Note also that no linear term in the material frame magnetic intensity $\mathcal{H}$ (16.6.10) depending on $\mathbf{P}^S$ can arise as did in the spatial frame magnetic intensity $\mathbf{H}$ (10.2.17).

The two stress terms needed for the center-of-mass equation may now be evaluated. First, the Piola-Kirchoff stress tensor can be found to the linear level to be

$$T_{iJ}=\frac{\partial \rho^0 \Sigma}{\partial E_{AJ}} x_{i,A}=\delta_{iA} c_{AJCD} u_{C,D}. \tag{17.2.8}$$

The other stress becomes

$$-\mathcal{E}_K \frac{\partial \mathcal{P}_K}{\partial x_{i,J}}=-\delta_{iA} e_{KAJ} \mathcal{E}_K. \tag{17.2.9}$$

The body force terms in (17.1.8) have no linear part and so may be dropped. By referring the displacement to the material coordinate system, $u_I=u_i\delta_{iI}$, the center-of-mass equation now becomes the linearized dynamic elasticity equation expressed in the material coordinate system,

$$\rho^0 \ddot{u}_I=(c_{IJCD} u_{C,D}-e_{KIJ} \mathcal{E}_K)_{,J}. \tag{17.2.10}$$

It follows from the reasoning of Section 11.1 that the electromagnetic equations needed to be solved with this equation in the quasi-electrostatic approximation reduce to

$$\nabla_X \cdot \mathfrak{D}=0 \tag{17.2.11}$$

with the linearized material frame electric displacement $\mathfrak{D}$, found from (16.6.11) and (17.2.7), given by

$$\mathfrak{D}_K=P_K^S+\epsilon_0 \kappa_{KN} \mathcal{E}_N+e_{KAB} u_{A,B} \tag{17.2.12}$$

and the material frame electric field given by

$$\mathcal{E}=-\nabla_X \varphi. \tag{17.2.13}$$

The last four equations were first given detailed justification as the material frame expression of piezoelectricity by Baumhauer and Tiersten [1973] in a study of polarized ferroelectric ceramics. The present derivation extends the proof to general pyroelectric crystals.

Note that the only difference between the linearized constitutive and dynamical equations of this section when applied to pyroelectrics as compared to piezoelectrics is the static spontaneous polarization term in $\mathcal{P}$ and $\mathfrak{D}$. No *linear* piezoelectric-like terms involving $\mathbf{P}^S$ appear in the material frame constitutive expressions for $\mathcal{P}$, $\mathfrak{D}$, or $\mathcal{H}$. This contrasts to the spatial frame expression of these equations where such terms do appear even though they are shown to disappear from the differential equations and boundary conditions and so are unobservable. The nonappearance of corresponding terms in the material frame equations does not contradict the necessity or the reality of the piezoelectric-like terms in the spatial frame equations; the material frame equations with material electric fields do, however, give an alternate, simpler, and thus more convenient set of equations with which to study piezoelectricity in pyroelectrics.

17.3 Material Frame Polarization to Bilinear Order

In preparation for studying three-field acoustic and electroacoustic interactions we obtain the material frame polarization to bilinear order in driving fields in this section. The procedure to be followed is rather similar to that just used for the linear material frame polarization; the main difference is that two stages of iteration are needed to accomplish the adiabatic elimination of the internal coordinates to bilinear order. The treatment applies, as usual, to piezoelectrics and pyroelectrics as well as ordinary dielectrics. Just as in the preceding section, we make two simplifying assumptions: (1) no spontaneous electric field if the crystal is pyroelectric,

$$\mathbf{E}^S=0, \tag{17.3.1}$$

and (2) no spontaneous stress,

$$\mathbf{t}^S=0. \tag{17.3.2}$$

The stored energy is now needed to trilinear order as given in (12.1.3). Substituting that form into the internal motion equations (17.1.4) and dropping the inertial term by the adiabatic assumption leads to

$$\begin{aligned}0=q^{\nu}X_{I,i}\mathcal{E}_I-X_{K,i}\Bigg(&2\sum_{\mu}{}^{20}M^{\nu\mu}_{KL}\Pi^{\mu}_L+{}^{11}M^{\nu}_{KAB}E_{AB}\\&+3\sum_{\mu\lambda}{}^{30}M^{\nu\mu\lambda}_{KLM}\Pi^{\mu}_L\Pi^{\lambda}_M+2\sum_{\mu}{}^{21}M^{\nu\mu}_{KLAB}\Pi^{\mu}_L E_{AB}\\&+{}^{12}M^{\nu}_{KABCD}E_{AB}E_{CD}\Bigg).\end{aligned} \tag{17.3.3}$$

Since third order terms in the stored energy are regarded as small compared to second order terms, we can solve this equation in two stages of iteration. First, we solve for $\mathbf{\Pi}^{\nu}$ from the linear term of (17.3.3) and then substitute the linear solution (17.2.5) into the bilinear terms. The procedure is similar to the iteration process employed in Sections 12.2 and 12.4 for three-field optical interactions. This gives us

$$\begin{aligned}
\Pi_J^\nu = \sum_\rho \Upsilon_{JK}^{\nu\rho}\Bigg[& q^\rho \mathcal{E}_K - {}^{11}M_{KAB}^{\rho} E_{AB} \\
& - 3\sum_{\mu\lambda} {}^{30}M_{KLM}^{\rho\mu\lambda}\Pi_L^\mu \Pi_M^\lambda - 2\sum_\mu {}^{21}M_{KLAB}^{\rho\mu}\Pi_L^\mu E_{AB} \\
& - {}^{12}M_{KABCD}^{\rho} E_{AB} E_{CD}\Bigg] \\
= \sum_\rho \Upsilon_{JK}^{\nu\rho} \Bigg\{ & q^\rho \mathcal{E}_K - {}^{11}M_{KAB}^{\rho} E_{AB} \\
& - 3\sum_{\mu\lambda\sigma\xi} {}^{30}M_{KLM}^{\rho\mu\lambda} \Upsilon_{LE}^{\mu\sigma} q^\sigma \Upsilon_{MF}^{\lambda\xi} q^\xi \mathcal{E}_E \mathcal{E}_F \\
& + \Bigg[6\sum_{\mu\lambda\sigma\xi} {}^{30}M_{KLM}^{\rho\mu\sigma} \Upsilon_{MF}^{\lambda\xi}\, {}^{11}M_{FAB}^{\xi} \Upsilon_{LE}^{\mu\sigma} q^\sigma \\
& - 2\sum_{\mu\sigma} {}^{21}M_{KLAB}^{\rho\mu} \Upsilon_{LE}^{\mu\sigma} q^\sigma \Bigg] E_{AB}\mathcal{E}_E \\
& + \Bigg[-3\sum_{\mu\lambda\sigma\xi} {}^{30}M_{KLM}^{\rho\mu\lambda} \Upsilon_{LE}^{\mu\sigma}\, {}^{11}M_{EAB}^{\sigma} \Upsilon_{MF}^{\lambda\xi}\, {}^{11}M_{FCD}^{\xi} \\
& + \sum_{\mu\sigma} {}^{21}M_{KLAB}^{\rho\mu} \Upsilon_{LE}^{\mu\sigma}\, {}^{11}M_{ECD}^{\sigma} \\
& + \sum_{\mu\sigma} {}^{21}M_{KLCD}^{\rho\mu} \Upsilon_{LE}^{\mu\sigma}\, {}^{11}M_{EAB}^{\sigma} - {}^{12}M_{KABCD}^{\rho} \Bigg] E_{AB} E_{CD} \Bigg\}. \qquad (17.3.4)
\end{aligned}$$

With the expression for $\boldsymbol{\Pi}^\nu$ to bilinear order the material frame polarization $\mathcal{P}$ (17.1.2) becomes

$$\begin{aligned}
\mathcal{P}_K &= P_K^S + \epsilon_0 \chi_{KL} \mathcal{E}_L + e_{KAB} E_{AB} + 2\epsilon_0 b_{KLM} \mathcal{E}_L \mathcal{E}_M \\
&\quad + \epsilon_0 l_{KLAB}^s \mathcal{E}_L E_{AB} + \tfrac{1}{2} e_{KABCD} E_{AB} E_{CD} \\
&= P_K^S + \epsilon_0 \chi_{KL} \mathcal{E}_L + e_{KAB} u_{A,B} + 2\epsilon_0 b_{KLM} \mathcal{E}_L \mathcal{E}_M \\
&\quad + \epsilon_0 l_{KLAB}^s \mathcal{E}_L u_{A,B} + \tfrac{1}{2}(e_{KABCD} + e_{KBD}\delta_{AC}) u_{A,B} u_{C,D}. \qquad (17.3.5)
\end{aligned}$$

Here χ_{KL} and b_{KLM} are the low frequency values of the linear electric susceptibility (9.1.13) and the optical mixing tensor (12.4.7) more appropriately called here the *electric field mixing tensor*; e_{KAB} is the piezoelectric stress tensor (10.2.2); e_{KABCD} is the *nonlinear piezoelectric tensor* defined by

$$\begin{aligned} e_{KABCD} \equiv & -6 \sum_{\substack{\nu\rho\mu \\ \lambda\sigma\xi}} q^{\nu} \Upsilon^{\nu\rho}_{KJ} {}^{30}M^{\rho\mu\lambda}_{JLM} \Upsilon^{\mu\sigma}_{LI} {}^{11}M^{\sigma}_{IAB} \Upsilon^{\lambda\xi}_{MH} {}^{11}M^{\xi}_{HCD} \\ & +2 \sum_{\nu\rho\mu\sigma} q^{\nu} \Upsilon^{\nu\rho}_{KJ} {}^{21}M^{\rho\mu}_{JLAB} \Upsilon^{\mu\sigma}_{LI} {}^{11}M^{\sigma}_{ICD} \\ & +2 \sum_{\nu\rho\mu\sigma} q^{\nu} \Upsilon^{\nu\rho}_{KJ} {}^{21}M^{\rho\mu}_{JLCD} \Upsilon^{\mu\sigma}_{LI} {}^{11}M^{\sigma}_{IAB} \\ & -2 \sum_{\nu\rho} q^{\nu} \Upsilon^{\nu\rho}_{KJ} {}^{12}M^{\rho}_{JABCD} \end{aligned} \tag{17.3.6}$$

and possesses the interchange symmetry

$$e_{KABCD} = e_{K(AB)(CD)} = e_{K(CD)(AB)}. \tag{17.3.7}$$

The form that crystal symmetry requires this tensor to take for each crystal class is given in the Appendix. The other tensor introduced in (17.3.5) is the *electrostrictive susceptibility* which is defined by

$$\begin{aligned} \epsilon_0 l^s_{IJKL} \equiv & -2 \sum_{\nu\mu\lambda\rho} q^{\mu} \Upsilon^{\mu\nu}_{IM} q^{\lambda} \Upsilon^{\lambda\rho}_{JN} {}^{21}M^{\nu\rho}_{MNKL} \\ & +6 \sum_{\substack{\nu\mu\lambda \\ \rho\xi\theta}} q^{\mu} \Upsilon^{\mu\nu}_{IM} q^{\lambda} \Upsilon^{\lambda\rho}_{JN} {}^{30}M^{\nu\rho\theta}_{MNP} \Upsilon^{\theta\xi}_{PQ} {}^{11}M^{\xi}_{QKL}. \end{aligned} \tag{17.3.8}$$

The electrostrictive susceptibility can be seen to have the same interchange symmetry,

$$l^s_{KLAB} = l^s_{(KL)(AB)}, \tag{17.3.9}$$

as that of the static elastooptic susceptibility that couples to strain.

17.4 Elasticity Equation to Bilinear Order

In this section we obtain the bilinearized dynamic elasticity equation by expanding the stress and body force terms in the general nonlinear equation (17.1.8).

The Piola-Kirchoff stress tensor T_{iJ} is found from the stored energy by (17.1.7). The internal coordinates may be adiabatically eliminated from the stored energy by substitution of $\boldsymbol{\Pi}^{\nu}$ from (17.3.4) into (12.1.3) with the result

$$\begin{aligned}\rho^0\Sigma=&\tfrac{1}{2}c_{ABCD}E_{AB}E_{CD}+\frac{\epsilon_0}{2}\chi_{KL}\mathcal{E}_K\mathcal{E}_L\\&+\frac{4\epsilon_0}{3}b_{KLM}\mathcal{E}_K\mathcal{E}_L\mathcal{E}_M+\frac{1}{6}c_{ABCDEF}E_{AB}E_{CD}E_{EF}\\&+\frac{1}{2}\epsilon_0 l^s_{KLAB}\mathcal{E}_K\mathcal{E}_L E_{AB}.\end{aligned}\tag{17.4.1}$$

The new tensor c_{ABCDEF} is the *third order stiffness tensor* and is defined by

$$\begin{aligned}c_{ABCDEF}\equiv&6\,{}^{03}M_{ABCDEF}-2\sum_{\nu\rho}{}^{12}M^{\nu}_{KABCD}\Upsilon^{\nu\rho}_{KJ}\,{}^{11}M^{\rho}_{JEF}\\&-2\sum_{\nu\rho}{}^{12}M^{\nu}_{KABEF}\Upsilon^{\nu\rho}_{KJ}\,{}^{11}M^{\rho}_{JCD}-2\sum_{\nu\rho}{}^{12}M^{\nu}_{KEFCD}\Upsilon^{\nu\rho}_{KJ}\,{}^{11}M^{\rho}_{JAB}\\&+2\sum_{\nu\mu\rho\sigma}{}^{21}M^{\nu\mu}_{KLAB}\Upsilon^{\nu\rho}_{KJ}\,{}^{11}M^{\rho}_{JCD}\Upsilon^{\mu\sigma}_{LI}\,{}^{11}M^{\sigma}_{IEF}\\&+2\sum_{\nu\mu\rho\sigma}{}^{21}M^{\nu\mu}_{KLCD}\Upsilon^{\nu\rho}_{KJ}\,{}^{11}M^{\rho}_{JAB}\Upsilon^{\mu\sigma}_{LI}\,{}^{11}M^{\sigma}_{IEF}\\&+2\sum_{\nu\mu\rho\sigma}{}^{21}M^{\nu\mu}_{KLEF}\Upsilon^{\nu\rho}_{KJ}\,{}^{11}M^{\rho}_{JCD}\Upsilon^{\mu\sigma}_{LI}\,{}^{11}M^{\sigma}_{IAB}\\&-6\sum_{\substack{\nu\mu\lambda\\\rho\sigma\xi}}{}^{30}M^{\nu\mu\lambda}_{KLM}\Upsilon^{\nu\rho}_{KJ}\,{}^{11}M^{\rho}_{JAB}\Upsilon^{\mu\sigma}_{LI}\,{}^{11}M^{\sigma}_{ICD}\Upsilon^{\lambda\xi}_{MH}\,{}^{11}M^{\xi}_{HEF}.\end{aligned}\tag{17.4.2}$$

It is apparent from either of the last two equations that the third order stiffness tensor possesses the interchange symmetry,

$$\begin{aligned}c_{ABCDEF}&=c_{(AB)(CD)(EF)}=c_{(AB)(EF)(CD)}=c_{(EF)(CD)(AB)}\\&=c_{(CD)(AB)(EF)}=c_{(EF)(AB)(CD)}=c_{(CD)(EF)(AB)}.\end{aligned}\tag{17.4.3}$$

The form that crystal symmetry requires this tensor to take for each crystal class is given in the Appendix.

The Piola-Kirchoff stress tensor can now be found to bilinear order as

$$\begin{aligned}T_{iJ}=\delta_{iI}\Big[&c_{IJAB}u_{A,B}+\tfrac{1}{2}(c_{IJABCD}+\delta_{IA}c_{BJCD}\\&+\delta_{IC}c_{DJAB}+\delta_{AC}c_{IJBD})u_{A,B}u_{C,D}\\&+\tfrac{1}{2}\epsilon_0 l^s_{LMIJ}\,\mathcal{E}_L\,\mathcal{E}_M\Big].\end{aligned}\tag{17.4.4}$$

Differentiation of (17.3.5) yields the other stress contribution

$$-\mathcal{E}_L \frac{\partial \mathcal{P}_L}{\partial x_{i,J}} = -\delta_{iI}\big[e_{LIJ} + \epsilon_0 l^s_{LMIJ}\mathcal{E}_M + (e_{LIJCD} + \delta_{IC}e_{LDJ})u_{C,D}\big]\mathcal{E}_L. \tag{17.4.5}$$

The electromagnetic body forces in the center-of-mass equation to bilinear order are

$$X_{I,i}\left(\mathcal{E} + \frac{\partial \mathbf{X}}{\partial t} \times \mathcal{B}\right)_I \mathcal{Q}^D = -\delta_{Ii}\mathcal{E}_I \mathcal{P}_{K,K} = -\delta_{Ii}\mathcal{E}_I(e_{JAB}u_{A,B} + \epsilon_0\chi_{JL}\mathcal{E}_L)_{,J}, \tag{17.4.6}$$

$$X_{I,i}\left[\left(\mathcal{J}^D + \mathcal{Q}^D\frac{\partial \mathbf{X}}{\partial t}\right) \times \mathcal{B}\right]_I = \delta_{Ii}\epsilon_{IJK}\dot{\mathcal{P}}_J\mathcal{B}_K = \delta_{Ii}\epsilon_{IJK}\big(e_{JAB}\dot{u}_{A,B} + \epsilon_0\chi_{JL}\dot{\mathcal{E}}_L\big)\mathcal{B}_K. \tag{17.4.7}$$

Substitution of the last four equations into the center-of-mass equation (17.1.8) produces

$$\begin{aligned}\rho^0\ddot{u}_I = \Big[& c_{IJAB}u_{A,B} - e_{KIJ}\mathcal{E}_K + \tfrac{1}{2}(c_{IJABCD} \\ & + \delta_{IA}c_{BJCD} + \delta_{IC}c_{DJAB} + \delta_{AC}c_{IJBD})u_{A,B}u_{C,D} \\ & - \tfrac{1}{2}\epsilon_0 l^s_{LMIJ}\mathcal{E}_L\mathcal{E}_M - (e_{LIJAB} + \delta_{IA}e_{LBJ})u_{A,B}\mathcal{E}_L\Big]_{,J} \\ & - \mathcal{E}_I(e_{JAB}u_{A,B} + \epsilon_0\chi_{JL}\mathcal{E}_L)_{,J} \\ & + \epsilon_{IJK}\big(e_{JAB}\dot{u}_{A,B} + \epsilon_0\chi_{JL}\dot{\mathcal{E}}_L\big)\mathcal{B}_K \end{aligned} \tag{17.4.8}$$

for the *bilinearized dynamic elasticity equation* expressed in the material frame. Note that the ordinary (second order) stiffness tensor c_{ABCD} contributes to the effective third order stiffness coefficients of $u_{A,B}u_{C,D}$. Note also that these additional terms, as a group, possess $IJ \leftrightarrow AB \leftrightarrow CD$ interchange symmetry but do not possess interchange symmetry within each of the three pairs as the third order stiffness tensor c_{ABCDEF} is shown to have. Thus these additional terms couple to rotation as well as to strain. It is also easily seen that $\delta_{IA}e_{LBJ}$ has lower interchange symmetry than the nonlinear piezoelectric tensor e_{LIJAB}. The $\delta_{IA}e_{LBJ}$ term possesses pair interchange symmetry, $AB \leftrightarrow IJ$, but does not possess interchange symmetry within A, B or within I,J. Thus it couples to rotation as well as to strain.

We want to consider the simplification of the bilinearized elasticity equation in the quasi-electrostatic approximation in a later section. In this approximation the momentum conservation law offers an alternate and more convenient form of the equation for the displacement. In anticipation of this we now add the bilinearized form of the electromagnetic momentum continuity equation (16.P.1),

$$\frac{d}{dt}[\epsilon_0(\mathcal{E}\times\mathcal{B})_I]=\left(\epsilon_0\mathcal{E}_I\mathcal{E}_J-\tfrac{1}{2}\epsilon_0\mathcal{E}_L\mathcal{E}_L\delta_{IJ}+\frac{1}{\mu_0}\mathcal{B}_I\mathcal{B}_J-\frac{1}{2\mu_0}\mathcal{B}_L\mathcal{B}_L\delta_{IJ}\right)_{,J}+\mathcal{E}_I\mathcal{P}_{K,K}-(\dot{\mathcal{P}}\times\mathcal{B})_I, \tag{17.4.9}$$

to (17.4.8) to obtain the bilinearized momentum conservation law. This cancels the body force terms while adding additional electric and magnetic stresses and also a time derivative term. The result is

$$\begin{aligned}\rho^0\ddot{u}_I+\frac{d}{dt}[\epsilon_0(\mathcal{E}\times\mathcal{B})_I]=\Big[&c_{IJAB}u_{A,B}-e_{KIJ}\mathcal{E}_K\\&+\tfrac{1}{2}(c_{IJABCD}+\delta_{IA}c_{BJCD}+\delta_{IC}c_{DJAB}\\&+\delta_{AC}c_{IJBD})u_{A,B}u_{C,D}-\tfrac{1}{2}\epsilon_0 l_{LMIJ}\mathcal{E}_L\mathcal{E}_M\\&-(e_{LIJAB}+\delta_{IA}e_{LBJ})u_{A,B}\mathcal{E}_L+\frac{1}{\mu_0}\mathcal{B}_I\mathcal{B}_J\\&-\frac{1}{2\mu_0}\mathcal{B}_L\mathcal{B}_L\delta_{IJ}\Big]_{,J}\end{aligned} \tag{17.4.10}$$

where the *electrostriction tensor* is defined by

$$l_{LMIJ}\equiv l^s_{LMIJ}-2\delta_{L(I}\delta_{J)M}+\delta_{LM}\delta_{IJ}, \tag{17.4.11}$$

which states that it is the sum of electrostrictive effects caused by the presence of the dielectric crystal plus those present in a vacuum.

A boundary condition on the stress $\mathcal{T}$, which is contained within the divergence in (17.4.10), can readily be obtained from this equation by the procedure used in Sections 7.2 and 8.4. If $\mathbf{N}$ denotes a unit material frame normal of a body surface and o and i represent the two sides of the surface, then we find

$$[\mathcal{T}^o-\mathcal{T}^i]\cdot\mathbf{N}=0. \tag{17.4.12}$$

Before considering the quasi-electrostatic approximation of this equation and the importance of the magnetic terms, we obtain the bilinearized Maxwell equations.

17.5 Material Form of Maxwell Equations to Bilinear Order

The material form of the Maxwell equations that are found in Section 16.1 may be written for a dielectric as

$$\nabla_X \times \mathcal{H} - \dot{\mathcal{D}} = 0, \tag{17.5.1}$$

$$\nabla_X \times \mathcal{E} + \dot{\mathcal{B}} = 0, \tag{17.5.2}$$

$$\nabla_X \cdot \mathcal{D} = 0, \tag{17.5.3}$$

$$\nabla_X \cdot \mathcal{B} = 0. \tag{17.5.4}$$

Bilinearizing these equations requires finding $\mathcal{D}$ and $\mathcal{H}$ to bilinear order. Since we have $\mathcal{P}$ to bilinear order in (17.3.5), we use the relation (16.6.11) between $\mathcal{D}$ and $\mathcal{P}$ to obtain

$$\begin{aligned}\mathcal{D}_K &= \epsilon_0 J(C^{-1})_{KL}\left(\mathcal{E} + \frac{\partial \mathbf{X}}{\partial t} \times \mathcal{B}\right)_L + \mathcal{P}_K \\ &= \epsilon_0(1+u_{M,M})(\delta_{KL} - 2u_{(K,L)})\left(\mathcal{E} + \frac{\partial \mathbf{X}}{\partial t} \times \mathcal{B}\right)_L + \mathcal{P}_K \\ &= P_K^S + \epsilon_0 \kappa_{KL}\mathcal{E}_L + e_{KAB}u_{A,B} + 2\epsilon_0 b_{KLM}\mathcal{E}_L\mathcal{E}_M \\ &\quad + \epsilon_0 l_{KLAB}\mathcal{E}_L u_{A,B} + \tfrac{1}{2}(e_{KABCD} + e_{KBD}\delta_{AC})u_{A,B}u_{C,D} \\ &\quad - \epsilon_0(\dot{\mathbf{u}} \times \mathcal{B})_K. \end{aligned} \tag{17.5.5}$$

The material frame magnetic intensity $\mathcal{H}$ for a nonmagnetic dielectric is obtained by setting the magnetization $\mathcal{M} = 0$ in (16.6.10). Bilinearizing $\mathcal{H}$ then yields

$$\begin{aligned}\mathcal{H}_K &= \frac{1}{\mu_0 J} C_{KL}\mathcal{B}_L + \epsilon_{KLM}\frac{\partial X_L}{\partial t}\epsilon_0 J(C^{-1})_{MN}\left(\mathcal{E} + \frac{\partial \mathbf{X}}{\partial t} \times \mathcal{B}\right)_N \\ &= \frac{1}{\mu_0}(\delta_{KL} - \delta_{KL}u_{P,P} + 2u_{(K,L)})\mathcal{B}_L - \epsilon_0\epsilon_{KLM}\dot{u}_L\mathcal{E}_M. \end{aligned} \tag{17.5.6}$$

The six equations of this section and (17.4.10) of the preceding section form a complete set of differential equations to bilinear order for electroelastic phenomena in dielectrics in the fully electrodynamic regime. We show in the next section that under commonly met conditions they may be greatly simplified.

17.6 Quasi-electrostatic Approximation

The derivation of the set of equations above does not consider the relative size of various terms. We show in this section that under typical conditions of electroelastic experiments a number of terms in these equations are small and may be dropped and that only a truncated set of the equations need be considered.

In Section 11.1 it is shown that for the study of linear piezoelectric phenomena all terms involving time derivatives except the $\rho^0\ddot{\mathbf{u}}$ term are smaller than the remaining terms by the factor $\eta \equiv \omega L/c$. For a crystal dimension of $L \lesssim 1$ cm and a frequency of $\omega/2\pi \lesssim 1$ GHz we have $\eta \ll 1$ and so we drop all such terms. We call this the quasi-electrostatic approximation. The scaling argument used in Section 11.1 can be applied to the bilinearized equations considered here with a similar dropping of time derivative terms by the quasi-electrostatic approximation.

One difference occurs in the present case in that some magnetic effects remain. Magnetostrictive terms bilinear in the material frame magnetic induction $\mathfrak{B}$ remain in the dynamic elasticity equation (17.4.10). A difference between $\mathfrak{H}$ and $\mathfrak{B}$ (apart from the constant μ_0) also remains, necessitating continued consideration of (17.5.1) (with $\mathfrak{D}$ dropped). This result is not surprising, since it merely indicates that the magnetostrictive effect (a bilinear effect) can be important even in a nonmagnetic dielectric if the *applied* magnetic field is sufficiently strong.

The important thing to be realized is that the magnetic field must be applied because in the zeroth order quasi-static approximation there is no coupling between electric and magnetic fields. Thus the magnetic field cannot be produced by an applied electric field. We now wish to invoke the additional assumption of no applied magnetic fields, as would be typical for studies of nonlinear electroelastic phenomena in dielectrics. We thus drop from consideration the magnetostrictive terms and the two equations (17.5.1) and (17.5.4) governing $\mathfrak{B}$ and $\mathfrak{H}$. We then speak of the resultant truncated set of equations and constitutive expressions as applying to the *quasi-electrostatic regime*.

In this regime a complete set of bilinearized differential equations consists of

$$\rho^0 \ddot{u}_I = \mathfrak{T}_{IJ,J}, \tag{17.6.1}$$

$$\mathfrak{D}_{K,K} = 0 \tag{17.6.2}$$

with the constitutive relations

$$\mathfrak{T}_{IJ}=c_{IJAB}u_{A,B}+e_{KIJ}\varphi_{,K}+\tfrac{1}{2}(c_{IJABCD}$$
$$+\delta_{IA}c_{BJCD}+\delta_{IC}c_{DJAB}+\delta_{AC}c_{IJBD})u_{A,B}u_{C,D}$$
$$-\tfrac{1}{2}\epsilon_0 l_{LMIJ}\varphi_{,L}\varphi_{,M}+(e_{LIJAB}+\delta_{IA}e_{LBJ})u_{A,B}\varphi_{,L}, \tag{17.6.3}$$

$$\mathfrak{D}_K=P_K^S-\epsilon_0\kappa_{KL}\varphi_{,L}+e_{KAB}u_{A,B}+2\epsilon_0 b_{KLM}\varphi_{,L}\varphi_{,M}$$
$$-\epsilon_0 l_{KLAB}\varphi_{,L}u_{A,B}+\tfrac{1}{2}(e_{KABCD}+e_{KBD}\delta_{AC})u_{A,B}u_{C,D} \tag{17.6.4}$$

where we use

$$\mathfrak{E}=-\nabla_X\varphi, \tag{17.6.5}$$

which is implied by (17.5.2) after $\dot{\mathfrak{B}}$ is dropped. Equations (17.6.1) and (17.6.2) consist of four differential equations in the four unknown functions **u** and φ. They were first obtained by Baumhauer and Tiersten [1973].

The boundary condition on the stress $\mathfrak{T}$ is readily found from a "pill-box" argument (see Sections 7.2 and 8.4) to be

$$[\mathfrak{T}^o_{IJ}-\mathfrak{T}^i_{IJ}]N_J=0 \tag{17.6.6}$$

where **N** is the unit outward body surface normal in the material frame. Continuity of the displacement and some of its derivatives is discussed in Section 16.3. The electrical boundary condition usually takes the form of specifying the potential

$$\varphi=\varphi(t) \tag{17.6.7}$$

on an equipotential surface S.

17.7 Relation of Electrostriction to Elastooptic Effect

Since both the electrostrictive susceptibility and the elastooptic susceptibility represent the interaction of two electric fields and an elastic deformation in a material medium, it is apparent that the two tensors are closely related. If, in fact, we were to speak of the mixing of two light waves to generate an acoustic wave as electrostriction (as is sometimes done [Caddes et al., 1967]), then the tensors governing the two effects would be identical. However, the term electrostriction has traditionally been used to denote the deformation induced by two *low* frequency electric fields and it is in this sense that we use the term in this chapter. Thus the electrostrictive susceptibility must be compared with the *low frequency limit* of the elastooptic susceptibility.

This comparison yields

$$\begin{aligned} l^s_{IJKL} &= 2\chi_{IJ(KL)} - \delta_{I(K}\chi_{L)J} - \delta_{J(K}\chi_{L)I} + \chi_{IJ}\delta_{KL} \\ &= -\kappa_{IM}\kappa_{JN}p_{MN(KL)} - \delta_{I(K}\chi_{L)J} - \delta_{J(K}\chi_{L)I} + \chi_{IJ}\delta_{KL} \end{aligned} \tag{17.7.1}$$

with the use of (13.2.12), (13.2.11), and (17.3.8). This result [Nelson, 1978a] states that the electrostrictive susceptibility differs from the low frequency elastooptic susceptibility that couples to strain by Maxwell stress terms arising from the presence of a polarizable medium. Note that the elastooptic susceptibility that couples to rotation does not enter the relation (17.7.1).

The cause of the difference between the electrostrictive susceptibility and the low frequency elastooptic susceptibility is the differing frames of reference in which each is defined. It is clear that elastooptic measurements detect spatial frame electric fields and that the elastooptic tensor should be defined in the spatial frame. It is also clear that electrostriction is usually studied with low frequency electric fields that are applied to the medium by electrodes attached to the medium. Thus the electric fields measured here are material frame electric fields and so the electrostrictive susceptibility is properly defined in the material coordinate system.

The relation (17.7.1) could be converted to a relation for the electrostriction tensor l_{IJKL} with the use of (17.4.11). We regard both the electrostriction tensor and the electrostrictive susceptibility as equally important. The former tensor is the quantity that enters the differential equations and represents the electrostrictive effects caused by the dielectric medium and those present in a vacuum. The latter tensor represents only the electrostrictive effects caused by the dielectric medium. The electrostrictive susceptibility, like the linear electric susceptibility, vanishes for a vacuum while the electrostriction tensor, like the dielectric tensor, does not.

The new terms involving the linear electric susceptibility in (17.7.1) are measurably large. Consider the l^s_{1111} component in $LiNbO_3$. If we assume the low frequency limit of p_{1111} is the same as its optical value, -0.026, and use $\kappa_{11}=84$ and thus $\chi_{11}=83$, we find that the last three terms in (17.7.1) are $\sim$45% of the entire right side.

17.8 Acoustic Wave Mixing

Acoustic wave mixing is the production of an output acoustic wave by two input acoustic waves. The generated wave can have a frequency which is either the sum or difference of the frequencies of the two input waves. Acoustic second harmonic generation is a special case of sum frequency generation.

We find a solution [Nelson, 1978b] by the method of free-plus-forced-waves analogous to that used in Section 13.5 for acoustooptic diffraction and applied to optical mixing in Section 15.2. We assume that all three acoustic waves are plane waves and that the strengths of the two input waves are undepleted by the nonlinear interaction. The latter approximation restricts the validity of the solution to the region of initial growth of the generated wave. We consider the solution only near or at phase matching, since only then is the interaction interestingly large.

The solution treats aspects of anisotropy of the medium quite generally. Any of the three interacting acoustic eigenmodes can be of any type —longitudinal, transverse, or hybrid. They may travel collinearly or noncollinearly in any direction (or directions) with respect to the crystal axes and the input surface with the proviso that they be near to or at phase matching.

The treatment of acoustic wave mixing applies to ordinary (nonpiezoelectric) dielectrics, to piezoelectrics, and to pyroelectrics. For the latter category of crystals we continue to assume that the spontaneous electric field and the spontaneous stress are zero, (17.3.1) and (17.3.2). Under these conditions no difference between pyroelectrics and piezoelectrics show up. The result can be specialized to ordinary (nonpiezoelectric) dielectrics by setting all odd rank tensors equal to zero.

We begin by making a Fourier expansion of the quasi-electrostatic equations (17.6.1) to (17.6.4) for sum frequency generation,

$$\omega_3 = \omega_1 + \omega_2, \tag{17.8.1}$$

using the method of Section 12.2. Amplitude designations of (1,0), (0,1), and (1,1) are associated with the frequencies ω_1, ω_2, and ω_3 respectively. The dynamical equations for the output field at frequency ω_3 are then

$$-\rho^0\omega_3^2 u_I(\mathbf{X};1,1) - c_{IJAB}u_{A,BJ}(\mathbf{X};1,1) - e_{KIJ}\varphi_{,KJ}(\mathbf{X};1,1) = G_I(\mathbf{X};1,1), \tag{17.8.2}$$

$$\epsilon_0\kappa_{KL}\varphi_{,LK}(\mathbf{X};1,1) - e_{KAB}u_{A,BK}(\mathbf{X};1,1) = Q_{K,K}(\mathbf{X};1,1) \tag{17.8.3}$$

where $\mathbf{G}$ includes all the bilinear terms in the stress (17.6.3) and $\nabla\cdot\mathbf{Q}$ includes all the bilinear terms in the divergence of the electric displacement

(17.6.4). They are given by

$$\begin{aligned} G_I(\mathbf{X};1,1) = \Big[& \tfrac{1}{2}(c_{IJABCD} + \delta_{IA}c_{BJCD} + \delta_{IC}c_{DJAB} \\ & + \delta_{AC}c_{IJBD})u_{A,B}(\mathbf{X};1,0)u_{C,D}(\mathbf{X};0,1) \\ & - \tfrac{1}{2}\epsilon_0 l_{LMIJ}\varphi_{,L}(\mathbf{X};1,0)\varphi_{,M}(\mathbf{X};0,1) \\ & + \tfrac{1}{2}(e_{LIJAB} + \delta_{IA}e_{LBJ})(u_{A,B}(\mathbf{X};1,0)\varphi_L(\mathbf{X};0,1) \\ & + u_{A,B}(\mathbf{X};0,1)\varphi_{,L}(\mathbf{X};1,0))\Big]_{,J}, \end{aligned} \tag{17.8.4}$$

$$\begin{aligned} Q_{K,K}(\mathbf{X};1,1) = \Big[& 2\epsilon_0 b_{KLM}\varphi_{,L}(\mathbf{X};1,0)\varphi_{,M}(\mathbf{X};0,1) \\ & - \tfrac{1}{2}\epsilon_0 l_{KLAB}(\varphi_{,L}(\mathbf{X};1,0)u_{A,B}(\mathbf{X};0,1) \\ & + \varphi_{,L}(\mathbf{X};0,1)u_{A,B}(\mathbf{X};1,0)) \\ & + \tfrac{1}{2}(e_{KABCD} + e_{KBD}\delta_{AC}) \\ & \times u_{A,B}(\mathbf{X};1,0)u_{C,D}(\mathbf{X};0,1)\Big]_{,K}. \end{aligned} \tag{17.8.5}$$

The assumptions of (1) plane waves and (2) no depletion of input waves allow us to write the two input acoustic waves as

$$\mathbf{u}(\mathbf{X};1,0) = U_1\mathbf{b}^{\alpha}e^{i\mathbf{K}_1^{\alpha}\cdot\mathbf{X}}, \tag{17.8.6}$$

$$\mathbf{u}(\mathbf{X};0,1) = U_2\mathbf{b}^{\beta}e^{i\mathbf{K}_2^{\beta}\cdot\mathbf{X}} \tag{17.8.7}$$

where U_1 and U_2 are amplitude constants, $\mathbf{b}$ is an orthonormal displacement eigenvector discussed in Section 10.8, $\mathbf{K}$ is an acoustic wave propagation vector, and α and β refer to the eigenmodes at frequencies ω_1 and ω_2. Because two acoustic waves are the input fields for this interaction, we wish to eliminate the piezoelectrically generated electric potentials $\varphi(\mathbf{X};1,0)$ and $\varphi(\mathbf{X};0,1)$ in favor of displacement fields in the driving functions (17.8.4) and (17.8.5). Since those functions are already bilinear, a linear replacement is adequate. The electric potential can thus be found from the linearized electric displacement equation,

$$\epsilon_0\kappa_{KL}\varphi_{,LK}(\mathbf{X};1,0) - e_{KAB}u_{A,BK}(\mathbf{X};1,0) = 0. \tag{17.8.8}$$

The plane wave displacement (17.8.6) produces by this equation a plane wave electric potential

$$\varphi(\mathbf{X};1,0) = \frac{s_K^{(1)}e_{KAB}b_A^{\alpha}s_B^{(1)}}{\epsilon_0\mathbf{s}^{(1)}\cdot\boldsymbol{\kappa}\cdot\mathbf{s}^{(1)}}U_1e^{i\mathbf{K}_1^{\alpha}\cdot\mathbf{X}} \tag{17.8.9}$$

where $\mathbf{s}^{(1)}$ is a unit vector in the direction of $\mathbf{K}_1^\alpha$. A similar result can be found for $\varphi(\mathbf{X};0,1)$. We find unit propagation vectors $\mathbf{s}^{(1)}$, $\mathbf{s}^{(2)}$, $\mathbf{s}^{(3)}$, and $\mathbf{s}^D$, defined by

$$\mathbf{K}_1^\alpha = K_1^\alpha \mathbf{s}^{(1)}, \tag{17.8.10}$$

$$\mathbf{K}_2^\beta = K_2^\beta \mathbf{s}^{(2)}, \tag{17.8.11}$$

$$\mathbf{K}_3^\gamma = K_3^\gamma \mathbf{s}^{(3)}, \tag{17.8.12}$$

$$\mathbf{K}^D \equiv \mathbf{K}_1^\alpha + \mathbf{K}_2^\beta = K^D \mathbf{s}^D \tag{17.8.13}$$

useful in expressing the interaction coefficient. Elimination of the electric potentials from the driving functions $\mathbf{G}$ and $\nabla \cdot \mathbf{Q}$ now yields

$$G_I(\mathbf{X};1,1) = -iK_1^\alpha K_2^\beta K^D g_{IJABCD} b_A^\alpha s_B^{(1)} b_C^\beta s_D^{(2)} s_J^D U_1 U_2 e^{i\mathbf{K}^D \cdot \mathbf{X}}, \tag{17.8.14}$$

$$Q_{J,J}(\mathbf{X};1,1) = -iK_1^\alpha K_2^\beta K^D q_{JABCD} b_A^\alpha s_B^{(1)} b_C^\beta s_D^{(2)} s_J^D U_1 U_2 e^{i\mathbf{K}^D \cdot \mathbf{X}} \tag{17.8.15}$$

where

$$\begin{aligned} g_{IJABCD} \equiv {} & \tfrac{1}{2}(c_{IJABCD} + \delta_{IA} c_{BJCD} + \delta_{IC} c_{DJAB} + \delta_{AC} c_{IJBD}) \\ & - \frac{l_{EFIJ} s_E^{(1)} s_F^{(2)} s_M^{(1)} e_{MAB} s_N^{(2)} e_{NCD}}{2\epsilon_0 (\mathbf{s}^{(1)} \cdot \boldsymbol{\kappa} \cdot \mathbf{s}^{(1)})(\mathbf{s}^{(2)} \cdot \boldsymbol{\kappa} \cdot \mathbf{s}^{(2)})} \\ & + \frac{(e_{LIJCD} + \delta_{IC} e_{LDJ}) s_L^{(1)} s_K^{(1)} e_{KAB}}{2\epsilon_0 (\mathbf{s}^{(1)} \cdot \boldsymbol{\kappa} \cdot \mathbf{s}^{(1)})} \\ & + \frac{(e_{LIJAB} + \delta_{IA} e_{LBJ}) s_L^{(2)} s_K^{(2)} e_{KCD}}{2\epsilon_0 (\mathbf{s}^{(2)} \cdot \boldsymbol{\kappa} \cdot \mathbf{s}^{(2)})}, \end{aligned} \tag{17.8.16}$$

$$\begin{aligned} q_{LABCD} \equiv {} & \frac{2 b_{LEF} s_E^{(1)} s_F^{(2)} s_P^{(1)} e_{PAB} s_Q^{(2)} e_{QCD}}{\epsilon_0 (\mathbf{s}^{(1)} \cdot \boldsymbol{\kappa} \cdot \mathbf{s}^{(1)})(\mathbf{s}^{(2)} \cdot \boldsymbol{\kappa} \cdot \mathbf{s}^{(2)})} \\ & - \frac{l_{LECD} s_E^{(1)} s_P^{(1)} e_{PAB}}{2(\mathbf{s}^{(1)} \cdot \boldsymbol{\kappa} \cdot \mathbf{s}^{(1)})} - \frac{l_{LEAB} s_E^{(2)} s_P^{(2)} e_{PCD}}{2(\mathbf{s}^{(2)} \cdot \boldsymbol{\kappa} \cdot \mathbf{s}^{(2)})} \\ & + \tfrac{1}{2}(e_{LABCD} + e_{LBD} \delta_{AC}). \end{aligned} \tag{17.8.17}$$

The general solution of (17.8.2) and (17.8.3) consists of a forced or driven wave solution of the inhomogeneous differential equations plus a free wave solution of the homogeneous differential equations. First, we find a driven wave of the form

$$\mathbf{u}^D(\mathbf{X};1,1)=\mathbf{u}^D(1,1)e^{i\mathbf{K}^D\cdot\mathbf{X}}, \tag{17.8.18}$$

$$\varphi^D(\mathbf{X};1,1)=\varphi^D(1,1)e^{i\mathbf{K}^D\cdot\mathbf{X}}. \tag{17.8.19}$$

The constant amplitudes, $\mathbf{u}^D(1,1)$ and $\varphi^D(1,1)$, then satisfy

$$\begin{aligned}-\rho^0\omega_3^2u_I^D(1,1)&+(K^D)^2s_B^Ds_J^Dc_{IJAB}u_A^D(1,1)\\&+(K^D)^2s_K^Ds_J^De_{KIJ}\varphi^D(1,1)\\&=-iK_1^\alpha K_2^\beta K^Dg_{IJABCD}b_A^\alpha s_B^{(1)}b_C^\beta s_D^{(2)}s_J^DU_1U_2,\end{aligned} \tag{17.8.20}$$

$$\begin{aligned}-\epsilon_0(K^D)^2s_K^D\kappa_{KL}s_L^D\varphi^D(1,1)+(K^D)^2s_K^Ds_B^De_{KAB}u_A^D(1,1)\\=-iK_1^\alpha K_2^\beta K^Dq_{JABCD}b_A^\alpha s_B^{(1)}b_C^\beta s_D^{(2)}s_J^DU_1U_2.\end{aligned} \tag{17.8.21}$$

Elimination of $\varphi^D(1,1)$ between these equations gives

$$\beta_{IA}u_A^D(1,1)=\frac{-iK_1^\alpha K_2^\beta}{\rho^0K^D}\bar{c}_{IJABCD}s_J^Db_A^\alpha s_B^{(1)}b_C^\beta s_D^{(2)}U_1U_2 \tag{17.8.22}$$

where

$$\beta_{IA}\equiv\frac{1}{\rho^0}\bar{c}_{IJAB}s_J^Ds_B^D-v_D^2\delta_{IA}, \tag{17.8.23}$$

$$\bar{c}_{IJAB}\equiv c_{IJAB}+\frac{s_K^De_{KIJ}s_L^De_{LAB}}{\epsilon_0\mathbf{s}^D\cdot\boldsymbol{\kappa}\cdot\mathbf{s}^D}, \tag{17.8.24}$$

$$v_D\equiv\frac{\omega_3}{K^D}, \tag{17.8.25}$$

$$\bar{c}_{IJABCD}\equiv g_{IJABCD}+\frac{s_K^De_{KIJ}s_L^Dq_{LABCD}}{\epsilon_0\mathbf{s}^D\cdot\boldsymbol{\kappa}\cdot\mathbf{s}^D}. \tag{17.8.26}$$

Using the inverse of $\boldsymbol{\beta}$ found in (10.8.12), we can obtain from (17.8.22) the driven wave displacement amplitude

$$u_K^D(1,1)=\frac{-iK_1^\alpha K_2^\beta}{\rho^0K^D}\sum_{\epsilon=1}^{3}\frac{b_K^\epsilon(\mathbf{s}^D,\omega_3)b_I^\epsilon(\mathbf{s}^D,\omega_3)}{v_\epsilon^2(\mathbf{s}^D)-v_D^2}\bar{c}_{IJABCD}s_J^Db_A^\alpha s_B^{(1)}b_C^\beta s_D^{(2)}U_1U_2. \tag{17.8.27}$$

The general free wave solution of a piezoelectrically stiffened vibration traveling into the nonlinear medium from the boundary is

$$\mathbf{u}^F(\mathbf{X};1,1)=\sum_{\epsilon=1}^{3} C^\epsilon \mathbf{b}^\epsilon(\mathbf{s}^{(3)},\omega_3)e^{i\mathbf{K}_3^\epsilon\cdot\mathbf{X}} \tag{17.8.28}$$

where $C^\epsilon(\epsilon=1,2,3)$ are integration constants. Thus the general solution is

$$\mathbf{u}(\mathbf{X};1,1)=\sum_{\epsilon=1}^{3} C^\epsilon \mathbf{b}^\epsilon(\mathbf{s}^{(3)},\omega_3)e^{i\mathbf{K}_3^\epsilon\cdot\mathbf{X}}+\mathbf{u}^D(1,1)e^{i\mathbf{K}^D\cdot\mathbf{X}}. \tag{17.8.29}$$

The interesting region of this solution is near phase matching,

$$\mathbf{K}_3^\gamma=\mathbf{K}^D\equiv\mathbf{K}_1^\alpha+\mathbf{K}_2^\beta, \tag{17.8.30}$$

since there the output intensity reaches a maximum, just as happens for acoustooptic diffraction and optical mixing. The particular free wave acoustic eigenmode that can be phase matched is denoted by γ. The understanding that we gained from obtaining the solution near phase matching from the exact general solution in the study of acoustooptic diffraction can be used here to obtain the desired solution by a simplified approximate procedure. This procedure is to drop those components of the solution proportional to eigenvectors of modes which cannot be phase matched ($\epsilon\neq\gamma$) and to determine the remaining integration constant C^γ by an approximate boundary condition which requires that the component of the displacement in the γ eigenmode at the sum frequency ω_3 vanish,

$$\mathbf{u}(\mathbf{X}^P;1,1)\cdot\mathbf{b}^\gamma(\mathbf{s}^{(3)},\omega_3)=0, \tag{17.8.31}$$

on the input plane surface given by

$$\mathbf{X}^P\cdot\mathbf{N}=0. \tag{17.8.32}$$

Here $\mathbf{N}$ is the unit inward normal to the surface and $\mathbf{X}^P$ is a coordinate vector to any point P in the plane from the origin of coordinates taken as lying in the plane. See Fig. 17.1.

Application of this boundary condition to the solution (17.8.29) gives

$$C^\gamma=-\mathbf{b}^\gamma(\mathbf{s}^{(3)},\omega_3)\cdot\mathbf{u}^D(1,1)e^{i\Delta\mathbf{K}\cdot\mathbf{X}^P} \tag{17.8.33}$$

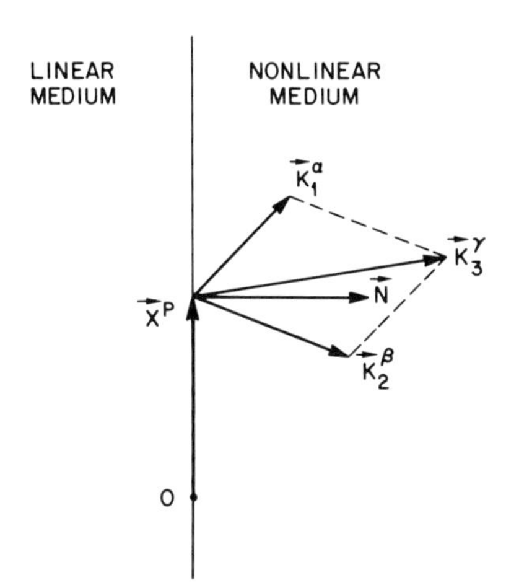

Fig. 17.1. Schematic diagram of interaction geometry for phase matched sum frequency generation.

where

$$\Delta\mathbf{K} \equiv \mathbf{K}_3^{\gamma} - \mathbf{K}^D = \mathbf{K}_3^{\gamma} - \mathbf{K}_1^{\alpha} - \mathbf{K}_2^{\beta} \tag{17.8.34}$$

is the wavevector mismatch. Since C^{γ} must be a constant independent of the location on the input plane, the components of $\Delta\mathbf{K}$ tangent to the plane must vanish,

$$(\Delta\mathbf{K})_T = 0. \tag{17.8.35}$$

Since the solution we desire is close to phase matching, $\mathbf{s}^D \cong \mathbf{s}^{(3)}$ and

$$\mathbf{b}^{\gamma}\left(\mathbf{s}^{(3)}, \omega_3\right) \cdot \mathbf{b}^{\epsilon}\left(\mathbf{s}^D, \omega_3\right) \cong \delta^{\gamma\epsilon}. \tag{17.8.36}$$

Thus the integration constant is

$$C^{\gamma} = \frac{iK_1^{\alpha}K_2^{\beta}\bar{c}U_1U_2}{\rho^0 K^D\left[v_{\gamma}^2(\mathbf{s}^D) - v_D^2\right]} \tag{17.8.37}$$

where

$$\bar{c} \equiv \bar{c}_{IJABCD} b_I^{\gamma} s_J^{(3)} b_A^{\alpha} s_B^{(1)} b_C^{\beta} s_D^{(2)}. \tag{17.8.38}$$

It is now apparent that the material interaction coefficient $\bar{c}$, the effective third order stiffness, governs the strength of acoustic wave mixing. Inserting the approximation $\mathbf{s}^D \cong \mathbf{s}^{(3)}$, good near phase matching, we obtain from (17.8.26), (17.8.16), and (17.8.17) a final expression for $\bar{c}$,

$$\bar{c}=\Bigg[\tfrac{1}{2}(c_{IJABCD}+\delta_{IA}c_{BJCD}+\delta_{IC}c_{DJAB}+\delta_{AC}c_{IJBD})$$
$$+\frac{(e_{LIJCD}+\delta_{IC}e_{LDJ})s_L^{(1)}s_K^{(1)}e_{KAB}}{2\epsilon_0(\mathbf{s}^{(1)}\cdot\boldsymbol{\kappa}\cdot\mathbf{s}^{(1)})}+\frac{(e_{LIJAB}+\delta_{IA}e_{LBJ})s_L^{(2)}s_K^{(2)}e_{KCD}}{2\epsilon_0(\mathbf{s}^{(2)}\cdot\boldsymbol{\kappa}\cdot\mathbf{s}^{(2)})}$$
$$+\frac{(e_{LABCD}+\delta_{AC}e_{LDB})s_L^{(3)}s_K^{(3)}e_{KIJ}}{2\epsilon_0(\mathbf{s}^{(3)}\cdot\boldsymbol{\kappa}\cdot\mathbf{s}^{(3)})}-\frac{l_{EFIJ}s_E^{(1)}s_F^{(2)}s_M^{(1)}e_{MAB}s_N^{(2)}e_{NCD}}{2\epsilon_0(\mathbf{s}^{(1)}\cdot\boldsymbol{\kappa}\cdot\mathbf{s}^{(1)})(\mathbf{s}^{(2)}\cdot\boldsymbol{\kappa}\cdot\mathbf{s}^{(2)})}$$
$$-\frac{l_{EFAB}s_E^{(2)}s_F^{(3)}s_M^{(2)}e_{MCD}s_N^{(3)}e_{NIJ}}{2\epsilon_0(\mathbf{s}^{(2)}\cdot\boldsymbol{\kappa}\cdot\mathbf{s}^{(2)})(\mathbf{s}^{(3)}\cdot\boldsymbol{\kappa}\cdot\mathbf{s}^{(3)})}-\frac{l_{EFCD}s_E^{(3)}s_F^{(1)}s_M^{(3)}e_{MIJ}s_N^{(1)}e_{NAB}}{2\epsilon_0(\mathbf{s}^{(3)}\cdot\boldsymbol{\kappa}\cdot\mathbf{s}^{(3)})(\mathbf{s}^{(1)}\cdot\boldsymbol{\kappa}\cdot\mathbf{s}^{(1)})}$$
$$+\frac{2b_{KLM}s_K^{(3)}s_L^{(1)}s_M^{(2)}s_N^{(3)}e_{NIJ}s_P^{(1)}e_{PAB}s_Q^{(2)}e_{QCD}}{\epsilon_0^2(\mathbf{s}^{(1)}\cdot\boldsymbol{\kappa}\cdot\mathbf{s}^{(1)})(\mathbf{s}^{(2)}\cdot\boldsymbol{\kappa}\cdot\mathbf{s}^{(2)})(\mathbf{s}^{(3)}\cdot\boldsymbol{\kappa}\cdot\mathbf{s}^{(3)})}\Bigg]b_I^{\gamma}s_J^{(3)}b_A^{\alpha}s_B^{(1)}b_C^{\beta}s_D^{(2)}. \tag{17.8.39}$$

This expression contains a great amount of information concerning the mechanisms of interaction between acoustic waves. First, we note that this expression is symmetric upon the interchange of any two acoustic waves, for example, $\mathbf{b}^{\alpha},\mathbf{s}^{(1)}\rightarrow\mathbf{b}^{\beta},\mathbf{s}^{(2)}$. Next, we note that $\bar{c}$ contains many contributions experimentally distinguishable either through their dependence on propagation directions or through their coupling to rotation. The first term in $\bar{c}$ is the third order stiffness tensor that couples three strain fields directly.

The next three terms are effective third order (nonlinear) stiffness contributions that involve only the second order (linear) stiffness tensor. They thus represent nonlinear elasticity *required* of a linear elastic medium. They arise because the second order stiffness tensor is a series expansion coefficient of the Green finite strain tensor in the stored energy and in the stress and because the Green finite strain tensor contains an inherent nonlinear part as seen in (3.10.5). These effective third order stiffness contributions can couple to *rotations* as well as to strains, a distinguishing feature compared to the true third order stiffness tensor.

The fifth through seventh terms in $\bar{c}$ represent two-step indirect effects. In these the longitudinal electric field, produced piezoelectrically by one acoustic wave, interacts with the strain of each of the other two acoustic waves through the direct nonlinear piezoelectric effect and also interacts with *both* the strain and rotation of the other two acoustic waves through an effective nonlinear piezoelectric effect which involves only the linear piezoelectric tensor. The latter effect occurs because e_{LBJ} is the series expansion coefficient of the Green finite strain tensor in the polarization (17.3.5). This last term represents the nonlinear piezoelectricity *required* of a linear piezoelectric medium. These terms have the form of piezoelectric stiffening for the stiffness tensors of terms two through four.

The eighth through tenth terms in $\bar{c}$ represent three-step indirect effects. In these the longitudinal electric fields, produced piezoelectrically by two of the acoustic waves, interact with the strain of the third acoustic wave through the electrostrictive effect.

Finally, the last term in $\bar{c}$ represents a four-step indirect effect. Here the longitudinal electric fields, produced piezoelectrically by all three acoustic waves, interact through the electric field mixing tensor b_{KLM}.

The solution for the displacement may now be written as

$$\mathbf{u}(\mathbf{X};1,1)=\frac{-2K_1^{\alpha}K_2^{\beta}\mathbf{b}^{\gamma}\bar{c}U_1U_2}{\rho^0K^D\left[v_{\gamma}^2(\mathbf{s}_D)-v_D^2\right]}\sin\left(\frac{\Delta K_N X}{2}\right)e^{i\mathbf{K}_3^{\gamma}\cdot\mathbf{X}} \tag{17.8.40}$$

where in view of (17.8.35) ΔK_N is defined by

$$\Delta\mathbf{K}=\Delta K_N\mathbf{N}, \tag{17.8.41}$$

a scalar coordinate,

$$X\equiv\mathbf{N}\cdot\mathbf{X}, \tag{17.8.42}$$

is introduced, and in the propagation factor we set $(\mathbf{K}^D+\mathbf{K}_3^{\gamma})/2\cong\mathbf{K}_3^{\gamma}$. Next it is useful to reexpress the denominator of $\mathbf{u}(\mathbf{X};1,1)$ in terms of ΔK_N by methods analogous to those used for the optical output field of acoustooptic diffraction in Section 13.5. The result, which we leave as an exercise, is

$$v_{\gamma}^2(\mathbf{s}^D)-v_D^2=\frac{-2v_{\gamma}^2\Delta K_N(\mathbf{N}\cdot\mathbf{v}_g^{\gamma})}{\omega_3} \tag{17.8.43}$$

where the expression (10.11.19) for the group velocity $\mathbf{v}_g^{\gamma}$ is used. The displacement of the output acoustic wave may now be expressed as

$$\mathbf{u}(\mathbf{X};1,1)=\frac{K_1^{\alpha}K_2^{\beta}\bar{c}X\mathbf{b}^{\gamma}}{2\rho^0v_{\gamma}\mathbf{N}\cdot\mathbf{v}_g^{\gamma}}\Phi^{1/2}\left(\frac{\Delta K_N X}{2}\right)U_1U_2e^{i\mathbf{K}_3^{\gamma}\cdot\mathbf{X}} \tag{17.8.44}$$

where the phase-matching function $\Phi(\phi)$ is defined by

$$\Phi(\phi) \equiv \frac{\sin^2 \phi}{\phi^2} \tag{17.8.45}$$

and so has a value of unity at exact phase matching. A plot of it versus ϕ is shown in Fig. 13.3.

The time-averaged energy flux vector of the output acoustic wave is

$$\begin{aligned} \langle S_J^\gamma \rangle &= \langle -\dot{u}_I(\mathbf{X},t)\left[c_{IJKL} u_{K,L}(\mathbf{X},t) - e_{KIJ} \mathcal{E}_K(\mathbf{X},t)\right] \rangle \\ &= \langle -\dot{u}_I(\mathbf{X},t) \bar{c}_{IJKL} u_{K,L}(\mathbf{X},t) \rangle. \end{aligned} \tag{17.8.46}$$

Substitution of the displacement vector into this gives the final result

$$\begin{aligned} \langle S_J^\gamma \rangle &= \frac{\omega_3 K_3^\gamma}{2} \left(\frac{D K_1^\alpha K_2^\beta \bar{c} X}{4\rho^0 v_\gamma (\mathbf{N} \cdot \mathbf{v}_g^\gamma)} \right)^2 \Phi\left(\frac{\Delta K_N X}{2} \right) |U_1|^2 |U_2|^2 b_I^\gamma \bar{c}_{IJKL} b_K^\gamma s_L \\ &= \frac{\cos\delta^\alpha \cos\delta^\beta \cos\delta^\gamma}{8\left(\rho^0 v_\alpha v_\beta v_\gamma\right)^3} \left(\frac{D\omega_3 \bar{c} X}{\mathbf{N} \cdot \mathbf{t}^\gamma} \right)^2 \Phi\left(\frac{\Delta K_N X}{2} \right) \langle S^\alpha \rangle \langle S^\beta \rangle t_J^\gamma \end{aligned} \tag{17.8.47}$$

where the time-averaged magnitudes of the input intensities, $\langle S^\alpha \rangle$ and $\langle S^\beta \rangle$, are introduced by way of (10.11.16), where the expression (10.11.13) for the unit vector $\mathbf{t}$ in the direction of energy flux is used, and where

$$\begin{aligned} D &= 2 \quad \text{if} \quad \omega_1 \neq \omega_2 \text{ and/or } \alpha \neq \beta \\ &= 1 \quad \text{if} \quad \omega_1 = \omega_2 \text{ and } \alpha = \beta. \end{aligned} \tag{17.8.48}$$

The factor of $(D/2)^2$ introduced in (17.8.47) makes it applicable to second harmonic generation also. It is needed because for the latter case an extra factor of $\frac{1}{2}$ enters the Fourier expansion of $\mathbf{u}(\mathbf{X}; 1, 1)$.

The expression (17.8.47) for the output acoustic flux applies to sum frequency generation (17.8.1) near or at phase matching (17.8.30). When it is applied to acoustic second harmonic generation, the input flux should be substituted for both $\langle S^\alpha \rangle$ and $\langle S^\beta \rangle$ and D should be set equal to 1. The expression also can be applied to difference frequency generation (with $\omega_1 \neq \omega_2, D=2$),

$$\omega_3 = \omega_1 - \omega_2, \tag{17.8.49}$$

provided that the wavevectors are at least close to satisfying the different phase-matching condition,

$$\mathbf{K}_3^{\gamma}=\mathbf{K}_1^{\alpha}-\mathbf{K}_2^{\beta}. \tag{17.8.50}$$

The expression (17.8.47) shows that the output acoustic flux at exact phase matching ($\Delta K_N=0, \Phi=1$) grows as the square of the coordinate X measured normal to the input surface. Deviation from this dependence occurs when significant depletion of the input waves occurs. The formula also shows that the output flux is proportional to the square of the nonlinear interaction strength $\bar{c}$ as well as the output frequency ω_3.

17.9 Phase Matching

Phase matching is conceptually the same in acoustic mixing as it is in optical mixing. However, in contrast to optical mixing phase matching in acoustic mixing is easily attained. The reason for this is that acoustic velocities in the long wavelength region are dispersionless, that is, they are independent of frequency.

The result of dispersionless propagation is that collinear acoustic second harmonic generation is always phase matched for a harmonic eigenmode type that is the same as the fundamental. This is true regardless of mode type, crystal symmetry, or propagation direction. For the same reason third and higher harmonics are also automatically phase matched in a collinear interaction. When the power at the fundamental frequency is sizable, power transfer to these higher harmonic frequencies must be considered. Since the harmonic and fundamental waves have the same mode type, energy walk-off that usually limits power transfer in collinear optical harmonic generation does not occur in collinear acoustic harmonic generation.

If we consider collinear sum frequency generation by different eigenmode types, (17.8.1) and (17.8.30) lead to a requirement on the ratio of the two input frequencies,

$$\frac{\omega_2}{\omega_1}=\left(\frac{v_\alpha-v_\gamma}{v_\gamma-v_\beta}\right)\frac{v_\beta}{v_\alpha}. \tag{17.9.1}$$

This equation has a solution only if the velocity v_γ of the output sum frequency wave is intermediate in size between the velocities v_α and v_β of the input waves. A similar solution of (17.8.49) and (17.8.50) for collinear difference frequency generation is

$$\frac{\omega_2}{\omega_1}=\frac{v_\gamma-v_\alpha}{v_\gamma-v_\beta}\frac{v_\beta}{v_\alpha}. \tag{17.9.2}$$

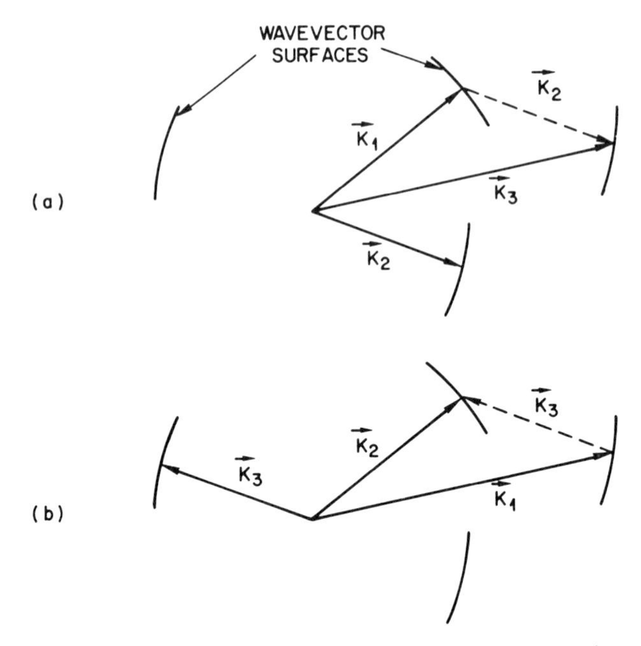

Fig. 17.2. Comparison of phase matching for (*a*) sum frequency generation and (*b*) difference frequency generation using the same three eigenmodes and same three frequencies.

In this case it can be seen that the difference frequency wave velocity v_γ must be either the largest or the smallest of the three eigenmode velocities. For noncollinear sum or difference frequency generation the solution of the wavevector and frequency conditions is quite complicated but easily grasped graphically. Figure 17.2 illustrates phase matching for both sum and difference frequency generation using the same three acoustic eigenmodes.

17.10 Three-Wave Acoustic Parametric Interaction

We now wish to study a three-wave acoustic parametric interaction which is analogous to the three-wave optical parametric interaction studied in Sections 15.5 to 15.8. Since the three waves considered are all freely propagating waves, the parametric interaction has output waves of significant intensity only near phase matching. It is thus advantageous to consider the parametric interaction by way of coupled mode equations derived in the slowly varying amplitude (SVA) approximation. Just as the understanding gained in a free-plus-forced-wave calculation in Section 13.5 on the acoustooptic interaction aided in obtaining the SVA equations in

Section 15.5 for a three-wave optical interaction, so also that understanding helps here.

The coupled mode equations to be derived apply to simple dielectrics, piezoelectrics, and pyroelectrics (for which the spontaneous electric field and spontaneous stress vanish). They apply to arbitrary crystal symmetry and to arbitrary orientations between the crystal axes, the input surface, and the propagation directions of the three waves (consistent with being close to or at phase matching).

The frequencies of the three waves to be considered once again satisfy

$$\omega_3 = \omega_1 + \omega_2. \tag{17.10.1}$$

The ω_3 wave is called the *pump wave*; ω_1 and ω_2 waves are *signal* and *idler waves*. We employ the Fourier expansion notation of Section 12.2, that is, amplitudes for the frequencies ω_1, ω_2, and ω_3 are denoted by $(1,0)$, $(0,1)$, and $(1,1)$. The acoustic eigenmodes at these three frequencies are denoted by α, β, and γ respectively.

Since we wish to consider temporal as well as spatial growth of the waves, the SVA treatment here allows the amplitude factors of the various waves to be slowly varying in time. By this we mean for the ω_1 field, for example,

$$\mathbf{b}^{\alpha}\cdot\frac{d\mathbf{u}(\mathbf{X},t;1,0)}{dt} \ll \omega_1 \mathbf{b}^{\alpha}\cdot\mathbf{u}(\mathbf{X},t;1,0). \tag{17.10.2}$$

The scalar product with the acoustic eigenvector $\mathbf{b}^{\alpha}$ is used to make the equation a condition on the scalar amplitude function of the mode.

The Fourier components at ω_1 of (17.6.1) to (17.6.4), which are the basis of this calculation, are

$$-\rho^0\omega_1^2 b_I^{\alpha} u_I(\mathbf{X},t;1,0) - 2i\rho^0\omega_1 b_I^{\alpha}\frac{du_I(\mathbf{X},t;1,0)}{dt}$$
$$- b_I^{\alpha} c_{IJAB} u_{A,BJ}(\mathbf{X},t;1,0) - b_I^{\alpha} e_{KIJ}\varphi_{,KJ}(\mathbf{X},t;1,0) = b_I^{\alpha} G_I(\mathbf{X},t;1,0), \tag{17.10.3}$$

$$\epsilon_0\kappa_{KL}\varphi_{,KL}(\mathbf{X},t;1,0) - e_{KAB}u_{A,BK}(\mathbf{X},t;1,0) = Q_{K,K}(\mathbf{X},t;1,0) \tag{17.10.4}$$

where $\mathbf{G}$ and $\nabla\cdot\mathbf{Q}$ are given by equations analogous to (17.8.4) and (17.8.5) and the second time derivative of $\mathbf{u}(\mathbf{X},t;1,0)$ is dropped because of (17.10.2). Equation (17.10.3) is a scalar product of the dynamic elasticity equation with the acoustic eigenvector because only this component of the vector equation can be satisfied in a SVA treatment, as is shown in Section 13.5.

In the SVA approximation the amplitude function is represented as

$$\mathbf{u}(\mathbf{X},t;1,0)=\mathbf{b}^{\alpha}U_1(X,t)e^{i\mathbf{K}_1^{\alpha}\cdot\mathbf{X}}, \tag{17.10.5}$$

where the eigenvector $\mathbf{b}^{\alpha}$ and the wavevector $\mathbf{K}_1^{\alpha}$ are those characteristic of the unperturbed mode, where the amplitude U_1 is a function of the scalar coordinate

$$X\equiv\mathbf{N}\cdot\mathbf{X} \tag{17.10.6}$$

measured along the unit inward normal $\mathbf{N}$ of the plane input surface, and where $U_1(X,t)$ is a slowly varying function of X,

$$\frac{\partial U_1}{\partial X}\ll K_1^{\alpha}U_1. \tag{17.10.7}$$

Further it must be realized that the eigenvector $\mathbf{b}^{\alpha}$ for a piezoelectric crystal is the eigenvector of the acoustic propagation equation (10.8.2) in which the electric potential is eliminated to produce a piezoelectrically stiffened stiffness tensor. Thus the equation for the amplitude U_1 should follow from the driven acoustic propagation equation formed from a combination of (17.10.3) and (17.10.4). The electric potential should be regarded as a plane wave driven by the acoustic wave displacement and the nonlinearity $\nabla\cdot\mathbf{Q}$. Thus the solution of (17.10.4) for the potential should be taken as

$$\varphi(\mathbf{X},t;1,0)=\frac{s_K^{(1)}e_{KLM}s_M^{(1)}u_L(\mathbf{X},t;1,0)}{\epsilon_0\mathbf{s}^{(1)}\cdot\boldsymbol{\kappa}\cdot\mathbf{s}^{(1)}}-\frac{Q_{K,K}(\mathbf{X},t;1,0)}{(K_1^{\alpha})^2\epsilon_0\mathbf{s}^{(1)}\cdot\boldsymbol{\kappa}\cdot\mathbf{s}^{(1)}} \tag{17.10.8}$$

where $\mathbf{s}^{(1)}$ is a unit vector in the direction of $\mathbf{K}_1^{\alpha}$ [see (17.8.10) to (17.8.12) for the unit vector notation that we use here]. The correctness of this equation can be checked by deducing it from the solution of Section 17.8. Equation (17.10.8) can be substituted into (17.10.3) to obtain

$$-\rho^0\omega_1^2b_I^{\alpha}u_I(\mathbf{X},t;1,0)-2i\rho^0\omega_1b_I^{\alpha}\frac{du_I(\mathbf{X},t;1,0)}{dt}$$
$$-b_I^{\alpha}\left(c_{IJAB}+\frac{e_{BIJ}s_K^{(1)}e_{KAM}s_M^{(1)}}{\epsilon_0\mathbf{s}^{(1)}\cdot\boldsymbol{\kappa}\cdot\mathbf{s}^{(1)}}\right)u_{A,BJ}(\mathbf{X},t;1,0)=b_I^{\alpha}\mathcal{G}_I(\mathbf{X},t;1,0) \tag{17.10.9}$$

where

$$\mathcal{G}_I(\mathbf{X},t;1,0)=G_I(\mathbf{X},t;1,0)-\frac{e_{KIJ}Q_{L,LKJ}(\mathbf{X},t;1,0)}{\epsilon_0\mathbf{s}^{(1)}\cdot\boldsymbol{\kappa}\cdot\mathbf{s}^{(1)}}. \quad (17.10.10)$$

Note that the two derivatives $\partial^2/\partial X_B\partial X_J$ acting on $\mathbf{u}$ in (17.10.9) make only that part of the tensor coefficient in parentheses which is symmetric upon interchange of the B and J indices measurable from this equation. Since this is the usual occurrence in the acoustic propagation equation, we use no special notation to indicate it.

Substitution of (17.10.5) into (17.10.9) with the use of (17.10.7) to drop second space derivatives of U_1 now yields

$$\left[-\rho^0\omega_1^2+(K_1^\alpha)^2\bar{c}_{IJAB}s_B^{(1)}s_J^{(1)}b_I^\alpha b_A^\alpha\right]U_1-2i\rho^0\omega_1\frac{dU_1}{dt}$$
$$-2iK_1^\alpha\left[c_{IJAB}+\frac{e_{BIJ}s_K^{(1)}e_{KAM}s_M^{(1)}}{\epsilon_0\mathbf{s}^{(1)}\cdot\boldsymbol{\kappa}\cdot\mathbf{s}^{(1)}}\right]s_J^{(1)}N_Bb_I^\alpha b_A^\alpha\frac{\partial U_1}{\partial X}=b_I^\alpha\mathcal{G}_I(\mathbf{X},t;1,0)e^{-i\mathbf{K}_1^\alpha\cdot\mathbf{X}}$$
$$(17.10.11)$$

where $\bar{c}_{IJAB}$ is the stiffened stiffness tensor given in (10.8.4) [with $\mathbf{E}^S=0$, see (10.3.13) and (10.3.14)]. The first bracketed term vanishes because it is the unperturbed eigenvector equation. The second bracketed term may be rewritten in terms of the group velocity (10.11.11) as

$$\left[c_{IJAB}+\frac{e_{BIJ}s_K^{(1)}e_{KAM}s_M^{(1)}}{\epsilon_0\mathbf{s}^{(1)}\cdot\boldsymbol{\kappa}\cdot\mathbf{s}^{(1)}}\right]s_J^{(1)}N_Bb_I^\alpha b_A^\alpha=N_J\bar{c}_{IJAB}b_I^\alpha b_A^\alpha s_B^{(1)}$$
$$=\rho^0v_\alpha\mathbf{N}\cdot\mathbf{v}_g^\alpha \quad (17.10.12)$$

where v_α is the phase velocity and $\mathbf{v}_g^\alpha$ is the group velocity. The equation for the amplitude U_1 is now

$$\frac{\partial U_1}{\partial X}+\frac{1}{\mathbf{N}\cdot\mathbf{v}_g^\alpha}\frac{dU_1}{dt}=\frac{i\mathbf{b}^\alpha\cdot\mathcal{G}(\mathbf{X},t;1,0)e^{-i\mathbf{K}_1^\alpha\cdot\mathbf{X}}}{2\rho^0\omega_1(\mathbf{N}\cdot\mathbf{v}_g^\alpha)}. \quad (17.10.13)$$

Evaluation of $\mathbf{b}^\alpha\cdot\mathcal{G}$ for displacement fields at frequencies ω_3 and ω_2 having the form of (17.10.5) leads to

$$\mathbf{b}^\alpha\cdot\mathcal{G}=iK_1^\alpha K_2^\beta K_3^\gamma\bar{c}U_3U_2^*e^{i(\mathbf{K}_3^\gamma-\mathbf{K}_2^\beta)\cdot\mathbf{X}} \quad (17.10.14)$$

where $\bar{c}$ is the material interaction coefficient given by (17.8.39). This expression when substituted into the previous equation shows that the phase mismatch $\Delta\mathbf{K}$,

$$\Delta\mathbf{K} \equiv \mathbf{K}_3^\gamma - \mathbf{K}_1^\alpha - \mathbf{K}_2^\beta, \tag{17.10.15}$$

enters the differential equation. Since input surface boundary conditions require the tangential components of this to vanish,

$$(\Delta\mathbf{K})_T = 0, \tag{17.10.16}$$

we may express $\Delta\mathbf{K}$ in terms of its component ΔK_N normal to the input surface,

$$\Delta\mathbf{K} = \Delta K_N \mathbf{N}. \tag{17.10.17}$$

The differential equation for the amplitude U_1 in the SVA approximation in its final form is now

$$\frac{\partial U_1}{\partial X} + \frac{1}{\mathbf{N}\cdot\mathbf{v}_g^\alpha}\frac{dU_1}{dt} = -\frac{K_1^\alpha K_2^\beta K_3^\gamma \bar{c}\, U_3 U_2^* e^{i\Delta K_N X}}{2\rho^0\omega_1(\mathbf{N}\cdot\mathbf{v}_g^\alpha)}. \tag{17.10.18}$$

Analogous derivations for U_2 and U_3 yield

$$\frac{\partial U_2}{\partial X} + \frac{1}{\mathbf{N}\cdot\mathbf{v}_g^\beta}\frac{dU_2}{dt} = -\frac{K_1^\alpha K_2^\beta K_3^\gamma \bar{c}\, U_3 U_1^* e^{i\Delta K_N X}}{2\rho^0\omega_2(\mathbf{N}\cdot\mathbf{v}_g^\beta)}, \tag{17.10.19}$$

$$\frac{\partial U_3}{\partial X} + \frac{1}{\mathbf{N}\cdot\mathbf{v}_g^\gamma}\frac{dU_3}{dt} = +\frac{K_1^\alpha K_2^\beta K_3^\gamma \bar{c}\, U_1 U_2 e^{-i\Delta K_N X}}{2\rho^0\omega_3(\mathbf{N}\cdot\mathbf{v}_g^\gamma)}. \tag{17.10.20}$$

These coupled mode equations are first order differential equations for the slowly varying time and space dependence of the amplitudes of three free acoustic waves interacting with one another through the material interaction coefficient $\bar{c}$ given in (17.8.39). Note that the factor $\mathbf{N}\cdot\mathbf{v}_g$ becomes simply the phase velocity if the phase velocity is normal to the surface.

A set of coupled mode equations is an adequate characterization of an interaction only if it includes all modes into which significant power is coupled. In practice this means that there should be no mode that obtains power by way of a phase matched interaction and is not represented in the set of equations. Thus care must be taken when applying equations (17.10.18) to (17.10.20) to collinearly propagating waves of the same mode type, since as discussed in Section 17.9 all harmonics are phase matched under this condition. In particular if frequencies ω_1 and $\omega_3 = \omega_1 + \omega_2$ are fed in, a wave at $\omega_4 = \omega_3 + \omega_1$ will be generated in a phase matched interaction as well as a wave at $\omega_2 = \omega_3 - \omega_1$. In fact, the output flux formula (17.8.47) shows that the former will be stronger in ratio to the latter by $(\omega_4/\omega_2)^2 = (\omega_3 + \omega_1)^2/(\omega_3 - \omega_1)^2$.

17.11 Manley-Rowe Relations

The Manley-Rowe relations (see Section 15.6) relate the growth and decay in the number of quanta in the three interacting waves. Acoustic quanta, called *phonons*, possess energies of $\hbar\omega$ just as photons do.

The Manley-Rowe relations can be derived by multiplying first (17.10.18) by $\rho^0\omega_1\mathbf{N}\cdot\mathbf{v}_g^\alpha U_1^*/2$ and then adding the result to its own complex conjugate. The result is

$$\frac{\partial}{\partial X}\left(\frac{\langle\mathbf{S}_1\rangle\cdot\mathbf{N}}{\omega_1}\right)+\frac{1}{\mathbf{N}\cdot\mathbf{v}_g^\alpha}\frac{d}{dt}\left(\frac{\langle\mathbf{S}_1\rangle\cdot\mathbf{N}}{\omega_1}\right)=\Gamma \tag{17.11.1}$$

where

$$\Gamma\equiv\Re\left\{-\tfrac{1}{2}K_1^\alpha K_2^\beta K_3^\gamma \bar{c}\, U_1^* U_2^* U_3 e^{+i\Delta K_N X}\right\} \tag{17.11.2}$$

and $\langle\mathbf{S}_1\rangle$ is the time-averaged energy flux vector given by (10.11.16). Use of the relation (10.11.1) for the acoustic group velocity allows (17.11.1) to be written as

$$\frac{\partial}{\partial X}\left(\frac{\langle\mathbf{S}_1\rangle\cdot\mathbf{N}}{\omega_1}\right)+\frac{d}{dt}\left(\frac{\langle H_1\rangle}{\omega_1}\right)=\Gamma \tag{17.11.3}$$

where $\langle H_1\rangle$ is the time-averaged energy density in the ω_1 wave. Analogous treatment of (17.10.19) and (17.10.20) leads to

$$\frac{\partial}{\partial X}\left(\frac{\langle\mathbf{S}_2\rangle\cdot\mathbf{N}}{\omega_2}\right)+\frac{d}{dt}\left(\frac{\langle H_2\rangle}{\omega_2}\right)=\Gamma, \tag{17.11.4}$$

$$\frac{\partial}{\partial X}\left(\frac{\langle\mathbf{S}_3\rangle\cdot\mathbf{N}}{\omega_3}\right)+\frac{d}{dt}\left(\frac{\langle H_3\rangle}{\omega_3}\right)=-\Gamma. \tag{17.11.5}$$

The Manley-Rowe relations are then

$$\begin{aligned}\frac{\partial}{\partial X}\left(\frac{\langle\mathbf{S}_1\rangle\cdot\mathbf{N}}{\omega_1}\right)+\frac{d}{dt}\left(\frac{\langle H_1\rangle}{\omega_1}\right)&=\frac{\partial}{\partial X}\left(\frac{\langle\mathbf{S}_2\rangle\cdot\mathbf{N}}{\omega_2}\right)+\frac{d}{dt}\left(\frac{\langle H_2\rangle}{\omega_2}\right)\\&=-\frac{\partial}{\partial X}\left(\frac{\langle\mathbf{S}_3\rangle\cdot\mathbf{N}}{\omega_3}\right)-\frac{d}{dt}\left(\frac{\langle H_3\rangle}{\omega_3}\right).\end{aligned} \tag{17.11.6}$$

If the relations are divided by $\hbar$, they may be interpreted in terms of numbers of phonons.

If we speak of the *growth* in the number of phonons as the sum of the increase in phonons flowing in the direction $\mathbf{N}$ and the increase in density of phonons with time, then the Manley-Rowe relations state that the growth of phonons in the ω_1 wave is equal to the growth of phonons in the ω_2 wave and equal to the decay (negative growth) of phonons in the ω_3 wave. The relations are closely related to energy conservation as expressed by (17.10.1) when multipled by $\hbar$. This can be seen if (17.11.3), (17.11.4), and (17.11.5) are multiplied respectively by ω_1, ω_2, and ω_3 and then added. Use of (17.10.1) to make the right side vanish then yields

$$\frac{\partial}{\partial X}\langle \mathbf{S}_1+\mathbf{S}_2+\mathbf{S}_3\rangle \cdot \mathbf{N}+\frac{d}{dt}\langle H_1+H_2+H_3\rangle=0, \qquad (17.11.7)$$

which is the energy conservation statement for the three interacting waves.

17.12 Parametric Interaction of an Electric Field and Two Acoustic Waves

Thompson and Quate [1971] in an interesting experiment produced two counterpropagating ultrasonic waves in piezoelectric $LiNbO_3$ both at half the frequency of a homogeneous microwave electric field pump applied perpendicularly to the propagation direction of the ultrasonic waves. In this section we derive the coupled mode equations that govern this interaction.

The coupled mode equations for the two acoustic waves in this interaction differ from those of Section 17.10 only in the form of the bilinear driving factor. That difference arises because the ω_3 (pump) electric field is assumed (1) to be homogeneous in the crystal,

$$\varphi(\mathbf{X},t;1,1)=-\mathbf{e}^{(3)}\cdot\mathbf{X}\mathcal{E}_3(t), \qquad (17.12.1)$$

and (2) to generate no significant acoustic wave at frequency ω_3,

$$\mathbf{u}(\mathbf{X},t;1,1)=0. \qquad (17.12.2)$$

As before, the acoustic waves at frequencies ω_1 and ω_2 (the signal and idler waves) are vibrations stiffened by piezoelectrically generated electric fields. As always in such interactions we consider solutions for which

$$\omega_3=\omega_1+\omega_2. \qquad (17.12.3)$$

For these reasons we see that (17.10.13) applies to the ω_1 wave here,

$$\frac{\partial U_1}{\partial X}+\frac{1}{\mathbf{N}\cdot\mathbf{v}_g^\alpha}\frac{dU_1}{dt}=\frac{i\mathbf{b}^\alpha\cdot\mathcal{G}(\mathbf{X},t;1,0)e^{-i\mathbf{K}_1^\alpha\cdot\mathbf{X}}}{2\rho^0\omega_1(\mathbf{N}\cdot\mathbf{v}_g^\alpha)}, \tag{17.12.4}$$

except that $\mathbf{b}^\alpha\cdot\mathcal{G}$ must be reevaluated for an acoustic wave input at ω_2 and an electric field input at ω_3. This factor is given in this interaction by

$$\begin{aligned}\mathbf{b}^\alpha\cdot\mathcal{G}(\mathbf{X},t;1,0)=&-b_I^\alpha\Big[-\tfrac{1}{2}\epsilon_0 l_{LMIJ}\varphi_{,L}^*(\mathbf{X},t;0,1)e_M^{(3)}\mathcal{E}_3(t)\\&+\tfrac{1}{2}(e_{LIJAB}+\delta_{IA}e_{LBJ})u_{A,B}^*(\mathbf{X},t;0,1)e_L^{(3)}\mathcal{E}_3(t)\Big]_{,J}\\&-\frac{b_I^\alpha e_{NIJ}}{(K_1^\alpha)^2\epsilon_0\mathbf{s}^{(1)}\cdot\boldsymbol{\kappa}\cdot\mathbf{s}^{(1)}}\Big[-2\epsilon_0 b_{KLM}\varphi_{,L}^*(\mathbf{X},t;0,1)e_M^{(3)}\mathcal{E}_3(t)\\&+\tfrac{1}{2}\epsilon_0 l_{KLAB}u_{A,B}^*(\mathbf{X},t;0,1)e_L^{(3)}\mathcal{E}_3(t)\Big]_{,KNJ}\\=&\,(K_2^\beta)^2\bar{h}\,U_2^*\mathcal{E}_3e^{-i\mathbf{K}_2^\beta\cdot\mathbf{X}}\end{aligned} \tag{17.12.5}$$

where the material interaction coefficient $\bar{h}$ may be called the effective nonlinear piezoelectric coefficient and is defined by

$$\begin{aligned}\bar{h}\equiv\Bigg[&\tfrac{1}{2}(e_{LIJAB}+\delta_{IA}e_{LBJ})-\frac{l_{LMIJ}s_Ms_Ne_{NAB}}{2(\mathbf{s}\cdot\boldsymbol{\kappa}\cdot\mathbf{s})}-\frac{l_{LMAB}s_Ms_Ne_{NIJ}}{2(\mathbf{s}\cdot\boldsymbol{\kappa}\cdot\mathbf{s})}\\&+\frac{2s_Kb_{KLM}s_Ms_Pe_{PIJ}s_Qe_{QAB}}{\epsilon_0(\mathbf{s}\cdot\boldsymbol{\kappa}\cdot\mathbf{s})^2}\Bigg]e_L^{(3)}b_I^\alpha s_Jb_A^\beta s_B.\end{aligned} \tag{17.12.6}$$

In expressing $\bar{h}$ we use $\mathbf{s}^{(1)}=-\mathbf{s}^{(2)}\equiv\mathbf{s}$ and $(K_2^\beta/K_1^\alpha)^2=1$, true as we see presently for exact phase matching and approximately true whenever the coupled mode equations are applicable.

The effective nonlinear piezoelectric coefficient $\bar{h}$ contains several experimentally distinguishable contributions to the interaction of an electric field and two acoustic waves. First, there is the direct interaction of the homogeneous electric field and the strain fields of the two acoustic waves by way of the fifth rank nonlinear piezoelectric stress tensor. Second, the homogeneous electric field and the acoustic waves can interact through an effective nonlinear piezoelectric effect that involves only the linear piezoelectric tensor. This term, discussed in Section 17.8, arises because the linear piezoelectric tensor appears as the coefficient of the Green finite strain tensor which contains a nonlinear part. The next two terms represent the two-step indirect interaction of the homogeneous electric field, the longitudinal electric field piezoelectrically created by one acoustic wave, and the strain of the other acoustic wave through the electrostrictive effect. The last term represents the three-step indirect interaction of the homogeneous electric field and both longitudinal electric fields piezoelectrically generated by the two acoustic waves by way of the electric field mixing tensor.

Since the ω_3 field has no propagation vector, we define the phase mismatch as

$$\Delta\mathbf{K} \equiv \mathbf{K}_1^\alpha + \mathbf{K}_2^\beta. \tag{17.12.7}$$

Since boundary conditions require the tangential components $(\Delta\mathbf{K})_T$ to vanish,

$$(\Delta\mathbf{K})_T = 0, \tag{17.12.8}$$

we may write

$$\Delta\mathbf{K} = \Delta K_N \mathbf{N} \tag{17.12.9}$$

where once again $\mathbf{N}$ is the unit inward normal of the input plane surface.

Phase matching ($\Delta\mathbf{K}=0$), which causes the interaction to attain its maximum strength, requires arranging $\Delta K_N = 0$. When this is accomplished, (17.12.7) implies $\mathbf{K}_1^\alpha = -\mathbf{K}_2^\beta$. This corresponds to counterpropagating waves, $\mathbf{s}^{(1)} = -\mathbf{s}^{(2)}$, having equal wavenumbers $K_1^\alpha = K_2^\beta$. If the two waves also have equal frequencies, $\omega_1 = \omega_2$, then their velocities must be equal. This is most easily accomplished if they are of the same mode type, $\alpha = \beta$, but would also occur for degenerate modes.

Combining (17.12.4), (17.12.5), (17.12.7), and (17.12.9) now yields

$$\frac{\partial U_1}{\partial X} + \frac{1}{\mathbf{N}\cdot\mathbf{v}_g^\alpha}\frac{dU_1}{dt} = \frac{-iK_1^\alpha K_2^\beta \bar{h}\, U_2^* \mathcal{E}_3 e^{-i\Delta K_N X}}{2\rho^0\omega_1(\mathbf{N}\cdot\mathbf{v}_g^\alpha)} \tag{17.12.10}$$

as the equation for the scalar amplitude U_1 of the acoustic wave at frequency ω_1. A completely analogous development for U_2 yields

$$\frac{\partial U_2}{\partial X} + \frac{1}{\mathbf{N}\cdot\mathbf{v}_g^\beta}\frac{dU_2}{dt} = \frac{-iK_1^\alpha K_2^\beta \bar{h}\, U_1^* \mathcal{E}_3 e^{-i\Delta K_N X}}{2\rho^0\omega_2(\mathbf{N}\cdot\mathbf{v}_g^\beta)}. \tag{17.12.11}$$

These two coupled mode equations are all that are needed to analyze the Thompson-Quate experiment, since the $\mathcal{E}_3$ field can be regarded as specified.

In the absence of a coupled mode equation for $\mathcal{E}_3$ the portion of the Manley-Rowe relations relating to this field must be found from a somewhat different basis. First of all, the coupled mode equations for U_1 and U_2 can be converted to energy continuity equations in the manner used for (17.11.1) and (17.11.3). This gives

$$\frac{\partial}{\partial X}\left(\frac{\langle \mathbf{S}_1\rangle\cdot\mathbf{N}}{\omega_1}\right)+\frac{d}{dt}\left(\frac{\langle H_1\rangle}{\omega_1}\right)=\Lambda, \tag{17.12.12}$$

$$\frac{\partial}{\partial X}\left(\frac{\langle \mathbf{S}_2\rangle\cdot\mathbf{N}}{\omega_2}\right)+\frac{d}{dt}\left(\frac{\langle H_2\rangle}{\omega_2}\right)=\Lambda \tag{17.12.13}$$

where

$$\Lambda\equiv\Re\left\{\frac{i}{2}K_1^{\alpha}K_2^{\beta}\bar{h}\,U_1U_2\mathcal{E}_3^{*}e^{i\Delta K_N X}\right\}. \tag{17.12.14}$$

To obtain the energy continuity equation for the $\mathcal{E}_3$ electric field we use Poynting's theorem,

$$\nabla_X\cdot(\mathcal{E}\times\mathcal{H})+\mathcal{E}\cdot\dot{\mathfrak{D}}+\mathcal{H}\cdot\dot{\mathfrak{B}}=0, \tag{17.12.15}$$

found from (17.5.1) and (17.5.2). Dropping all magnetic field terms as is appropriate in the quasi-electrostatic regime leaves

$$\mathcal{E}\cdot\dot{\mathfrak{D}}=0. \tag{17.12.16}$$

Use of (17.6.4) for $\mathfrak{D}$ and the usual Fourier expansion in time for obtaining that part of the equation relating to the ω_3 field gives

$$\frac{d}{dt}\left(\frac{\epsilon_0}{4}\varphi_{,K}(\mathbf{X},t;1,1)\kappa_{KL}\varphi_{,L}^{*}(\mathbf{X},t;1,1)\right)$$
$$=-\omega_3\Re\left\{i\varphi_{,K}^{*}(\mathbf{X},t;1,1)\mathfrak{D}_K^{NL}(\mathbf{X},t;1,1)\right\} \tag{17.12.17}$$

with

$$\begin{aligned}\mathfrak{D}_K^{NL}(\mathbf{X},t;1,1)=&\,\epsilon_0 b_{KLM}\varphi_{,L}(\mathbf{X},t;1,0)\varphi_{,M}(\mathbf{X},t;0,1)\\&+\tfrac{1}{4}\epsilon_0 l_{KLAB}\big[\varphi_{,L}(\mathbf{X},t;1,0)u_{A,B}(\mathbf{X},t;0,1)\\&+\varphi_{,L}(\mathbf{X},t;0,1)u_{A,B}(\mathbf{X},t;1,0)\big]\\&-\tfrac{1}{4}(e_{KABCD}+e_{KBD}\delta_{AC})u_{A,B}(\mathbf{X},t;1,0)u_{C,D}(\mathbf{X},t;0,1).\end{aligned} \tag{17.12.18}$$

Recognizing the enclosed quantity on the left as the energy density H_3 and simplifying the right side yields

$$\frac{d}{dt}\left(\frac{H_3}{\omega_3}\right) = -\Lambda. \qquad (17.12.19)$$

The Manley-Rowe relations,

$$\frac{\partial}{\partial X}\left(\frac{\langle \mathbf{S}_1\rangle\cdot\mathbf{N}}{\omega_1}\right)+\frac{d}{dt}\left(\frac{\langle H_1\rangle}{\omega_1}\right)=\frac{\partial}{\partial X}\left(\frac{\langle \mathbf{S}_2\rangle\cdot\mathbf{N}}{\omega_2}\right)+\frac{d}{dt}\left(\frac{\langle H_2\rangle}{\omega_2}\right)$$
$$= -\frac{d}{dt}\left(\frac{\langle H_3\rangle}{\omega_3}\right), \qquad (17.12.20)$$

follow immediately. Their interpretation is quite analogous to that discussed in Section 17.11 (except that $\mathbf{S}_3=0$ here) and so is not repeated.

17.13 Temporal Growth of Acoustic Parametric Oscillation

The treatment of optical parametric interactions in Chapter 15 does not consider time dependence of the wave amplitudes and so applies only to steady state phenomena. This is reasonable because temporal growth of optical oscillations in an optical parametric oscillator are very fast and so are not usually observed. The speed of acoustic waves, however, is typically 10^5 times slower than the speed of electromagnetic waves and so the temporal growth of acoustic oscillations is easily observable. For this reason we include a slowly varying time dependence of the displacement amplitude in the coupled mode equations of Section 17.10 and 17.12. It thus is interesting to solve for the temporal growth of acoustic oscillations in the Thompson-Quate [1971] experiment.

Let us consider a special case, considered by Thompson and Quate, of the general coupled mode equations (17.12.10) and (17.12.11) in which the signal and idler acoustic waves have the same frequency, $\omega_1=\omega_2=\omega_3/2$, are the same mode type, $\alpha=\beta$, and propagate normal to the end surfaces. Further let us consider exact phase matching, $\Delta K_N=0$. The signal and idler waves are thus counterpropagating. We denote $\mathbf{s}\equiv\mathbf{s}^{(1)}=-\mathbf{s}^{(2)}$, $K\equiv K_1^\alpha = K_2^\beta$, $v\equiv v_\alpha = v_\beta$, and $\mathbf{v}_g\equiv\mathbf{v}_g^\alpha=-\mathbf{v}_g^\beta$. We define a coupling constant g by

$$g\equiv\frac{K^2\bar{h}|\mathcal{E}_3|}{\rho^0\omega_3 v} \qquad (17.13.1)$$

where the homogeneous electric field pump amplitude is expressed as

$$\mathcal{E}_3 = |\mathcal{E}_3| e^{-i\varphi_P}. \tag{17.13.2}$$

We also introduce phenomenologically a loss term αU_1 on the left side of the U_1 equation and a loss term $-\alpha U_2$ on the left side of the U_2 equation, the minus sign accounting for the reverse propagation direction of the U_2 wave. Thus the coupled mode equations can be written

$$\frac{\partial U_1}{\partial X} + \frac{1}{v}\frac{dU_1}{dt} + \alpha U_1 = -ige^{-i\varphi_P} U_2^*, \tag{17.13.3}$$

$$\frac{\partial U_2^*}{\partial X} + \frac{1}{v}\frac{dU_2^*}{dt} - \alpha U_2^* = -ige^{+i\varphi_P} U_1. \tag{17.13.4}$$

We desire a temporally growing solution of the form

$$U_1 = V_1(X) e^{\eta t}, \tag{17.13.5}$$

$$U_2 = V_2(X) e^{\eta t} \tag{17.13.6}$$

where η is real and positive. If we denote $f \equiv \alpha + \eta / v$, we obtain

$$\frac{\partial V_1}{\partial X} + fV_1 = -ige^{-i\varphi_P} V_2^*, \tag{17.13.7}$$

$$\frac{\partial V_2^*}{\partial X} + fV_2^* = -ige^{+i\varphi_P} V_1. \tag{17.13.8}$$

An equation for V_1 alone may be obtained by differentiating the first of these,

$$\frac{\partial^2 V_1}{\partial X^2} + f\frac{\partial V_1}{\partial X} = -ige^{-i\varphi_P}\frac{\partial V_2^*}{\partial X}, \tag{17.13.9}$$

multiplying the first equation by $-f$ and the second equation by $-ig\exp(-i\varphi_P)$ and then adding the last three equations together. The result is

$$\frac{\partial^2 V_1}{\partial X^2} = -\Gamma^2 V_1 \tag{17.13.10}$$

where

$$\Gamma^2 \equiv g^2 - f^2. \tag{17.13.11}$$

We may adopt a solution of the form

$$V_1 = (A\sin\Gamma X + B\cos\Gamma X)e^{-i\varphi} \tag{17.13.12}$$

with A, B, and φ real, that is, both coefficients of the V_1 wave solution have the same phase. Substitution of this solution into (17.13.8) and integration with the use of an integrating factor yields

$$V_2^* = -\frac{i}{g}e^{i(\varphi_P-\varphi)}\big[(-f\sin\Gamma X - \Gamma\cos\Gamma X)A + (-f\cos\Gamma X + \Gamma\sin\Gamma X)B\big] + Ce^{fX}. \tag{17.13.13}$$

Substitution of the two solutions into (17.13.7) shows that the integration constant $C=0$.

For boundary conditions we relate the displacement by a reflectivity factor r at each end of the crystal located at $X=0$ and $X=L$,

$$V_1(0) = rV_2(0), \tag{17.13.14}$$

$$rV_1(L)e^{+iKL} = V_2(L)e^{-iKL}. \tag{17.13.15}$$

Actually, since we are considering a single eigenmode reflecting at normal incidence, $r=1$ as shown in Prob. 10.7. For the moment, however, we allow for a possible nonunit reflectivity, though we regard it as real. We make the simplifying assumption that the length of the crystal is acoustically resonant for the ω_1 frequency,

$$2KL = 2n\pi \qquad (n \text{ integral}). \tag{17.13.16}$$

The two boundary conditions lead to

$$Be^{-i\varphi} = \frac{ir}{g}e^{-i(\varphi_P-\varphi)}(-\Gamma A - fB), \tag{17.13.17}$$

$$r(A\sin\Gamma L + B\cos\Gamma L)e^{-i\varphi} = \frac{i}{g}e^{-i(\varphi_P-\varphi)}\big[(-f\sin\Gamma L - \Gamma\cos\Gamma L)A + (-f\cos\Gamma L + \Gamma\sin\Gamma L)B\big]. \tag{17.13.18}$$

The phase in these equations must satisfy

$$-ie^{-i(\varphi_P-2\varphi)} = \pm 1. \tag{17.13.19}$$

The plus sign should be chosen here, since it makes Λ in the Manley-Rowe relations of Section 17.12 positive and thus corresponds to power transfer from the electric field pump to the signal and idler waves. This leads to the relation

$$\varphi_P - 2\varphi = -\frac{\pi}{2} \tag{17.13.20}$$

between the phases of the pump, signal, and idler waves.

The two equations for A and B have a solution provided that the determinant of their coefficients vanishes. This yields

$$\tan \Gamma L = -\frac{\Gamma(1-r^2)}{f(1+r^2)-2gr} \tag{17.13.21}$$

as the condition on Γ. It may be reexpressed as

$$\left[\tan\frac{\Gamma L}{2} + \frac{g-f}{\Gamma}\left(\frac{1+r}{1-r}\right)\right]\left[\tan\frac{\Gamma L}{2} - \frac{g+f}{\Gamma}\left(\frac{1-r}{1+r}\right)\right] = 0, \tag{17.13.22}$$

which shows that, contained in the previous equation, there are two families of solutions corresponding to even and odd solutions of the symmetric geometry.

From (17.13.17) we have

$$B = \frac{\Gamma r A}{g - fr} \tag{17.13.23}$$

and so the final solutions take the form

$$U_1 = \frac{A}{g-fr}\left[(g-fr)\sin\Gamma X + \Gamma r \cos\Gamma X\right]e^{\eta t - i\varphi}, \tag{17.13.24}$$

$$U_2 = \frac{A}{g-fr}\left[(f-gr)\sin\Gamma X + \Gamma \cos\Gamma X\right]e^{\eta t - i\varphi}. \tag{17.13.25}$$

The temporal growth constant η is given by

$$\eta = v\left[(g^2 - \Gamma^2)^{1/2} - \alpha\right] \tag{17.13.26}$$

with Γ found from (17.13.21). If $r = 1$, $\Gamma = 0$ and we have

$$\eta = v(g - \alpha) \tag{17.13.27}$$

which gives us a minimum condition, $g > \alpha$, for temporal growth of acoustic oscillations. An initial condition on the displacement fields is necessary to determine the constant A in the solution. In reality the temporal growth of the acoustic oscillations begins from thermal noise present in the eigenmode.

In the Thompson-Quate experiment $v=4.76\times10^5$ cm/sec and $\alpha=0.3$ cm^{-1}. If we assume the pump field $\mathcal{E}_3$ is strong enough to give $g=2\alpha$, which is a substantial gain, then $\eta=(7.0\ \mu\text{sec})^{-1}$. With this growth rate it is apparent that the approach to a steady state oscillation can take tens of microseconds.

PROBLEMS

17.1. Show that (17.1.6) can be obtained by transformation of (8.2.3).

17.2. Show that (17.1.8) can be obtained by transformation of (7.5.11).

17.3. Find the material frame energy conservation law in terms of material frame electromagnetic fields from (8.7.10).

17.4. Beginning from (17.10.20) find the generated sum frequency flux in the absence of input wave depletion and show that it agrees with (17.8.47).

17.5. Find a steady state solution to (17.13.3) and (17.13.4) subject to the boundary conditions

$$U_1(0)=|U|e^{-i\varphi}, \tag{17.P.1}$$

$$U_2(L)=0. \tag{17.P.2}$$

What is the instability condition for this solution corresponding to an oscillation threshold? How does it compare to the condition implied by (17.13.27)?

17.6. Derive (17.8.43).

17.7. Show by the transformation of the relevant terms in the material frame polarization (17.3.5) to the spatial frame that the low frequency limit of the elastooptic susceptibility, including both coupling to strain and rotation, is obtained.

CHAPTER 18

Higher Order Interactions

Having finished consideration of three-field interactions, we wish in this last chapter to consider two examples of higher order interactions, third order optical mixing and a four-wave acoustooptic interaction. The study of the latter, acoustically induced optical harmonic generation (AIOHG) [Boyd, Nash, and Nelson, 1970], in fact, stimulated the development of the theoretical framework on which this book is based. AIOHG is an interaction in which an input optical wave interacts with itself and with an input acoustic wave to produce an output optical wave at a frequency displaced from the optical second harmonic frequency by the acoustic frequency.

As with all wave interactions the strength of AIOHG is large only when it is near or at phase matching. The analysis of AIOHG led to a further development of the phase-matching concept. It showed that indirect contributions to AIOHG could be doubly phase matched and so become even stronger than the singly phase matched direct contribution [Nelson and Lax, 1971a]. Double phase matching of all-optical interactions was observed about the same time [Andrews, Rabin, and Tang, 1970]. Triple phase matching of a five-wave acoustooptic interaction was later demonstrated [Nelson and Mikulyak, 1972].

We begin by deriving the nonlinear polarization created by either three input optical waves or two input optical waves and one input acoustic wave. This is done in a manner analogous to that used for three-field interactions in Chapter 12. Before considering the acoustooptic interaction application is made to optical third harmonic generation, first observed by Terhune, Maker, and Savage [1962].

18.1 Four-Wave Optical Interactions

The output optical wave from a nonlinear interaction is found from the driven wave equation (12.3.14) with a nonlinear polarization $\mathscr{P}$. The object of this section is to derive the constitutive form of $\mathscr{P}$ for the case of two input optical waves and a third wave that may be either acoustic or optical and so may have either a low or high frequency compared to optic mode resonant frequencies.

We begin from the internal motion equation expressed in the material frame,

$$m^{\nu}\ddot{y}_i^{\nu} = q^{\nu}\mathcal{E}_i - \frac{\partial \rho^0 \Sigma}{\partial \Pi_C^{\nu}} X_{C,i}, \tag{18.1.1}$$

where the stored energy is given by (12.1.3) with the addition of one more term,

$$\sum_{\nu\mu\lambda} {}^{31}M_{CDEAB}^{\nu\mu\lambda}\Pi_C^{\nu}\Pi_D^{\mu}\Pi_E^{\lambda}E_{AB}. \tag{18.1.2}$$

Other fourth order energy terms need not be considered because they cannot affect the nonlinear polarization of the processes discussed. The first order energy terms may be dropped by the assumptions of no spontaneous electric field and no spontaneous stress. The terms in the internal motion equation that are dropped in Section 12.1 as being negligibly small can be dropped here also. They are the correction terms transforming the material time derivative to a spatial time derivative and the magnetic term. The result is

$$m^{\nu}\frac{\partial^2 y_i^{\nu}}{\partial t^2} - q^{\nu}E_i + 2\sum_{\mu} {}^{20}M_{id}^{\nu\mu}(y_d^{\mu} - u_{d,e}Y_e^{\mu}) + {}^{11}M_{iab}^{\nu}u_{a,b} = F_i^{\nu} \tag{18.1.3}$$

where the nonlinear terms are collected in

$$\begin{aligned} F_i^{\nu} \equiv\ & 2\sum_{\mu} {}^{20}M_{cd}^{\nu\mu}(\delta_{ci}y_j^{\mu}u_{d,j} + y_d^{\mu}u_{c,i}) \\ & -3\sum_{\mu\lambda} {}^{30}M_{ide}^{\nu\mu\lambda}\left[(y_d^{\mu} - u_{d,f}Y_f^{\mu})(y_e^{\lambda} - u_{e,g}Y_g^{\lambda}) - 2y_d^{\mu}y_g^{\lambda}u_{e,g}\right] \\ & +3\sum_{\mu\lambda} {}^{30}M_{cde}^{\nu\mu\lambda}y_d^{\mu}y_e^{\lambda}u_{c,i} - 4\sum_{\mu\lambda\rho} {}^{40}M_{idef}^{\nu\mu\lambda\rho}y_d^{\mu}y_e^{\lambda}(y_f^{\rho} - u_{f,g}Y_g^{\rho}) \\ & -2\sum_{\mu} {}^{21}M_{idab}^{\nu\mu}y_d^{\mu}u_{a,b} - 3\sum_{\mu\lambda} {}^{31}M_{ideab}^{\nu\mu\lambda}y_d^{\mu}y_e^{\lambda}u_{a,b}. \end{aligned} \tag{18.1.4}$$

Next we perform a Fourier expansion of the internal motion equation in the manner of Section 12.2 (except that we have three input frequencies here) and select the components at the output frequency ω_C of the interaction. We take the output frequency as being equal to the sum of the two input optical frequencies ω_1, ω_2 and the third frequency ω_A, which we allow to be any frequency between zero and optical frequencies,

$$\omega_C = \omega_1 + \omega_2 + \omega_A. \tag{18.1.5}$$

It is also convenient to denote the intermediate frequencies by

$$\omega_{12} = \omega_1 + \omega_2, \tag{18.1.6}$$

$$\omega_{1A} = \omega_1 + \omega_A, \tag{18.1.7}$$

$$\omega_{2A} = \omega_2 + \omega_A. \tag{18.1.8}$$

We designate the amplitudes associated with each of these frequencies by the frequency itself.

If all the terms in the Fourier expansion of (18.1.3) with a displacement gradient oscillating at an optical frequency are dropped as being negligibly small, the result is

$$-m^\nu \omega_C^2 y_i^\nu(\omega_C) + 2\sum_\mu {}^{20}M_{id}^{\nu\mu} y_d^\mu(\omega_C) - q^\nu E_i(\omega_C) = F_i^\nu(\omega_C) \tag{18.1.9}$$

with

$$\begin{aligned}
F_i^\nu(\omega_C) = {} & \sum_\mu {}^{20}M_{cd}^{\nu\mu}\big(\delta_{ci} y_j^\mu(\omega_{12}) u_{d,j}(\omega_A) + y_d^\mu(\omega_{12}) u_{c,i}(\omega_A)\big) \\
& - \tfrac{3}{2}\sum_{\mu\lambda} {}^{30}M_{ide}^{\nu\mu\lambda}\big[2y_d^\mu(\omega_{1A}) y_e^\lambda(\omega_2) + 2y_d^\mu(\omega_{2A}) y_e^\lambda(\omega_1) \\
& + 2y_d^\mu(\omega_{12})\big(y_e^\lambda(\omega_A) - u_{e,g}(\omega_A) Y_g^\lambda\big) - y_d^\mu(\omega_1) y_g^\lambda(\omega_2) u_{e,g}(\omega_A) \\
& - y_d^\mu(\omega_2) y_g^\lambda(\omega_1) u_{e,g}(\omega_A)\big] \\
& + \tfrac{3}{2}\sum_{\mu\lambda} {}^{30}M_{cde}^{\nu\mu\lambda} y_d^\mu(\omega_1) y_e^\lambda(\omega_2) u_{c,i}(\omega_A) \\
& - 6\sum_{\mu\lambda\rho} {}^{40}M_{idef}^{\nu\mu\lambda\rho} y_d^\mu(\omega_1) y_e^\lambda(\omega_2)\big(y_f^\rho(\omega_A) - u_{f,g}(\omega_A) Y_g^\rho\big) \\
& - \sum_\mu {}^{21}M_{idab}^{\nu\mu} y_d^\mu(\omega_{12}) u_{a,b}(\omega_A) \\
& - \tfrac{3}{2}\sum_{\mu\lambda} {}^{31}M_{ideab}^{\nu\mu\lambda} y_d^\mu(\omega_1) y_e^\lambda(\omega_2) u_{a,b}(\omega_A).
\end{aligned} \tag{18.1.10}$$

The nonlinear driving polarization active in the wave equation $\mathscr{P}(\omega_C)$ is shown in Section 12.3 to be given in terms of $\mathbf{F}^\nu(\omega_C)$ and the nonlinear current $\mathbf{I}(\omega_C)$, defined in (12.3.5), by

$$\mathscr{P}_i(\omega_C) = \sum_{\sigma\nu} q^\sigma \Upsilon_{ij}^{\sigma\nu}(\omega_C) F_j^\nu(\omega_C) + \frac{i}{\omega_C} I_i(\omega_C). \tag{18.1.11}$$

Just as is found in Section 12.3, the only term of significant size in $\mathbf{I}(\omega_C)$ is

$$I_i(\omega_C)=\frac{i\omega_C}{2}u_{l,l}(\omega_A)\sum_\nu q^\nu y_i^\nu(\omega_{12}). \tag{18.1.12}$$

We must now combine the last three equations with (9.1.9) used for $\mathbf{y}^\mu(\omega_1)$ and $\mathbf{y}^\mu(\omega_2)$, with (12.4.2) used for $\mathbf{y}^\lambda(\omega_A)$, with (12.2.16) used for $\mathbf{y}^\rho(\omega_{1A})$ and $\mathbf{y}^\rho(\omega_{2A})$, and with

$$y_i^\sigma(\omega_{12})=\sum_\zeta \Upsilon_{ij}^{\sigma\zeta}(\omega_{12})\left[q^\zeta E_j(\omega_{12})+F_j^\zeta(\omega_{12})\right], \tag{18.1.13}$$

$$F_l^\zeta(\omega_{12})=-3\sum_{\mu\lambda} {}^{30}M_{lde}^{\zeta\mu\lambda}\Upsilon_{dj}^{\mu\rho}(\omega_1)q^\rho\Upsilon_{ek}^{\lambda\eta}(\omega_2)q^\eta E_j(\omega_1)E_k(\omega_2) \tag{18.1.14}$$

for $\mathbf{y}^\sigma(\omega_{12})$. The result may be written as

$$\begin{aligned}\mathscr{P}_i^{\omega_C}(\mathbf{z})=&\,\epsilon_0 D_1\chi_{i\ j\ k\ lm}^{\omega_C\omega_1\omega_2\omega_A}E_j^{\omega_1}(\mathbf{z})E_k^{\omega_2}(\mathbf{z})u_{l,m}^{\omega_A}(\mathbf{z})\\ &+\epsilon_0 D_2 e_{i\ j\ k\ l}^{\omega_C\omega_1\omega_2\omega_A}E_j^{\omega_1}(\mathbf{z})E_k^{\omega_2}(\mathbf{z})E_l^{\omega_A}(\mathbf{z})\\ &+\epsilon_0 D_3\chi_{i\ j\ lm}^{\omega_C\omega_{12}\omega_A}E_j^{\omega_{12}}(\mathbf{z})u_{l,m}^{\omega_A}(\mathbf{z})\\ &+\epsilon_0 D_4 b_{i\ j\ k}^{\omega_C\omega_{12}\omega_A}E_j^{\omega_{12}}(\mathbf{z})E_k^{\omega_A}(\mathbf{z})\\ &+\epsilon_0 D_5 b_{i\ j\ k}^{\omega_C\omega_{1A}\omega_2}E_j^{\omega_{1A}}(\mathbf{z})E_k^{\omega_2}(\mathbf{z})\\ &+\epsilon_0 D_6 b_{i\ j\ k}^{\omega_C\omega_{2A}\omega_1}E_j^{\omega_{2A}}(\mathbf{z})E_k^{\omega_1}(\mathbf{z})\end{aligned} \tag{18.1.15}$$

where the space dependence is now explicitly indicated and the frequency designations of the amplitudes are shown as superscripts. The degeneracy factors are defined by

$$D_1=\left\{\begin{array}{lll}1 & \text{if} & \omega_1=\omega_2,\omega_A\neq 0\\ 2 & \text{if} & \omega_1=\omega_2,\omega_A=0\\ 2 & \text{if} & \omega_1\neq\omega_2,\omega_A\neq 0\\ 4 & \text{if} & \omega_1\neq\omega_2,\omega_A=0\end{array}\right\}, \tag{18.1.16}$$

$$D_2=\left\{\begin{array}{lll}1 & \text{if} & \omega_1=\omega_2=\omega_A\\ 3 & \text{if} & \omega_1=\omega_2\neq\omega_A\neq 0\\ 6 & \text{if} & \omega_1=\omega_2,\omega_A=0\\ 6 & \text{if} & \omega_1\neq\omega_2\neq\omega_A\neq 0\\ 12 & \text{if} & \omega_1\neq\omega_2,\omega_A=0\end{array}\right\}, \tag{18.1.17}$$

$$D_3=\left\{\begin{array}{lll}1 & \text{if} & \omega_A\neq 0\\ 2 & \text{if} & \omega_A=0\end{array}\right\}, \tag{18.1.18}$$

$$\left.\begin{cases} D_4=2, \quad D_5=D_6=0 & \text{if} \quad \omega_1=\omega_2=\omega_A\neq 0 \\ D_4=D_5=2, \quad D_6=0 & \text{if} \quad \omega_1=\omega_2\neq\omega_A\neq 0 \\ D_4=4, \quad D_5=2, \quad D_6=0 & \text{if} \quad \omega_1=\omega_2, \omega_A=0 \\ D_4=D_5=D_6=2 & \text{if} \quad \omega_1\neq\omega_2\neq\omega_A\neq 0 \\ D_4=4, \quad D_5=D_6=2 & \text{if} \quad \omega_1\neq\omega_2, \omega_A=0 \end{cases}\right\}. \tag{18.1.19}$$

This procedure gives a defining relation for both the *acoustically induced optical mixing tensor* $\chi_{i\,j\,k\,lm}^{\omega_C\omega_1\omega_2\omega_A}$ [called the *acoustically induced optical harmonic generation* (AIOHG) *tensor* when $\omega_1=\omega_2$] and the *third order optical mixing tensor* $e_{i\,j\,k\,l}^{\omega_C\omega_1\omega_2\omega_A}$. *They are*

$$\begin{aligned}
\epsilon_0\chi_{i\,j\,k\,lm}^{\omega_C\omega_1\omega_2\omega_C} = -\tfrac{3}{4}\sum_{\substack{\nu\mu\lambda\\ \alpha\beta\gamma}} & q^\nu q^\mu q^\lambda \Upsilon_{ia}^{\nu\alpha}(\omega_C)\Upsilon_{jb}^{\mu\beta}(\omega_1)\Upsilon_{kc}^{\lambda\gamma}(\omega_2) \\
\times\Bigg[& {}^{31}M_{abclm}^{\alpha\beta\gamma} - 4\sum_{\delta\epsilon} {}^{40}M_{abcd}^{\alpha\beta\gamma\delta}\Upsilon_{de}^{\delta\epsilon}(\omega_A)\,{}^{11}M_{elm}^{\epsilon} \\
& -2\sum_{\delta\epsilon}\big({}^{30}M_{abd}^{\alpha\beta\delta}\Upsilon_{de}^{\delta\epsilon}(\omega_{2A})\,{}^{21}M_{eclm}^{\epsilon\gamma} \\
& +{}^{30}M_{acd}^{\alpha\gamma\delta}\Upsilon_{de}^{\delta\epsilon}(\omega_{1A})\,{}^{21}M_{eblm}^{\epsilon\beta} + {}^{30}M_{bcd}^{\beta\gamma\delta}\Upsilon_{de}^{\delta\epsilon}(\omega_{12})\,{}^{21}M_{ealm}^{\epsilon\alpha}\big) \\
& +6\sum_{\delta\epsilon\zeta\eta}\Big({}^{30}M_{abd}^{\alpha\beta\delta}\Upsilon_{de}^{\delta\epsilon}(\omega_{2A})\,{}^{30}M_{ecf}^{\epsilon\gamma\zeta}\Upsilon_{fg}^{\zeta\eta}(\omega_A)\,{}^{11}M_{glm}^{\eta} \\
& +{}^{30}M_{acd}^{\alpha\gamma\delta}\Upsilon_{de}^{\delta\epsilon}(\omega_{1A})\,{}^{30}M_{ebf}^{\epsilon\beta\zeta}\Upsilon_{fg}^{\zeta\eta}(\omega_A)\,{}^{11}M_{glm}^{\eta} \\
& +{}^{30}M_{bcd}^{\beta\gamma\delta}\Upsilon_{de}^{\delta\epsilon}(\omega_{12})\,{}^{30}M_{eaf}^{\epsilon\alpha\zeta}\Upsilon_{fg}^{\zeta\eta}(\omega_A)\,{}^{11}M_{glm}^{\eta}\Big) \\
& +\sum_{\delta\epsilon}\big(2\,{}^{30}M_{ace}^{\alpha\gamma\epsilon}\Upsilon_{em}^{\epsilon\delta}(\omega_{1A})\,{}^{20}M_{lb}^{\delta\beta} + 2\,{}^{30}M_{bce}^{\beta\gamma\epsilon}\Upsilon_{em}^{\epsilon\delta}(\omega_{12})\,{}^{20}M_{la}^{\delta\alpha} \\
& +2\,{}^{30}M_{abe}^{\alpha\beta\epsilon}\Upsilon_{em}^{\epsilon\delta}(\omega_{2A})\,{}^{20}M_{lc}^{\delta\gamma} + 2\,{}^{30}M_{bcd}^{\beta\gamma\delta}\Upsilon_{de}^{\delta\epsilon}(\omega_{12})\,{}^{20}M_{el}^{\epsilon\alpha}\delta_{ma} \\
& +2\,{}^{30}M_{cad}^{\gamma\alpha\delta}\Upsilon_{de}^{\delta\epsilon}(\omega_{1A})\,{}^{20}M_{el}^{\epsilon\beta}\delta_{mb} + 2\,{}^{30}M_{abd}^{\alpha\beta\delta}\Upsilon_{de}^{\delta\epsilon}(\omega_{2A})\,{}^{20}M_{el}^{\epsilon\gamma}\delta_{mc} \\
& -{}^{30}M_{bcl}^{\beta\gamma\alpha}\delta_{ma} - {}^{30}M_{cal}^{\gamma\alpha\beta}\delta_{mb} - {}^{30}M_{abl}^{\alpha\beta\gamma}\delta_{mc}\big)\Bigg] - \tfrac{1}{2}\epsilon_0 b_{i\,j\,k}^{\omega_{12}\omega_1\omega_2}\delta_{lm},
\end{aligned} \tag{18.1.20}$$

$$\begin{aligned}
\epsilon_0 e_{i\,j\,k\,l}^{\omega_C\omega_1\omega_2\omega_A} = -\sum_{\substack{\nu\mu\lambda\rho\\ \alpha\beta\gamma\delta}} & q^\nu q^\mu q^\lambda q^\rho\, \Upsilon_{ia}^{\nu\alpha}(\omega_C)\Upsilon_{jb}^{\mu\beta}(\omega_1)\Upsilon_{kc}^{\lambda\gamma}(\omega_2)\Upsilon_{ld}^{\rho\delta}(\omega_A) \\
\times\Bigg[& {}^{40}M_{abcd}^{\alpha\beta\gamma\delta} - \tfrac{3}{2}\sum_{\zeta\epsilon}\big({}^{30}M_{abf}^{\alpha\beta\zeta}\Upsilon_{fe}^{\zeta\epsilon}(\omega_{1A})\,{}^{30}M_{ecd}^{\epsilon\gamma\delta} \\
& +{}^{30}M_{acf}^{\alpha\gamma\zeta}\Upsilon_{fe}^{\zeta\epsilon}(\omega_{2A})\,{}^{30}M_{ebd}^{\epsilon\beta\delta} + {}^{30}M_{adf}^{\alpha\delta\zeta}\Upsilon_{fe}^{\zeta\epsilon}(\omega_{12})\,{}^{30}M_{ebc}^{\epsilon\beta\gamma}\big)\Bigg].
\end{aligned} \tag{18.1.21}$$

The third order optical mixing tensor produces effects that have a variety of names. When $\omega_1 = \omega_2 = \omega_A$, the process is called *third harmonic generation*. When $\omega_A = 0$ and $\omega_1 = \omega_2$, the process is called *electric field induced second harmonic generation* [Terhune, Maker, and Savage, 1962]. When $\omega_2 = \omega_A$ and both are low frequencies, the process is called the *Kerr effect*. If this tensor is used to produce a mixing $\omega_C = \omega + \omega - \omega = \omega$, the result is an intensity dependent refractive index that leads to *self-focusing* of a light beam [Chiao, Garmire, and Townes, 1964]. Four-wave mixing of the form $\omega_C = 2\omega_1 - \omega_2$, called *coherent anti-Stokes Raman scattering*, is a very useful spectroscopic tool to study material resonances at $\omega_1 - \omega_2$.

Since the third order optical mixing tensor is a fourth rank tensor, it exists for all materials. It can be seen to possess permutation symmetry, that is, the interchange of any pair of tensor indices provided the associated frequencies are also interchanged. This is analogous to the symmetry possessed by the second order optical mixing tensor. If there is substantial frequency dispersion between each of the four frequencies, then $e_{i\ j\ k\ l}^{\omega_C \omega_1 \omega_2 \omega_A}$ possesses no interchange symmetry of its tensor indices alone. If any two (or three) frequencies are the same, then there is interchange symmetry between the associated tensor indices. Lastly, the tensor possesses approximate interchange symmetry between any pair of tensor indices if all four frequencies fall in an essentially dispersionless region. This can be called Kleinman symmetry in analogy to the similar situation with the second order optical mixing tensor. Similar remarks could be made concerning the first three tensor indices of the acoustically induced optical mixing tensor. The form that crystal symmetry requires the AIOHG tensor ($\omega_1 = \omega_2$) to take for each crystal class is given in the Appendix.

18.2 Optical Third Harmonic Generation

Let us consider optical third harmonic generation, first observed by Terhune, Maker, and Savage [1962], as an example of the action of the nonlinear driving polarization (18.1.15). In such an interaction

$$\omega_1 = \omega_2 = \omega_A \equiv \omega_O, \tag{18.2.1}$$

$$\omega_{12} = \omega_{1A} = \omega_{2A} = 2\omega_O \equiv \omega_H, \tag{18.2.2}$$

$$\omega_C = 3\omega_O. \tag{18.2.3}$$

Thus $D_2 = 1, D_4 = 2, D_5 = D_6 = 0$. Since a displacement gradient at an optical frequency is negligibly small, the terms containing it in (18.1.15) may be dropped. We then have

$$\mathcal{P}_i^{\omega_C}(\mathbf{z}) = \epsilon_0 e_{i\ j\ k\ l}^{\omega_C \omega_O \omega_O \omega_O} E_j^{\omega_O}(\mathbf{z}) E_k^{\omega_O}(\mathbf{z}) E_l^{\omega_O}(\mathbf{z})$$

$$+ 2\epsilon_0 b_{i\ j\ k}^{\omega_C \omega_H \omega_O} E_j^{\omega_H}(\mathbf{z}) E_k^{\omega_O}(\mathbf{z}). \tag{18.2.4}$$

The driving polarization must be expressed only in terms of the input field

$$\mathbf{E}^{\omega_O}(\mathbf{z}) = \mathbf{e}^{\beta} E_O e^{i\mathbf{k}_O \cdot \mathbf{z}} \tag{18.2.5}$$

where $\mathbf{e}^{\beta}$ is a unit electric field vector of the β eigenmode. Thus $\mathbf{E}^{\omega_H}(\mathbf{z})$ must be found from

$$\left(\frac{c}{\omega_H}\right)^2 [\nabla \times (\nabla \times \mathbf{E}^{\omega_H}(\mathbf{z}))]_i - \kappa_{ij} E_j^{\omega_H}(\mathbf{z}) = b_{i\ j\ k}^{\omega_H \omega_O \omega_O} e_j^{\beta} e_k^{\beta} (E_O)^2 e^{2i\mathbf{k}_O \cdot \mathbf{z}}. \tag{18.2.6}$$

We assume that the generation of the intermediate field $\mathbf{E}^{\omega_H}(\mathbf{z})$ is far from being phase matched, the most likely situation. We return to a discussion of phase matching of intermediate fields (multiple phase matching) in Section 18.5. In generating $\mathbf{E}^{\omega_H}(\mathbf{z})$ far from phase matching we may ignore the free wave of this field and include only the forced wave. This simplification is justified in Section 18.6. A forced wave solution of the wave equation (18.2.6) is

$$\mathbf{E}^{\omega_H}(\mathbf{z}) = \sum_{\alpha=1}^{3} \frac{\boldsymbol{\mathcal{E}}^{\alpha}(\mathbf{s}_D, \omega_H) \mathcal{E}_n^{\alpha}(\mathbf{s}_D, \omega_H) b_{n\ j\ k}^{\omega_H \omega_O \omega_O} e_j^{\beta} e_k^{\beta} (E_O)^2 e^{2i\mathbf{k}_O \cdot \mathbf{z}}}{\dfrac{|2\mathbf{k}_O|^2}{|\mathbf{k}_H^{\alpha}(\mathbf{s}_D)|^2} - 1}, \tag{18.2.7}$$

a result analogous to (13.5.13). Here the unit wavevector $\mathbf{s}_D$ of the driven wave is in the direction of $2\mathbf{k}_O$ and so is just $\mathbf{s}_O$. The wavevector $\mathbf{k}_H^{\alpha}(\mathbf{s}_D)$ is that of the eigenmode α at frequency ω_H and in the direction $\mathbf{s}_D$. For the longitudinal mode ($\alpha = 3$), $\mathbf{k}_H^{\alpha}(\mathbf{s}_D) = \infty$ and so the denominator becomes just -1. When the second harmonic wave is far from being phase matched, all *three* eigenmodes ($\alpha = 1, 2, 3$) can contribute a comparable amount under appropriate conditions of crystal symmetry and field orientations. This illustrates the importance of the three-eigenvector approach to crystal optics introduced in Chapter 9.

The driving polarization is now given by

$$\mathscr{P}_i^{\omega_C}(\mathbf{z})=\epsilon_0\left[e_{i\,j\,k\,l}^{\omega_C\omega_O\omega_O\omega_O}+2b_{i\,m\,k}^{\omega_C\omega_H\omega_O}\sum_{\alpha=1}^{3}\frac{\mathcal{E}_m^{\alpha}\mathcal{E}_n^{\alpha}b_{n\,j\,l}^{\omega_H\omega_O\omega_O}}{\dfrac{|2\mathbf{k}_O|^2}{|\mathbf{k}_H^{\alpha}(\mathbf{s}_D)|^2}-1}\right]$$

$$\times e_j^{\beta}e_k^{\beta}e_l^{\beta}(E_O)^3e^{3i\mathbf{k}_O\cdot\mathbf{z}}$$

$$\equiv \mathscr{P}_i^{\omega_C}e^{3i\mathbf{k}_O\cdot\mathbf{z}} \tag{18.2.8}$$

where the functional dependences shown in the preceding equation are dropped for brevity. We note that third harmonic generation can occur through the direct effect represented by $e_{i\,j\,k\,l}^{\omega_C\omega_O\omega_O\omega_O}$ in any material. The two-step indirect effect that proceeds through second harmonic generation and optical mixing, however, can contribute only in piezoelectric crystals.

We now must solve the wave equation with the driving polarization (18.2.8),

$$\left(\frac{c}{\omega_C}\right)^2\nabla\times(\nabla\times\mathbf{E}^{\omega_C}(\mathbf{z}))-\boldsymbol{\kappa}\cdot\mathbf{E}^{\omega_C}(\mathbf{z})=\frac{\mathscr{P}^{\omega_C}}{\epsilon_0}e^{3i\mathbf{k}_O\cdot\mathbf{z}}. \tag{18.2.9}$$

An equation of this form is solved in Section 13.5 under the approximation that the input field amplitudes are essentially constant and so that result can be used here. In particular the solution near phase matching is approximated in (13.5.40) and the resultant time-averaged Poynting vector is found in (13.5.46). That result is

$$\langle\mathbf{S}_C^{\xi}\rangle=\frac{\omega_C^2z^2|\mathbf{e}^{\xi}\cdot\mathscr{P}^{\omega_C}|^2\Phi(\Delta k_n z/2)\mathbf{t}_C^{\xi}}{8\epsilon_0 cn^{\xi}\cos\delta^{\xi}(\mathbf{n}\cdot\mathbf{t}_C^{\xi})^2}. \tag{18.2.10}$$

Here ξ denotes the output free wave eigenmode that can be phase matched in the geometry considered, $z=\mathbf{n}\cdot\mathbf{z}$ is the direction of growth of the third harmonic wave which is along the normal $\mathbf{n}$ to the input surface, Φ is the phase-matching function defined in (13.5.42), Δk_n is the normal component of phase mismatch given here by

$$\Delta k_n\equiv\Delta\mathbf{k}\cdot\mathbf{n}=(\mathbf{k}_C-3\mathbf{k}_O)\cdot\mathbf{n}, \tag{18.2.11}$$

$\mathbf{t}_C^{\xi}$ is the unit vector in the direction of the group velocity, and δ^{ξ} is the angle between this direction and the output wavevector $\mathbf{k}_C$ direction. If the input time-averaged intensity is denoted by $\langle S_O^\beta \rangle$ and the material interaction coefficient (the effective susceptibility component) e is defined by

$$e \equiv e_i^\xi \left[e_{i\ j\ k\ l}^{\omega_C \omega_O \omega_O \omega_O} + 2 b_{i\ m\ k}^{\omega_C \omega_H \omega_O} \sum_{\alpha=1}^{3} \frac{\mathcal{E}_m^\alpha \mathcal{E}_n^\alpha b_{n\ j\ l}^{\omega_H \omega_O \omega_O}}{\dfrac{|2\mathbf{k}_O|^2}{|\mathbf{k}_H^\alpha(\mathbf{s}_D)|^2} - 1} \right] e_j^\beta e_k^\beta e_l^\beta, \quad (18.2.12)$$

then the output Poynting vector may be expressed as

$$\langle \mathbf{S}_C^\xi \rangle = \left(\frac{\omega_C z e}{\epsilon_0 c^2 \mathbf{n} \cdot \mathbf{t}_C^\xi} \right)^2 \frac{\langle S_O^\beta \rangle^3 \Phi(\Delta k_n z/2) \mathbf{t}_C^\xi}{(n^\beta)^3 n^\xi \cos^3 \delta^\beta \cos \delta^\xi}. \quad (18.2.13)$$

This applies near phase matching and for interaction lengths z such that no appreciable depletion of $\langle S_O^\beta \rangle$ occurs. The dependence on the cube of the input intensity makes focusing the input power into a smaller area very effective. The other dependences are similar to those of other nonlinear optical interactions that are discussed in Chapters 13, 14, and 15.

18.3 Acoustically Induced Optical Harmonic Generation

Acoustically induced optical harmonic generation (AIOHG) is the interaction of an input light wave with itself and with an input acoustic wave to produce an output light wave at a frequency displaced from the optical second harmonic by the acoustic frequency. In this section we consider this interaction [Nelson and Lax, 1971c] near or at (single) phase matching,

$$\mathbf{k}_C = 2\mathbf{k}_O + \mathbf{k}_A. \quad (18.3.1)$$

Double phase matching of the interaction is considered in Section 18.5.

To apply the nonlinear driving polarization (18.1.15) to this process we set

$$\omega_1 = \omega_2 \equiv \omega_O, \quad (18.3.2)$$

$$\omega_{1A} = \omega_{2A} = \omega_O + \omega_A \equiv \omega_B, \quad (18.3.3)$$

$$\omega_{12} = 2\omega_O \equiv \omega_H, \quad (18.3.4)$$

$$\omega_C = 2\omega_O + \omega_A. \quad (18.3.5)$$

Thus $D_1=1$, $D_2=3$, $D_3=1$, $D_4=D_5=2$, and $D_6=0$. Therefore

$$\mathscr{P}_i^{\omega_C}(\mathbf{z})=\epsilon_0\chi_{i\ j\ k\ lm}^{\omega_C\omega_O\omega_O\omega_A}E_j^{\omega_O}(\mathbf{z})E_k^{\omega_O}(\mathbf{z})u_{l,m}^{\omega_A}(\mathbf{z})+3\epsilon_0 e_{i\ j\ k\ l}^{\omega_C\omega_O\omega_O\omega_A}E_j^{\omega_O}(\mathbf{z})E_k^{\omega_O}(\mathbf{z})E_l^{\omega_A}(\mathbf{z})$$
$$+\epsilon_0\chi_{i\ j\ lm}^{\omega_C\omega_H\omega_A}E_j^{\omega_H}(\mathbf{z})u_{l,m}^{\omega_A}(\mathbf{z})+2\epsilon_0 b_{i\ j\ k}^{\omega_C\omega_B\omega_O}E_j^{\omega_B}(\mathbf{z})E_k^{\omega_O}(\mathbf{z}). \tag{18.3.6}$$

This driving polarization must be expressed in terms of the input fields $E_j^{\omega_O}$ and $u_{l,m}^{\omega_A}$. We assume that neither of the electric fields at the intermediate frequencies ω_H and ω_B is near phase matching. Thus they can be represented as forced waves only (see Section 18.6). The electric field at ω_H is given by (18.2.7), that at ω_B is given by an analogous expression obtained from (13.5.13) and (13.1.2),

$$\mathbf{E}^{\omega_B}(\mathbf{z})=\sum_{\xi=1}^{3}\frac{\boldsymbol{\mathcal{E}}^{\xi}(\mathbf{s}_D,\omega_B)\mathcal{E}_n^{\xi}(\mathbf{s}_D,\omega_B)}{\dfrac{|\mathbf{k}_O+\mathbf{k}_A|^2}{|\mathbf{k}_B^{\xi}(\mathbf{s}_D)|^2}-1}\left(\chi_{n\ j\ lm}^{\omega_B\omega_O\omega_A}-\frac{2b_{n\ j\ p}^{\omega_B\omega_O\omega_A}s_p^A s_q^A e_{qlm}}{\epsilon_0\mathbf{s}^A\cdot\boldsymbol{\kappa}(\omega_A)\cdot\mathbf{s}^A}\right)$$
$$\times u_{l,m}^{\omega_A}(\mathbf{z})E_j^{\omega_O}(\mathbf{z}) \tag{18.3.7}$$

where $\mathbf{s}_D$ and $\mathbf{s}^A$ are unit wavevectors in the directions of $\mathbf{k}_O+\mathbf{k}_A$ and $\mathbf{k}_A$ respectively, and the electric field at ω_A is given by (10.7.21) as

$$\mathbf{E}^{\omega_A}(\mathbf{z})=-\frac{\mathbf{s}^A s_n^A e_{nlm}u_{l,m}^{\omega_A}(\mathbf{z})}{\epsilon_0\mathbf{s}^A\cdot\boldsymbol{\kappa}(\omega_A)\cdot\mathbf{s}^A}. \tag{18.3.8}$$

Combining the equations now yields

$$\mathbf{e}^{\eta}\cdot\mathscr{P}^{\omega_C}(\mathbf{z})=i\epsilon_0 k_A\chi_S(E_O)^2Ue^{i(2\mathbf{k}_O+\mathbf{k}_A)\cdot\mathbf{z}} \tag{18.3.9}$$

with the material interaction coefficient χ_S given by

$$\chi_S\equiv e_i^{\eta}e_j^{\alpha}e_k^{\alpha}b_l^{\beta}s_m^A\left[\chi_{i\ j\ k\ lm}^{\omega_C\omega_O\omega_O\omega_A}-\frac{3e_{i\ j\ k\ p}^{\omega_C\omega_O\omega_O\omega_A}s_p^A s_n^A e_{nlm}}{\epsilon_0\mathbf{s}^A\cdot\boldsymbol{\kappa}(\omega_A)\cdot\mathbf{s}^A}\right.$$
$$+\left(\chi_{i\ p\ lm}^{\omega_C\omega_H\omega_A}-\frac{2b_{i\ p\ q}^{\omega_C\omega_H\omega_A}s_q^A s_r^A e_{rlm}}{\epsilon_0\mathbf{s}^A\cdot\boldsymbol{\kappa}(\omega_A)\cdot\mathbf{s}^A}\right)\sum_{\xi=1}^{3}\frac{\mathcal{E}_p^{\xi}(\mathbf{s}_O,\omega_H)\mathcal{E}_n^{\xi}(\mathbf{s}_O,\omega_H)b_{n\ j\ k}^{\omega_H\omega_O\omega_O}}{\dfrac{|2\mathbf{k}_O|^2}{|\mathbf{k}_H^{\xi}(\mathbf{s}_O)|^2}-1}$$
$$\left.+2b_{i\ p\ k}^{\omega_C\omega_B\omega_O}\sum_{\xi=1}^{3}\frac{\mathcal{E}_p^{\xi}(\mathbf{s}_D,\omega_B)\mathcal{E}_n^{\xi}(\mathbf{s}_D,\omega_B)}{\dfrac{|\mathbf{k}_O+\mathbf{k}_A|^2}{|\mathbf{k}_B^{\xi}(\mathbf{s}_D)|^2}-1}\left(\chi_{n\ j\ lm}^{\omega_B\omega_O\omega_A}-\frac{2b_{n\ j\ p}^{\omega_B\omega_O\omega_A}s_p^A s_q^A e_{qlm}}{\epsilon_0\mathbf{s}^A\cdot\boldsymbol{\kappa}(\omega_A)\cdot\mathbf{s}^A}\right)\right] \tag{18.3.10}$$

and with the amplitudes of the input optical electric field and input acoustic displacement given by E_O and U. Here $\mathbf{e}^\alpha$ and $\mathbf{e}^\eta$ are unit input and output electric field vectors respectively and $\mathbf{b}^\beta$ is a unit acoustic displacement vector. The eigenmodes of the three waves are denoted by α, η, and β. The output time-averaged Poynting vector for singly phase matched AIOHG can now be obtained by combining (18.2.10) and (18.3.9) with the result

$$\langle \mathbf{S}_C^\eta \rangle = \left(\frac{\omega_C z \chi_S}{\mathbf{n}\cdot\mathbf{t}_C^\eta} \right)^2 \frac{\cos\delta^\beta \langle S_O^\alpha \rangle^2 \langle S_A^\beta \rangle \Phi(\Delta k_n z/2)\mathbf{t}_C^\eta}{\epsilon_0 c^3 \rho^0 v_\beta^3 n^\eta \cos\delta^\eta (n^\alpha)^2 \cos^2\delta^\alpha} \tag{18.3.11}$$

where the normal component of phase mismatch Δk_n is defined by

$$\Delta k_n \equiv \Delta\mathbf{k}\cdot\mathbf{n} = (\mathbf{k}_C - 2\mathbf{k}_O - \mathbf{k}_A)\cdot\mathbf{n}, \tag{18.3.12}$$

$\mathbf{n}$ being the inward unit normal to the input surface.

The material interaction coefficient χ_S is the effective susceptibility governing the strength of AIOHG. It is interestingly complex, consisting of the direct effect, three two-step indirect effects, and two three-step indirect effects.

The direct effect is represented by the fifth rank AIOHG susceptibility. It can couple to both strain and rotation in a manner analogous to the elastooptic tensor. Its coupling to rotation is discussed in the next section.

The two-step indirect effects are of three types. In the first one the two optical electric fields interact through third order optical mixing with the electric field generated through the piezoelectric effect by the input acoustic wave. In the second one the two optical electric fields generate a second harmonic wave by way of second order optical mixing; this second harmonic wave then mixes with the input acoustic wave by way of acoustooptic diffraction. The latter process can couple to both the strain and the rotation of the acoustic wave. In the third two-step indirect effect these two processes are reversed. First, the input optical electric field mixes with the acoustic wave by way of acoustooptic diffraction; then the resultant light wave at frequency ω_B combines with the input light wave through optical mixing.

The two three-step indirect effects also involve the same interaction tensors but acting in different order. In one the electrooptic effect combines a low frequency electric field generated from the acoustic wave by the piezoelectric effect with an electric field of a light wave produced by second harmonic generation. In the other three-step effect the electrooptic effect combines an input optical electric field with a low frequency electric field generated by the piezoelectric effect from the input acoustic wave; the resultant electric field at frequency ω_B then undergoes optical mixing with the input optical electric field.

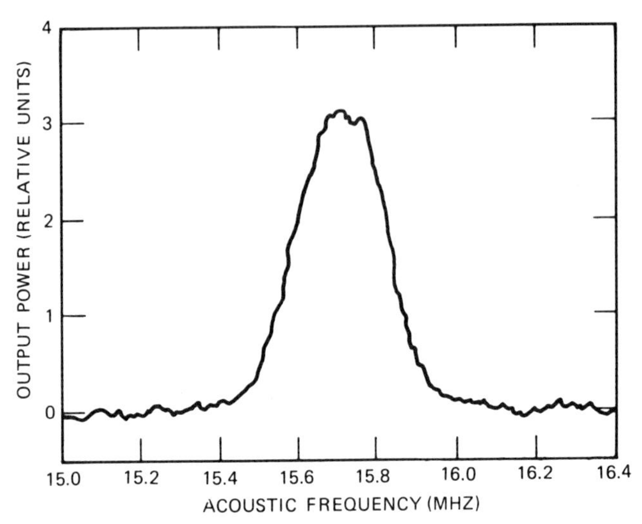

Fig. 18.1. Single phase matching of AIOHG output near 5.3 μm wavelength as seen by varying the acoustic wave frequency [reprinted by permission from Boyd et al., 1970].

As with all indirect effects, those here are physically distinguishable from the direct interaction by their differing dependences on frequency, on wavevectors, and on their coupling or lack of coupling to rotations. Note that either of two denominators involving wavevectors in indirect effect terms in (18.3.10) could approach zero and so make one of those terms dominant. This observation, in fact, led to the idea of double phase matching of AIOHG [Nelson and Lax, 1971a], since the wavevector relation needed to make either denominator vanish is the phase-matching condition for the intermediate optical wave at either ω_H or ω_B. Double phase matching is explored further in Section 18.5. In the present section we specifically exclude the double phase-matching case.

AIOHG was first observed in GaAs crystals with a laser beam at 10.6 μm wavelength and an acoustic shear wave at 15.7 MHz frequency [Boyd et al., 1970]. Collinear propagation of the beams along the [001] direction was employed, since no ordinary second harmonic generation can occur in this orientation. The AIOHG process, however, is allowed for this orientation. When the acoustic frequency was adjusted to produce phase matching, an output wave at 5.3 μm was detected as shown in Fig. 18.1.

18.4 Coupling to Rotation

It is shown in Chapter 13 that the rotational part of a displacement gradient produces a contribution to the elastooptic effect by rotating the dielectric tensor. By analogy we can expect a rotation to produce a contribution to AIOHG by rotating the second order optical mixing tensor. We can verify this surmise by obtaining the part of the AIOHG susceptibility that is antisymmetric upon interchange of its last two (elastic) indices.

From (18.1.20) we find

$$\begin{aligned}
\epsilon_0 \chi_{i\ j\ k\ [lm]}^{\omega_C \omega_1 \omega_2 \omega_A} = {} & \frac{\epsilon_0}{2}\,\delta_{i[l}\, b_{m]j\ k}^{\omega_{12}\omega_1\omega_2} + \frac{\epsilon_0}{2}\,\delta_{j[l}\, b_{m]k\ i}^{\omega_{1A}\omega_2\omega_C} + \frac{\epsilon_0}{2}\,\delta_{k[l}\, b_{m]i\ j}^{\omega_{2A}\omega_C\omega_1} \\
& + \tfrac{3}{4}\left(\omega_C^2 - \omega_{12}^2\right) \sum_{\substack{\nu\mu\lambda \\ \beta\gamma\delta\epsilon}} q^\nu q^\mu q^\lambda \Upsilon_{jb}^{\mu\beta}(\omega_1) \Upsilon_{kc}^{\lambda\gamma}(\omega_2)\, {}^{30}M_{bce}^{\beta\gamma\epsilon} \Upsilon_{e\ [l}^{\epsilon\delta}(\omega_{12})\, m^\delta \Upsilon_{m]i}^{\delta\nu}(\omega_C) \\
& - \tfrac{3}{4}\left(\omega_{1A}^2 - \omega_1^2\right) \sum_{\substack{\nu\mu\lambda \\ \alpha\gamma\delta\epsilon}} q^\nu q^\mu q^\lambda \Upsilon_{ia}^{\nu\alpha}(\omega_C) \Upsilon_{kc}^{\lambda\gamma}(\omega_2)\, {}^{30}M_{ace}^{\alpha\gamma\epsilon} \Upsilon_{e\ [l}^{\epsilon\delta}(\omega_{1A})\, m^\delta \Upsilon_{m]j}^{\delta\mu}(\omega_1) \\
& - \tfrac{3}{4}\left(\omega_{2A}^2 - \omega_2^2\right) \sum_{\substack{\nu\mu\lambda \\ \alpha\beta\delta\epsilon}} q^\nu q^\mu q^\lambda \Upsilon_{ia}^{\nu\alpha}(\omega_C) \Upsilon_{jb}^{\mu\beta}(\omega_1)\, {}^{30}M_{abe}^{\alpha\beta\epsilon} \Upsilon_{e\ [l}^{\epsilon\delta}(\omega_{2A})\, m^\delta \Upsilon_{m]k}^{\delta\lambda}(\omega_2).
\end{aligned} \tag{18.4.1}$$

The last three terms have coefficients that are the differences of the squares of two optical frequencies and can be shown to be smaller than the first three terms in the ratio of the acoustic frequency to one of the optical frequencies. They thus may be dropped with the result

$$\chi_{i\ j\ k\ [lm]}^{\omega_C \omega_1 \omega_2 \omega_A} = \tfrac{1}{2}\,\delta_{i[l}\, b_{m]j\ k}^{\omega_{12}\omega_1\omega_2} + \tfrac{1}{2}\,\delta_{j[l}\, b_{m]k\ i}^{\omega_{1A}\omega_2\omega_C} + \tfrac{1}{2}\,\delta_{k[l}\, b_{m]i\ j}^{\omega_{2A}\omega_C\omega_1}. \tag{18.4.2}$$

If the frequency dependence of the terms is ignored, the right side can be shown to be just one half of the change in the second order optical mixing tensor b_{ijk} arising from an infinitesimal rotation. The factor of one half enters because the AIOHG tensor on the left side is defined for one (of two) Fourier components arising from the finite frequency of the rotation. We leave the proof of this assertion for an exercise.

The antisymmetric part of the AIOHG tensor that couples to rotation can be expected to have a magnitude comparable to the symmetric part of the AIOHG tensor that couples to strain. Equation (18.4.2) predicts, for example,

$$\chi_{122[23]} = \tfrac{1}{2} b_{123} = +0.65 \times 10^{-10}\ m/V \tag{18.4.3}$$

for GaAs in the infrared. Measurement of the entire component that couples to both strain and rotation [Boyd et al., 1970] yielded for this situation

$$\chi_{12223} = (+0.8 \text{ or } +1.1) \times 10^{-10}\, m/V, \tag{18.4.4}$$

showing that the coupling to rotation in this case constituted either 60 or 80% of the total coupling. The two possibilities result from the measurement only of the absolute value of the effective AIOHG susceptibility that includes both the calculable indirect and the unknown direct contributions.

18.5 Doubly Phase Matched Interaction

Double phase matching of AIOHG is the phase matching of either of two three-wave processes that combine to produce a two-step indirect contribution to AIOHG. Two successive phase matchings guarantee the phase matching of the overall AIOHG interaction. The indirect effect that is doubly phase matched is much larger than the direct effect and the other indirect effects that are only singly phase matched. The susceptibilities of the latter may thus be dropped from the nonlinear polarization.

As can be seen from the possible vanishing of denominators in the effective susceptibility χ_S (18.3.10) for single phase matching, there are two possibilities for double phase matching. One corresponds to phase matching acoustooptic diffraction first,

$$\mathbf{k}_B = \mathbf{k}_O + \mathbf{k}_A, \tag{18.5.1}$$

and then optical mixing,

$$\mathbf{k}_C = \mathbf{k}_B + \mathbf{k}_O, \tag{18.5.2}$$

where the subscripts have the meaning introduced in Section 18.3. Combining these two equations,

$$\mathbf{k}_C = 2\mathbf{k}_O + \mathbf{k}_A, \tag{18.5.3}$$

shows that they imply phase matching of the overall AIOHG process. The other possibility is to phase match optical second harmonic generation first,

$$\mathbf{k}_H = 2\mathbf{k}_O, \tag{18.5.4}$$

and then acoustooptic diffraction,

$$\mathbf{k}_C = \mathbf{k}_H + \mathbf{k}_A, \tag{18.5.5}$$

which in combination also imply overall phase matching (18.5.3). Note that the piezoelectric step that enters some of the indirect contributions cannot be phase matched because as shown in Section 10.7 the piezoelectrically created electric field is a longitudinal forced wave and so has no free wave analogue.

Let us consider the first example of double phase matching mentioned above. We then need to consider only the last term in the nonlinear polarization in (18.3.6),

$$\mathscr{P}_i^{\omega_C}(\mathbf{z}) = 2\epsilon_0 b_{i\,j\,k}^{\omega_C\omega_B\omega_O} E_j^{\omega_B}(\mathbf{z}) E_k^{\omega_O}(\mathbf{z}). \tag{18.5.6}$$

We consider the wave generated by this polarization at frequency ω_C as well as the intermediate wave at frequency ω_B to be close to or at phase matching. This means we must include free waves as well as forced waves at these frequencies. The electric field of the intermediate acoustooptic diffracted wave near phase matching is given in Section 13.5. Combining (13.5.40) and (13.5.36) gives this field as

$$\mathbf{E}^{\omega_B}(\mathbf{z}) = \frac{\boldsymbol{\mathcal{E}}^{\xi}(\mathbf{s}_D, \omega_B)\boldsymbol{\mathcal{E}}^{\xi}(\mathbf{s}_D, \omega_B)\cdot \mathscr{P}^{\omega_B}\left(e^{i(\mathbf{k}_O+\mathbf{k}_A)\cdot\mathbf{z}} - e^{i\mathbf{k}_B^{\xi}\cdot\mathbf{z}}\right)}{\epsilon_0\left(\dfrac{|\mathbf{k}_O+\mathbf{k}_A|^2}{|\mathbf{k}_B^{\xi}(\mathbf{s}_D)|^2} - 1\right)} \tag{18.5.7}$$

where $\boldsymbol{\mathcal{E}}^{\xi}$ is the electric field eigenvector and $\mathbf{k}_B^{\xi}$ is the wavevector of the ξ eigenmode, which is phase matchable, $\mathbf{s}_D$ is a unit vector in the direction of $\mathbf{k}_O + \mathbf{k}_A$, and $\mathscr{P}^{\omega_B}$ is the nonlinear driving polarization for acoustooptic diffraction given by (13.1.2). With the use of the latter equation the nonlinear polarization becomes

$$\mathscr{P}^{\omega_C}(\mathbf{z}) = \frac{\mathbf{P}^{\omega_C}}{\dfrac{|\mathbf{k}_O+\mathbf{k}_A|^2}{|\mathbf{k}_B^{\xi}(\mathbf{s}_D)|^2} - 1}\left(e^{i(2\mathbf{k}_O+\mathbf{k}_A)\cdot\mathbf{z}} - e^{i(\mathbf{k}_B^{\xi}+\mathbf{k}_O)\cdot\mathbf{z}}\right) \tag{18.5.8}$$

where the coefficient is given by

$$P_i^{\omega_C} \equiv 2ik_A\epsilon_0 b_{i\,r\,k}^{\omega_C\omega_B\omega_O}\mathcal{E}_r^{\xi}(\mathbf{s}_D, \omega_B)\mathcal{E}_n^{\xi}(\mathbf{s}_D, \omega_B)\Bigg(\chi_{n\,j\,lm}^{\omega_B\omega_O\omega_A}$$

$$-\frac{2b_{n\,j\,p}^{\omega_B\omega_O\omega_A}s_p^A s_q^A e_{qlm}}{\epsilon_0 \mathbf{s}^A\cdot\kappa(\omega_A)\cdot\mathbf{s}^A}\Bigg)b_l^{\beta}s_m^A e_j^{\alpha}e_k^{\alpha}(E_0)^2 U. \tag{18.5.9}$$

The two input fields are expressed as

$$\mathbf{E}^{\omega_O}(\mathbf{z}) = E_O \mathbf{e}^{\alpha} e^{i\mathbf{k}_O \cdot \mathbf{z}}, \tag{18.5.10}$$

$$u_{l,m}^{\omega_A}(\mathbf{z}) = ik_A U b_l^{\beta} s_m^A e^{i\mathbf{k}_A \cdot \mathbf{z}} \tag{18.5.11}$$

with $\mathbf{e}^{\alpha}$, $\mathbf{b}^{\beta}$, $\mathbf{s}^A$ being unit vectors of the α optical eigenmode, the β acoustic eigenmode, and the acoustic propagation direction respectively.

In Section 13.5 it is shown that near phase matching only the free wave and the component of the forced wave that is in the phase matchable mode need be considered. Thus we may take the output electric field solution to be

$$\mathbf{E}^{\omega_C}(\mathbf{z}) = C^{\eta} \boldsymbol{\mathcal{E}}^{\eta}(\mathbf{s}_C, \omega_C) e^{i\mathbf{k}_C^{\eta} \cdot \mathbf{z}} + \frac{1}{\dfrac{|\mathbf{k}_O + \mathbf{k}_A|^2}{|\mathbf{k}_B^{\xi}(\mathbf{s}_D)|^2} - 1} \left\{ \frac{\boldsymbol{\mathcal{E}}^{\eta}(\mathbf{s}_F, \omega_C) \boldsymbol{\mathcal{E}}^{\eta}(\mathbf{s}_F, \omega_C) \cdot \mathbf{P}^{\omega_C} e^{i(2\mathbf{k}_O + \mathbf{k}_A) \cdot \mathbf{z}}}{\dfrac{|2\mathbf{k}_O + \mathbf{k}_A|^2}{|\mathbf{k}_C^{\eta}(\mathbf{s}_F)|^2} - 1} - \frac{\boldsymbol{\mathcal{E}}^{\eta}(\mathbf{s}_G, \omega_C) \boldsymbol{\mathcal{E}}^{\eta}(\mathbf{s}_G, \omega_C) \cdot \mathbf{P}^{\omega_C} e^{i(\mathbf{k}_B^{\xi} + \mathbf{k}_O) \cdot \mathbf{z}}}{\dfrac{|\mathbf{k}_B^{\xi} + \mathbf{k}_O|^2}{|\mathbf{k}_C^{\eta}(\mathbf{s}_G)|^2} - 1} \right\}. \tag{18.5.12}$$

Here η denotes the optical mode at frequency ω_C that is phase matchable, $\mathbf{k}_C^{\eta}$ is its wavevector, C^{η} is an integration constant, and $\mathbf{s}_F$ and $\mathbf{s}_G$ are unit propagation vectors in the directions of $2\mathbf{k}_O + \mathbf{k}_A$ and $\mathbf{k}_B^{\xi} + \mathbf{k}_O$ respectively.

It is also shown in Section 13.5 that near phase matching an approximate boundary condition may be used that requires the component of the output electric field in the η mode to vanish on the input plane surface. If we put the origin of coordinates in this plane, denote the coordinate vector to any point in the plane by $\mathbf{z}_p$, and denote the inward surface normal by $\mathbf{n}$, then the plane is described by

$$\mathbf{n} \cdot \mathbf{z}_p = 0 \tag{18.5.13}$$

and the approximate boundary condition can be expressed as

$$\boldsymbol{\mathfrak{D}}^{\eta}(\mathbf{s}_C, \omega_C) \cdot \mathbf{E}^{\omega_C}(\mathbf{z}_p) = 0. \tag{18.5.14}$$

In applying this condition we may ignore the slight difference between $\mathbf{s}_C$,

$\mathbf{s}_F$, and $\mathbf{s}_G$ in the arguments of $\mathcal{E}^\eta$ when using the biorthonormality of the $\mathcal{D}^\eta$ and $\mathcal{E}^\eta$ vectors (9.3.9). Satisfaction of the boundary condition at all points on the surface requires the tangential components of the three wavevectors to be equal,

$$(\mathbf{k}_C^\eta)_t = (2\mathbf{k}_O + \mathbf{k}_A)_t = (\mathbf{k}_B^\xi + \mathbf{k}_O)_t, \tag{18.5.15}$$

and the integration constant C^η to be given by

$$C^\eta = -\frac{\mathcal{E}^\eta\cdot\mathbf{P}^{\omega_C}}{\epsilon_0\left(\dfrac{|\mathbf{k}_O+\mathbf{k}_A|^2}{|\mathbf{k}_B^\xi(\mathbf{s}_D)|^2}-1\right)}\left[\frac{1}{\dfrac{|2\mathbf{k}_O+\mathbf{k}_A|^2}{|\mathbf{k}_C^\eta(\mathbf{s}_F)|^2}-1}-\frac{1}{\dfrac{|\mathbf{k}_B^\xi+\mathbf{k}_O|^2}{|\mathbf{k}_C^\eta(\mathbf{s}_G)|^2}-1}\right]. \tag{18.5.16}$$

The output electric field can now be obtained by substituting into it the expression for C^η and by reexpressing the three denominators in C^η by (13.5.36). The result is

$$\mathbf{E}^{\omega_C}(\mathbf{z}) = -\frac{i\mathcal{E}^\eta\,\mathcal{E}^\eta\cdot\mathbf{P}^{\omega_C}\omega_B\omega_C n_B^\xi n_C^\eta\cos\delta^\xi\cos\delta^\eta}{2\epsilon_0 c^2\Delta k_{Bn}(\mathbf{n}\cdot\mathbf{t}_B^\xi)(\mathbf{n}\cdot\mathbf{t}_C^\eta)}$$
$$\times\left(\frac{\sin(\Delta k_{Cn}z/2)}{\Delta k_{Cn}}e^{i(\mathbf{k}_C^\eta+2\mathbf{k}_O+k_A)\cdot\mathbf{z}/2} - \frac{\sin(\Delta k_{Dn}z/2)}{\Delta k_{Dn}}e^{i(\mathbf{k}_C^\eta+\mathbf{k}_B^\xi+\mathbf{k}_O)\cdot\mathbf{z}/2}\right) \tag{18.5.17}$$

where the phase mismatches are defined by

$$\Delta\mathbf{k}_B \equiv \mathbf{k}_B^\xi - \mathbf{k}_O - \mathbf{k}_A, \tag{18.5.18}$$

$$\Delta\mathbf{k}_D \equiv \mathbf{k}_C^\eta - \mathbf{k}_B^\xi - \mathbf{k}_O, \tag{18.5.19}$$

$$\Delta\mathbf{k}_C \equiv \mathbf{k}_C^\eta - 2\mathbf{k}_O - \mathbf{k}_A = \Delta\mathbf{k}_B + \Delta\mathbf{k}_D \tag{18.5.20}$$

and the subscript n on them indicates the component normal to the input surface, the tangential component having vanished by (18.5.15). The scalar coordinate z is defined as $\mathbf{n}\cdot\mathbf{z}$. The electric field may be more conveniently expressed in terms of the effective susceptibility χ_D of doubly phase matched AIOHG defined by

$$\chi_D \equiv 2e_i^\eta b_i^{\omega_C}{}_r^{\omega_B}{}_k^{\omega_O} e_r^\xi e_n^\xi\left(\chi_n^{\omega_B}{}_j^{\omega_O}{}_{lm}^{\omega_A} - \frac{2b_n^{\omega_B}{}_j^{\omega_O}{}_p^{\omega_A} S_p^A s_q^A e_{qlm}}{\epsilon_0\mathbf{s}^A\cdot\boldsymbol{\kappa}(\omega_A)\cdot\mathbf{s}^A}\right)e_j^\alpha e_k^\alpha b_l^\beta s_m^A. \tag{18.5.21}$$

The result is

$$\mathbf{E}^{\omega_C}(\mathbf{z}) = -\frac{\mathbf{e}^{\eta} k_A \omega_B \omega_C z^2 \chi_D (E_O)^2 U}{8c^2 n_C^{\eta} \cos\delta^{\eta} n_B^{\xi} \cos\delta^{\xi} (\mathbf{n}\cdot\mathbf{t}_B^{\xi})(\mathbf{n}\cdot\mathbf{t}_C^{\eta})}$$
$$\times \frac{1}{\Delta k_{Bn} z/2}\left[\Phi^{1/2}\left(\frac{\Delta k_{Dn} z}{2}\right) e^{i\Delta k_{Bn} z/4} - \Phi^{1/2}\left(\frac{\Delta k_{Cn} z}{2}\right) e^{-i\Delta k_{Bn} z/4}\right] e^{i\mathbf{k}_C^{\eta}\cdot\mathbf{z}} \qquad (18.5.22)$$

where Φ is the single phase-matching function defined in (13.5.42). In finding the magnetic field that accompanies this electric field the spatial dependence brought about by the phase mismatch may be ignored and only the spatial dependence of the propagating factor need be considered.

The time-averaged Poynting vector of the output wave is now easily found to be

$$\langle \mathbf{S}_C^{\eta} \rangle = \frac{(\omega_B \omega_C z^2 \chi_D)^2 \cos\delta^{\beta} \langle S_O^{\alpha} \rangle^2 \langle S_A^{\beta} \rangle \Psi(\Delta k_{Bn} z/2, \Delta k_{Dn} z/2) \mathbf{t}_C^{\eta}}{16 c^5 \epsilon_0 \rho^0 v_{\beta}^3 n_C^{\eta} \cos\delta^{\eta} (n_B^{\xi} \cos\delta^{\xi})^2 (n_O^{\alpha} \cos\delta^{\alpha})^2 (\mathbf{n}\cdot\mathbf{t}_B^{\xi})^2 (\mathbf{n}\cdot\mathbf{t}_C^{\eta})^2} \qquad (18.5.23)$$

where $\mathbf{t}_C^{\eta}$ is a unit vector in the direction of energy propagation and Ψ is the double phase-matching function defined by

$$\Psi(\theta,\varphi) \equiv \frac{1}{\theta^2}\left[\Phi(\varphi) + \Phi(\theta+\varphi) - 2\Phi^{1/2}(\varphi)\Phi^{1/2}(\theta+\varphi)\cos\theta\right] \qquad (18.5.24)$$

where θ, φ, and $\theta+\varphi$ are respectively the phase mismatches of the first interaction, the second interaction, and the overall two-step interaction. We have normalized this function so that

$$\Psi(0,0) = 1. \qquad (18.5.25)$$

The most important result of the Poynting vector expression is its dependence on z^4, where z is the interaction distance measured normal to the input surface. This dependence on z is a signature of a doubly phase matched interaction. By comparison the Poynting vector has a z^2 dependence for a singly phase matched interaction. In both cases the dependence persists only for distances into the nonlinear medium for which input wave depletion is negligible. The faster growth of doubly phase matched AIOHG compared to singly phase matched AIOHG indicates its greater conversion efficiency. In fact, the Poynting vector for double phase matching is $(\omega_B z/4c)^2$ times larger than that for the direct interaction (assuming $\chi_S \approx \chi_D$) and $(\Delta k_{B_n} z/2)^2$ times larger than that for the corresponding two-step indirect interaction when single phase matching is employed. For an interaction length of 1 mm and a wavelength of 1 μm each of these factors can be of order 10^6.

18.6 Double Phase-Matching Function

The form of the double phase-matching function (18.5.24) deserves further examination. Cursory inspection of $\Psi(\theta,\varphi)$ seems to indicate that its dependences on θ and φ are different and that its only invariance is under the transformation $\theta \to -\theta$, $\varphi \to -\varphi$. However, the use of several trigonometric identities allows $\Psi(\theta,\varphi)$ to be reexpressed in two other forms,

$$\Psi(\theta,\varphi) = \frac{1}{\varphi^2}\left[\Phi(\theta) + \Phi(\theta+\varphi) - 2\Phi^{1/2}(\theta)\Phi^{1/2}(\theta+\varphi)\cos\varphi\right] \quad (18.6.1a)$$

$$= \frac{1}{(\theta+\varphi)^2}\left[\Phi(\theta) + \Phi(\varphi) - 2\Phi^{1/2}(\theta)\Phi^{1/2}(\varphi)\cos(\theta+\varphi)\right]. \quad (18.6.1b)$$

Invariance to the transformations $\theta \to \varphi$, $\varphi \to \theta$ and $\theta \to -\varphi$, $\varphi \to -\theta$ is now also apparent. Equations (18.5.24) and (18.6.1) together show a symmetrical dependence on θ, φ, and $\theta+\varphi$, the phase mismatches of the first step, the second step, and the overall interaction respectively. Thus if only one exact phase matching can be accomplished, there is no preference between the three possibilities.

Let us next examine $\Psi(\theta,\varphi)$ under the three special conditions: (1) phase matching of the first step, $\theta = 0$, $\varphi = \eta$; (2) phase matching of the second step, $\varphi = 0$, $\theta = \eta$; (3) phase matching of the overall interaction, $\theta = -\varphi = \eta$. In each case we find the same result,

$$\Psi = \frac{1}{\eta^2}\left[1 + \Phi(\eta) - 2\Phi^{1/2}(\eta)\cos\eta\right]. \quad (18.6.2)$$

Note also that in none of these cases does Ψ reduce to the single phase-matching function Φ.

A geometric interpretation of the double phase-matching function is suggested by the numerators of the expressions (18.5.24) and (18.6.1),

$$a^2 \equiv \Phi(\varphi) + \Phi(\theta+\varphi) - 2\Phi^{1/2}(\varphi)\Phi^{1/2}(\theta+\varphi)\cos\theta, \qquad (18.6.3\text{a})$$

$$b^2 \equiv \Phi(\theta) + \Phi(\theta+\varphi) - 2\Phi^{1/2}(\theta)\Phi^{1/2}(\theta+\varphi)\cos\varphi, \qquad (18.6.3\text{b})$$

$$c^2 \equiv \Phi(\theta) + \Phi(\varphi) - 2\Phi^{1/2}(\theta)\Phi^{1/2}(\varphi)\cos(\theta+\varphi), \qquad (18.6.3\text{c})$$

each of which has the form of a cosine law of a (different) triangle. In terms of these quantities the double phase-matching function is given by

$$\Psi(\theta,\varphi) = \left(\frac{a}{\theta}\right)^2 = \left(\frac{b}{\varphi}\right)^2 = \left(\frac{c}{\theta+\varphi}\right)^2. \qquad (18.6.4)$$

Figure 18.2 depicts the three triangles implied by (18.6.3). Note that two of the triangles are right triangles and are enclosed by the third triangle. An interpretation of the double phase-matching function corresponding to each of the expressions in (18.6.4) can be obtained from the associated triangle. For example, consider the enclosing triangle. The lengths of the lower and upper sides of that triangle represent the strengths of the electric fields of the first and second steps due to the phase-matching effects in the individual steps. Their enclosed angle, $\theta+\varphi$, is the sum of the phase mismatches of the two steps. The side c opposite this angle when divided by the angle $\theta+\varphi$ represents the strength of the output electric field of the two-step interaction due to phase-matching effects of the entire process. Analogous interpretations hold for the other two triangles. The upper enclosed triangle corresponds to the electric field (18.5.17), since the field contains the factor $\Phi^{1/2}(\theta+\varphi) - \Phi^{1/2}(\varphi)$ exp $i\theta$ which shows that it is the difference of two components having a relative phase angle of θ.

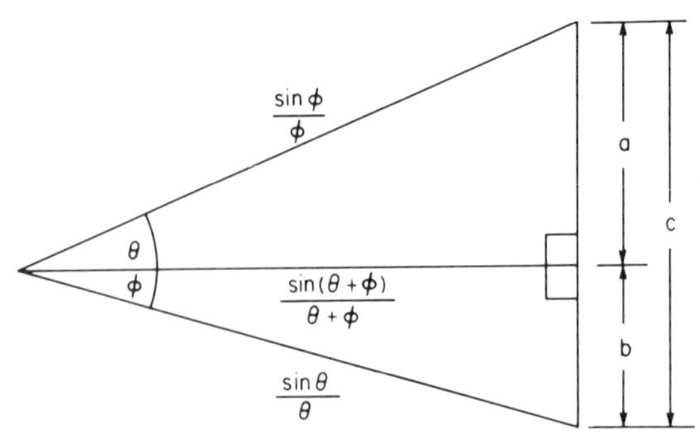

Fig. 18.2. Geometric interpretation of double phase matching in a two-step interaction [reprinted from Nelson, 1978d].

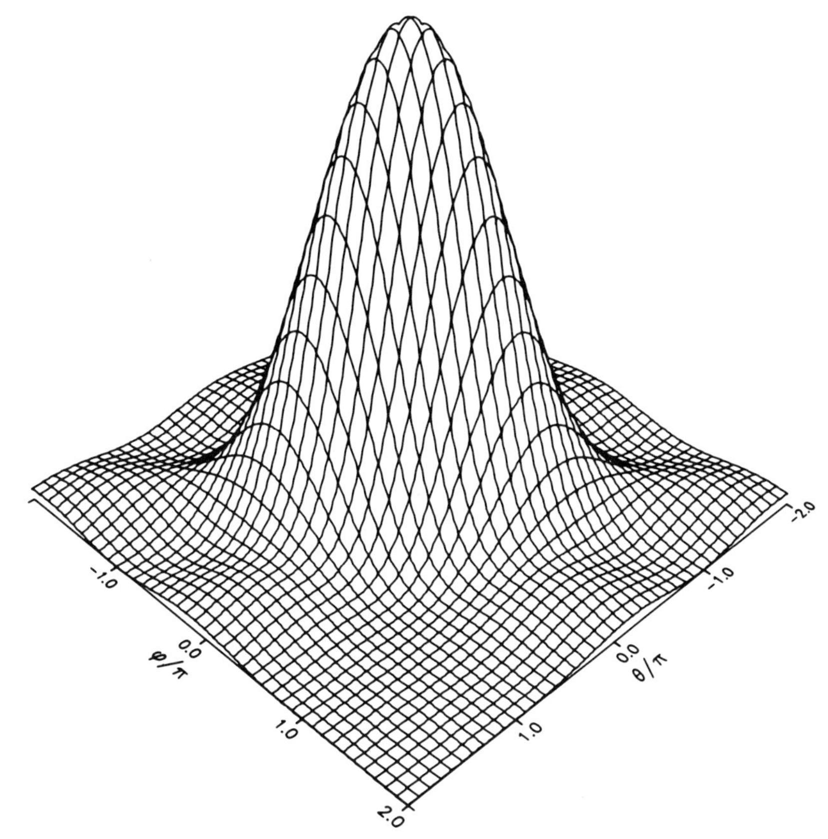

Fig. 18.3. Double phase-matching function $\Psi(\theta,\phi)$ versus θ and ϕ [reprinted from Nelson, 1978d].

A perspective view of the double phase-matching function is shown in Fig. 18.3. Exact phase matching of either of the two steps is represented by the values of the function along either axis, θ or φ. Phase matching of the overall interaction is represented by the values along the $\theta=-\varphi$ axis. Note that Ψ has no zeroes along any of these three axes except at $\pm\infty$. This is in sharp contrast to the single phase-matching function $\Phi(\varphi)$ (see Fig. 13.3) which has zeroes at $n\pi$ where n is a nonzero integer. The function $\Psi(\theta,\varphi)$ does have zeroes at

$$\theta=m\pi \qquad \text{and} \qquad \varphi=n\pi \tag{18.6.5a}$$

provided that the integers n and m satisfy

$$m\neq 0, n\neq 0 \qquad \text{and} \qquad m+n\neq 0. \tag{18.6.5b}$$

It is interesting to examine the double phase-matching function $\Psi(\theta,\varphi)$ in the limit called single phase-matching in Section 18.3. This limit corresponds to regarding $\theta+\varphi$, the mismatch of the overall interaction, as small in magnitude compared to the magnitude of either θ or φ. It is easy to see from either (18.5.24) or (18.6.1a) that in this limit

$$\Psi(\theta,\varphi)\rightarrow\frac{1}{\theta^2}\Phi(\theta+\varphi) \tag{18.6.6}$$

except very near $\Phi(\theta+\varphi)=0$ where the deleted term of order θ^{-3} contributes. If this is combined with the double phase-matching result (18.5.23), we obtain the single phase-matching result (18.3.11) provided, of course, that we drop all susceptibility terms from the latter except the two-step interaction considered in this section. This agreement justifies the approximation used in Section 18.3 of ignoring the free wave of the intermediate field that is generated by an interaction far from phase matching.

18.7 Multiple Phase Matching

It is apparent from the discussion of double phase matching of a four-wave interaction in the preceding two sections that the concept can be extended to multiple phase matching of higher order interactions. In general, n phase matchings of an n-step contribution to an $(n+2)$-wave interaction are possible. Such multiple phase matchings should allow higher order interactions to generate waves about as intense as result from three-wave interactions. The output wave from an interaction employing n phase matchings should have an initial growth proportional to the $2n$ power of the interaction length.

Triple phase matching of a collinear five-wave acoustooptic interaction in $LiNbO_3$ has been observed [Nelson and Mikulyak, 1972]. In this experiment an input extraordinary light wave at $2\pi c/\omega_O=1.06$ μm wavelength and a $\omega_A/2\pi=291$ MHz ultrasonic shear wave with its displacement along the X axis were propagated along the Y axis. These generated an ordinary light wave at the sum frequency ω_B by way of noncritically phase matched collinear acoustooptic diffraction (see Fig. 13.8). This light wave then interacted with itself to produce an optical second harmonic light wave of frequency ω_C and extraordinary polarization by a noncritically phase-matched collinear interaction (see Fig. 15.8). Thus the frequency conditions were

$$\omega_B=\omega_O+\omega_A, \tag{18.7.1}$$

used twice, and

$$\omega_C = 2\omega_B = 2\omega_O + 2\omega_A. \tag{18.7.2}$$

The latter equation shows the five-wave character of the interaction. The phase-matching conditions were

$$\mathbf{k}_B^o = \mathbf{k}_O^e + \mathbf{k}_A, \tag{18.7.3}$$

used twice, and

$$\mathbf{k}_C^e = 2\mathbf{k}_B^o = 2\mathbf{k}_O^e + 2\mathbf{k}_A. \tag{18.7.4}$$

Here o and e denote the ordinary and extraordinary propagation modes. Figure 18.4 shows the phase-matching curve versus ultrasonic frequency found in this experiment. The width of the curve was determined by the spread of wavevectors caused by focusing the input light beam and was about 10 times wider than a plane wave analysis, such as that presented above, would give.

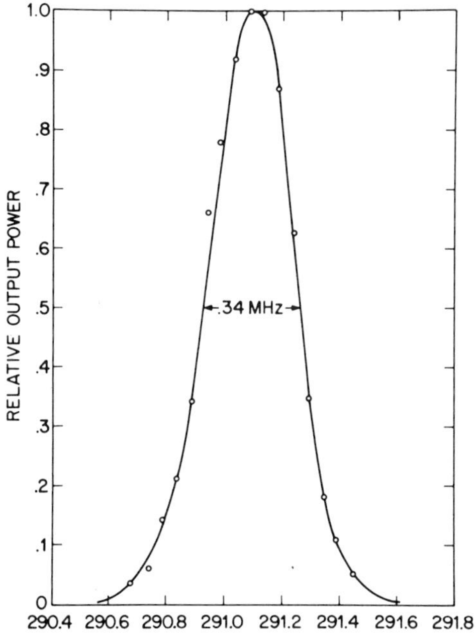

Fig. 18.4. Output power from a triply phase matched five-wave acoustooptic interaction versus the acoustic wave frequency [reprinted by permission from Nelson and Mikulyak, 1972].

PROBLEMS

18.1. Show that the change in the optical mixing tensor induced by a static body rotation is twice the right side of (18.4.2).

18.2. Derive an expression for the Poynting vector of a doubly phase matched third harmonic output wave analogous to the singly phase matched form in (18.2.13).

18.3. Show that (18.5.24) can be put into the form of (18.6.1b).

18.4. Consider the case that all the phase mismatch in both steps of the interaction considered in Section 18.5 is caused by the input optical wave, that is, $\Delta k_{Bn} = \Delta k_{Dn} = \Delta k_{On}$. Find the form of the double phase-matching function. Does it have zeroes?

18.5. Show that (18.6.6) in conjunction with the Poynting vector expression (18.5.23) agrees with the part of the singly phase matched Poynting vector (18.3.11) arising from the same two-step interaction.

18.6. Show that the third order optical mixing tensor may be reexpressed with the use of the normal coordinates defined in (9.P.7) to (9.P.10) as

$$e_{i\ j\ k\ l}^{\omega_C \omega_1 \omega_2 \omega_A} = \frac{1}{6\epsilon_0} \sum_{klmn} \frac{c_i^k c_j^l c_k^m c_l^n}{(\Omega_k^2 - \omega_C^2)(\Omega_l^2 - \omega_1^2)(\Omega_m^2 - \omega_2^2)(\Omega_n^2 - \omega_A^2)}$$
$$\times \left\{ \sum_p \left[\frac{{}^3R^{kpl\,3}R^{pmn}}{\Omega_p^2 - \omega_{2A}^2} + \frac{{}^3R^{kpm\,3}R^{pln}}{\Omega_p^2 - \omega_{1A}^2} + \frac{{}^3R^{kpn\,3}R^{pml}}{\Omega_p^2 - \Omega_{12}^2} \right] + {}^4R^{klmn} \right\} \tag{18.P.1}$$

where the frequency notation of (18.1.5) to (18.1.8) is used and

$${}^3R^{klm} \equiv -3 \sum_{\nu\mu\lambda} \frac{{}^{30}M_{pqr}^{\nu\mu\lambda} i_p^{k\nu} i_q^{l\mu} i_r^{m\lambda}}{(m^\nu m^\mu m^\lambda)^{1/2}}, \tag{18.P.2}$$

$${}^4R^{klmn} \equiv -6 \sum_{\nu\mu\lambda\rho} \frac{{}^{40}M_{pqrs}^{\nu\mu\lambda\rho} i_p^{k\nu} i_q^{l\mu} i_r^{m\lambda} i_s^{n\rho}}{(m^\nu m^\mu m^\lambda m^\rho)^{1/2}}. \tag{18.P.3}$$

18.7. The *second order Raman interaction* is the interaction between two light waves and two optic modes of a medium. It can be observed by *second order Raman scattering* in which an output light wave is observed at a frequency displaced from the frequency of an input light wave by the sum of the frequencies of two optic modes. The direct second order Raman interaction with the m and n optic modes can be characterized by a nonlinear polarization

$$\mathcal{P}_i^{\omega_C} = \epsilon_0 \chi_{i\ j}^{\omega_C \omega_1 mn} E_j^{\omega_1} \eta_{\omega_2}^m \eta_{\omega_A}^n \tag{18.P.4}$$

where η^m is a normal coordinate [see (9.P.7) to (9.P.10)] and the frequency notation of (18.1.5) to (18.1.8) is used. Show that the direct *second order Raman susceptibility* $\chi_{i\ j}^{\omega_C \omega_1 mn}$ is given by

$$\chi_{i\ j}^{\omega_C \omega_1 mn} \equiv \frac{1}{2\epsilon_0} \sum_{kl} \frac{c_i^k c_j^l}{(\Omega_k^2 - \omega_C^2)(\Omega_l^2 - \omega_1^2)} \left\{ \sum_p \left[\frac{{}^3R^{kpl\,3}R^{pmn}}{\Omega_p^2 - \omega_{2A}^2} + \frac{{}^3R^{kpm\,3}R^{pln}}{\Omega_p^2 - \omega_{1A}^2} + \frac{{}^3R^{kpn\,3}R^{pml}}{\Omega_p^2 - \omega_1^2} \right] + {}^4R^{klmn} \right\} \tag{18.P.5}$$

where the notation of Prob. 18.6 is used and the m and n optic modes include both ionic and electronic modes.

18.8. Show that the third order optical mixing tensor and the second order Raman susceptibility of the m and n optic modes are related by

$$3 e_{i\ j\ k\ l}^{\omega_C \omega_1 \omega_2 \omega_A} = \sum_{mn} \frac{\chi_{i\ j}^{\omega_C \omega_1 mn} c_k^m c_l^n}{(\Omega_m^2 - \omega_2^2)(\Omega_n^2 - \omega_A^2)} \tag{18.P.6}$$

where the frequency notation of (18.1.5) to (18.1.8) is used and the summation includes both ionic and electronic modes. What can be said from this relation about the optic modes that contribute to third order optical mixing?

Appendix

Crystal point group symmetry requires some components of various interaction tensors to vanish and requires relations between other components. We record here in matrix notation (see Section 10.5) the various tensors encountered in the text. They are listed under the 7 crystal systems (plus isotropic for even rank tensors) and 32 point groups (crystal classes). The latter are denoted by their symbol in the Hermann-Mauguin (International) notation. Translation between this and the Schoenflies notation is given in Table 1 along with whether or not the class is pyroelectric and, if so, the allowed direction of the spontaneous polarization $\mathbf{P}^S$. The notation in some of the tables is patterned after Nye [1957]. The orientation of the rectangular crystallographic axes to which the tensor components are referred conforms with the *Standards on Piezoelectric Crystals* [Bond et al., 1949].

The following notation applies to Tables 2 to 7:

- · Zero component.
- ● Nonzero component.
- ●—● Equal components.
- ●—○ Components equal in magnitude, opposite in sign.

Other special notation is given in the individual tables.

TABLE 1

Crystal System	Crystal Class		Pyroelectricity
	Hermann-Mauguin	Schoenflies	
Triclinic	1	C_1	$\mathbf{P}^S=[P_x^S, P_y^S, P_z^S]$
	$\bar{1}$	$C_i = S_2$	No
Monoclinic	2	C_2	$\mathbf{P}^S=[0, P_y^S, 0]$
	m	$C_s = C_{1h}$	$\mathbf{P}^S=[P_x^S, 0, P_z^S]$
	$2/m$	C_{2h}	No

TABLE 1 (*Continued*)

Crystal System	Crystal Class		Pyroelectricity
	Hermann-Mauguin	Schoenflies	
Orthorhombic	222	$D_2 = V$	No
	$mm2$	C_{2v}	$\mathbf{P}^S = [0,0,P_z^S]$
	mmm	$D_{2h} = V_h$	No
Tetragonal	4	C_4	$\mathbf{P}^S = [0,0,P_z^S]$
	$\bar{4}$	S_4	No
	$4/m$	C_{4h}	No
	422	D_4	No
	$4mm$	C_{4v}	$\mathbf{P}^S = [0,0,P_z^S]$
	$\bar{4}2m$	$D_{2d} = V_d$	No
	$4/mmm$	D_{4h}	No
Trigonal	3	C_3	$\mathbf{P}^S = [0,0,P_z^S]$
	$\bar{3}$	$C_{3i} = S_6$	No
	32	D_3	No
	$3m$	C_{3v}	$\mathbf{P}^S = [0,0,P_z^S]$
	$\bar{3}m$	D_{3d}	No
Hexagonal	6	C_6	$\mathbf{P}^S = [0,0,P_z^S]$
	$\bar{6}$	C_{3h}	No
	$6/m$	C_{6h}	No
	622	D_6	No
	$6mm$	C_{6v}	$\mathbf{P}^S = [0,0,P_z^S]$
	$\bar{6}m^2$	D_{3h}	No
	$6/mmm$	D_{6h}	No
Cubic	23	T	No
	$m3$	T_h	No
	432	O	No
	$\bar{4}3m$	T_d	No
	$m3m$	O_h	No

TABLE 2
Symmetric Second Rank Tensor

The dielectric tensor ($\kappa_{ij} = \kappa_{ji}$) and the linear electric susceptibility ($\chi_{ij} = \chi_{ji}$) are two examples.

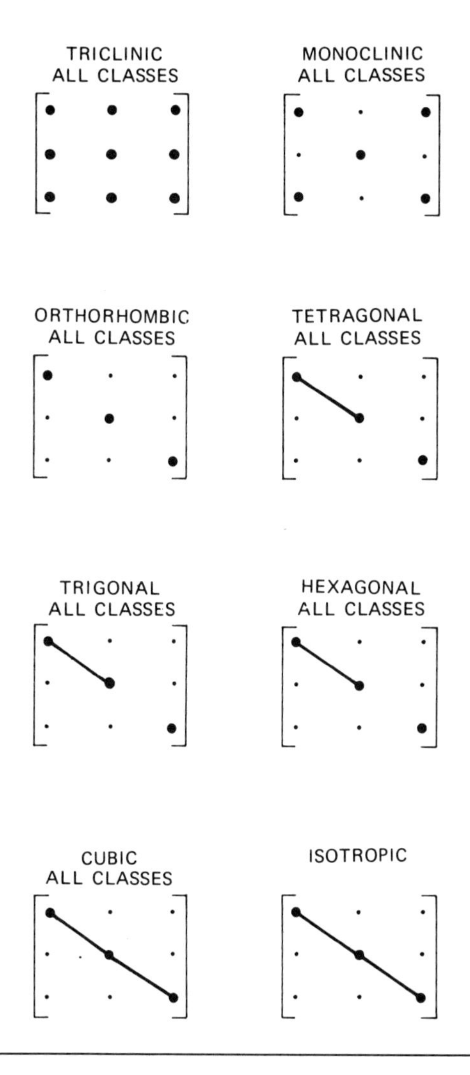

TABLE 3
Partially Symmetric Third Rank Tensor

The piezoelectric stress tensor ($e_{ijk}=e_{ikj}$), the piezoelectric strain tensor ($d_{ijk}=d_{ikj}$), the electrooptic susceptibility tensor ($b_i^{\omega_1}{}_j^{\omega_1}{}_k^{0}=b_j^{\omega_1}{}_i^{\omega_1}{}_k^{0}$), the Pockels strain-free electrooptic tensor ($r_{ijk}^S=r_{jik}^S$), and the optical second harmonic generation tensor $b_i^{\omega_H}{}_j^{\omega_O}{}_k^{\omega_O}$ are examples.

The pair of symmetric indices in each case is contracted by way of the association (10.5.7). This allows presentation of a 3×6 array of components in the table. For e_{in}, d_{in}, and $b_i^{\omega_H}{}_n^{\omega_O\omega_O}$ the second index is six-dimensioned; for $b_n^{\omega_1\omega_1}{}_k^{0}$ and r_{nk}^S the first index is six-dimensioned. Among these tensors only d_{in} incorporates factors of 2 in its contracted notation definition (10.5.11). When matrix multiplication is used in (15.1.1) for harmonic generation,

$$\mathscr{P}_i=\epsilon_0 b_{im}E_m^2, \tag{A.1}$$

the six-dimensioned quantity E_m^2 representing the product of the two input electric fields is defined by

$$
\begin{aligned}
E_jE_k &= E_m^2 \qquad (m=1,2,3)\\
&= \tfrac{1}{2}E_m^2 \qquad (m=4,5,6)
\end{aligned} \tag{A.2}
$$

with the use of the relation (10.5.7).

The following notation is used in Table 3:

◉ A component that is twice the solid dot component joined to it for the piezoelectric strain tensor d_{in} or is equal to the solid dot component joined to it for the other tensors listed above.

Since third rank tensors vanish for all centrosymmetric crystal classes, only acentric crystal classes are listed. The interchange symmetry prevents acentric class 432 from having any nonzero piezoelectric tensor components.

The X axis in the $\bar{6}m2$ group is chosen along a twofold rotation axis in accordance with Table 1 of Bond and co-workers, [1949]. Some workers have written the matrix for this class in a form that assumes that the Y axis is chosen along a twofold rotation axis while still stating that they are in conformance with the standard.

TRICLINIC

CLASS 1

$$\begin{bmatrix} \bullet & \bullet & \bullet & \bullet & \bullet & \bullet \\ \bullet & \bullet & \bullet & \bullet & \bullet & \bullet \\ \bullet & \bullet & \bullet & \bullet & \bullet & \bullet \end{bmatrix}$$

TABLE 3 (*Continued*)

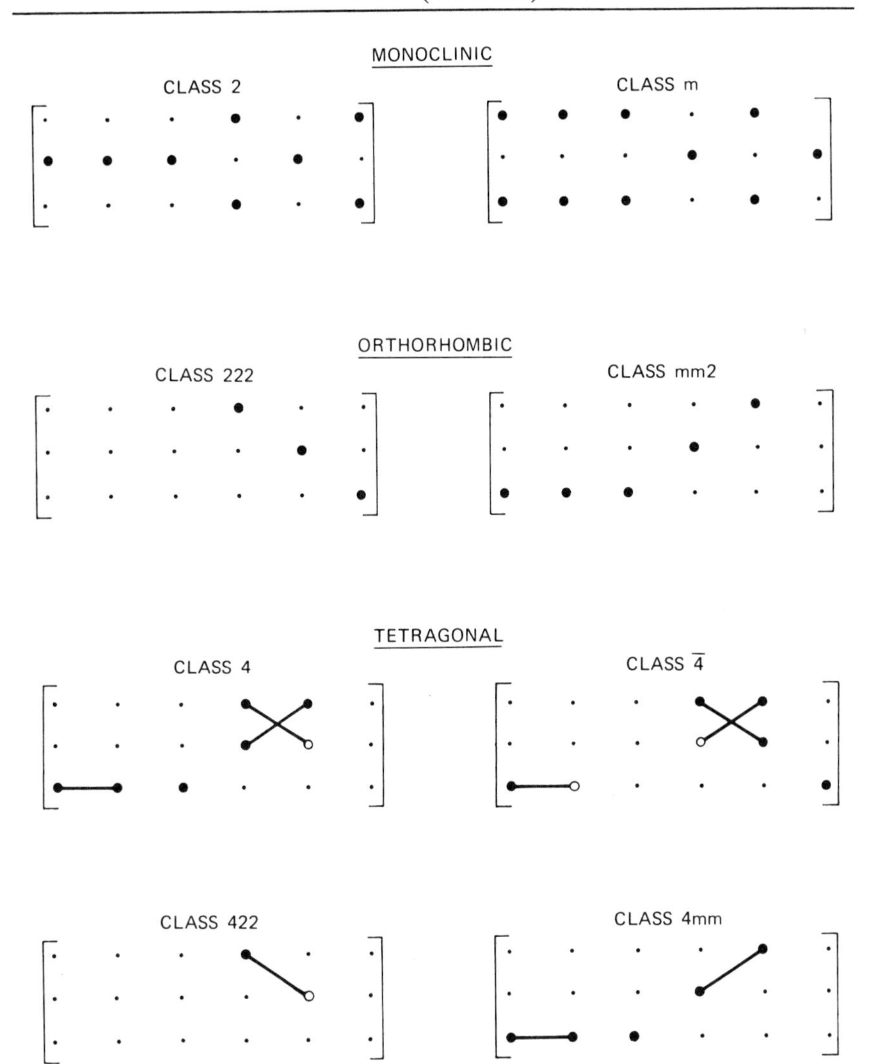

TABLE 3 (*Continued*)

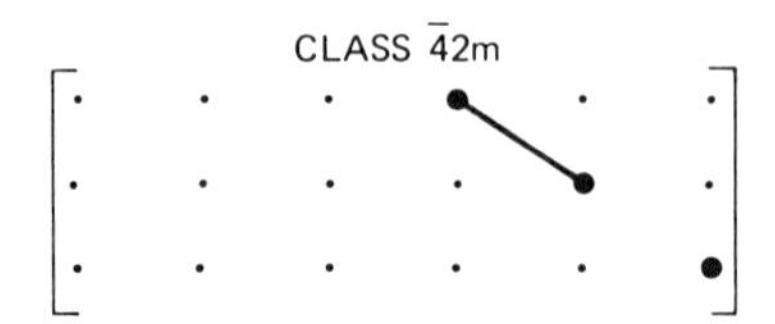

TRIGONAL

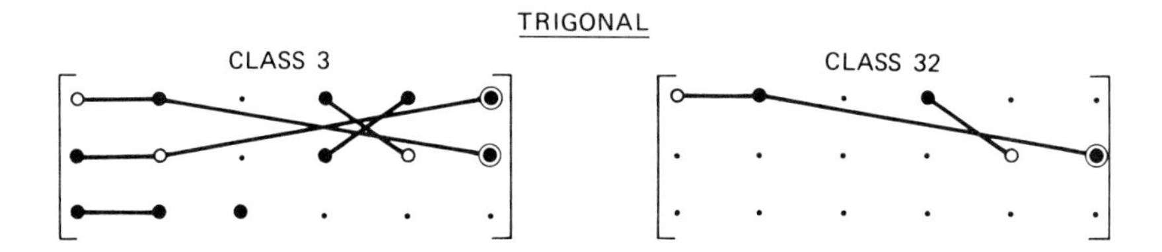

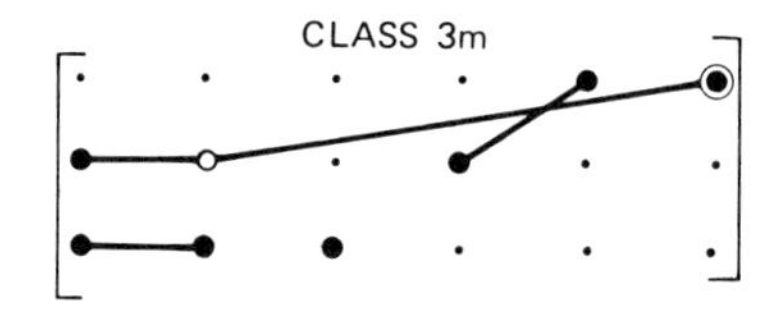

HEXAGONAL

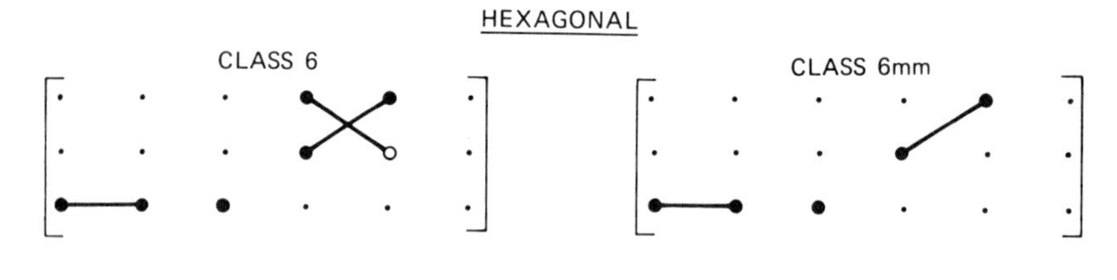

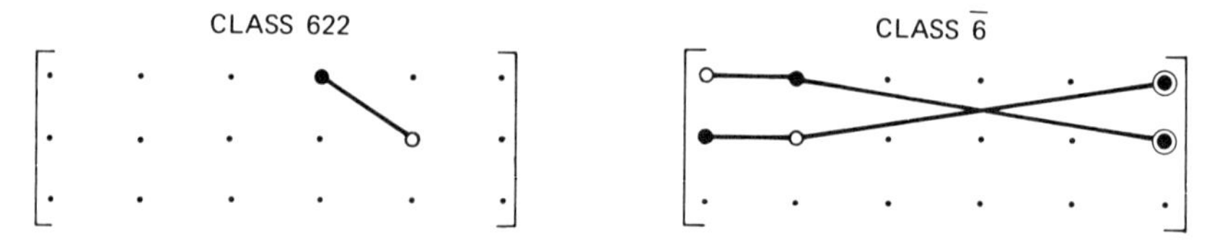

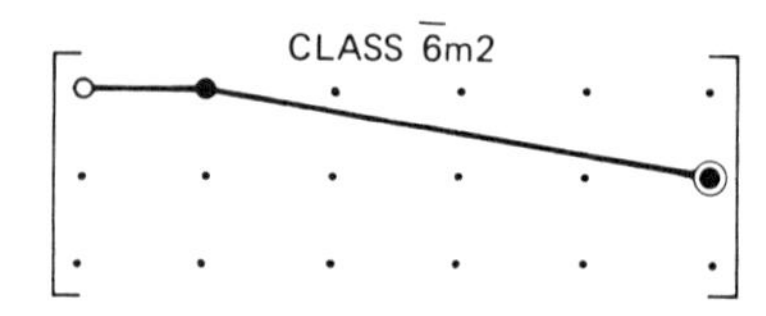

TABLE 3 (*Continued*)

CUBIC

CLASS 432

$$\begin{bmatrix} \cdot & \cdot & \cdot & \cdot & \cdot & \cdot \\ \cdot & \cdot & \cdot & \cdot & \cdot & \cdot \\ \cdot & \cdot & \cdot & \cdot & \cdot & \cdot \end{bmatrix}$$

CLASSES 23 and $\bar{4}3m$

$$\begin{bmatrix} \cdot & \cdot & \cdot & \bullet & \cdot & \cdot \\ \cdot & \cdot & \cdot & \cdot & \bullet & \cdot \\ \cdot & \cdot & \cdot & \cdot & \cdot & \bullet \end{bmatrix}$$

(the three filled dots are joined by a line)

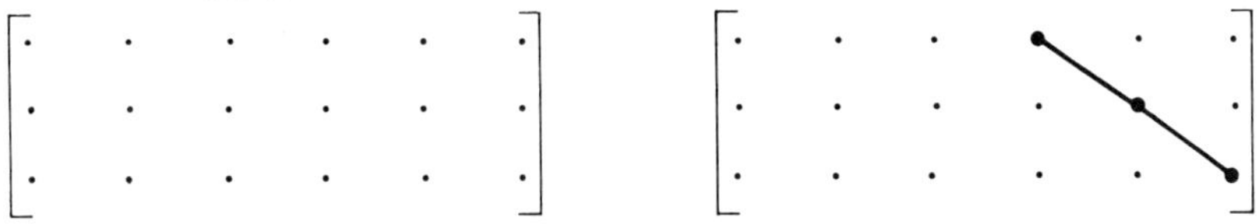

TABLE 4
Partially Antisymmetric Third Rank Tensor

The optical mixing tensor $b_{i\,j\,k}^{\omega_S\omega_1\omega_2}$ has no interchange symmetry when ω_S, ω_1, and ω_2 are all sufficiently separated that significant frequency dispersion occurs between them. Thus it may be divided into symmetric and antisymmetric parts,

$$b_{i\,j\,k}^{\omega_S\omega_1\omega_2} = b_{i\,(jk)}^{\omega_S\omega_1\omega_2} + b_{i\,[jk]}^{\omega_S\omega_1\omega_2}. \tag{A.3}$$

We may contract the notation for an antisymmetric tensor in the following manner [note the difference with relation (10.5.7)]. We associate values of j, k, and m by

$$\begin{matrix} j,k \rightarrow & 1,1 & 2,2 & 3,3 & 2,3 & 3,1 & 1,2 \\ m \rightarrow & 1 & 2 & 3 & 4 & 5 & 6 \end{matrix}. \tag{A.4}$$

An antisymmetric tensor is then contracted by

$${}^{A}T_m = T_{[jk]} = \tfrac{1}{2}(T_{jk} - T_{kj}). \tag{A.5}$$

Note that this convention may be applied to a symmetric tensor by

$${}^{S}T_m = T_{(jk)} = \tfrac{1}{2}(T_{jk} + T_{kj}) \tag{A.6}$$

with the same result as obtained by the convention (10.5.7). We then have

$$T_{jk} \equiv T_{(jk)} + T_{[jk]} = {}^{S}T_m + {}^{A}T_m \equiv T_m. \tag{A.7}$$

The symmetric part of the third rank tensor

$${}^{S}b_{im} = b_{i\,(jk)}^{\omega_S\omega_1\omega_2} \tag{A.8}$$

is listed in Table 3; we list below the antisymmetric part

$${}^{A}b_{im} = b_{i\,[jk]}^{\omega_S\omega_1\omega_2} \tag{A.9}$$

TABLE 4 (*Continued*)

for the acentric crystal classes [Giordmaine, 1965]. $^{A}b_{im}$ vanishes for centrosymmetric crystal classes.

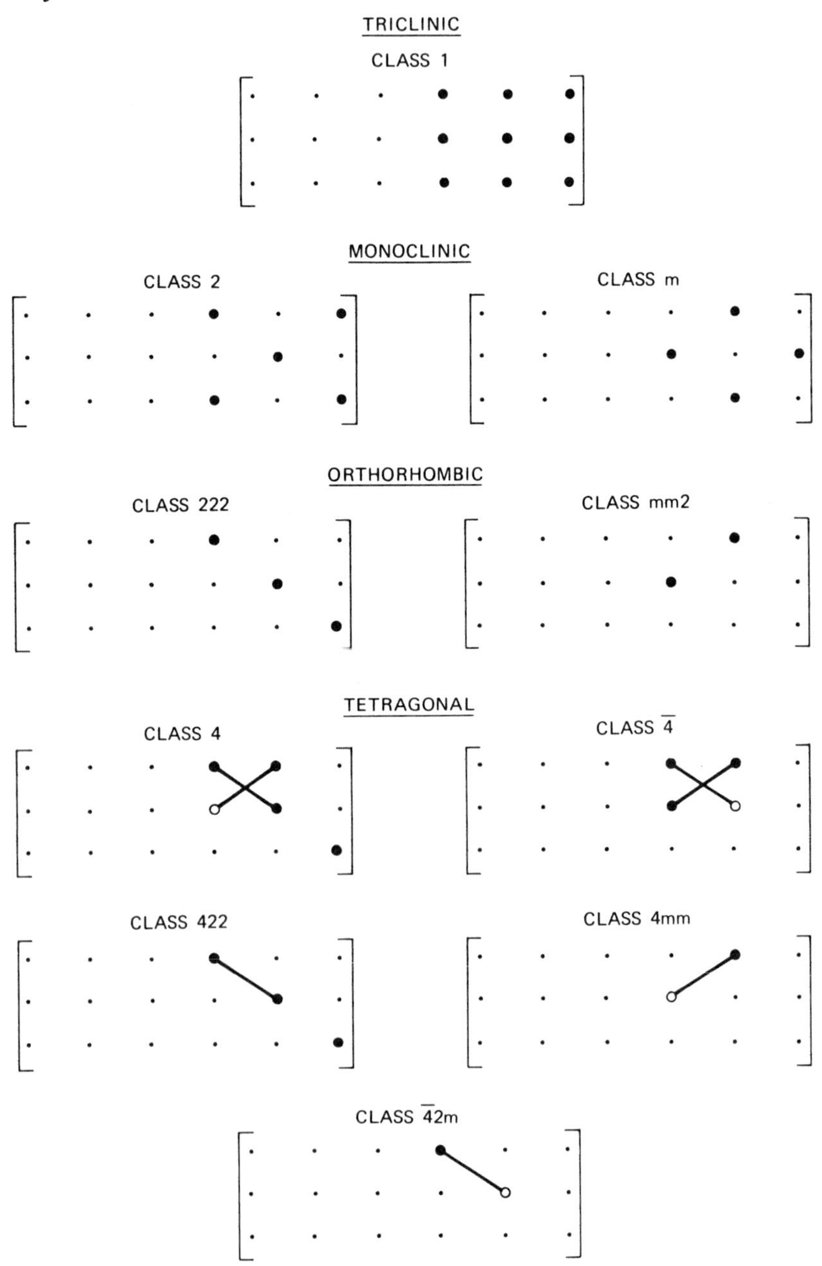

TABLE 4 (*Continued*)

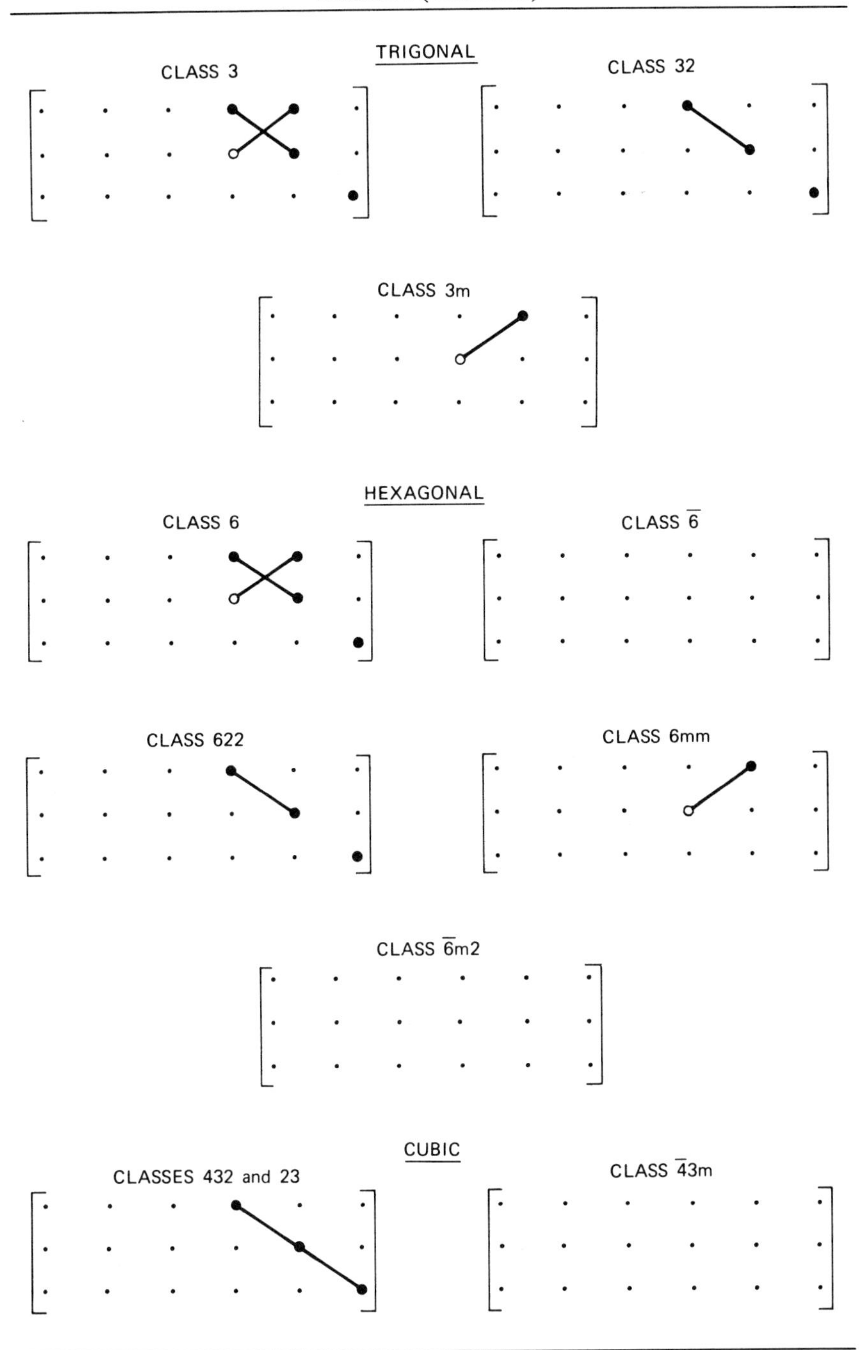

TABLE 5
Fully Symmetric Third Rank Tensor

The optical mixing tensor is fully symmetric (Kleinman symmetry),

$$b_{i\,j\,k}^{\omega_s\omega_1\omega_2} = b_{i\,k\,j}^{\omega_s\omega_1\omega_2} = b_{j\,k\,i}^{\omega_s\omega_1\omega_2} = b_{j\,i\,k}^{\omega_s\omega_1\omega_2} = b_{k\,i\,j}^{\omega_s\omega_1\omega_2} = b_{k\,j\,i}^{\omega_s\omega_1\omega_2}, \tag{A.10}$$

when all three frequencies ω_1, ω_2, $\omega_3 = \omega_1 + \omega_2$ are in a dispersionless region. The second and third indices are contracted into a six-dimensional index as in Table 3.

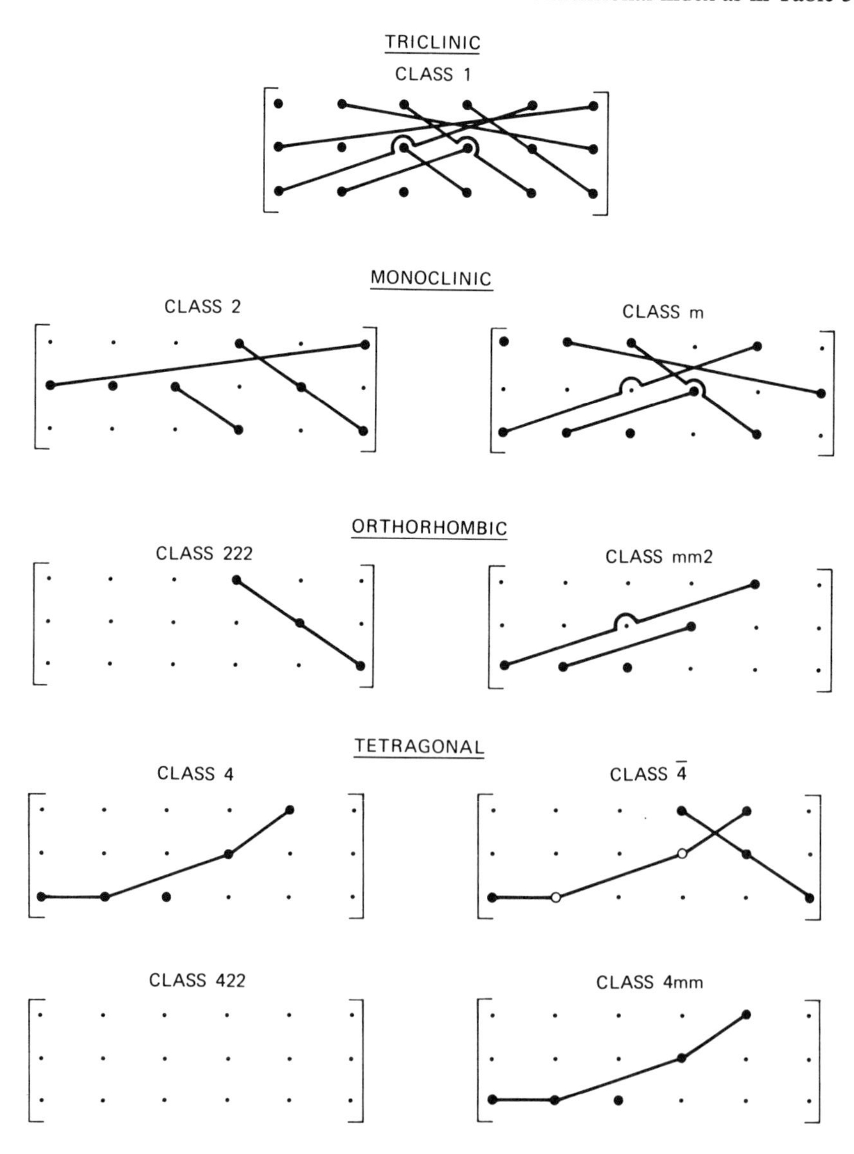

TABLE 5 (*Continued*)

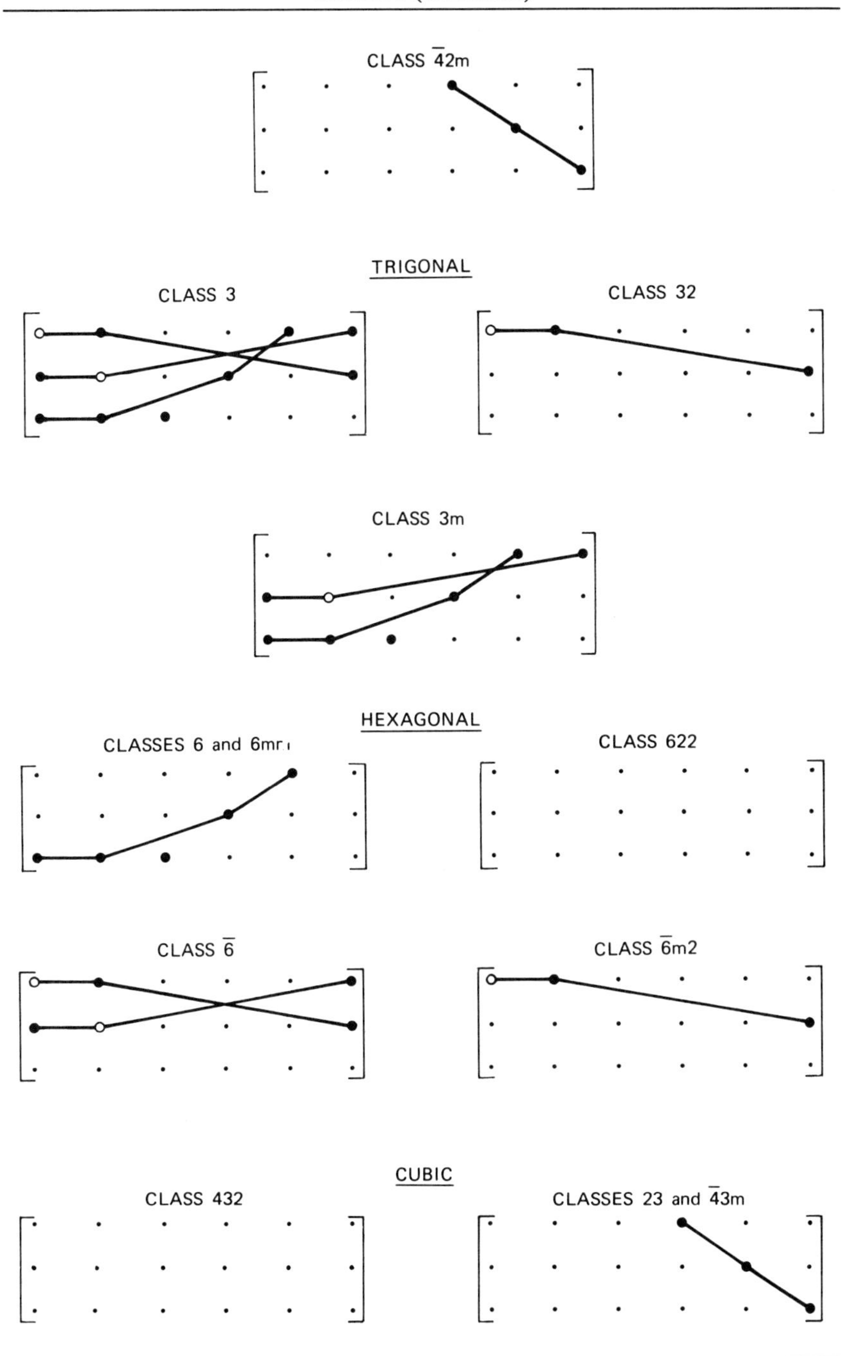

TABLE 6
Fourth Rank Elasticity Tensors

The elastic stiffness tensor c_{ijkl} and the elastic compliance tensor s_{ijkl} possess the same interchange symmetry

$$c_{ijkl} = c_{(ij)(kl)} = c_{(kl)(ij)}, \tag{A.11}$$

$$s_{ijkl} = s_{(ij)(kl)} = s_{(kl)(ij)}. \tag{A.12}$$

These tensors may then be represented in contracted notation by 6×6 arrays that are symmetric about the leading diagonal. The contracted notation for the two tensors differs by some factors of 2 as seen in (10.5.12) and (10.5.13) necessitating the following notation:

◉ A component that is twice the solid dot component joined to it for s_{mn} or is equal to the solid dot component joined to it for c_{mn}.

× A compliance component $= 2(s_{11} - s_{12})$ or a stiffness component $= \frac{1}{2}(c_{11} - c_{12})$.

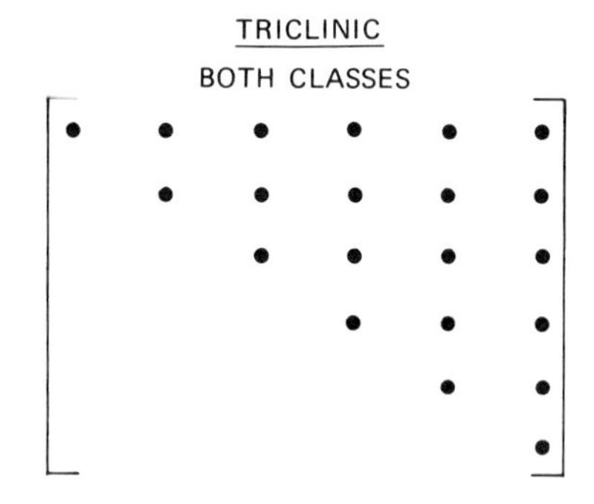

TABLE 6 (*Continued*)

MONOCLINIC

ALL CLASSES

ORTHORHOMBIC

ALL CLASSES

TETRAGONAL

CLASSES 4, $\bar{4}$, 4/m

CLASSES 4mm, $\bar{4}$2m, 422, 4/mmm

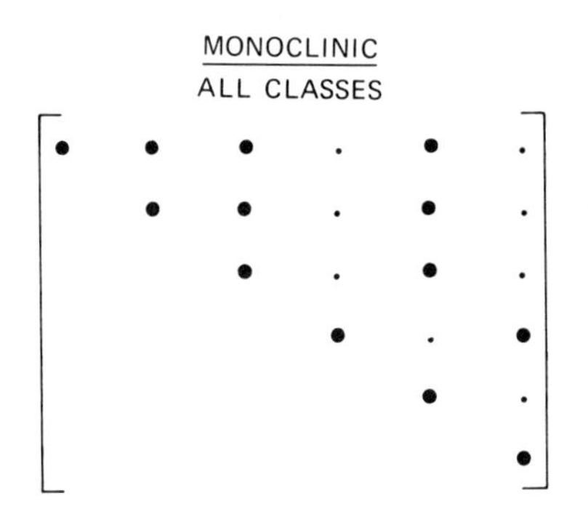

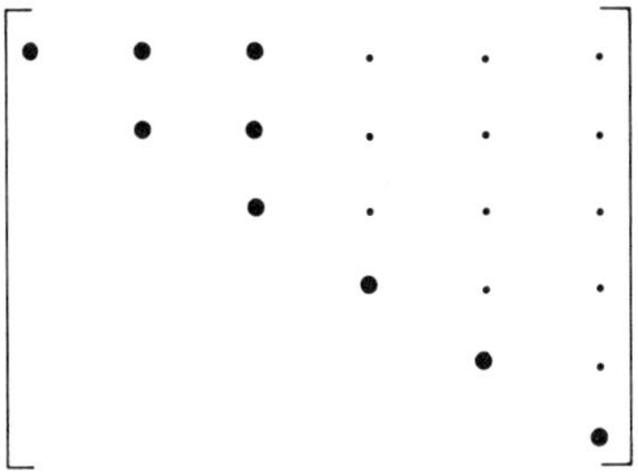

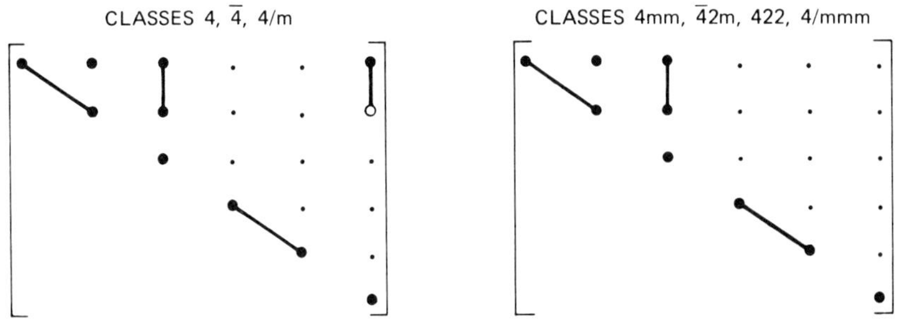

TABLE 6 (*Continued*)

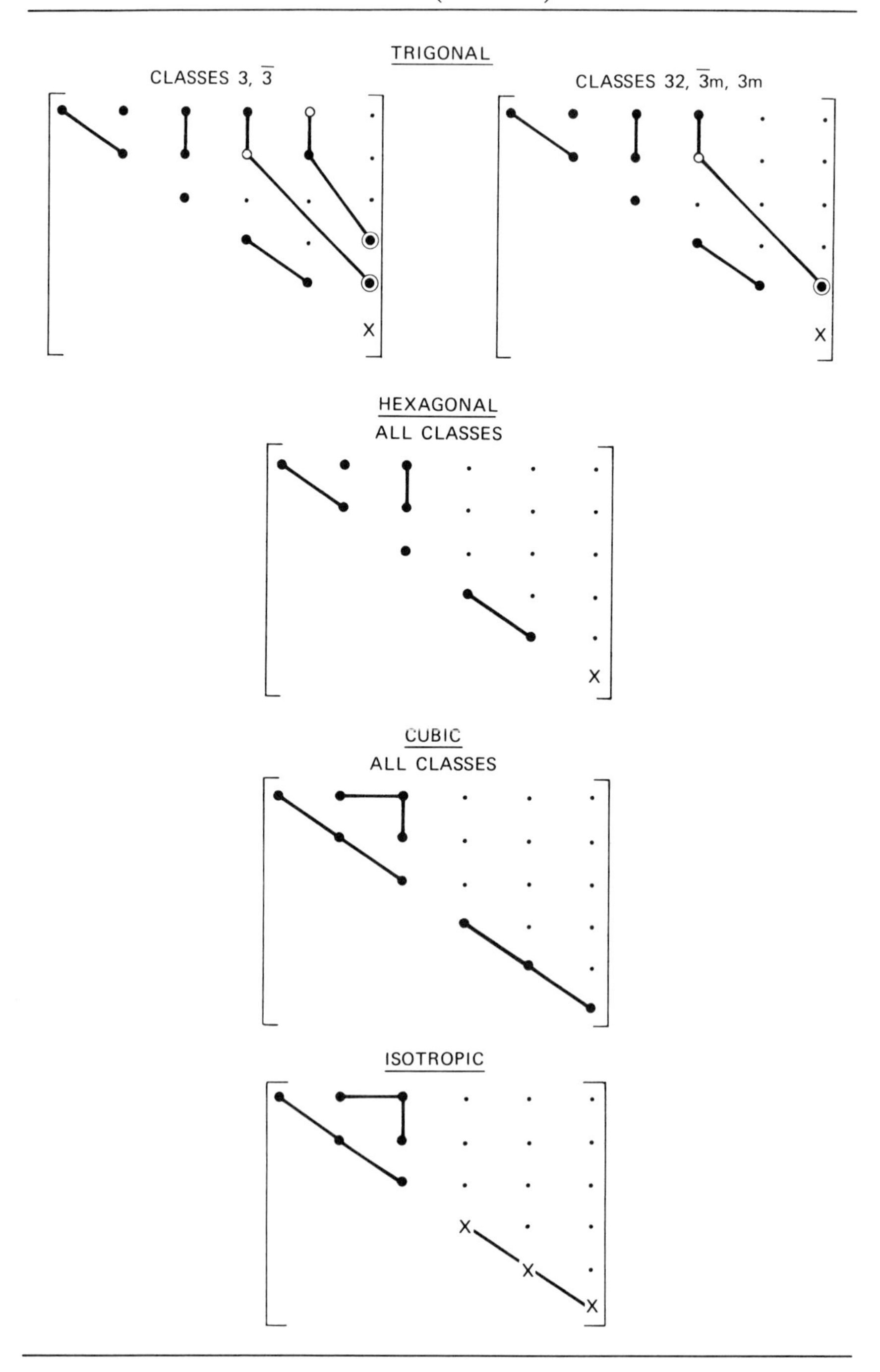

TABLE 7
Fourth Rank Elastooptic Tensors

The elastooptic susceptibility $\chi_{i\,j\,kl}^{\omega_s\omega_1\omega_2}$ that couples to strain (12.4.5), the Pockels elastooptic tensor p_{ijkl} (13.2.10), and the Pockels piezooptic tensor q_{ijkl} (13.7.13) possess the symmetry

$$T_{ijkl} = T_{(ij)(kl)} \tag{A.13}$$

provided that the frequency dispersion between the input and output optical frequencies is negligible [see discussion concerning (13.2.2)]. The rotooptic susceptibility that couples to rotation and possesses antisymmetry on its last pair of indices is given for the various crystal systems in Section 13.3. Contracted notation for the elastooptic tensors that couple to strain may be defined by

$$\Delta\kappa_m = \chi_{mn} S_n, \tag{A.14}$$

$$(\Delta\kappa^{-1})_m = p_{mn} S_n, \tag{A.15}$$

$$(\Delta\kappa^{-1})_m = q_{mn} t_n \tag{A.16}$$

where

$$\Delta\kappa_m = \Delta\kappa_{(ij)} \qquad (m = 1, 2, \ldots 6), \tag{A.17}$$

$$(\Delta\kappa^{-1})_m = (\Delta\kappa^{-1})_{(ij)} \qquad (m = 1, 2, \ldots 6), \tag{A.18}$$

$$\chi_{mn} = \chi_{(ij)\,(kl)}^{\omega_s\,\omega_1\omega_2} \qquad (m, n = 1, 2, \ldots 6), \tag{A.19}$$

$$p_{mn} = p_{(ij)(kl)} \qquad (m, n = 1, 2, \ldots 6), \tag{A.20}$$

$$q_{mn} = q_{(ij)(kl)} \qquad (m = 1, 2, \ldots 6; n = 1, 2, 3) \tag{A.21}$$

$$= 2q_{(ij)(kl)} \qquad (m = 1, 2, \ldots 6; n = 4, 5, 6) \tag{A.22}$$

and where the contracted notation for the strain and stress tensors is given in (10.5.8) and (10.5.9). The following notation is used in this table:

◉ A component that is equal to the solid dot component joined to it for either χ_{mn} or p_{mn} or is twice the solid dot component joined to it for q_{mn}.

◎ A component that is equal to minus the solid dot component joined to it for either χ_{mn} or p_{mn} or is minus twice the solid dot component joined to it for q_{mn}.

× A component equal to $\frac{1}{2}(p_{11} - p_{12})$ or $\frac{1}{2}(\chi_{11} - \chi_{12})$ or $(q_{11} - q_{12})$.

TABLE 7 (*Continued*)

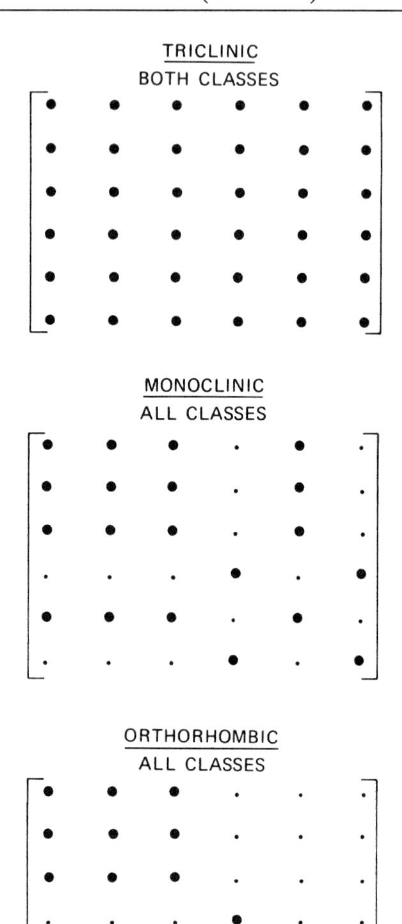

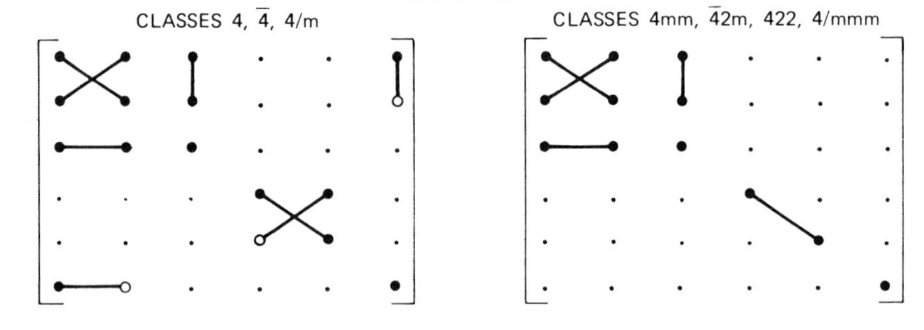

TABLE 7 (*Continued*)

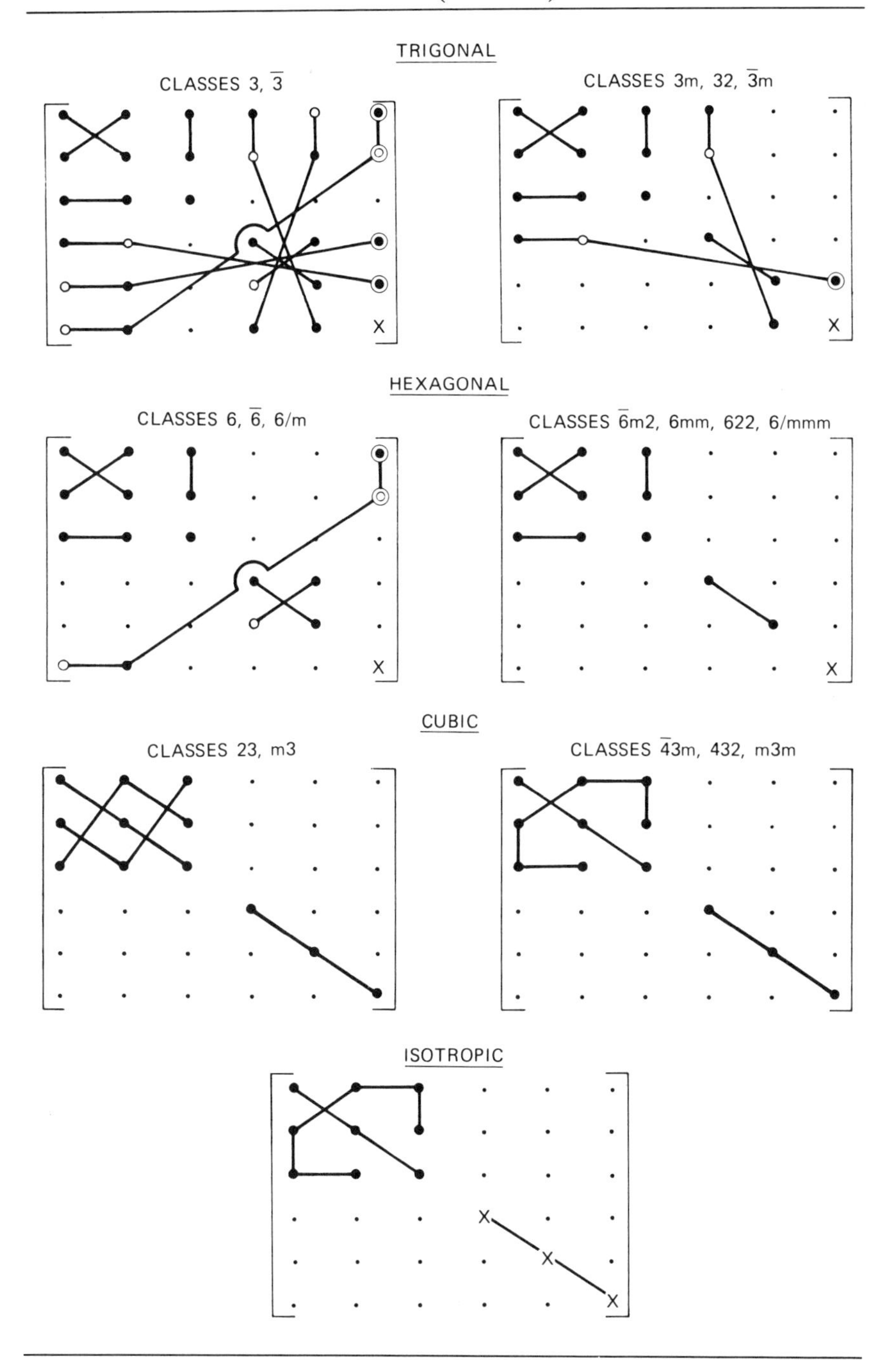

TABLE 8
Fifth Rank AIOHG Tensor

The portion of the susceptibility for AIOHG that couples to strain has the interchange symmetry

$$\chi_{i\,j\,k\,lm}^{\omega_C \omega_1 \omega_1 \omega_A} = \chi_{i\,(jk)(lm)}^{\omega_C \omega_1 \omega_1 \omega_A}. \tag{A.23}$$

With the notation of (10.5.7) this fifth rank tensor can be written as

$$\chi_{i(jk)(lm)} = \chi_{iab} \qquad (a \text{ and } b = 1,2,3,4,5,6). \tag{A.24}$$

The nonlinear polarization $\mathscr{P}$ may be written as a matrix product,

$$\mathscr{P}_i = \epsilon_0 \chi_{imn} E_m^2 S_n, \tag{A.25}$$

with the strain S_n defined in (10.5.8) and the product of the input electric fields defined in (A.2). The AIOHG tensor (A.24) is nonzero for the 21 acentric crystal classes. The symmetry of the portion of the AIOHG susceptibility that couples to rotation is discussed in Section 18.4.

In the following table zero elements are indicated by a dot; nonzero elements by letters; related elements are expressed in terms of the same letters except when only X appears, for which cases no relation exists between the nonzero elements. The matrix for χ_{imn} is shown in the table as three 6×6 arrays. The left 6×6 array corresponds to $i=1$, the middle one to $i=2$, and the right one to $i=3$. Within each 6×6 array m numbers the rows and n numbers the columns. The table is taken from Nelson and Lax [1971c] except that the matrices for classes $3m$, 32, and $\bar{6}m2$ are changed here to conform with standard axes [Bond et al., 1949].

TRICLINIC

CLASS 1

$$\begin{vmatrix} X & X & X & X & X & X \\ X & X & X & X & X & X \\ X & X & X & X & X & X \\ X & X & X & X & X & X \\ X & X & X & X & X & X \\ X & X & X & X & X & X \end{vmatrix} \qquad \begin{vmatrix} X & X & X & X & X & X \\ X & X & X & X & X & X \\ X & X & X & X & X & X \\ X & X & X & X & X & X \\ X & X & X & X & X & X \\ X & X & X & X & X & X \end{vmatrix} \qquad \begin{vmatrix} X & X & X & X & X & X \\ X & X & X & X & X & X \\ X & X & X & X & X & X \\ X & X & X & X & X & X \\ X & X & X & X & X & X \\ X & X & X & X & X & X \end{vmatrix}$$

MONOCLINIC

CLASS 2

$$
\begin{vmatrix}
\cdot & \cdot & \cdot & X & \cdot & X \\
\cdot & \cdot & \cdot & X & \cdot & X \\
\cdot & \cdot & \cdot & X & \cdot & X \\
X & X & X & \cdot & X & \cdot \\
\cdot & \cdot & \cdot & X & \cdot & X \\
X & X & X & \cdot & X & \cdot
\end{vmatrix}
\qquad
\begin{vmatrix}
X & X & X & \cdot & X & \cdot \\
X & X & X & \cdot & X & \cdot \\
X & X & X & \cdot & X & \cdot \\
\cdot & \cdot & \cdot & X & \cdot & X \\
X & X & X & \cdot & X & \cdot \\
\cdot & \cdot & \cdot & X & \cdot & X
\end{vmatrix}
\qquad
\begin{vmatrix}
\cdot & \cdot & \cdot & X & \cdot & X \\
\cdot & \cdot & \cdot & X & \cdot & X \\
\cdot & \cdot & \cdot & X & \cdot & X \\
X & X & X & \cdot & X & \cdot \\
\cdot & \cdot & \cdot & X & \cdot & X \\
X & X & X & \cdot & X & \cdot
\end{vmatrix}
$$

CLASS *m*

$$
\begin{vmatrix}
X & X & X & \cdot & X & \cdot \\
X & X & X & \cdot & X & \cdot \\
X & X & X & \cdot & X & \cdot \\
\cdot & \cdot & \cdot & X & \cdot & X \\
X & X & X & \cdot & X & \cdot \\
\cdot & \cdot & \cdot & X & \cdot & X
\end{vmatrix}
\qquad
\begin{vmatrix}
\cdot & \cdot & \cdot & X & \cdot & X \\
\cdot & \cdot & \cdot & X & \cdot & X \\
\cdot & \cdot & \cdot & X & \cdot & X \\
X & X & X & \cdot & X & \cdot \\
\cdot & \cdot & \cdot & X & \cdot & X \\
X & X & X & \cdot & X & \cdot
\end{vmatrix}
\qquad
\begin{vmatrix}
X & X & X & \cdot & X & \cdot \\
X & X & X & \cdot & X & \cdot \\
X & X & X & \cdot & X & \cdot \\
\cdot & \cdot & \cdot & X & \cdot & X \\
X & X & X & \cdot & X & \cdot \\
\cdot & \cdot & \cdot & X & \cdot & X
\end{vmatrix}
$$

ORTHORHOMBIC

CLASS 222

$$
\begin{vmatrix}
\cdot & \cdot & \cdot & X & \cdot & \cdot \\
\cdot & \cdot & \cdot & X & \cdot & \cdot \\
\cdot & \cdot & \cdot & X & \cdot & \cdot \\
X & X & X & \cdot & \cdot & \cdot \\
\cdot & \cdot & \cdot & \cdot & \cdot & X \\
\cdot & \cdot & \cdot & \cdot & X & \cdot
\end{vmatrix}
\qquad
\begin{vmatrix}
\cdot & \cdot & \cdot & \cdot & X & \cdot \\
\cdot & \cdot & \cdot & \cdot & X & \cdot \\
\cdot & \cdot & \cdot & \cdot & X & \cdot \\
\cdot & \cdot & \cdot & \cdot & \cdot & X \\
X & X & X & \cdot & \cdot & \cdot \\
\cdot & \cdot & \cdot & X & \cdot & \cdot
\end{vmatrix}
\qquad
\begin{vmatrix}
\cdot & \cdot & \cdot & \cdot & \cdot & X \\
\cdot & \cdot & \cdot & \cdot & \cdot & X \\
\cdot & \cdot & \cdot & \cdot & \cdot & X \\
\cdot & \cdot & \cdot & \cdot & X & \cdot \\
\cdot & \cdot & \cdot & X & \cdot & \cdot \\
X & X & X & \cdot & \cdot & \cdot
\end{vmatrix}
$$

CLASS *mm*2

$$
\begin{vmatrix}
\cdot & \cdot & \cdot & \cdot & X & \cdot \\
\cdot & \cdot & \cdot & \cdot & X & \cdot \\
\cdot & \cdot & \cdot & \cdot & X & \cdot \\
\cdot & \cdot & \cdot & \cdot & \cdot & X \\
X & X & X & \cdot & \cdot & \cdot \\
\cdot & \cdot & \cdot & X & \cdot & \cdot
\end{vmatrix}
\qquad
\begin{vmatrix}
\cdot & \cdot & \cdot & X & \cdot & \cdot \\
\cdot & \cdot & \cdot & X & \cdot & \cdot \\
\cdot & \cdot & \cdot & X & \cdot & \cdot \\
X & X & X & \cdot & \cdot & \cdot \\
\cdot & \cdot & \cdot & \cdot & \cdot & X \\
\cdot & \cdot & \cdot & \cdot & X & \cdot
\end{vmatrix}
\qquad
\begin{vmatrix}
X & X & X & \cdot & \cdot & \cdot \\
X & X & X & \cdot & \cdot & \cdot \\
X & X & X & \cdot & \cdot & \cdot \\
\cdot & \cdot & \cdot & X & \cdot & \cdot \\
\cdot & \cdot & \cdot & \cdot & X & \cdot \\
\cdot & \cdot & \cdot & \cdot & \cdot & X
\end{vmatrix}
$$

TABLE 8 (*Continued*)

TETRAGONAL

CLASS 4

$$\begin{vmatrix} \cdot & \cdot & \cdot & I & M & \cdot \\ \cdot & \cdot & \cdot & J & N & \cdot \\ \cdot & \cdot & \cdot & K & O & \cdot \\ A & B & C & \cdot & \cdot & D \\ E & F & G & \cdot & \cdot & H \\ \cdot & \cdot & \cdot & L & P & \cdot \end{vmatrix} \quad \begin{vmatrix} \cdot & \cdot & \cdot & N & -J & \cdot \\ \cdot & \cdot & \cdot & M & -I & \cdot \\ \cdot & \cdot & \cdot & O & -K & \cdot \\ F & E & G & \cdot & \cdot & -H \\ -B & -A & -C & \cdot & \cdot & D \\ \cdot & \cdot & \cdot & -P & L & \cdot \end{vmatrix} \quad \begin{vmatrix} Q & R & S & \cdot & \cdot & Z \\ R & Q & S & \cdot & \cdot & -Z \\ T & T & U & \cdot & \cdot & \cdot \\ \cdot & \cdot & \cdot & V & W & \cdot \\ \cdot & \cdot & \cdot & -W & V & \cdot \\ Y & -Y & \cdot & \cdot & \cdot & X \end{vmatrix}$$

CLASS $\bar{4}$

$$\begin{vmatrix} \cdot & \cdot & \cdot & I & M & \cdot \\ \cdot & \cdot & \cdot & J & N & \cdot \\ \cdot & \cdot & \cdot & K & W & \cdot \\ A & B & C & \cdot & \cdot & D \\ E & F & G & \cdot & \cdot & H \\ \cdot & \cdot & \cdot & L & P & \cdot \end{vmatrix} \quad \begin{vmatrix} \cdot & \cdot & \cdot & -N & J & \cdot \\ \cdot & \cdot & \cdot & -M & I & \cdot \\ \cdot & \cdot & \cdot & -W & K & \cdot \\ -F & -E & -G & \cdot & \cdot & H \\ B & A & C & \cdot & \cdot & -D \\ \cdot & \cdot & \cdot & P & -L & \cdot \end{vmatrix} \quad \begin{vmatrix} Q & R & S & \cdot & \cdot & Z \\ -R & -Q & -S & \cdot & \cdot & Z \\ T & -T & \cdot & \cdot & \cdot & X \\ \cdot & \cdot & \cdot & V & \cdot & \cdot \\ \cdot & \cdot & \cdot & \cdot & -V & \cdot \\ Y & Y & U & \cdot & \cdot & \cdot \end{vmatrix}$$

CLASS 422

$$\begin{vmatrix} \cdot & \cdot & \cdot & I & \cdot & \cdot \\ \cdot & \cdot & \cdot & J & \cdot & \cdot \\ \cdot & \cdot & \cdot & K & \cdot & \cdot \\ A & B & C & \cdot & \cdot & \cdot \\ \cdot & \cdot & \cdot & \cdot & \cdot & H \\ \cdot & \cdot & \cdot & \cdot & P & \cdot \end{vmatrix} \quad \begin{vmatrix} \cdot & \cdot & \cdot & \cdot & -J & \cdot \\ \cdot & \cdot & \cdot & \cdot & -I & \cdot \\ \cdot & \cdot & \cdot & \cdot & -K & \cdot \\ \cdot & \cdot & \cdot & \cdot & \cdot & -H \\ -B & -A & -C & \cdot & \cdot & \cdot \\ \cdot & \cdot & \cdot & -P & \cdot & \cdot \end{vmatrix} \quad \begin{vmatrix} \cdot & \cdot & \cdot & \cdot & \cdot & Z \\ \cdot & \cdot & \cdot & \cdot & \cdot & -Z \\ \cdot & \cdot & \cdot & \cdot & \cdot & \cdot \\ \cdot & \cdot & \cdot & \cdot & W & \cdot \\ \cdot & \cdot & \cdot & -W & \cdot & \cdot \\ Y & -Y & \cdot & \cdot & \cdot & \cdot \end{vmatrix}$$

CLASS 4*mm*

$$\begin{vmatrix} \cdot & \cdot & \cdot & \cdot & M & \cdot \\ \cdot & \cdot & \cdot & \cdot & N & \cdot \\ \cdot & \cdot & \cdot & \cdot & W & \cdot \\ \cdot & \cdot & \cdot & \cdot & \cdot & D \\ E & F & G & \cdot & \cdot & \cdot \\ \cdot & \cdot & \cdot & L & \cdot & \cdot \end{vmatrix} \quad \begin{vmatrix} \cdot & \cdot & \cdot & N & \cdot & \cdot \\ \cdot & \cdot & \cdot & M & \cdot & \cdot \\ \cdot & \cdot & \cdot & W & \cdot & \cdot \\ F & E & G & \cdot & \cdot & \cdot \\ \cdot & \cdot & \cdot & \cdot & \cdot & D \\ \cdot & \cdot & \cdot & \cdot & L & \cdot \end{vmatrix} \quad \begin{vmatrix} Q & R & S & \cdot & \cdot & \cdot \\ R & Q & S & \cdot & \cdot & \cdot \\ T & T & U & \cdot & \cdot & \cdot \\ \cdot & \cdot & \cdot & V & \cdot & \cdot \\ \cdot & \cdot & \cdot & \cdot & V & \cdot \\ \cdot & \cdot & \cdot & \cdot & \cdot & X \end{vmatrix}$$

CLASS 42m

$$
\left|\begin{array}{cccccc}
\cdot & \cdot & \cdot & I & \cdot & \cdot \\
\cdot & \cdot & \cdot & J & \cdot & \cdot \\
\cdot & \cdot & \cdot & K & \cdot & \cdot \\
A & B & C & \cdot & \cdot & \cdot \\
\cdot & \cdot & \cdot & \cdot & \cdot & H \\
\cdot & \cdot & \cdot & \cdot & P & \cdot
\end{array}\right|
\qquad
\left|\begin{array}{cccccc}
\cdot & \cdot & \cdot & \cdot & J & \cdot \\
\cdot & \cdot & \cdot & \cdot & I & \cdot \\
\cdot & \cdot & \cdot & \cdot & K & \cdot \\
\cdot & \cdot & \cdot & \cdot & \cdot & H \\
B & A & C & \cdot & \cdot & \cdot \\
\cdot & \cdot & \cdot & P & \cdot & \cdot
\end{array}\right|
\qquad
\left|\begin{array}{cccccc}
\cdot & \cdot & \cdot & \cdot & \cdot & Z \\
\cdot & \cdot & \cdot & \cdot & \cdot & Z \\
\cdot & \cdot & \cdot & \cdot & \cdot & X \\
\cdot & \cdot & \cdot & \cdot & W & \cdot \\
\cdot & \cdot & \cdot & W & \cdot & \cdot \\
Y & Y & U & \cdot & \cdot & \cdot
\end{array}\right|
$$

TRIGONAL

CLASS 3

$D \equiv -(A+B+C), m \equiv -(j+k+l)$

$$
\left|\begin{array}{cccccc}
2m & 2l & -n & 2d & 2I & (A-B) \\
2k & 2j & n & 2e & 2J & (C-D) \\
-z & z & \cdot & f & K & -F \\
2a & 2b & c & p & -M & (G-H) \\
2G & 2H & L & -M & -p & (b-a) \\
(A-C) & (B-D) & -E & (I-J) & (e-d) & (k+l)
\end{array}\right|
\left|\begin{array}{cccccc}
-2D & -2C & -E & 2J & -2e & (j-k) \\
-2B & -2A & E & 2I & -2d & (l-m) \\
-F & F & \cdot & K & -f & z \\
2H & 2G & L & M & p & (a-b) \\
-2b & -2a & -c & p & -M & (G-H) \\
(j-l) & (k-m) & n & (d-e) & (I-J) & (-B-C)
\end{array}\right|
\left|\begin{array}{cccccc}
2N & 2P & Q & -U & -r & t \\
2P & 2N & Q & U & r & -t \\
R & R & S & \cdot & \cdot & \cdot \\
-Z & Z & \cdot & T & s & q \\
-q & q & \cdot & -s & T & -Z \\
-t & t & \cdot & r & -U & (N-P)
\end{array}\right|
$$

CLASS 3m

$D \equiv -(A+B+C)$

$$
\left|\begin{array}{cccccc}
\cdot & \cdot & \cdot & \cdot & 2I & (A-B) \\
\cdot & \cdot & \cdot & \cdot & 2J & (C-D) \\
\cdot & \cdot & \cdot & \cdot & K & -F \\
\cdot & \cdot & \cdot & \cdot & -M & (G-H) \\
2G & 2H & L & -M & \cdot & \cdot \\
(A-C) & (B-D) & -E & (I-J) & \cdot & \cdot
\end{array}\right|
\left|\begin{array}{cccccc}
-2D & -2C & -E & 2J & \cdot & \cdot \\
-2B & -2A & E & 2I & \cdot & \cdot \\
-F & F & \cdot & K & \cdot & \cdot \\
2H & 2G & L & M & \cdot & \cdot \\
\cdot & \cdot & \cdot & \cdot & -M & (G-H) \\
\cdot & \cdot & \cdot & \cdot & (I-J) & (-B-C)
\end{array}\right|
\left|\begin{array}{cccccc}
2N & 2P & Q & -U & \cdot & \cdot \\
2P & 2N & Q & U & \cdot & \cdot \\
R & R & S & \cdot & \cdot & \cdot \\
-Z & Z & \cdot & T & \cdot & \cdot \\
\cdot & \cdot & \cdot & \cdot & T & -Z \\
\cdot & \cdot & \cdot & \cdot & -U & (N-P)
\end{array}\right|
$$

CLASS 32

$m \equiv -(j+k+l)$

$$
\left|\begin{array}{cccccc}
2m & 2l & -n & 2d & \cdot & \cdot \\
2k & 2j & n & 2e & \cdot & \cdot \\
-z & z & \cdot & f & \cdot & \cdot \\
2a & 2b & c & p & \cdot & \cdot \\
\cdot & \cdot & \cdot & \cdot & -p & (b-a) \\
\cdot & \cdot & \cdot & \cdot & (e-d) & (k+l)
\end{array}\right|
\left|\begin{array}{cccccc}
\cdot & \cdot & \cdot & \cdot & -2e & (j-k) \\
\cdot & \cdot & \cdot & \cdot & -2d & (l-m) \\
\cdot & \cdot & \cdot & \cdot & f & z \\
\cdot & \cdot & \cdot & \cdot & p & (a-b) \\
-2b & -2a & -c & p & \cdot & \cdot \\
(j-l) & (k-m) & n & (d-e) & \cdot & \cdot
\end{array}\right|
\left|\begin{array}{cccccc}
\cdot & \cdot & \cdot & \cdot & -r & t \\
\cdot & \cdot & \cdot & \cdot & r & -t \\
\cdot & \cdot & \cdot & \cdot & \cdot & \cdot \\
\cdot & \cdot & \cdot & \cdot & s & q \\
-q & q & \cdot & -s & \cdot & \cdot \\
-t & t & \cdot & r & \cdot & \cdot
\end{array}\right|
$$

TABLE 8 (*Continued*)

HEXAGONAL

CLASS 6

$$\begin{vmatrix} \cdot & \cdot & \cdot & 2D & 2J & \cdot \\ \cdot & \cdot & \cdot & 2E & 2K & \cdot \\ \cdot & \cdot & \cdot & F & L & \cdot \\ 2A & 2B & C & \cdot & \cdot & (G-H) \\ 2G & 2H & I & \cdot & \cdot & (B-A) \\ \cdot & \cdot & \cdot & (J-K) & (E-D) & \cdot \end{vmatrix} \quad \begin{vmatrix} \cdot & \cdot & \cdot & 2K & -2E & \cdot \\ \cdot & \cdot & \cdot & 2J & -2D & \cdot \\ \cdot & \cdot & \cdot & L & -F & \cdot \\ 2H & 2G & T & \cdot & \cdot & (A-B) \\ -2B & -2A & -C & \cdot & \cdot & (G-H) \\ \cdot & \cdot & \cdot & (D-E) & (J-K) & \cdot \end{vmatrix} \quad \begin{vmatrix} 2M & 2N & O & \cdot & \cdot & T \\ 2N & 2M & O & \cdot & \cdot & -T \\ P & P & Q & \cdot & \cdot & \cdot \\ \cdot & \cdot & \cdot & R & S & \cdot \\ \cdot & \cdot & \cdot & -S & R & \cdot \\ -T & T & \cdot & \cdot & \cdot & (M-N) \end{vmatrix}$$

CLASS $\bar{6}$

$D \equiv -(A+B+C), M \equiv -(J+K+L)$

$$\begin{vmatrix} 2A & 2B & E & \cdot & \cdot & (L-M) \\ 2C & 2D & -E & \cdot & \cdot & (J-K) \\ F & -F & \cdot & \cdot & \cdot & O \\ \cdot & \cdot & \cdot & G & P & \cdot \\ \cdot & \cdot & \cdot & P & -G & \cdot \\ (K-M) & (J-L) & N & \cdot & \cdot & (B+C) \end{vmatrix} \quad \begin{vmatrix} 2J & 2K & N & \cdot & \cdot & (D-C) \\ 2L & 2M & -N & \cdot & \cdot & (B-A) \\ O & -O & \cdot & \cdot & \cdot & -F \\ \cdot & \cdot & \cdot & -P & G & \cdot \\ \cdot & \cdot & \cdot & G & P & \cdot \\ (D-B) & (C-A) & -E & \cdot & \cdot & (K+L) \end{vmatrix} \quad \begin{vmatrix} \cdot & \cdot & \cdot & R & -I & \cdot \\ \cdot & \cdot & \cdot & -R & I & \cdot \\ \cdot & \cdot & \cdot & \cdot & \cdot & \cdot \\ Q & -Q & \cdot & \cdot & \cdot & H \\ -H & H & \cdot & \cdot & \cdot & Q \\ \cdot & \cdot & \cdot & I & R & \cdot \end{vmatrix}$$

CLASS 622

$$\begin{vmatrix} \cdot & \cdot & \cdot & 2D & \cdot & \cdot \\ \cdot & \cdot & \cdot & 2E & \cdot & \cdot \\ \cdot & \cdot & \cdot & F & \cdot & \cdot \\ 2A & 2B & C & \cdot & \cdot & \cdot \\ \cdot & \cdot & \cdot & \cdot & \cdot & (B-A) \\ \cdot & \cdot & \cdot & \cdot & (E-D) & \cdot \end{vmatrix} \quad \begin{vmatrix} \cdot & \cdot & \cdot & \cdot & -2E & \cdot \\ \cdot & \cdot & \cdot & \cdot & -2D & \cdot \\ \cdot & \cdot & \cdot & \cdot & -F & \cdot \\ \cdot & \cdot & \cdot & \cdot & \cdot & (A-B) \\ -2B & -2A & -C & \cdot & \cdot & \cdot \\ \cdot & \cdot & \cdot & (D-E) & \cdot & \cdot \end{vmatrix} \quad \begin{vmatrix} \cdot & \cdot & \cdot & \cdot & \cdot & T \\ \cdot & \cdot & \cdot & \cdot & \cdot & -T \\ \cdot & \cdot & \cdot & \cdot & \cdot & \cdot \\ \cdot & \cdot & \cdot & \cdot & S & \cdot \\ \cdot & \cdot & \cdot & -S & \cdot & \cdot \\ -T & T & \cdot & \cdot & \cdot & \cdot \end{vmatrix}$$

CLASS 6*mm*

$$\begin{vmatrix} \cdot & \cdot & \cdot & \cdot & 2J & \cdot \\ \cdot & \cdot & \cdot & \cdot & 2K & \cdot \\ \cdot & \cdot & \cdot & \cdot & L & \cdot \\ \cdot & \cdot & \cdot & \cdot & \cdot & (G-H) \\ 2G & 2H & I & \cdot & \cdot & \cdot \\ \cdot & \cdot & \cdot & (J-K) & \cdot & \cdot \end{vmatrix} \quad \begin{vmatrix} \cdot & \cdot & \cdot & 2K & \cdot & \cdot \\ \cdot & \cdot & \cdot & 2J & \cdot & \cdot \\ \cdot & \cdot & \cdot & L & \cdot & \cdot \\ 2H & 2G & I & \cdot & \cdot & \cdot \\ \cdot & \cdot & \cdot & \cdot & \cdot & (G-H) \\ \cdot & \cdot & \cdot & \cdot & (J-K) & \cdot \end{vmatrix} \quad \begin{vmatrix} 2M & 2N & O & \cdot & \cdot & \cdot \\ 2N & 2M & O & \cdot & \cdot & \cdot \\ P & P & Q & \cdot & \cdot & \cdot \\ \cdot & \cdot & \cdot & R & \cdot & \cdot \\ \cdot & \cdot & \cdot & \cdot & R & \cdot \\ \cdot & \cdot & \cdot & \cdot & \cdot & (M-N) \end{vmatrix}$$

CLASS $\bar{6}m2$

$M \equiv -(J+K+L)$

$$
\left|\begin{array}{cccccc}
2M & 2L & -N & \cdot & \cdot & \cdot \\
2K & 2J & N & \cdot & \cdot & \cdot \\
-Z & Z & \cdot & \cdot & \cdot & \cdot \\
\cdot & \cdot & \cdot & P & \cdot & \cdot \\
\cdot & \cdot & \cdot & \cdot & P & \cdot \\
\cdot & \cdot & \cdot & \cdot & \cdot & (K+L)
\end{array}\right|
\qquad
\left|\begin{array}{cccccc}
\cdot & \cdot & \cdot & \cdot & \cdot & (J-K) \\
\cdot & \cdot & \cdot & \cdot & \cdot & (L-M) \\
\cdot & \cdot & \cdot & \cdot & \cdot & Z \\
\cdot & \cdot & \cdot & \cdot & P & \cdot \\
\cdot & \cdot & \cdot & P & \cdot & \cdot \\
(J-L) & (K-M) & N & \cdot & \cdot & \cdot
\end{array}\right|
\qquad
\left|\begin{array}{cccccc}
\cdot & \cdot & \cdot & \cdot & -R & \cdot \\
\cdot & \cdot & \cdot & \cdot & R & \cdot \\
\cdot & \cdot & \cdot & \cdot & \cdot & \cdot \\
\cdot & \cdot & \cdot & \cdot & \cdot & Q \\
-Q & Q & \cdot & \cdot & \cdot & \cdot \\
\cdot & \cdot & \cdot & R & \cdot & \cdot
\end{array}\right|
$$

CUBIC

CLASS 23

$$
\left|\begin{array}{cccccc}
\cdot & \cdot & \cdot & I & \cdot & \cdot \\
\cdot & \cdot & \cdot & J & \cdot & \cdot \\
\cdot & \cdot & \cdot & K & \cdot & \cdot \\
A & B & C & \cdot & \cdot & \cdot \\
\cdot & \cdot & \cdot & \cdot & \cdot & H \\
\cdot & \cdot & \cdot & \cdot & P & \cdot
\end{array}\right|
\qquad
\left|\begin{array}{cccccc}
\cdot & \cdot & \cdot & \cdot & K & \cdot \\
\cdot & \cdot & \cdot & \cdot & I & \cdot \\
\cdot & \cdot & \cdot & \cdot & J & \cdot \\
\cdot & \cdot & \cdot & \cdot & \cdot & P \\
C & A & B & \cdot & \cdot & \cdot \\
\cdot & \cdot & \cdot & H & \cdot & \cdot
\end{array}\right|
\qquad
\left|\begin{array}{cccccc}
\cdot & \cdot & \cdot & \cdot & \cdot & J \\
\cdot & \cdot & \cdot & \cdot & \cdot & K \\
\cdot & \cdot & \cdot & \cdot & \cdot & I \\
\cdot & \cdot & \cdot & \cdot & H & \cdot \\
\cdot & \cdot & \cdot & P & \cdot & \cdot \\
B & C & A & \cdot & \cdot & \cdot
\end{array}\right|
$$

CLASS 432

$$
\left|\begin{array}{cccccc}
\cdot & \cdot & \cdot & \cdot & \cdot & \cdot \\
\cdot & \cdot & \cdot & J & \cdot & \cdot \\
\cdot & \cdot & \cdot & -J & \cdot & \cdot \\
\cdot & B & -B & \cdot & \cdot & \cdot \\
\cdot & \cdot & \cdot & \cdot & \cdot & H \\
\cdot & \cdot & \cdot & \cdot & -H & \cdot
\end{array}\right|
\qquad
\left|\begin{array}{cccccc}
\cdot & \cdot & \cdot & \cdot & -J & \cdot \\
\cdot & \cdot & \cdot & \cdot & \cdot & \cdot \\
\cdot & \cdot & \cdot & \cdot & J & \cdot \\
\cdot & \cdot & \cdot & \cdot & \cdot & -H \\
-B & \cdot & B & \cdot & \cdot & \cdot \\
\cdot & \cdot & \cdot & H & \cdot & \cdot
\end{array}\right|
\qquad
\left|\begin{array}{cccccc}
\cdot & \cdot & \cdot & \cdot & \cdot & J \\
\cdot & \cdot & \cdot & \cdot & \cdot & -J \\
\cdot & \cdot & \cdot & \cdot & \cdot & \cdot \\
\cdot & \cdot & \cdot & \cdot & H & \cdot \\
\cdot & \cdot & \cdot & -H & \cdot & \cdot \\
B & -B & \cdot & \cdot & \cdot & \cdot
\end{array}\right|
$$

CLASS $\bar{4}3m$

$$
\left|\begin{array}{cccccc}
\cdot & \cdot & \cdot & I & \cdot & \cdot \\
\cdot & \cdot & \cdot & J & \cdot & \cdot \\
\cdot & \cdot & \cdot & J & \cdot & \cdot \\
A & B & B & \cdot & \cdot & \cdot \\
\cdot & \cdot & \cdot & \cdot & \cdot & H \\
\cdot & \cdot & \cdot & \cdot & H & \cdot
\end{array}\right|
\qquad
\left|\begin{array}{cccccc}
\cdot & \cdot & \cdot & \cdot & J & \cdot \\
\cdot & \cdot & \cdot & \cdot & I & \cdot \\
\cdot & \cdot & \cdot & \cdot & J & \cdot \\
\cdot & \cdot & \cdot & \cdot & \cdot & H \\
B & A & B & \cdot & \cdot & \cdot \\
\cdot & \cdot & \cdot & H & \cdot & \cdot
\end{array}\right|
\qquad
\left|\begin{array}{cccccc}
\cdot & \cdot & \cdot & \cdot & \cdot & J \\
\cdot & \cdot & \cdot & \cdot & \cdot & J \\
\cdot & \cdot & \cdot & \cdot & \cdot & I \\
\cdot & \cdot & \cdot & \cdot & H & \cdot \\
\cdot & \cdot & \cdot & H & \cdot & \cdot \\
B & B & A & \cdot & \cdot & \cdot
\end{array}\right|
$$

TABLE 9
Fifth Rank Nonlinear Piezoelectric Tensor

The nonlinear piezoelectric tensor e_{KABCD} has the interchange symmetry

$$e_{KABCD} = e_{K(AB)(CD)} = e_{K(CD)(AB)} \tag{A.26}$$

and so has the symmetry of the AIOHG tensor *plus* symmetry about the leading diagonal of each of the three 6×6 arrays. No factors of 2 need be introduced in the matrix notation e_{KMN}. Here M and N are six-dimensional indices obeying the relation (10.5.7). From (17.3.5) the polarization resulting from this tensor in matrix notation is

$$\mathcal{P}_K = \tfrac{1}{2} e_{KMN} S_M S_N \tag{A.27}$$

with S_M given by (10.5.8). The notation in this table is the same as in Table 8.

TRICLINIC

CLASS 1

$$\left|\begin{array}{cccccc} X & X & X & X & X & X \\ & X & X & X & X & X \\ & & X & X & X & X \\ & & & X & X & X \\ & & & & X & X \\ & & & & & X \end{array}\right| \qquad \left|\begin{array}{cccccc} X & X & X & X & X & X \\ & X & X & X & X & X \\ & & X & X & X & X \\ & & & X & X & X \\ & & & & X & X \\ & & & & & X \end{array}\right| \qquad \left|\begin{array}{cccccc} X & X & X & X & X & X \\ & X & X & X & X & X \\ & & X & X & X & X \\ & & & X & X & X \\ & & & & X & X \\ & & & & & X \end{array}\right|$$

MONOCLINIC

CLASS 2

$$\left|\begin{array}{cccccc} \cdot & \cdot & \cdot & X & \cdot & X \\ & \cdot & \cdot & X & \cdot & X \\ & & \cdot & X & \cdot & X \\ & & & \cdot & X & \cdot \\ & & & & \cdot & X \\ & & & & & \cdot \end{array}\right| \qquad \left|\begin{array}{cccccc} X & X & X & \cdot & X & \cdot \\ & X & X & \cdot & X & \cdot \\ & & X & \cdot & X & \cdot \\ & & & X & \cdot & X \\ & & & & X & \cdot \\ & & & & & X \end{array}\right| \qquad \left|\begin{array}{cccccc} \cdot & \cdot & \cdot & X & \cdot & X \\ & \cdot & \cdot & X & \cdot & X \\ & & \cdot & X & \cdot & X \\ & & & \cdot & X & \cdot \\ & & & & \cdot & X \\ & & & & & \cdot \end{array}\right|$$

CLASS m

$$\left|\begin{array}{cccccc} X & X & X & \cdot & X & \cdot \\ & X & X & \cdot & X & \cdot \\ & & X & \cdot & X & \cdot \\ & & & X & \cdot & X \\ & & & & X & \cdot \\ & & & & & X \end{array}\right| \qquad \left|\begin{array}{cccccc} \cdot & \cdot & \cdot & X & \cdot & X \\ & \cdot & \cdot & X & \cdot & X \\ & & \cdot & X & \cdot & X \\ & & & \cdot & X & \cdot \\ & & & & \cdot & X \\ & & & & & \cdot \end{array}\right| \qquad \left|\begin{array}{cccccc} X & X & X & \cdot & X & \cdot \\ & X & X & \cdot & X & \cdot \\ & & X & \cdot & X & \cdot \\ & & & X & \cdot & X \\ & & & & X & \cdot \\ & & & & & X \end{array}\right|$$

ORTHORHOMBIC

CLASS 222

$$\left|\begin{array}{cccccc} \cdot & \cdot & \cdot & X & \cdot & \cdot \\ & \cdot & \cdot & X & \cdot & \cdot \\ & & \cdot & X & \cdot & \cdot \\ & & & \cdot & \cdot & \cdot \\ & & & & \cdot & X \\ & & & & & \cdot \end{array}\right| \qquad \left|\begin{array}{cccccc} \cdot & \cdot & \cdot & \cdot & X & \cdot \\ & \cdot & \cdot & \cdot & X & \cdot \\ & & \cdot & \cdot & X & \cdot \\ & & & \cdot & \cdot & X \\ & & & & \cdot & \cdot \\ & & & & & \cdot \end{array}\right| \qquad \left|\begin{array}{cccccc} \cdot & \cdot & \cdot & \cdot & \cdot & X \\ & \cdot & \cdot & \cdot & \cdot & X \\ & & \cdot & \cdot & \cdot & X \\ & & & \cdot & X & \cdot \\ & & & & \cdot & \cdot \\ & & & & & \cdot \end{array}\right|$$

CLASS $mm2$

$$\left|\begin{array}{cccccc} \cdot & \cdot & \cdot & \cdot & X & \cdot \\ & \cdot & \cdot & \cdot & X & \cdot \\ & & \cdot & \cdot & X & \cdot \\ & & & \cdot & \cdot & X \\ & & & & \cdot & \cdot \\ & & & & & \cdot \end{array}\right| \qquad \left|\begin{array}{cccccc} \cdot & \cdot & \cdot & X & \cdot & \cdot \\ & \cdot & \cdot & X & \cdot & \cdot \\ & & \cdot & X & \cdot & \cdot \\ & & & \cdot & \cdot & \cdot \\ & & & & \cdot & X \\ & & & & & \cdot \end{array}\right| \qquad \left|\begin{array}{cccccc} X & X & X & \cdot & \cdot & \cdot \\ & X & X & \cdot & \cdot & \cdot \\ & & X & \cdot & \cdot & \cdot \\ & & & X & \cdot & \cdot \\ & & & & X & \cdot \\ & & & & & X \end{array}\right|$$

TABLE 9 (*Continued*)

TETRAGONAL

CLASS 4

$$
\left|\begin{array}{cccccc}
\cdot & \cdot & \cdot & I & E & \cdot \\
 & \cdot & \cdot & J & F & \cdot \\
 & & \cdot & K & G & \cdot \\
 & & & \cdot & \cdot & D \\
 & & & & \cdot & H \\
 & & & & & \cdot
\end{array}\right|
\qquad
\left|\begin{array}{cccccc}
\cdot & \cdot & \cdot & F & -J & \cdot \\
 & \cdot & \cdot & E & -I & \cdot \\
 & & \cdot & G & -K & \cdot \\
 & & & \cdot & \cdot & -H \\
 & & & & \cdot & D \\
 & & & & & \cdot
\end{array}\right|
\qquad
\left|\begin{array}{cccccc}
Q & R & S & \cdot & \cdot & Z \\
 & Q & S & \cdot & \cdot & -Z \\
 & & U & \cdot & \cdot & \cdot \\
 & & & V & \cdot & \cdot \\
 & & & & V & \cdot \\
 & & & & & X
\end{array}\right|
$$

CLASS $\bar{4}$

$$
\left|\begin{array}{cccccc}
\cdot & \cdot & \cdot & I & M & \cdot \\
 & \cdot & \cdot & J & N & \cdot \\
 & & \cdot & K & W & \cdot \\
 & & & \cdot & \cdot & D \\
 & & & & \cdot & H \\
 & & & & & \cdot
\end{array}\right|
\qquad
\left|\begin{array}{cccccc}
\cdot & \cdot & \cdot & -N & J & \cdot \\
 & \cdot & \cdot & -M & I & \cdot \\
 & & \cdot & -W & K & \cdot \\
 & & & \cdot & \cdot & H \\
 & & & & \cdot & -D \\
 & & & & & \cdot
\end{array}\right|
\qquad
\left|\begin{array}{cccccc}
Q & \cdot & S & \cdot & \cdot & Z \\
 & -Q & -S & \cdot & \cdot & Z \\
 & & \cdot & \cdot & \cdot & X \\
 & & & V & \cdot & \cdot \\
 & & & & -V & \cdot \\
 & & & & & \cdot
\end{array}\right|
$$

CLASS 422

$$
\left|\begin{array}{cccccc}
\cdot & \cdot & \cdot & I & \cdot & \cdot \\
 & \cdot & \cdot & J & \cdot & \cdot \\
 & & \cdot & K & \cdot & \cdot \\
 & & & \cdot & \cdot & \cdot \\
 & & & & \cdot & H \\
 & & & & & \cdot
\end{array}\right|
\qquad
\left|\begin{array}{cccccc}
\cdot & \cdot & \cdot & \cdot & -J & \cdot \\
 & \cdot & \cdot & \cdot & -I & \cdot \\
 & & \cdot & \cdot & -K & \cdot \\
 & & & \cdot & \cdot & -H \\
 & & & & \cdot & \cdot \\
 & & & & & \cdot
\end{array}\right|
\qquad
\left|\begin{array}{cccccc}
\cdot & \cdot & \cdot & \cdot & \cdot & Z \\
 & \cdot & \cdot & \cdot & \cdot & -Z \\
 & & \cdot & \cdot & \cdot & \cdot \\
 & & & \cdot & \cdot & \cdot \\
 & & & & \cdot & \cdot \\
 & & & & & \cdot
\end{array}\right|
$$

CLASS 4*mm*

$$
\left|\begin{array}{cccccc}
\cdot & \cdot & \cdot & \cdot & M & \cdot \\
 & \cdot & \cdot & \cdot & N & \cdot \\
 & & \cdot & \cdot & W & \cdot \\
 & & & \cdot & \cdot & D \\
 & & & & \cdot & \cdot \\
 & & & & & \cdot
\end{array}\right|
\qquad
\left|\begin{array}{cccccc}
\cdot & \cdot & \cdot & N & \cdot & \cdot \\
 & \cdot & \cdot & M & \cdot & \cdot \\
 & & \cdot & W & \cdot & \cdot \\
 & & & \cdot & \cdot & \cdot \\
 & & & & \cdot & D \\
 & & & & & \cdot
\end{array}\right|
\qquad
\left|\begin{array}{cccccc}
Q & R & S & \cdot & \cdot & \cdot \\
 & Q & S & \cdot & \cdot & \cdot \\
 & & U & \cdot & \cdot & \cdot \\
 & & & V & \cdot & \cdot \\
 & & & & V & \cdot \\
 & & & & & X
\end{array}\right|
$$

CLASS $\bar{4}2m$

$$
\left|\begin{array}{cccccc}
\cdot & \cdot & \cdot & I & \cdot & \cdot \\
 & \cdot & \cdot & J & \cdot & \cdot \\
 & & \cdot & K & \cdot & \cdot \\
 & & & \cdot & \cdot & \cdot \\
 & & & & \cdot & H \\
 & & & & & \cdot
\end{array}\right|
\qquad
\left|\begin{array}{cccccc}
\cdot & \cdot & \cdot & \cdot & J & \cdot \\
 & \cdot & \cdot & \cdot & I & \cdot \\
 & & \cdot & \cdot & K & \cdot \\
 & & & \cdot & \cdot & H \\
 & & & & \cdot & \cdot \\
 & & & & & \cdot
\end{array}\right|
\qquad
\left|\begin{array}{cccccc}
\cdot & \cdot & \cdot & \cdot & \cdot & Z \\
 & \cdot & \cdot & \cdot & \cdot & Z \\
 & & \cdot & \cdot & \cdot & X \\
 & & & \cdot & W & \cdot \\
 & & & & \cdot & \cdot \\
 & & & & & \cdot
\end{array}\right|
$$

TRIGONAL

CLASS 3

$D \equiv -(A+2B),\ m \equiv -(j+2k)$

$$
\left|\begin{array}{cccccc}
2m & 2k & -n & 2a & 2I & (A-B) \\
 & 2j & n & 2b & 2J & (B-D) \\
 & & \cdot & f & K & -E \\
 & & & p & -M & (I-J) \\
 & & & & -p & (b-a) \\
 & & & & & 2k
\end{array}\right|
\qquad
\left|\begin{array}{cccccc}
-2D & -2B & -E & 2J & -2b & (j-k) \\
 & -2A & E & 2I & -2a & (k-m) \\
 & & \cdot & K & -f & n \\
 & & & M & p & (a-b) \\
 & & & & -M & (I-J) \\
 & & & & & -2B
\end{array}\right|
\qquad
\left|\begin{array}{cccccc}
2N & 2P & Q & -U & -r & \cdot \\
 & 2N & Q & U & r & \cdot \\
 & & S & \cdot & \cdot & \cdot \\
 & & & T & \cdot & r \\
 & & & & T & -U \\
 & & & & & (N-P)
\end{array}\right|
$$

CLASS $3m$

$D \equiv -(A-2B)$

$$
\left|\begin{array}{cccccc}
\cdot & \cdot & \cdot & \cdot & 2I & (A-B) \\
 & \cdot & \cdot & \cdot & 2J & (B-D) \\
 & & \cdot & \cdot & K & -E \\
 & & & \cdot & -M & (I-J) \\
 & & & & \cdot & \cdot \\
 & & & & & \cdot
\end{array}\right|
\qquad
\left|\begin{array}{cccccc}
-2D & -2B & -E & 2J & \cdot & \cdot \\
 & -2A & E & 2I & \cdot & \cdot \\
 & & \cdot & K & \cdot & \cdot \\
 & & & M & \cdot & \cdot \\
 & & & & -M & (I-J) \\
 & & & & & -2B
\end{array}\right|
\qquad
\left|\begin{array}{cccccc}
2N & 2P & Q & -U & \cdot & \cdot \\
 & 2N & Q & U & \cdot & \cdot \\
 & & S & \cdot & \cdot & \cdot \\
 & & & T & \cdot & \cdot \\
 & & & & T & -U \\
 & & & & & (N-P)
\end{array}\right|
$$

TABLE 9 (*Continued*)

TRIGONAL

CLASS 32

$m \equiv -(j+2k)$

$$
\begin{vmatrix}
2m & 2k & -n & 2a & \cdot & \cdot \\
 & 2j & n & 2b & \cdot & \cdot \\
 & & \cdot & f & \cdot & \cdot \\
 & & & p & \cdot & \cdot \\
 & & & & -p & (b-a) \\
 & & & & & 2k
\end{vmatrix}
\qquad
\begin{vmatrix}
\cdot & \cdot & \cdot & \cdot & -2b & (j-k) \\
 & \cdot & \cdot & \cdot & -2a & (k-m) \\
 & & \cdot & \cdot & -f & n \\
 & & & \cdot & p & (a-b) \\
 & & & & \cdot & \cdot \\
 & & & & & \cdot
\end{vmatrix}
\qquad
\begin{vmatrix}
\cdot & \cdot & \cdot & \cdot & -r & \cdot \\
 & \cdot & \cdot & \cdot & r & \cdot \\
 & & \cdot & \cdot & \cdot & \cdot \\
 & & & \cdot & \cdot & r \\
 & & & & \cdot & \cdot \\
 & & & & & \cdot
\end{vmatrix}
$$

HEXAGONAL

CLASS $\overline{6}$

$$
\begin{vmatrix}
\cdot & \cdot & \cdot & 2D & 2J & \cdot \\
 & \cdot & \cdot & 2E & 2K & \cdot \\
 & & \cdot & F & L & \cdot \\
 & & & \cdot & \cdot & J-K \\
 & & & & \cdot & E-D \\
 & & & & & \cdot
\end{vmatrix}
\qquad
\begin{vmatrix}
\cdot & \cdot & \cdot & 2K & -2E & \cdot \\
 & \cdot & \cdot & 2J & -2D & \cdot \\
 & & \cdot & L & -F & \cdot \\
 & & & \cdot & \cdot & (D-E) \\
 & & & & \cdot & (J-K) \\
 & & & & & \cdot
\end{vmatrix}
\qquad
\begin{vmatrix}
2M & 2N & P & \cdot & \cdot & \cdot \\
 & 2M & P & \cdot & \cdot & \cdot \\
 & & Q & \cdot & \cdot & \cdot \\
 & & & R & \cdot & \cdot \\
 & & & & R & \cdot \\
 & & & & & (M-N)
\end{vmatrix}
$$

CLASS $\overline{6}$

$D \equiv -(A+2B),\ M \equiv -(J+2K)$

$$
\begin{vmatrix}
2A & 2B & E & \cdot & \cdot & (K-M) \\
 & 2D & -E & \cdot & \cdot & (J-K) \\
 & & \cdot & \cdot & \cdot & N \\
 & & & G & P & \cdot \\
 & & & & -G & \cdot \\
 & & & & & 2B
\end{vmatrix}
\qquad
\begin{vmatrix}
2J & 2K & N & \cdot & \cdot & (D-B) \\
 & 2M & -N & \cdot & \cdot & (B-A) \\
 & & \cdot & \cdot & \cdot & -E \\
 & & & -P & G & \cdot \\
 & & & & P & \cdot \\
 & & & & & 2K
\end{vmatrix}
\qquad
\begin{vmatrix}
\cdot & \cdot & \cdot & R & -I & \cdot \\
 & \cdot & \cdot & -R & I & \cdot \\
 & & \cdot & \cdot & \cdot & \cdot \\
 & & & \cdot & \cdot & I \\
 & & & & \cdot & R \\
 & & & & & \cdot
\end{vmatrix}
$$

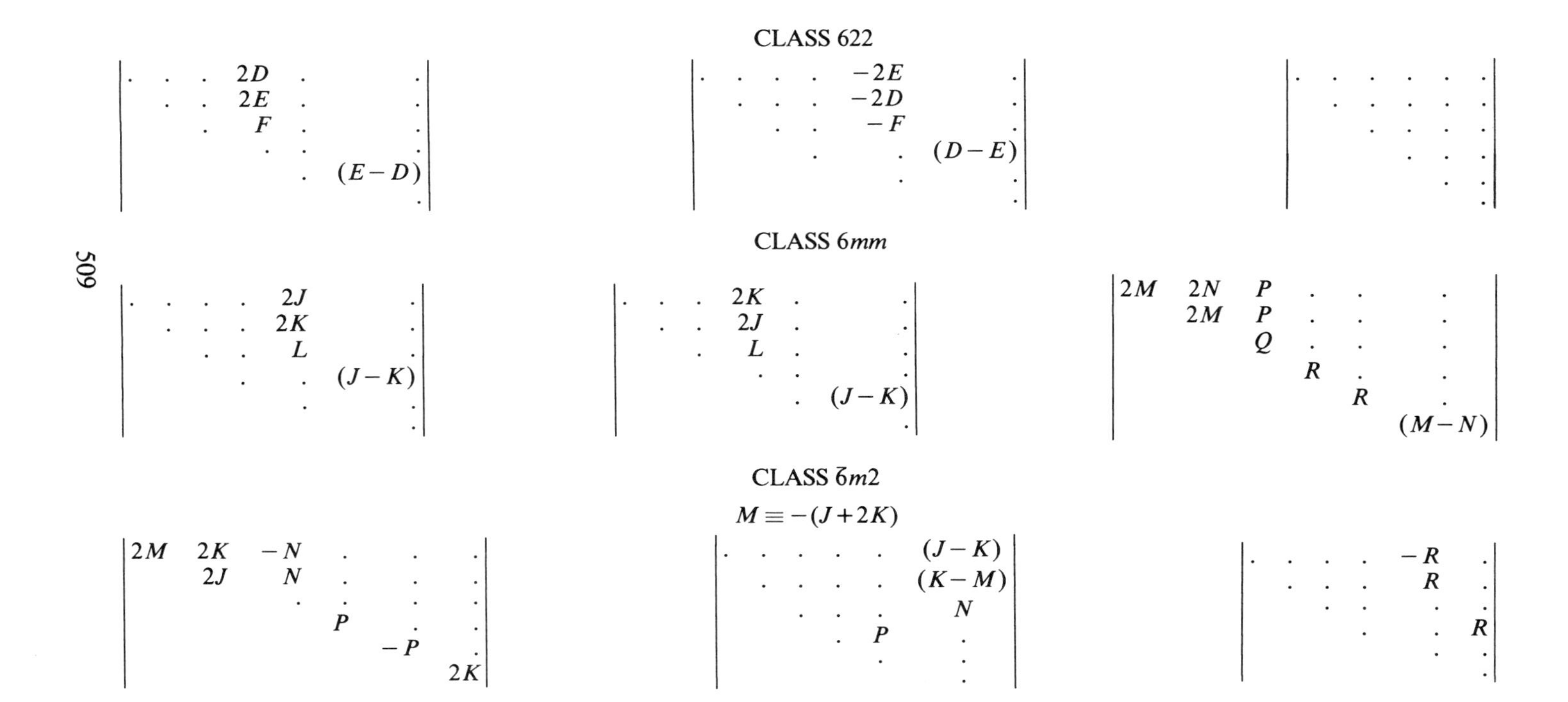

CLASS 622

$$\begin{vmatrix} \cdot & \cdot & \cdot & 2D & \cdot & \cdot \\ & \cdot & \cdot & 2E & \cdot & \cdot \\ & & \cdot & F & \cdot & \cdot \\ & & & \cdot & \cdot & \cdot \\ & & & & \cdot & (E-D) \\ & & & & & \cdot \end{vmatrix} \quad \begin{vmatrix} \cdot & \cdot & \cdot & \cdot & -2E & \cdot \\ & \cdot & \cdot & \cdot & -2D & \cdot \\ & & \cdot & \cdot & -F & \cdot \\ & & & \cdot & \cdot & (D-E) \\ & & & & \cdot & \cdot \\ & & & & & \cdot \end{vmatrix} \quad \begin{vmatrix} \cdot & \cdot & \cdot & \cdot & \cdot & \cdot \\ & \cdot & \cdot & \cdot & \cdot & \cdot \\ & & \cdot & \cdot & \cdot & \cdot \\ & & & \cdot & \cdot & \cdot \\ & & & & \cdot & \cdot \\ & & & & & \cdot \end{vmatrix}$$

CLASS 6*mm*

$$\begin{vmatrix} \cdot & \cdot & \cdot & \cdot & 2J & \cdot \\ & \cdot & \cdot & \cdot & 2K & \cdot \\ & & \cdot & \cdot & L & \cdot \\ & & & \cdot & \cdot & (J-K) \\ & & & & \cdot & \cdot \\ & & & & & \cdot \end{vmatrix} \quad \begin{vmatrix} \cdot & \cdot & \cdot & 2K & \cdot & \cdot \\ & \cdot & \cdot & 2J & \cdot & \cdot \\ & & \cdot & L & \cdot & \cdot \\ & & & \cdot & \cdot & \cdot \\ & & & & \cdot & (J-K) \\ & & & & & \cdot \end{vmatrix} \quad \begin{vmatrix} 2M & 2N & P & \cdot & \cdot & \cdot \\ & 2M & P & \cdot & \cdot & \cdot \\ & & Q & \cdot & \cdot & \cdot \\ & & & R & \cdot & \cdot \\ & & & & R & \cdot \\ & & & & & (M-N) \end{vmatrix}$$

CLASS $\bar{6}m2$

$M \equiv -(J+2K)$

$$\begin{vmatrix} 2M & 2K & -N & \cdot & \cdot & \cdot \\ & 2J & N & \cdot & \cdot & \cdot \\ & & \cdot & \cdot & \cdot & \cdot \\ & & & P & \cdot & \cdot \\ & & & & -P & \cdot \\ & & & & & 2K \end{vmatrix} \quad \begin{vmatrix} \cdot & \cdot & \cdot & \cdot & \cdot & (J-K) \\ & \cdot & \cdot & \cdot & \cdot & (K-M) \\ & & \cdot & \cdot & \cdot & N \\ & & & \cdot & P & \cdot \\ & & & & \cdot & \cdot \\ & & & & & \cdot \end{vmatrix} \quad \begin{vmatrix} \cdot & \cdot & \cdot & \cdot & -R & \cdot \\ & \cdot & \cdot & \cdot & R & \cdot \\ & & \cdot & \cdot & \cdot & \cdot \\ & & & \cdot & \cdot & R \\ & & & & \cdot & \cdot \\ & & & & & \cdot \end{vmatrix}$$

TABLE 9 (*Continued*)

CUBIC

CLASS 23

$$
\left|\begin{array}{cccccc}
\cdot & \cdot & \cdot & I & \cdot & \cdot \\
 & \cdot & \cdot & J & \cdot & \cdot \\
 & & \cdot & K & \cdot & \cdot \\
 & & & \cdot & \cdot & \cdot \\
 & & & & \cdot & H \\
 & & & & & \cdot
\end{array}\right|
\qquad
\left|\begin{array}{cccccc}
\cdot & \cdot & \cdot & \cdot & K & \cdot \\
 & \cdot & \cdot & \cdot & I & \cdot \\
 & & \cdot & \cdot & J & \cdot \\
 & & & \cdot & \cdot & H \\
 & & & & \cdot & \cdot \\
 & & & & & \cdot
\end{array}\right|
\qquad
\left|\begin{array}{cccccc}
\cdot & \cdot & \cdot & \cdot & \cdot & J \\
 & \cdot & \cdot & \cdot & \cdot & K \\
 & & \cdot & \cdot & \cdot & I \\
 & & & \cdot & H & \cdot \\
 & & & & \cdot & \cdot \\
 & & & & & \cdot
\end{array}\right|
$$

CLASS 432

$$
\left|\begin{array}{cccccc}
\cdot & \cdot & \cdot & \cdot & \cdot & \cdot \\
 & \cdot & \cdot & J & \cdot & \cdot \\
 & & \cdot & -J & \cdot & \cdot \\
 & & & \cdot & \cdot & \cdot \\
 & & & & \cdot & \cdot \\
 & & & & & \cdot
\end{array}\right|
\qquad
\left|\begin{array}{cccccc}
\cdot & \cdot & \cdot & \cdot & -J & \cdot \\
 & \cdot & \cdot & \cdot & \cdot & \cdot \\
 & & \cdot & \cdot & J & \cdot \\
 & & & \cdot & \cdot & \cdot \\
 & & & & \cdot & \cdot \\
 & & & & & \cdot
\end{array}\right|
\qquad
\left|\begin{array}{cccccc}
\cdot & \cdot & \cdot & \cdot & \cdot & J \\
 & \cdot & \cdot & \cdot & \cdot & -J \\
 & & \cdot & \cdot & \cdot & \cdot \\
 & & & \cdot & \cdot & \cdot \\
 & & & & \cdot & \cdot \\
 & & & & & \cdot
\end{array}\right|
$$

CLASS $\bar{4}3m$

$$
\left|\begin{array}{cccccc}
\cdot & \cdot & \cdot & I & \cdot & \cdot \\
 & \cdot & \cdot & J & \cdot & \cdot \\
 & & \cdot & J & \cdot & \cdot \\
 & & & \cdot & \cdot & \cdot \\
 & & & & \cdot & H \\
 & & & & & \cdot
\end{array}\right|
\qquad
\left|\begin{array}{cccccc}
\cdot & \cdot & \cdot & \cdot & J & \cdot \\
 & \cdot & \cdot & \cdot & I & \cdot \\
 & & \cdot & \cdot & J & \cdot \\
 & & & \cdot & \cdot & H \\
 & & & & \cdot & \cdot \\
 & & & & & \cdot
\end{array}\right|
\qquad
\left|\begin{array}{cccccc}
\cdot & \cdot & \cdot & \cdot & \cdot & J \\
 & \cdot & \cdot & \cdot & \cdot & J \\
 & & \cdot & \cdot & \cdot & I \\
 & & & \cdot & H & \cdot \\
 & & & & \cdot & \cdot \\
 & & & & & \cdot
\end{array}\right|
$$

TABLE 10
Sixth Rank Elasticity Tensor

The third order stiffness tensor is a sixth rank tensor possessing the interchange symmetry given in (17.4.3). That symmetry allows it to be written as

$$c_{(AB)(CD)(EF)} = c_{KLM} \qquad (K, L, M = 1,2,3,4,5,6) \tag{A.28}$$

with the use of the relation (10.5.7). Equation (17.4.3) implies

$$c_{KLM} = c_{KML} = c_{LMK} = c_{LKM} = c_{MKL} = c_{MLK}. \tag{A.29}$$

The table is adapted from Brugger [1965]. In the table tensor components are denoted by their subscripts. A negative sign indicates that the component belonging in that position is equal to the negative of the component indicated.

Triclinic	Monoclinic	Orthorhombic	Tetragonal		Trigonal		Hexagonal		Cubic		Isotropic
1	2	222	4	422	3	32	6	622	23	432	
$\bar{1}$	m	$mm2$	$\bar{4}$	$4mm$	$\bar{3}$	$3m$	$\bar{6}$	$6mm$	$m3$	$\bar{4}3m$	
	$2/m$	mmm	$4/m$	$\bar{4}2m$		$\bar{3}m$	$6/m$	$\bar{6}m2$		$m3m$	
				$4/mmm$				$6/mm$			
111	111	111	111	111	111	111	111	111	111	111	111
112	112	112	112	112	112	112	112	112	112	112	112
113	113	113	113	113	113	113	113	113	113	112	112
114	0	0	0	0	114	114	0	0	0	0	0
115	115	0	0	0	115	0	0	0	0	0	0
116	0	0	116	0	116	0	116	0	0	0	0
122	122	122	112	112	122[a]	122[a]	122[a]	122[a]	113	112	112
123	123	123	123	123	123	123	123	123	123	123	123
124	0	0	0	0	124	124	0	0	0	0	0

TABLE 10 (*Continued*)

Triclinic	Monoclinic	Orthorhombic	Tetragonal		Trigonal		Hexagonal		Cubic		Isotropic
1	2	222	4	422	3	32	6	622	23	432	
$\bar{1}$	m	$mm2$	$\bar{4}$	$4mm$	$\bar{3}$	$3m$	$\bar{6}$	$6mm$	$m3$	$\bar{4}3m$	
	$2/m$	mmm	$4/m$	$\bar{4}2m$		$\bar{3}m$	$6/m$	$\bar{6}m2$		$m3m$	
				$4/mmm$				$6/mm$			
125	125	0	0	0	125	0	0	0	0	0	0
126	0	0	0	0	−116	0	−116	0	0	0	0
133	133	133	133	133	133	133	133	133	112	112	112
134	0	0	0	0	134	134	0	0	0	0	0
135	135	0	0	0	135	0	0	0	0	0	0
136	0	0	136	0	0	0	0	0	0	0	0
144	144	144	144	144	144	144	144	144	144	144	144[1]
145	0	0	145	0	145	0	145	0	0	0	0
146	146	0	0	0	146[b]	0	0	0	0	0	0
155	155	155	155	155	155	155	155	155	155	155	155[m]
156	0	0	0	0	156[c]	156[c]	0	0	0	0	0
166	166	166	166	166	166[d]	166[d]	166[d]	166[d]	166	155	155[m]
222	222	222	111	111	222	222	222	222	111	111	111
223	223	223	113	113	113	113	113	113	112	112	112
224	0	0	0	0	224[e]	224[e]	0	0	0	0	0
225	225	0	0	0	225[f]	0	0	0	0	0	0
226	0	0	−116	0	116	0	116	0	0	0	0
233	233	233	133	133	133	133	133	133	113	112	112
234	0	0	0	0	−134	−134	0	0	0	0	0
235	235	0	0	0	−135	0	0	0	0	0	0
236	0	0	−136	0	0	0	0	0	0	0	0
244	244	244	155	155	155	155	155	155	166	155	155[m]
245	0	0	−145	0	−145	0	−145	0	0	0	0
246	246	0	0	0	246[g]	0	0	0	0	0	0

255	255	255	144	144	144	144	144	144	144	144	144[l]
256	0	0	0	0	256[h]	256[h]	0	0	0	0	0
266	266	266	166	166	266[i]	266[i]	266[i]	266[i]	155	155	155[m]
333	333	333	333	333	333	333	333	333	111	111	111
334	0	0	0	0	0	0	0	0	0	0	0
335	335	0	0	0	0	0	0	0	0	0	0
336	0	0	0	0	0	0	0	0	0	0	0
344	344	344	344	344	344	344	344	344	155	155	155[m]
345	0	0	0	0	0	0	0	0	0	0	0
346	346	0	0	0	−135	0	0	0	0	0	0
355	355	355	344	344	344	344	344	344	166	155	155[m]
356	0	0	0	0	134	134	0	0	0	0	0
366	366	366	366	366	366[j]	366[j]	366[j]	366[j]	144	144	144[l]
444	0	0	0	0	444	444	0	0	0	0	0
445	445	0	0	0	445	0	0	0	0	0	0
446	0	0	446	0	145	0	145	0	0	0	0
455	0	0	0	0	−444	−444	0	0	0	0	0
456	456	456	456	456	456[k]	456[k]	456[k]	456[k]	456	456	456[n]
466	0	0	0	0	124	124	0	0	0	0	0
555	555	0	0	0	−445	0	0	0	0	0	0
556	0	0	−446	0	−145	0	−145	0	0	0	0
566	566	0	0	0	125	0	0	0	0	0	0
666	0	0	0	0	−116	0	−116	0	0	0	0

[a] $c_{122}=c_{111}+c_{112}-c_{222}$.
[b] $c_{146}=\frac{1}{2}(-c_{115}-3c_{125})$.
[c] $c_{156}=\frac{1}{2}(c_{114}+3c_{124})$.
[d] $c_{166}=\frac{1}{4}(-2c_{111}-c_{112}+3c_{222})$.
[e] $c_{224}=-c_{114}-2c_{124}$.
[f] $c_{225}=-c_{115}-2c_{125}$.
[g] $c_{246}=\frac{1}{2}(-c_{115}+c_{125})$.
[h] $c_{256}=\frac{1}{2}(c_{114}-c_{124})$.
[i] $c_{266}=\frac{1}{4}(2c_{111}-c_{112}-c_{222})$.
[j] $c_{366}=\frac{1}{2}(c_{113}-c_{123})$.
[k] $c_{456}=\frac{1}{2}(-c_{144}+c_{155})$.
[l] $c_{144}=\frac{1}{2}(c_{112}-c_{123})$.
[m] $c_{155}=\frac{1}{4}(c_{111}-c_{112})$.
[n] $c_{456}=\frac{1}{8}(c_{111}-3c_{112}+2c_{123})$.

References

Akhmanov, S. A., A. I. Kovrigin, V. A. Kolosov, A. S. Piskarskas, V. V. Fadeev, and R. V. Khokhlov, *JETP Pis'ma Red.* **3**, 372 (1966) [Engl. transl.: *JETP Letters* **3**, 241 (1966)].

Andrews, R. A., H. Rabin, and C. L. Tang, *Phys. Rev. Lett.* **25**, 605 (1970).

Armstrong, J. A., N. Bloembergen, J. Ducuing, and P. S. Pershan, *Phys. Rev.* **127**, 1918 (1962).

Ashkin, A., G. D. Boyd, and J. M. Dziedzic, *Phys. Rev. Lett.* **11**, 14 (1963).

Bass, M., P. A. Franken, A. E. Hill, C. W. Peters, and G. Weinreich, *Phys. Rev. Lett.* **8**, 18 (1962a).

Bass, M., P. A. Franken, J. F. Ward, and G. Weinreich, *Phys. Rev. Lett.* **9**, 446 (1962b).

Baumhauer, J. C., and H. F. Tiersten, *J. Acoust. Soc. Am.* **54**, 1017 (1973).

Berk, A. D., *IRE Trans. Antennas Propag.* **AP4**, 104 (1956).

Bloembergen, N., and P. S. Pershan, *Phys. Rev.* **128**, 606 (1962).

Bond, W. L., et al., *Proc. IRE* **37**, 1378 (1949).

Boyd, G. D., A. Ashkin, J. M. Dziedzic, and D. A. Kleinman, *Phys. Rev.* **137** A1305 (1965).

Boyd, G. D., and D. A. Kleinman, *J. Appl. Phys.* **39**, 3597 (1968).

Boyd, G. D., F. R. Nash, and D. F. Nelson, *Phys. Rev. Lett.* **24**, 1298 (1970).

Boyd, G. D., H. M. Kasper, and J. H. McFee, *IEEE J. Quant. Electron.* **QE7**, 563 (1971).

Boyd, G. D., and M. A. Pollack, *Phys. Rev. B* **7**, 5345 (1973).

Brillouin, L., *C. R. Acad. Sci.* **158**, 1331 (1914).

Brillouin, L., *Ann. Phys. (Paris)* **17**, 88 (1922).

Brugger, K., *J. Appl. Phys.* **36**, 759 (1965).

Caddes, D. E., C. F. Quate, and C. D. W. Wilkinson, in *Modern Optics*, J. Fox, Ed. (Polytechnic Press, Brooklyn, N.Y., 1967), p. 219.

Chang, I. C., *Appl. Phys. Lett.* **25**, 370 (1974).

Chapelle, J., and L. Taurel, *C. R. Acad. Sci.* **240**, 743 (1955).

Chiao, R. Y., C. H. Townes, and B. P. Stoicheff, *Phys. Rev. Lett.* **12**, 592 (1964).

Chiao, R. Y., E. Garmire, and C. H. Townes, *Phys. Rev. Lett.* **13** 479 (1964).

Collins, R. J., D. F. Nelson, A. L. Schawlow, W. Bond, C. G. B. Garrett, and W. Kaiser, *Phys. Rev. Lett.* **5**, 303 (1960).

Cummins, H. Z., and P. E. Schoen, in *Laser Handbook*, F. T. Arecchi and E. O. Schulz-Dubois, Eds. (North-Holland, Amsterdam, 1972), p. 1029.

Debye, P., and F. W. Sears, *Proc. Natl. Acad. Sci. U.S.* **18**, 409 (1932).

Dixon, R. C., and A. C. Eringen, *Int. J. Eng. Sci.* **3**, 359 (1965).

Dixon, R. W., *IEEE J. Quant. Electron.* **QE3**, 85 (1967).

Eckhardt, G., R. W. Hellwarth, F. J. McClung, S. E. Schwarz, D. Weiner, and E. J. Woodbury, *Phys. Rev. Lett.* **9**, 455 (1962).

Eringen, A. C., *Mechanics of Continua* (Wiley, New York, 1967).

Falk, J., and J. E. Murray, *Appl. Phys. Lett.* **14**, 245 (1969).

Faust, W. L., and C. H. Henry, *Phys. Rev. Lett.* **17**, 1265 (1966).

Faust, W. L., C. H. Henry, and R. H. Eick, *Phys. Rev.* **173**, 781 (1968).

Feinberg, G., in *Atomic Physics*, B. Bederson, V. W. Cohen, and F. M. J. Pichanick, Eds. (Plenum, New York, 1969), p. 1.

Franken, P. A., A. E. Hill, C. W. Peters, and G. Weinreich, *Phys. Rev. Lett.* **7**, 118 (1961).

Garwin, R. L., L. M. Lederman, and M. Weinrich, *Phys. Rev.* **105**, 1415 (1957).

Geusic, J. E., H. J. Levinstein, S. Singh, R. G. Smith, and L. G. Van Uitert, *Appl. Phys. Lett.* **12**, 306 (1968).

Giordmaine, J. A., *Phys. Rev. Lett.* **8**, 19 (1962).

Giordmaine, J. A., *Phys. Rev.* **138**, A1599 (1965).

Giordmaine, J. A., and R. C. Miller, *Phys. Rev. Lett.* **14**, 973 (1965).

Giordmaine, J. A., and W. Kaiser, *Phys. Rev.* **144**, 676 (1966).

Giordmaine, J. A., and P. M. Rentzepis, *J. Chim. Phys.* **67**, 215 (1967).

Gross, E., *Nature* **126**, 201, 400, 603 (1930).

Harris, S. E., and R. W. Wallace, *J. Opt. Soc. Am.* **59**, 744 (1969).

Harris, S. E., S. T. K. Nieh, and R. S. Feigelson, *Appl. Phys. Lett.* **17**, 223 (1970).

Hearmon, R. F. S., and D. F. Nelson, in *Landolt-Börnstein*, New Series, Vol. III/11, K. H. Hellwege, Ed. (Springer, Berlin, 1979), Sec. 5.2.3, p. 530.

Henry, C. H., and J. J. Hopfield, *Phys. Rev. Lett.* **15**, 964 (1965).

Hill, E. L., *Rev. Mod. Phys.* **23**, 253 (1951).

Huang, K., *Proc. Roy. Soc. A* **208**, 352 (1951).

Joffrin, J., and A. Levelut, *Solid State Commun.* **8**, 1573 (1970).

Kaminow, I. P., in *Ferroelectricity*, E. F. Weller, Ed. (Elsevier, Amsterdam, 1967), p. 183.

Kleinman, D. A., *Phys. Rev.* **126**, 1977 (1962a).

Kleinman, D. A., *Phys. Rev.* **128**, 1761 (1962b).

Kleinman, D. A., A. Ashkin, and G. D. Boyd, *Phys. Rev.* **145**, 338 (1966).

Kleinman, D. A., *Phys. Rev.* **174**, 1027 (1968).

Kleinman, D. A., in *Laser Handbook*, F. T. Arecchi and E. O. Schulz-Dubois, Eds. (North-Holland, Amsterdam, 1972), p. 1229.

Kreutzer, L. B., *Appl. Phy. Lett.* **10**, 33 (1967).

Kurosawa, T., *J. Phys. Soc. Jap.* **16**, 1298 (1961).

Kurtz, S. K., and J. A. Giordmaine, *Phys. Rev. Lett.* **22**, 192 (1969).

Landsberg, G., and L. Mandelstam, *Naturwiss.* **16**, 557, 772 (1928).

Lax, M., and D. F. Nelson, *Phys. Rev. B* **4**, 3694 (1971).

Lax, M., and D. F. Nelson, in *Atomic Structure and Properties of Solids*, E. Burstein, Ed. (Academic Press, New York, 1972), p. 105.

Lax, M., and D. F. Nelson, *Phys. Rev. B* **13**, 1759 (1976a).

Lax, M., and D. F. Nelson, *Phys. Rev. B* **13**, 1777 (1976b).

Lee, T. D., and C. N. Yang, *Phys. Rev.* **104**, 254 (1956).

Levine, B. F., *Phys. Rev. B* **7**, 2600 (1973).

Lighthill, M. J., *J. Inst. Math. Appl.* **1**, 1 (1965).

Lucas, R., and P. Biquard, *J. Phys. Radium* **3**, 464 (1932).

Lyddane, R. H., R. G. Sachs, and E. Teller, *Phys. Rev.* **59**, 673 (1941).

Maiman, T. H., *Nature* **187**, 493 (1960); *Brit. Commun. Electron.* **7**, 674 (1960).

Maker, P. D., R. W. Terhune, M. Nisenoff, and C. M. Savage, *Phys. Rev. Lett.* **8**, 21 (1962).

Manley, J. M., and H. E. Rowe, *Proc. IRE* **44**, 904 (1956).

Mason, M., and W. Weaver, *The Electromagnetic Field* (University of Chicago Press, Chicago, 1929; republished by Dover, New York, undated).

McKenna, J., and F. K. Reinhart, *J. Appl. Phys.* **47**, 2069 (1976).

Miller, R. C., *Appl. Phys. Lett.* **5**, 17 (1964).

Miller, R. C., G. D. Boyd, and A. Savage, *Appl. Phys. Lett.* **6**, 77 (1965).

Nath, G., H. Mehmanesch, and M. Gsanger, *Appl. Phys. Lett.* **17**, 286 (1970).

Nelson, D. F., and F. K. Reinhart, *Appl. Phys. Lett.* **7**, 148 (1964).

Nelson, D. F., and J. McKenna, *J. Appl. Phys.* **38**, 4057 (1967).

Nelson, D F., and E. H. Turner, *J. Appl. Phys.* **39**, 3337 (1968).

Nelson, D. F., and M. Lax, *Phys. Rev. Lett.* **24**, 379 (1970).

Nelson, D. F., and P. D. Lazay, *Phys. Rev. Lett.* **17**, 1187, 1638 (1970).

Nelson, D. F., and M. Lax, *Appl. Phys. Lett.* **18**, 10 (1971a).

Nelson, D. F., and M. Lax, *Phys. Rev. B* **3**, 2778 (1971b).

Nelson, D. F., and M. Lax, *Phys. Rev. B* **3**, 2795 (1971c).

Nelson, D. F., and R. M. Mikulyak, *Phys. Rev. Lett.* **28**, 1574 (1972).

Nelson, D. F., P. D. Lazay, and M. Lax, *Phys. Rev. B* **6**, 3109 (1972).

Nelson, D. F., and R. M. Mikulyak, *Appl. Phys. Lett.* **27**, 548 (1975).

Nelson, D. F., *J. Opt. Soc. Am.* **65**, 1144 (1975).

Nelson, D. F., and M. Lax, *Phys. Rev B* **13**, 1770 (1976a).

Nelson, D. F., and M. Lax, *Phys. Rev B* **13**, 1785 (1976b).

Nelson, D. F., *Am. J. Phys.* **45**, 1187 (1977).

Nelson, D. F., *J. Acoust. Soc. Am.* **63**, 1738 (1978a).

Nelson, D. F., *J. Acoust. Soc. Am.* **64**, 652 (1978b).

Nelson, D. F., *J. Acoust. Soc. Am.* **64**, 891 (1978c).

Nelson, D. F., *J. Opt. Soc. Am.* **68**, 1780 (1978d).

Noether, E. *Nachr. kgl. Ges. Wiss. Göttingen* **(1918)** 235.

Nye, J. F., *Physical Properties of Crystals* (Clarendon Press, Oxford, 1957).

Pine, A. S., *Phys. Rev. B* **2**, 2049 (1970).

Schinke, D. P., *IEEE J. Quant. Electronics* **QE8**, 86 (1972).

Smith, R. G., J. E. Geusic, H. J. Levinstein, J. J. Rubin, S. Singh, and L. G. Van Uitert, *Appl. Phys. Lett.* **12**, 308 (1968).

Smith, R. G., ir *Laser Handbook*, F. T. Arecchi and E. O. Schulz-Dubois, Eds. (North-Holland, Amsterdam, 1972), p. 837.

Terhune, R. W., P. D. Maker, and C. M. Savage, *Phys. Rev. Lett.* **8**, 404 (1962).

Thompson, R. B., and C. F. Quate, *J. Appl. Phys.* **42**, 907 (1971).

Thurston, R. N., and K. Brugger, *Phys. Rev.* **133**, A1604 (1964).

Thurston, R. N., and M. J. Shapiro, *J. Acoust. Soc. Am.* **41**, 1112 (1967).

Thurston, R. N., in *Handbuch der Physik*, Vol. VIa/4, S. Flügge, Ed. (Springer, Berlin, 1974), p. 109.

Tiersten, H. F., *J. Acoust. Soc. Am.* **35**, 53 (1963).

Tiersten, H. F., *Linear Piezoelectric Plate Vibrations* (Plenum, New York, 1969).

Tiersten, H. F., and J. C. Baumhauer, *J. Appl. Phys.* **45**, 4272 (1974).

Toupin R. A., *J. Rational Mech. Anal.* **5**, 849 (1956).

Toupin, R. A., and B. Bernstein, *J. Acoust. Soc. Am.* **33**, 216 (1961).

Toupin, R. A., *Arch. Ratl. Mech. Anal.* **11**, 385 (1962).

Toupin, R. A., *Int. J. Eng. Sci.* **1**, 101 (1963).

Truesdell, C., and R. A. Toupin, in *Handbuch der Physik*, Vol. III/1, S. Flügge, Ed. (Springer, Berlin, 1960), p. 226.

Walker, J. B., A. C. Pipkin, and R. S. Rivlin, *Accad. Naz. Lin., Ser 8*, **38**, 674 (1965).

Warner, A. W., M. Onoe, and G. A. Coquin, *J. Acoust. Soc. Am.* **42**, 1223 (1967).

Wemple, S. H., and M. DiDomenico, Jr., *Phys. Rev. Lett.* **23**, 1156 (1969).

Wemple, S. H., and M. DiDomenico, Jr., *Phys. Rev. B* **1**, 193 (1970).

Whitham, G. B., *Linear and Nonlinear Waves* (Wiley-Interscience, New York, 1974).

Woodbury, E. J., and W. K. Ng, *Proc. IRE* **50**, 2367 (1962).

Wu, C. S., E. Ambler, R. W. Hayward, D. D. Hoppes, and R. P. Hudson, *Phys. Rev.* **105**, 1413 (1957).

Yamada, T., and N. Niizeki, *Rev. Elec. Commun. Lab. (Tokyo)* **19**, 705 (1971).

Bibliography

Acoustics and Elasticity of Solids

Achenbach, J. D., *Wave Propagation in Elastic Solids* (North-Holland, Amsterdam, 1973).

Auld, B. A., *Acoustic Fields and Waves in Solids* (Wiley-Interscience, New York, 1973), 2 vols.

Dieulesaint, E., and D. Royer, *Elastic Waves in Solids* (Wiley-Interscience, New York), in press.

Federov, F. I., *Theory of Elastic Waves in Crystals* (Plenum, New York, 1968).

Graff, K. F., *Wave Motion in Elastic Solids* (Ohio State University, Columbus, 1975).

Green, A. E., and J. E. Adkins, *Large Elastic Deformations*, 2nd ed. (Clarendon, Oxford, 1970).

Green, A. E., and W. Zerna, *Theoretical Elasticity*, 2nd ed. (Clarendon, Oxford, 1968).

Hearmon, R. F. S., *Introduction to Applied Anisotropic Elasticity* (Oxford University, London, 1961).

Musgrave, M. J. P., *Crystal Acoustics* (Holden Day, San Francisco, 1970).

Rudenko, O. V., and S. I. Soluyan, *Theoretical Foundations of Nonlinear Acoustics* (Consultants Bureau, New York, 1977).

Thurston, R. N., "Waves in Solids," in *Handbuch der Physik*, Vol. VIa/4, S. Flügge, Ed., (Springer, Berlin, 1974), p. 109.

Tucker, J. W., and V. W. Rampton, *Microwave Ultrasonics in Solid State Physics* (North-Holland, Amsterdam, 1972).

Continuum Mechanics of Solids

Born, M., and K. Huang, *Dynamical Theory of Crystal Lattices* (Clarendon, Oxford, 1954).

Eringen, A. C., *Mechanics of Continua* (Wiley, New York, 1967).

Eringen, A. C., *Nonlinear Theory of Continuum Mechanics* (McGraw-Hill, New York, 1962).

Haar, D. ter, *Elements of Hamiltonian Mechanics* (North-Holland, Amsterdam, 1961).

Jaunzemis, W., *Continuum Mechanics* (Macmillan, New York, 1967).

Lanczos, C., *The Variational Principles of Mechanics*, 2nd ed. (University of Toronto, Toronto, 1962).

Landau, L. D., and E. M. Lifshitz, *The Classical Theory of Fields*, 4th ed. (Pergamon, Oxford, 1975).

Logan, J. D., *Invariant Variational Principles* (Academic, New York, 1977).

Soper, D. E., *Classical Field Theory* (Wiley-Interscience, New York, 1976).

Truesdell, C., *A First Course in Rational Continuum Mechanics* (Academic, New York, 1977).

Truesdell, C., and R. A. Toupin, "The Classical Field Theories," in *Handbuch der Physik*, Vol. III/1, S. Flügge, Ed. (Springer, Berlin, 1960), p. 226.

Yourgrau, W., and S. Mandelstam, *Variational Principles in Dynamics and Quantum Theory*, 3rd ed. (Saunders, Philadelphia, 1968).

Crystal Symmetry

Bhagavantam, S., *Crystal Symmetry and Physical Properties* (Academic, New York, 1966).

Birman, J. L., "Theory of Crystal Space Groups and Infrared and Raman Lattice Processes in Insulating Crystals," in *Handbuch der Physik*, Vol. XXV/2b, S. Flügge, Ed. (Springer, Berlin, 1974).

Burns, G., *Introduction to Group Theory with Applications* (Academic, New York, 1977).

Lax, M., *Symmetry Principles in Solid State and Molecular Physics* (Wiley, New York, 1974).

Lyubarskii, G. Ya., *The Application of Group Theory in Physics* (Pergamon, Oxford, 1960).

Nye, J. F., *Physical Properties of Crystals* (Clarendon, Oxford, 1957).

Electromagnetism in Solids

Böttcher, C. J. F., O. C. van Belle, P. Bordewuk, and A. Rip, *Theory of Electric Polarization* (Elsevier Scientific, Amsterdam, 1973).

Groot, S. R. de, *The Maxwell Equations* (North-Holland, Amsterdam, 1969).

Groot, S. R. de, and L. G. Suttorp, *Foundations of Electrodynamics* (North-Holland, Amsterdam, 1972).

Landau, L. D., and E. M. Lifshitz, *Electrodynamics of Continuous Media* (Pergamon, Oxford, 1960).

Penfield, P., Jr., and H. A. Haus, *Electrodynamics of Moving Media* (MIT Press, Cambridge, 1967).

Robinson, F. N. H., *Macroscopic Electromagnetism* (Pergamon, Oxford, 1973).

Optics in Solids

Akhmanov, S. A., and R. V. Khokhlov, *Problems of Nonlinear Optics* (Gordon and Breach, New York, 1972).

Arecchi, F. T., and E. O. Schulz-Dubois, Eds., *Laser Handbook* (North-Holland, Amsterdam, 1972), 2 vols.

Bloembergen, N., *Nonlinear Optics* (Benjamin, New York, 1965).

Born, M., and E. Wolf, *Principles of Optics*, 5th ed. (Pergamon, Oxford, 1975).

Butcher, P. N., *Nonlinear Optical Phenomena* (Ohio State University, Columbus, 1965).

Rabin, H., and C. L. Tang, Eds., *Quantum Electronics*, Vol. I, *Nonlinear Optics*, Parts A and B (Academic, New York, 1975).

Ramachadran, G. N., and S. Ramaseshan, "Crystal Optics," in *Handbuch der Physik*, Vol. XXV/1, S. Flügge, Ed. (Springer, Berlin, 1961), p. 1.

Symbols

Roman Letters

$\mathbf{A}$	Spatial frame vector potential
$\mathfrak{A}$	Material frame vector potential
$\mathbf{B}$	Spatial frame magnetic induction field
$\mathfrak{B}$	Material frame magnetic induction field
$\mathbf{b}^{\alpha}$	Acoustic displacement eigenvector ($\alpha=1,2,3$)
b_{ABC}	Electric field mixing tensor
$b_{i\ j\ k}^{\omega_S\omega_1\omega_2}$	Optical mixing susceptibility
C	Capacitance
C_{MN}	Green deformation tensor
c	Velocity of light in vacuum
c_{mn}	Cauchy deformation tensor (Chapter 3), contracted stiffness tensor (Chapter 11 and Appendix)
c_{ABCD}	Second order stiffness tensor
c_{ABCD}^{F}	Second order stiffness tensor in presence of a spontaneous electric field
c_{ABCDEF}	Third order stiffness tensor
$\mathbf{D}$	Spatial frame electric displacement vector
$\mathfrak{D}$	Material frame electric displacement vector
$\mathfrak{D}^{\alpha}$	Spatial frame electric displacement eigenvector ($\alpha=1,2,3$)
$\mathbf{d}^{\alpha}$	Unit spatial frame electric displacement eigenvector ($\alpha=1,2,3$)
dA	Scalar element of area in material frame
$d\mathbf{A}$	Vector element of area in material frame
da	Scalar element of area in spatial frame
$d\mathbf{a}$	Vector element of area in spatial frame
dS	Length of element of arc in material frame
ds	Length of element of arc in spatial frame

$d\mathbf{X}$	Vector element of arc in material frame
$d\mathbf{x}$	Vector element of arc in spatial frame
dV	Element of volume in material frame
dv	Element of volume in spatial frame
d_{ABC}	Piezoelectric strain tensor
E	Young's modulus
$\mathbf{E}$	Spatial frame electric field
$\mathbf{E}^S$	Spatial frame spontaneous electric field
$\mathfrak{E}$	Material frame electric field
$\mathscr{E}$	Effective spatial frame electric field felt by matter $(\mathbf{E}+\dot{\mathbf{x}}\times\mathbf{B})$
$\mathfrak{E}^\alpha$	Spatial frame electric field eigenvector ($\alpha=1,2,3$)
E_{AB}	Green finite strain tensor
e^α	Electric charge of particle of type α ($\alpha=1,2,\ldots N$)
$\mathbf{e}^\alpha$	Unit spatial frame electric field eigenvector ($\alpha=1,2,3$)
e_{ABC}	Piezoelectric stress tensor
e^F_{ABC}	Piezoelectric stress tensor in the presence of a spontaneous electric field
$e_{i\ j\ k\ l}^{\omega_C\omega_1\omega_2\omega_A}$	Third order optical mixing tensor
e_{KABCD}	Nonlinear piezoelectric tensor
f^C_{ij}	Flux of canonical angular momentum in spatial frame
$\mathbf{g}$	Total momentum density in spatial frame
$\mathbf{g}^C$	Canonical momentum density in spatial frame
g_{iJ}	Parallel displacement tensor or shifter
H	Spatial frame energy density
$\mathbf{H}$	Spatial frame magnetic intensity vector
$\mathcal{H}$	Material frame magnetic intensity vector
I	Action integral
$\mathbf{i}^{k\mu}$	Normal mode eigenvector of an internal vibration ($k=1,2,\ldots 3N-3;\mu=1,2,\ldots N-1$)
$J(\mathbf{x}/\mathbf{X})$	Jacobian of deformation transformation
$\mathfrak{J}$	Total material frame current density
$\mathfrak{J}^c$	Material frame conduction current density
$\mathfrak{J}^D$	Material frame dielectric current density
$j(\mathbf{X}/\mathbf{x})$	Jacobian of inverse deformation transformation
$\mathbf{j}$	Total spatial frame current density

$\mathbf{j}^c$	Spatial frame conduction current density
$\mathbf{j}^D$	Spatial frame dielectirc current density
$\mathbf{j}^f$	Total spatial frame free charge current density
K^α	Electromechanical coupling factor of α mode ($\alpha = 1,2,3$) (Chapter 11); magnitude of acoustic wavevector of α mode ($\alpha = 1,2,3$) (Chapter 17)
$\mathbf{K}^\alpha$	Acoustic wavevector (propagation vector) of α mode ($\alpha = 1,2,3$)
$\mathbf{K}^c$	Material frame surface conduction current
$\mathbf{k}^c$	Spatial frame surface conduction current
$\mathbf{k}^\alpha$	Optical wavevector (propagation vector) of α mode ($\alpha = 1,2,3$)
k_{ij}	Flux of total angular momentum in spatial frame
L	Total Lagrangian
L_F	Field Lagrangian
L_I	Interaction Lagrangian
L_M	Matter Lagrangian
$\mathfrak{L}_M$	Total Lagrangian density in material frame
$\mathfrak{L}_S$	Total Lagrangian density in spatial frame
$\mathfrak{L}_{FM}$	Field Lagrangian density in material frame
$\mathfrak{L}_{FS}$	Field Lagrangian density in spatial frame
$\mathfrak{L}_{IM}$	Interaction Lagrangian density in material frame
$\mathfrak{L}_{IS}$	Interaction Lagrangian density in spatial frame
$\mathfrak{L}_{MM}$	Matter Lagrangian density in material frame
$\mathfrak{L}_{MS}$	Matter Lagrangian density in spatial frame
$\mathbf{l}$	Spatial frame internal angular momentum density
l_{ABCD}	Electrostriction tensor
l^s_{ABCD}	Electrostrictive susceptibility
$\mathbf{M}$	Spatial frame magnetization vector
M_{i_j}	Spatial frame magnetization tensor
M_{i_J}	Mixed frame electromagnetic stress tensor
${}^{mn}M^{\mu\cdots}_{K\cdots AB\cdots}$	Material descriptor
$\mathfrak{M}$	Material frame magnetization vector
m^α	Mass of particle of type α ($\alpha = 1,2,\ldots N$)
m^μ	Mass associated with internal motion of type μ ($\mu = 1,2,\ldots N-1$)

m_{ij}	Spatial frame Maxwell vacuum field stress tensor
N	Number of particles in primitive unit cell
$\mathbf{N}$	Unit surface normal in material frame
n	Index of refraction
$\mathbf{n}$	Unit surface normal in spatial frame
$\mathbf{P}$	Spatial frame polarization vector
$\mathbf{P}^S$	Spontaneous polarization vector
$\boldsymbol{\mathcal{P}}$	Material frame polarization vector
$\mathscr{P}$	Spatial frame nonlinear polarization appearing in the electric field wave equation
p_{ijkl}	Pockels elastooptic tensor
Q	Total surface charge
Q_{ij}	Spatial frame quadrupolarization tensor
$\mathfrak{Q}$	Total material frame charge density
$\mathfrak{Q}^D$	Material frame dielectric charge density
$\mathfrak{Q}^f$	Material frame free charge density
q	Total spatial frame charge density
q^f	Spatial frame free charge density
q^α	Material frame charge density of particles of type α $(\alpha=1,2,\ldots N)$
q^μ	Material frame charge density associated with internal motion of type μ $(\mu=1,2,\ldots N-1)$
q_k	Four-dimensional coordinate vector (Chapter 5) in which q_1, q_2, q_3 are either spatial or material coordinates and q_4 is the time
q_{ijkl}	Pockels piezooptic tensor
R_{iJ}	Finite rotation tensor
r_{ijk}^S	Pockels strain-free electrooptic tensor
r_{ijk}^T	Pockels stress-free electrooptic tensor
$\mathbf{S}$	Spatial frame flux of energy
S_{AB}	Infinitesimal strain tensor
$\mathbf{s}$	Unit wavevector
s_{ABCD}	Compliance tensor
T_{iJ}	Piola-Kirchoff stress tensor
$\mathfrak{T}_{IJ}$	Bilinearized material frame stress tensor

$\mathbf{t}$	Unit vector in direction of group velocity
t_{ij}^C	Spatial frame canonical stress tensor
t_{ij}^E	Spatial frame elastic stress tensor
t_{AB}^S	Spontaneous stress tensor
t_{ij}^T	Total spatial frame stress tensor
t_{ij}^y	Spatial frame local stress tensor
$\mathbf{u}$	Spatial frame displacement vector
$\mathbf{u}^\alpha$	Spatial frame internal displacement of particle of type α ($\alpha = 1,2,\ldots N$)
$u_{i,K}$	Displacement gradient
V	Potential energy (Chapters 1 and 4); applied voltage (Chapter 11); volume in material frame (elsewhere)
v	Phase velocity; volume in spatial frame
$\mathbf{v}_g$	Group velocity
$\mathbf{v}_r$	Ray velocity or energy propagation velocity
$\mathbf{X}$	Material frame position vector of a matter point
$\partial\mathbf{X}/\partial t$	Flow of matter
$X_{J,i}$	Deformation gradient
$\mathbf{x}$	Spatial frame position vector of a matter point
$\dot{\mathbf{x}}$	Velocity of matter
$\ddot{\mathbf{x}}$	Acceleration of matter
$\mathbf{x}^{\alpha n}$	Spatial frame position of particle of type α in the n primitive unit cell ($\alpha = 1,2,\ldots N$; $n = 1,2,\ldots$)
$x_{i,J}$	Deformation gradient
$\mathbf{Y}^\mu$	Material frame internal coordinate in natural state ($\mu = 1,2,\ldots N-1$)
$\mathbf{y}^\mu$	Spatial frame internal coordinate relative to natural state ($\mu = 1,2,\ldots N-1$)
$\mathbf{y}^{T\mu}$	Spatial frame internal coordinate ($\mu = 1,2,\ldots N-1$)
$\mathbf{z}$	Spatial frame position vector of a space point

Greek Letters

β	Propagation constant of a waveguide mode
δ	Electrooptic perturbation (Chapter 14); angle between wavevector and group velocity (elsewhere)

$\delta(\mathbf{z})$	Three-dimensional Dirac delta function
$\delta(S)$	One-dimensional Dirac delta function
δ_{ij}	Kronecker delta
ϵ_0	Permittivity of free space
ϵ_{ABC}	Material frame permutation symbol
ϵ_{ijk}	Spatial frame permutation symbol
η^k	Normal coordinate of an internal vibration ($k=1,2,\ldots 3N-3$)
$\kappa_{ij}(\omega)$	Dielectric tensor
λ	Stretch (Chapter 3); Lamé coefficient (Chapter 10); wavelength (elsewhere)
μ	Lamé coefficient
μ_0	Permeability of free space
ν	Poisson's ratio (Chapter 10); circular frequency (elsewhere)
Π_C^{ν}	Rotationally invariant measure of $\mathbf{y}^{T\nu}$ relative to the natural state value $\Pi_C^{S\nu}$ ($\nu=1,2,\ldots N-1$)
$\Pi_C^{S\nu}$	Natural state value of $\Pi_C^{T\nu}$ ($\nu=1,2,\ldots N-1$)
$\Pi_C^{T\nu}$	Rotationally invariant measure of $\mathbf{y}^{T\nu}$ ($\nu=1,2,\ldots N-1$)
ρ	Spatial frame (deformed) mass density
ρ^0	Material frame (undeformed) mass density
ρ^{ν}	Spatial frame mass density associated with the ν internal motion ($\nu=1,2,\ldots N-1$)
Σ	Potential energy per unit undeformed mass
Σ^f	Material frame surface free charge density
σ^f	Spatial frame surface free charge density
$\Upsilon_{ij}^{\mu\nu}(\omega)$	Mechanical admittance
$\Upsilon_{AB}^{\mu\nu}$	Low frequency limit of mechanical admittance
Φ	Spatial frame scalar potential
$\Phi(\varphi)$	Single phase-matching function
φ	Material frame scalar potential (Chapters 16 and 17); phase angle (elsewhere)
$\chi_{ij}(\omega)$	Linear electric susceptibility
$\chi_{i\ j\ kl}^{\omega_S\omega_1\omega_2}$	Elastooptic susceptibility
$\chi_{i\ j\ [kl]}^{\omega_S\omega_1\omega_2}$	Rotooptic susceptibility
$\chi_{i\ j\ k\ lm}^{\omega_C\omega_1\omega_2\omega_A}$	Acoustically induced optical mixing susceptibility

$\Psi(\theta,\phi)$	Double phase-matching function
Ω_0	Primitive unit cell volume in material frame
ω	Angular frequency
ω_L	Longitudinal optic mode frequency
ω_T	Transverse optic mode frequency
$\boldsymbol{\omega}$	Total spatial frame angular momentum density
$\boldsymbol{\omega}^C$	Canonical angular momentum density in spatial frame

AUTHOR INDEX

SUBJECT INDEX